D0026128

DESIGN OF AGRICULTURAL MACHINERY

Gary Krutz
Purdue University

Lester Thompson
Sundstrand Corporation

Paul Claar
Iowa State University

John Wiley & Sons
New York Chichester Brisbane Toronto Singapore

DEDICATED TO OUR FAMILIES

Cover photo: Dave Repp, Photo Researchers, Inc.

Library of Congress in Publication Data:

Krutz, Gary.
Design of agricultural machinery.

Includes index.
1. Agricultural machinery–Design and construction.
I. Thompson, Lester. II. Claar, Paul. III. Title.
TJ1480.K73 1984 681'.763 83-23251
ISBN 0-471-08672-X

Printed in the United States of America

10 9 8 7 6 5 4 3 2 1

PREFACE

This book, intended for use by college seniors and practicing design engineers, presents the state of art in engineering design of agricultural machinery. Because of time limitations, however, not all of the topics involved in design in the outdoor environment, such as variability of the soil and crops harvested, are included.

The book covers combined stresses, assuming a prior course in strength of materials, and the design topics selected as most appropriate for the future success of practicing engineers. These topics include fatigue analysis, finite element analysis, statistical tolerance design, power transmission, and hydraulics.

The state of the art in engineering is constantly changing, and the reader is encouraged to be aware of new techniques that can improve design efficiency. Continuing education is important as the competition for qualified engineers increases. This book provides a solid basis for continuing education in the design of agricultural machinery.

We thank Ph.D. candidates Larry Gaultney, Mike Mailander, and John Schueller, who made significant contributions to Chapters 5 and 9. We are indebted to our wives and families, who sacrificed their time for our effort, and our special thanks for secretarial, artistic, and editorial assistance goes to Denise Evans, Connie Harth, RoxAndra Evans, Tammy Endres, Holly Wheaton, Dawn Lehman, Jane Nelson, and Randy Weyhrich. Also, those who influenced our design philosophy—former professors, engineering associates, department heads, and deans—should feel that a part of this book is attributed to them. Finally, our appreciation is extended to Deborah Herbert, who copy edited the manuscript, and to our reviewers, Leonard L. Bashford, Carroll E. Goering, and Stephen J. Marley, for their interest and help throughout this publishing process.

PURDUE UNIVERSITY
W. Lafayette, Indiana — *Gary Krutz*

SUNDSTRAND CORP.
Ames, Iowa — *Lester Thompson*

IOWA STATE UNIVERSITY
Ames, Iowa — *Paul Claar*

CONTENTS

CHAPTER EIGHT

POWER TRANSMISSIONS

CHAPTER NINE

LINKAGES ON FARM MACHINERY

CHAPTER TEN

HYDRAULIC POWER SYSTEMS

CHAPTER ONE

PHILOSOPHY OF DESIGN

The quality that makes good designers is the capability to develop their own philosophy of design. The philosophy of design includes organizing the methodology needed to proceed from concept to final product, determining which technological resources, such as mathematical relationships, to use, and considering economics, timeliness, reliability, and safety during the design process. In developing a design philosophy, design engineers can add to skills already learned and those included in this book to create products that benefit humanity and improve the quality of life now and for future generations.

Gasson defines designers as "those who employ the agents of technological skill and facility and make use of their creative energies to bring about some sought advantage useful to man." Design is a highly innovative cross-disciplinary process that uses the most acceptable, aesthetic, efficient, and economical means to satisfy people's needs. As technology changes, methods and philosophy of design change. For example, the advent of computer graphics has enabled designers to quickly design more efficient, lower cost components.

For engineering students the art of design requires gradual progression from the level of intellectual development that considers answers either right or wrong to a higher level of creativity (i.e., using an equation to get the correct answer might require additional information on the validity of that equation, such as how it was derived or whether it requires metric units). Only after reaching the peak of intellectual development can a person design products. This ability requires that the student first look at all possible solutions, then choose a course, stick with it, and continually modify and improve it until all conditions are satisfied.

Most engineering students are accustomed to simply plugging facts into equations. A design course provides a culture shock, which ultimately enables students to formulate their own philosophy on how to make assumptions, gather data, and use mathematical relationships to devise (design) some new structure, machine, or system.

1.1 FORMULATING A PROCEDURE

Most engineering courses deal with evaluation and analysis of loads, stresses, and motion after a design has been established (function and geometry are assumed). Richey provides an excellent description of design.

> Design requires that a configuration be devised and created to perform a function. At the same time, the configuration should be evaluated for adequate strength and minimum cost. Often this cannot be done until the machine is built and loads measured, but it is better to make assumptions and calculate major loads than to proceed without approximate calculations.
>
> Experience can, to some extent, be substituted for calculations as is evidenced by farmers, mechanics, and cross road shops. Usually, however, their designs can be further improved by an engineer's ability to evaluate.
>
> Note! Engineers do not have a monopoly on inventiveness and ingenuity required of a good designer. In fact, engineering training might inhibit these qualities because a young engineer might not want to proceed until he or she has all the information, whereas an educated mechanic would proceed by trial and error.

This chapter is intended to formulate a basic process for designing agricultural machines. The concept of design involves past experience of analysis, synthesis, and judgement. Some problems are so complex and vary so much with time and the environment that all the equations available cannot provide a solution. Catalogs based on experience and data are therefore needed to supplement the design process, and this book provides such a practical, nut-and-bolt approach.

1.2 THREE TYPES OF DESIGN

1. Original Design. Originality is judged on the degree of copying. The less a design resembles existing designs, the more original it is (patented designs).
2. Transitional Design. A sound, basic design is improved by using detailed refinements, such as finite element techniques and modeling.
3. Extensional Design. In an extensional design, an extrapolative procedure is used to increase the capacity of proven designs.

An example of original design is designing an air planter; an example of transitional design is reducing the weight of a tractor front-end loader; and an example of extensional design is designing a 200-kW tractor with configurations similar to those of the 100-kW unit.

Corporate management usually adds input to the design process. For example, sales, research, and marketing people may suggest ideas for new products, or current products might be found to be deficient in some way and therefore need to be redesigned. A typical flow of the design process is shown in Fig. 1.1. Once given guidelines, the engineer usually sketches the idea, reviews constraints and standards, compares the idea with those of competitors, analyzes

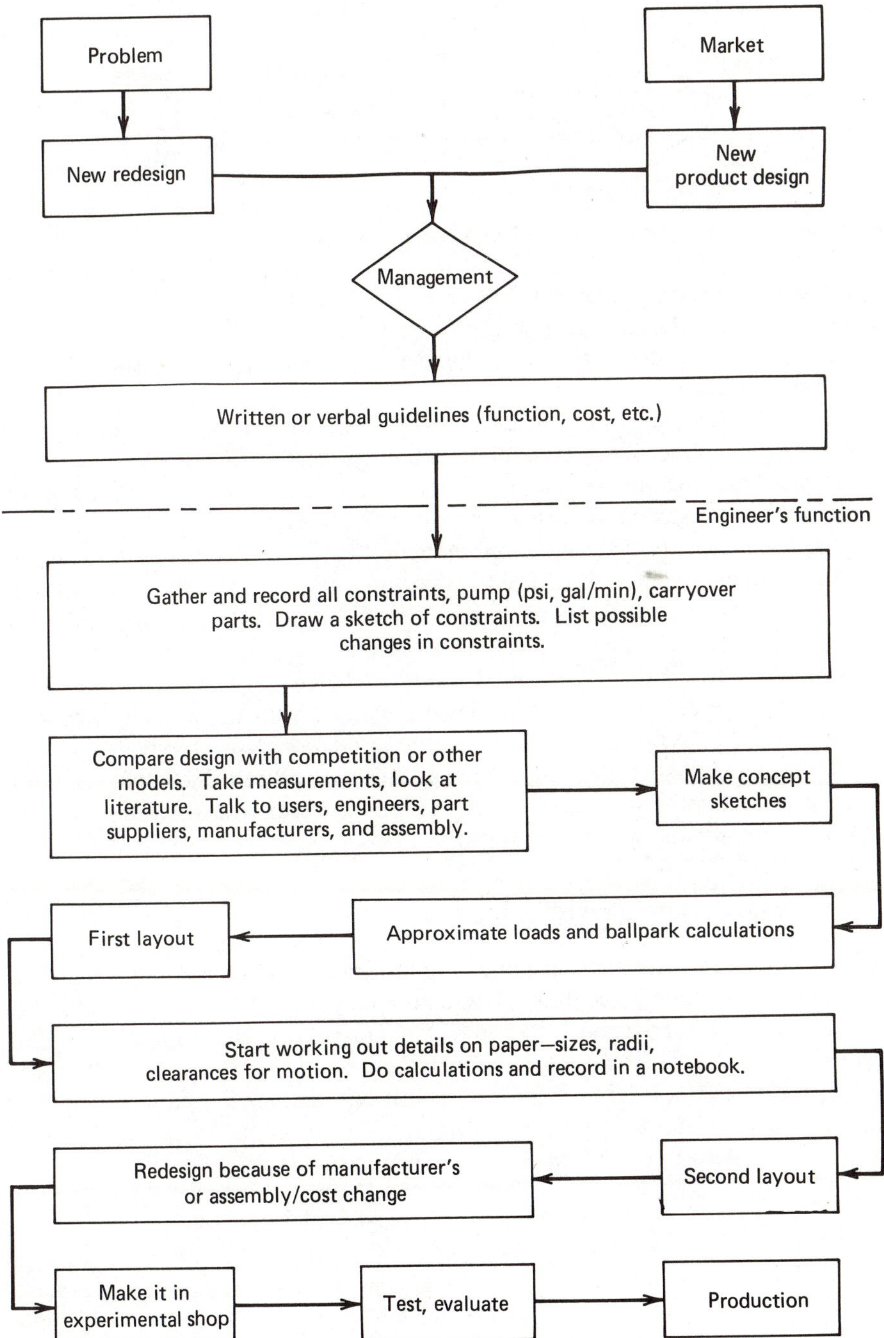

Figure 1.1 Typical organization of a design process.

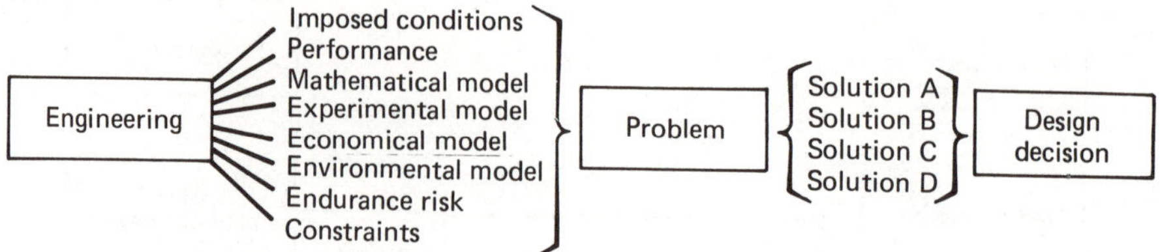

Figure 1.2 Making the right engineering decision.

the structure, creates drawings, checks manufacturing practices, builds, tests, and finally sees the idea go into production. After 1 to 3 years of hard work, this final step, production, gives the designer a feeling of satisfaction and accomplishment.

The aim of engineering education is to prepare students by providing examples of design philosophy that will aid in their success as professional engineers. The problems encountered by engineers do not have a simple correct answer and consist of many constraints, as shown in Fig. 1.2.

Engineers are not expected to remember all the analytical methods taught in the classroom, but they should remember the topics and be able to look them up, review them and their assumptions from original derivations, and then apply them to a design. General concepts, such as energy conversion processes, must be quickly recalled when a relevant problem arises (see Fig. 1.3). The analytical relationship between length, velocity, acceleration, and force should be remembered. Some of these relationships are given in Table 1.1. Other physical relationships, such as thermal and centripetal effects, thermal stress, steam pressure, thermal expansion, and coriolis acceleration, should also be kept in mind while designing.

1.3 COMMUNICATING IDEAS

The process of design requires good communication between the engineer, the prototype build shop, assembly plant, and management. If ideas are not accurately and fully understood, the project might be canceled and a good idea shelved. Engineers should therefore take the time to communicate their design to management, not only for self-promotion but also for the project's success.

***Table 1.1* A Designer's List of Principles of Conversion to Consider**

	Outcome			
Input	*Length, Volume*	*Velocity*	*Acceleration*	*Force*
Force, pressure, mechanical energy	Hooke's law Poisson's ratio Bouyancy principle Coulomb's law	Energy conservation Bernoulli's theorem Impulse Stokes's law	Newton's law	Lever Wedge Hydrostatics Friction

	CHEMICAL	ELASTIC	ELECTRICAL	GRAVITATIONAL	HEAT	KINETIC	RADIATION
CHEMICAL			Storage battery Cell Fuel cell Bacteria cell	(Flight)	Combustion Heat of solution Metabolism		Cold emmision of light (Fish and insects)
ELASTIC			Piezoelectric crystal	"Pogo stick"	Cooling by expansion Compression ignition	Clock (spring) Air rifle	
ELECTRICAL	Electrolysis Electroplating Storage battery		Transformer		Electric heater Thermoelectric heater and cooler Spark	Electric motor Loudspeaker Particle accelerator Electrostatic tweeter	Electronic flash Radio Radar X-Rays, Spark Fluorescent light
GRAVITATIONAL		"Pogo stick"	Hydroelectric generator			Falling object Grandfather clock Pendulum Tides	
HEAT	Endothermic reactions (Reduction of ore)	Thermal stress	Thermoelectric generator Thermionic convertor	Convection		Heat engines Wind Ocean currents Rocket	Radiation from a hot surface Electric light
KINETIC	Muscle (Rocket) (Explosion)	Recoil spring Tire pump Watch spring	Electric generator Microphone	Trajectory Pendulum	Friction Impact Mechanical refrigerator	(Collision) Wind mill	
NUCLEAR					Nuclear reactor Fission Fusion		Nuclear explosion The sun's radiation
RADIATION	Photosynthesis Photographic process		Photocell Ion emission Solar battery		Absorption at a surface Radio-frequency heating Infrared heating		Flurescence

Figure 1.3 Energy conversion process chart.

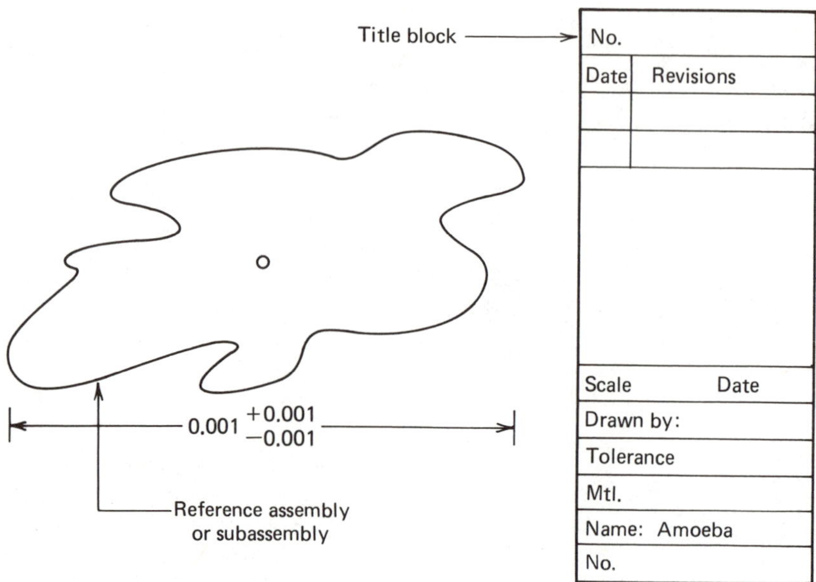

Figure 1.4 A sample engineering layout.

Discussion and engineering drawings are both part of this communication process.

Communication of a design begins with the drawing. A good layout needs to be drawn so it can easily be reproduced (see Fig. 1.4). Techniques and drawing systems vary from company to company. Some engineers make their own layouts, whereas computer graphic systems generate drawings for others (see Fig. 1.5). Computer-aided-design (CAD) is the common term used for these computer drafting systems.

Figure 1.5 Using computer graphics to design a hydraulic system.

A list of items needed for transmitting ideas to the build shop are as follows.

1. Layouts should be made so that major dimensions can be determined. Areas of part interference or clearance should be shown. Two orthographic projection views or a single view with sections are needed. CAD machines provide three-dimensional additions to a drawing, which can aid in comprehension. All critical dimensions should be calculated from a layout; they should never be scaled, because of paper stretch.
2. For tractors, a general 0,0,0 dimension is specified as the center of the crankshaft at the front face of the engine block. For automobiles and trucks 0,0,0 is at the center of the crankshaft on the top of the frame at the front face of the engine block. All other critical dimensions (e.g., the intersection of the rear axle and transmission, as shown in Fig. 1.6) should be calculated once this location is established. Critical joining dimensions, such as bolts, holes, and welds, must also be specified.
3. Reference drawings that control other drawings should be identified. "This drawing controlled by layout LN-3462."
4. All part numbers, torques, materials, and tolerances should be specified on the master layout.
5. A parts list can be constructed and should include quantities, material stock sizes, and material type.
6. The nameplate on the right-hand side usually includes the machine part name, drawing identification number, material used, scale, date, and name of designer. Revisions that accurately describe changes made during the design process should be listed chronologically.

Other pertinent information should be stored in a bound book for patent protection and for future reference. Such information includes sketches and concepts dated and witnessed by two or more competent persons. The witnesses' statement on the concept sketch might read: Understood and witnessed by me, *Don Jones*, date *May 2, 1982*. Information on how to file for a

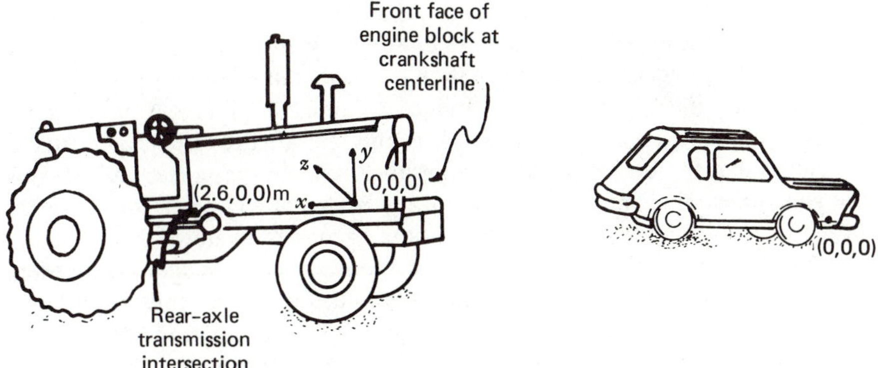

Figure 1.6 Standard organization of vehicle drawings.

patent, including necessary forms and drawing specifications, is available in *Code of Federal Regulations 37 Patents, Trademarks and Copyrights.*

The design notebook should include all free body diagrams and assumptions used (reference drawings, textbooks, etc.). Calculations should be detailed, and the designer should remember that almost all parts are subjected to fatigue loads and stress concentrations at geometry changes.

The source of standard equations, such as moments of inertia for common cross sections, is important to practicing designers. The Appendix provides a starting point for finding various commonly used inertia equations previously taught in statics and strength-of-materials courses. Many advanced texts and engineering handbooks provide standard solutions for more-advanced structural loading cases. It is advisable to consider all sources of information on engineering practice when beginning a design.

1.4 ETHICAL ENGINEERING ISSUES IN THE FUTURE

Today there are many significant dilemmas concerning safety in engineering practices, as expressed by Smith. These dilemmas result from the wording of a portion of the Fundamental Principles of Professional Engineering Practice.

> The Engineer, to uphold and advance the honor and dignity of the engineering profession and in keeping with high standards of ethical conduct:
>
> I. Will be honest and impartial, and will serve with devotion his employer, his clients, and the public;
> II. Will strive to increase the competence and prestige of the engineering profession;
> III. Will use his knowledge and skill for the advancement of human welfare.
>
> ***Relations With the Public***
> 1.1 The Engineer will have proper regard for the safety, health and welfare of the public in the performance of his professional duties . . . etc....

Difficult decisions arise when engineers are confronted with conflicts in the areas listed in section 1.1 of the Code of Ethics. If the public is endangered, the engineer is supposed to notify the proper authority and withdraw from further service on the project. A sample situation is considered in Smith's paper.

During an investigation of a bridge collapse, Engineer A investigates another, similar bridge and finds that it is only marginally safe. The engineer concludes that the bridge might collapse in certain circumstances and informs the governmental agency responsible for the bridge of a concern for the safety of the structure. Engineer A is told that the agency is aware of this situation, and has planned to provide for repairs in the next year's budget. Until then, the bridge must remain open to traffic. Without this bridge, emergency vehicles, such as police and fire apparatus, would have to use an alternate route, which would increase the response time by about 20 minutes. Because the agency is confident that the bridge is safe, Engineer A is asked to say nothing about the condition of the bridge.

What course should the engineer take? The code of ethics requires that the engineer protect the public safety, but which public, the public that needs emergency service or the public that uses the bridge daily? Smith states that the concept of safety needs to be redefined because of its complexity and its interactions with risk factors. There is always some risk in designing products; the issue of how much risk an engineer should allow in a design is unresolved. This issue can be partially resolved by putting some financial value on safety and after doing an economic evaluation before proceeding with the design. Nevertheless, the question remains: Are these risks acceptable with current technology and engineering practices?

How safe must a product be? The degree of safety is questioned in the cases that follow.

Case 1.

In designing airplane components, considerations of safety must be balanced against considerations of weight. In the first DC-10s, for example, an electronic closing device was used for a cargo hatch, rather than a heavier hydraulic closing device, even though the latter was safer. As a result, the cargo hatch could appear to be securely closed when it wasn't and it could blow out during a flight. The first crash of a fully loaded jumbo jet was caused by the opening of the cargo door in flight. The crash resulted in the death of 346 people.[1] Was this a reasonably safe design?

Case 2.

Can a product be too safe? Consider the following. An engineer accepts a design assignment, even though unsure of his or her ability to do the job. Because of these doubts, the engineer overdesigns the project to ensure its safety. The completed project is about 30% more expensive than it could have been. Should the engineer have admitted his or her shortcomings, or is the production of a safe, workable design sufficient, even though it is unduly expensive?

Conflicting values lead to another dilemma between safety and economic or other societal values.

Case 3.

An archaeologist, a member of the staff of a highway design firm, discovers some significant Indian artifacts while evaluating a possible highway route. This discovery may necessitate a change to an alternate route, which for several reasons could prove to be slightly less safe and less desirable. Any route changes would also stall the design work.

The engineers decide to suppress the archaeological find, on the basis that it might lead to a somewhat less safe design. They are concerned that human

[1]Paul Eddy, Elaine Potter, and Bruce Page, *Destination Disaster From the Tri-Motor to the DC 10: The Risk of Flying*, as excerpted in Baum and Flores *Ethical Problems in Engineering*, Troy, N.Y., pp. 248-261, 1978.

lives might be lost in the future in order to protect a possibly significant archaeological area. Even though this procedure conflicts with standard rules and regulations of the Department of Transportation, the engineers are convinced that it is the right thing to do. The staff archaeologist, who is worried about keeping the job, agrees with the decision.

How safe should products be? What risk level should be acceptable? How does safety relate to the job and duty to the employer? These questions must be resolved in every design process. As technology advances and new analytical methods become available, engineers must continually educate themselves in these areas so that they can make correct decisions. Some of the ethical dilemmas previously addressed will have to be decided in the courts.

1.5 FACTOR OF SAFETY (Safety Index)

An engineering judgment on how safe a product is can be made by calculating the factor of safety. For structural cases, this factor is expressed as

$$FS = \frac{\text{yield stress}}{\text{actual stress}} \tag{1.1}$$

with a numerical value between 1 and 4 desirable in the worst loaded condition. In actuality, loads may vary; thus a greater factor of safety should be used than that discussed here. The formulas used to calculate stress are based on several assumptions. All formulas presented in this book obey Hooke's law of linear elastic relationships, and the authors are not responsible for misuse of the formulas. It is assumed that engineers will use the latest design tools and techniques and will upgrade their technical competence as design practices change over time.

Variations and nonuniformity of the material's strength must be considered when deciding on a reasonable safety factor. Engineers use the term factor of safety to include uncertainties (risks) in a design. If the magnitude and direction of various forces are not known, especially on original designs, it is difficult to evaluate this term. Time elements, such as corrosion, lack of lubrication, loading, temperature, and history, can also affect the life of a part.

Knowledge of engineering materials must be acquired when designing. Some knowledge can be obtained from testing specimens or from suppliers. The choice of the factor of safety might relate to the amount of time available for engineering and testing. A large factor of safety might be uneconomical, whereas a failure that involves some slight loss of time or inconvenience might be acceptable. Safety of human life should definitely have a high value applied to it during design. Detailed testing could raise the product's cost and may or may not be justified.

Experience, accumulated by designing over a long period of time, is the best method in choosing the correct factor of safety in engineering design. There are not many reliable rules for determining a suitable value for factor of safety. The criterion of Eq. 1.1, assuming yield as the state at which failure occurs, is not valid in a tractor rollover bar design in which operator safety is achieved by yielding the part and absorbing energy. A new relationship exists in this case.

$$\text{FS} = \frac{\text{ultimate stress}}{\text{actual stress}} \qquad [1.2]$$

or

$$\text{FS} = \frac{\text{total energy absorbed}}{\text{minimum required energy absorption per ASAE Standards}}$$

A different factor-of-safety relationship should be derived for each design and its calculated value included in the engineer's notebook.

1.6 RELIABLITY

Excessive strength is usually wasteful in engineering design. A practice of accurately evaluating load and strength using a small factor of safety has been developed; it is called reliability design. Reliability uses statistical evaluations to determine the chance of success for a design. Because nothing is 100% safe, some products will fail under extreme conditions. Some conditions, such as environment, aging, and temperature, occur over time. It is good practice to make a product 99.9% reliable by considering replacing one out of every thousandth component that becomes defective thus keeping the design economical when human safety is not a factor.

Values for reliability are between zero and 1, values for the factor of safety range from zero to infinity.

Components attached in series will have an overall factor of safety equivalent to the weakest link. Therefore, the system factor of safety equals the smallest factor of safety of all the components. Because reliability is a random occurrence and not necessarily a function of load (e.g., sand and dirt can affect bearing life), the overall reliability of a system of components R_S in series is

$$R_s = R_1 \times R_2 \times R_3 \ldots \qquad [1.3]$$

where $R_{1,} R_2$ = reliability of each component
R_s = reliability of a system in series
R_p = reliabilty of a system in parallel

For a system with components in parallel, the probability of the system failure, F_p, is the product of the individual components' probabilities of failure, where $F_p = 1 - R_p$. That is, a system with 30% probability of failure has a reliability of 70% (see Fig. 1.7).

The only disadvantage of parallel systems is the added cost for redundancy. An example of parallel systems is the dual, parallel hydraulic systems on airplanes that improve the system's overall reliability; these systems approach an almost fail-safe design that prevent accidents in case of a single component failure. This system is more costly, but necessary because of the human lives involved.

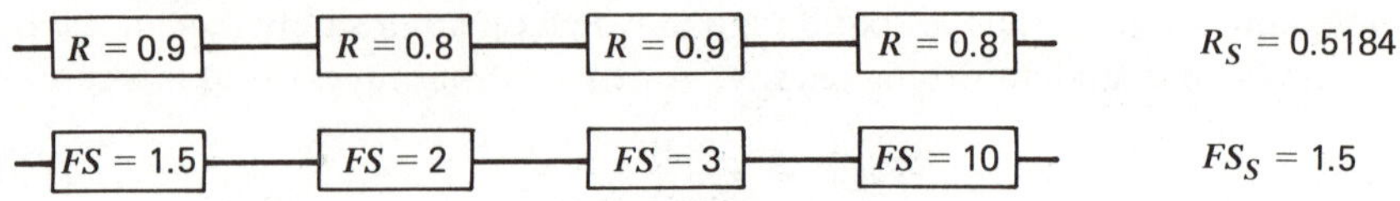

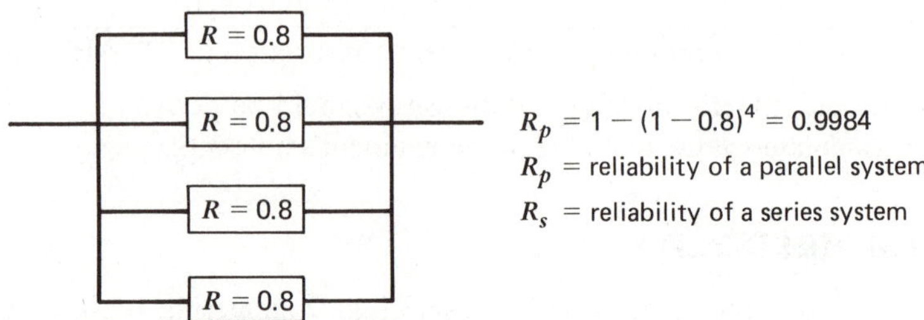

Figure 1.7 Reliability of series and parallel systems.

Assume a distribution of material strength and load as illustrated in Fig. 1.8 and assume they are both normally distributed.

From interference theory the mean interference, μ_Z, is

$$\mu_Z = \mu_S - \mu_L \qquad [1.4]$$

where μ_L = mean load
μ_S = mean strength

and the standard deviation, S_Z, is

$$S_Z = (S_s^2 + S_L^2)^{1/2} \qquad [1.5]$$

where S_S = standard deviation of strength
S_L = standard deviation of load

A conversion from component reliability to factor of safety can now be made. If values for S_S/μ_S and S_L/μ_L are known, the following formula gives values of factor of safety. This factor of safety (FS) nomenclature is currently called the safety index by those using Eq. 1.6.

$$FS = W\left[\left(\frac{S_S}{\mu_S}\right)^2 + \left(\frac{S_L}{\mu_L}\right)^2\right]^{1/2} + 1 \qquad [1.6]$$

where W = weighting factor for reliability levels R
$W = 1.28$ at $R = 0.9$
$W = 2.33$ at $R = 0.99$
$W = 3.08$ at $R = 0.999$

Evaluating and optimizing reliability of mixed systems of parallel and series components can be accomplished by using techniques called dynamic programming. The design goal of minimizing overall cost requires balancing the cost of reliability and repair costs.

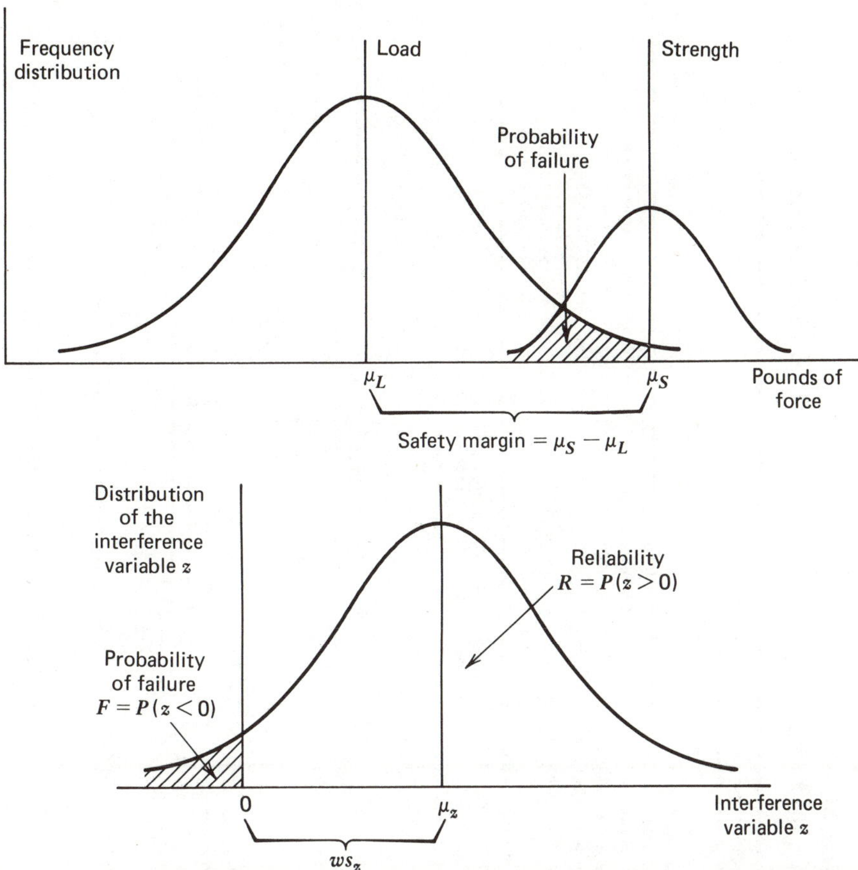

Figure 1.8 Natural frequency distribution of load and material strength. Reprinted with permission © William Wai Chong Chow, Cost Reduction in Product Design, Von Nostrand Reinhold Co. Publishers.

1.7 STANDARDS

The safety portion of the code of ethics is fulfilled partially by following product standards written and supervised by an appropriate engineering governing body. Currently, four groups, Society of Automotive Engineers (SAE), American Society of Agricultural Engineers (ASAE), National Fluid Power Association (NFPA), and American Society of Mechanical Engineers (ASME), provide the major source of standards for the design of agricultural vehicles. Other standards for factors such as materials, electronic components, and so on also exist.

Standards are technical reports approved and recommended for practice to engineers. The use of standards is voluntary and no liability is assumed by the sponsoring association for their misuse. Legal issues of product liability are currently concerned with defective design, product misuse, product alterations,

THREE-POINT FREE-LINK HITCH ATTACHMENT OF IMPLEMENTS TO AGRICULTURAL WHEELED TRACTORS—SAE J715 MAY80

SAE Standard

Report of the Tractor Technical Committee, approved April 1959, last revised July 1975, editorial change June 1977, reaffirmed without change May 1980. Conforms to report of FIEI Advisory Engineering Committee, Chicago, Illinois. Corresponds to ASAE S217.5.

1. Scope—This specification sets forth requirements for the attachment of three-point hitch implements or equipment to the rear of agricultural wheeled tractors by means of a three-point free-link in association with a power lift.

In order to assure proper performance of certain implements, standard dimensions for mast height, mast pitch adjustment and implement leveling adjustment are included. Location of link attachment points is not restricted and is, therefore, left to the discretion of the tractor designer.

If draft links are used for trailing power take-off implements, a means shall be included for locking the draft links in a fixed position, and a draw-bar hitch point shall be positioned in conformance with power take-off standards.

Dimensions comprising the standard specifications are divided into three categories:

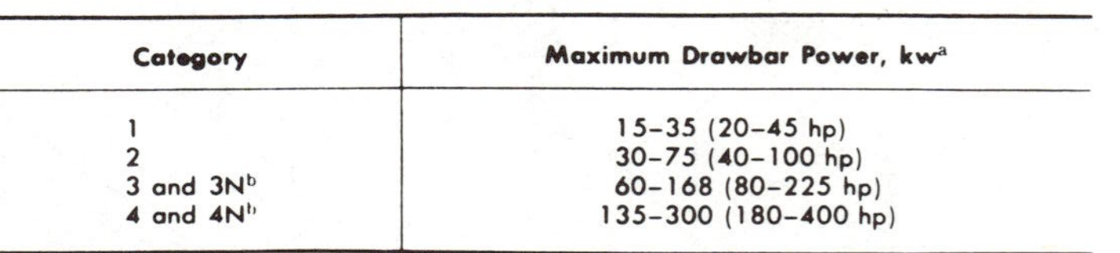

Category	Maximum Drawbar Power, kw[a]
1	15–35 (20–45 hp)
2	30–75 (40–100 hp)
3 and 3N[b]	60–168 (80–225 hp)
4 and 4N[b]	135–300 (180–400 hp)

[a]Based on SAE J708, paragraph 2.5.
[b]Refer to Special Hitch Categories, Section 3.

2. Definition of Terms

2.1 Linkage—The combination of one upper link and two lower links, each articulated to the tractor and the implement at opposite ends in order to connect the implement to the tractor.

2.2 Upper Link, Lower Link—Element in the linkage.

2.3 Hitch Point—The articulated connection between a link and the implement. For geometrical analysis, the hitch point is established as the center of the articulated connection between a link and the implement.

2.4 Link Point—The articulated connection between a link and the tractor. For geometrical analysis, the link point is established as the center of the articulated connection between a link and the tractor.

2.5 Upper Hitch Point—The articulated connection between the upper link and the implement.

2.6 Upper Link Point—The articulated connection between the upper link and the tractor.

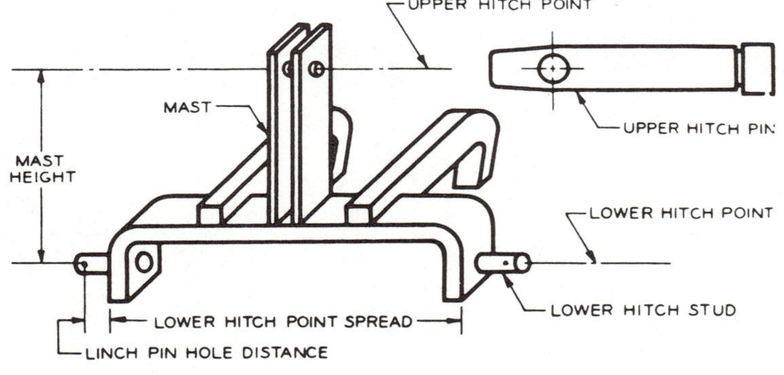

FIG. 2—DIMENSIONS ASSOCIATED WITH IMPLEMENT

Figure 1.9 SAE standard for tractor three-point hitches. Reprinted with permission © 1982 Society of Automotive Engineers, Inc.

2.7 Lower Hitch Point—The articulated connection between a lower link and the implement.

2.8 Lower Link Point—The articulated connection between a lower link and the tractor.

2.9 Upper Hitch Pin—The pin that connects the upper link to the implement.

2.10 Upper Link Pin—The pin that connects the upper link to the tractor.

2.11 Lower Hitch Stud or Pin—The stud or pin, attached to the implement, on which a lower link is secured.

2.12 Linchpin—The retaining pin used in the hitch pins or studs.

2.13 Mast—The member that provides attachment of the upper link to the implement.

2.14 Mast Height—The perpendicular distance between the upper hitch point and common axis of the lower hitch points.

TABLE 1—DIMENSIONS ASSOCIATED WITH IMPLEMENT

	Category I				Category II				Category III[e]				Category IV[e]			
	mm		in		mm		in		mm		in		mm		in	
	Min	Max	Min	Max	Min	Max	Min	Max	Min	Max	Min	Max	Min	Max	Min	Max
Upper Hitch Point																
Width inside	44.5	—	1.75	—	52.3	—	2.06	—	52.3	—	2.06	—	65	—	2.56	—
Width outside	—	85.9	—	3.38	—	95.3	—	3.75	—	95.3	—	3.75	—	132	—	5.20
Clearance radius for upper link[a,b]	57.2	—	2.25	—	57.2	—	2.25	—	57.2	—	2.25	—	76.2	—	3.00	—
Hitch pin hole diameter	19.3	19.56	0.76	0.77	26.65	25.91	1.01	1.02	32.0	32.26	1.26	1.27	45.2	45.5	1.78	1.79
Lower Hitch Point																
Stud diameter	21.84	22.10	0.86	0.87	28.19	28.45	1.11	1.12	36.32	36.58	1.43	1.44	49.7	50.8	1.96	2.00
Linchpin hole distance[a,b]	38.86	—	1.53	—	48.52	—	1.91	—	48.52	—	1.91	—	68	—	2.68	—
Linchpin hole diameter	11.68	12.19	0.46	0.48	11.68	12.19	0.46	0.48	11.68	12.19	0.46	0.48	17.5	18	0.69	0.71
Lower hitch point spread	681.0	684.3	26.81	26.94	822.5	825.5	32.38	32.50	963.7	966.7	37.94	38.06	1165	1168	45.87	45.99
Clearance radius for lower link[a,b]	63.5	—	2.50	—	73.2	—	2.88	—	82.6	—	3.25	—	82.6	—	3.25	—
Implement encroachment in front of lower hitch point if implement extends laterally behind tire	—	12.7	—	0.5	—	12.7	—	0.5	—	12.7	—	0.5	—	12.7	—	0.5
Implement Mast Height[c,d]	457		18		483		19		559		22		686		27	

Figure 1.9 (Cont.)

or lack of safety warning statements. If design engineers use standards published by the societies (committees with years of experience), they can minimize their companies' potential liability in designing products. Standards also provide a mechanism for part interchangeability, which benefits the consumer by resulting in lower costs. Most engineering standards are now being adopted worldwide under the direction of the International Standards Organization (ISO).

An organization establishing standards is not concerned with all ISO standards. NFPA standards, for example, include ISO graphic symbols for hydraulic systems, basic terminology, filter performance testing, fitting-connection dimensions, and testing procedures for high-pressure components. ASME, on the other hand, is mainly concerned with high pressure vessels in stationary plants.

One typical standard commonly used in designing agricultural equipment is ASAE S217.10 (SAE J715f), "Three-Point Free-Link Attachment for Hitching Implements to Agricultural Wheel Tractors." This standard, like several vehicle standards, is a mutual standard established by ASAE and SAE. This standard defines terminology and sets basic category sizes (see Fig. 1.9). SAE provides additional standards that are used to design agricultural vehicles. Some include materials, bolts, lighting systems, brakes, tires, operator controls, roll-over bars, hydraulic fittings, and operator sight patterns. Figure 1.10 illustrates a portion of SAE STD J941e, Motor Vehicle Driver's Eye Range in a Rear Seat Position, which defines the graphical methods used in constructing the driver's field of view (SAE STD J1050). Figure 1.11, from ISO STD 5713 on plow bolts, provides sizes, materials, and thread pitch used when designing plows. The checklist in Table 1.2 will help an engineer be successful.

Table 1.2 **Checklist of Factors in a Design Process**

Governmental regulations
Voluntary and involuntary standards
Costs
Environments of use
Operator skill, knowledge, background
Operator position and station
Nonoperator position and station
Repair and service requirements
Starting and stopping or shutdown
Control of product during operation
Instructions for operation, maintenance, and repair
Shielding and guarding
Inherent hazards
Modes of usage
General safety of all modes of usage and operations
Possible modes of failure and their consequences
Current state of knowledge and capabilities

MOTOR VEHICLE DRIVER'S EYE RANGE—SAE J941 MAR81

SAE Recommended Practice

Report of the Body Engineering Committee, approved November 1965, last revised by the Human Factors Engineering and Automotive Safety Committees February 1975, editorial change March 1977.

1. Scope—This SAE Recommended Practice establishes two-dimensional Eyellipses representative of 90th, 95th, and 99th percentile increments of driver eye locations for use in passenger cars, trucks, buses, and multipurpose passenger vehicles. A uniform method for describing and measuring the driver's direct and indirect fields of view using the Eyellipse is established in the recommended practice, Describing and Measuring the Driver's Field of View—SAE J1050a.

2. Background—The Eyellipse contours were developed by the statistical analysis of photogrammetric data of driver eye locations[1] and represent a population mix, primarily of United States licensed drivers, with a male-to-female ratio of one-to-one. The Eyellipse templates are the perimeters of envelopes formed by an infinite number of planes dividing the eye positions so that (P) % of the eyes are on one side of the plane and (100 – P) % are on the other. It should be noted that the 95th Eyellipse does not include 95% of the driver eye locations. For example, if a plane seen as a straight line in the side view is drawn tangent to the upper edge of the 95th percentile Eyellipse, then 95% of the driver eye locations, both inside and outside of the ellipse, will be below the line and 5% of the driver eye locations will be above the line (Fig. 1). Conversely, if a plane seen as a straight line in the side view is drawn tangent to the lower edge of the 95th percentile Eyellipse, then 95% of the driver eye locations, both inside and outside of the ellipse, will be above the line and 5% of the driver eye locations will be below the line. These planes or sight planes are drawn from the object in the driver's field of view tangent to the Eyellipse contour.

This recommended practice is based on an original study, involving drivers with a straight-ahead viewing task without head turning. A subsequent study[2] has provided a method of accounting for driver viewing targets that are located at extreme lateral angles from the forward line of sight which accommodates head movement up to 60 deg after an eye movement of up to 30 deg. An Eyellipse locator line has been developed in a third study,[3] to position the Eyellipse in the driver work space for back angles ranging 5–40 deg in 1 deg increments.

[1] J. F. Meldrum, "Automobile Driver Eye Position," SAE Transactions, Vol. 74 (1966), Paper 650464.

[2] W. A. Devlin and R. W. Roe, "The Eyellipse and Consideration in the Driver's Forward Field of View," Paper 680105 presented at SAE Automotive Engineering Congress, Detroit, 1968.

[3] R. W. Roe and D. C. Hammond, "Driver Head and Eye Position," Paper 720200 presented at SAE Automotive Engineering Congress, Detroit, January 1972.

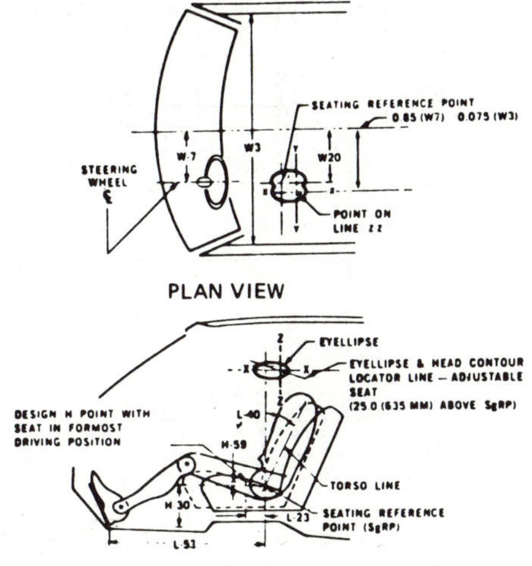

FIG. 6—EYELLIPSE TEMPLATE LOCATION

TABLE A-2—EYELLIPSE MAJOR AXIS LENGTHS—PLAN AND SIDE VIEW

L23		Percent Tangent Cutoff					
		90		95		99	
in.	mm	in.	mm	in.	mm	in.	mm
4.0	102	4.3	109	5.8	147	8.5	216
4.5	114	4.8	122	6.3	160	9.0	229
5.0	127	5.3	135	6.8	173	9.5	241
5.5	140	5.8	147	7.3	185	10.0	254
6.0	152	6.1	155	7.6	193	10.3	262
6.5	165	6.3	160	7.8	198	10.5	267

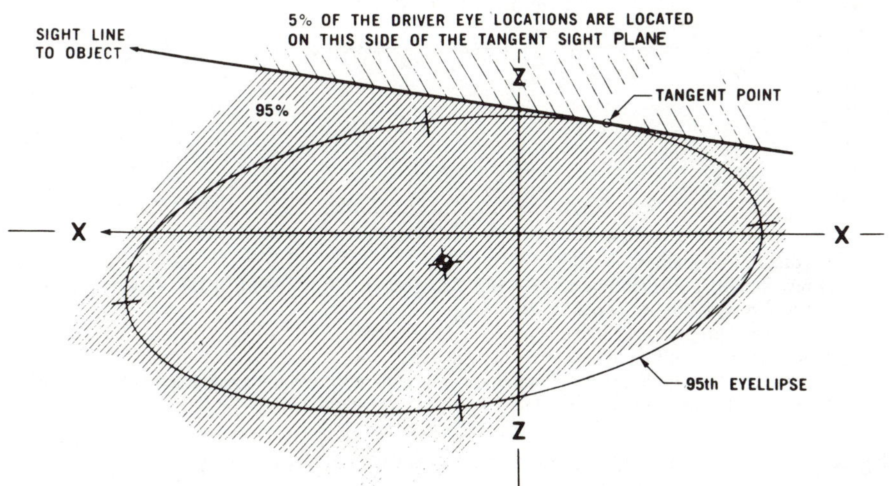

Figure 1.10 SAE standard on motor-vehicle driver eye range. Reprinted with permission © 1982 Society of Automotive Engineers, Inc.

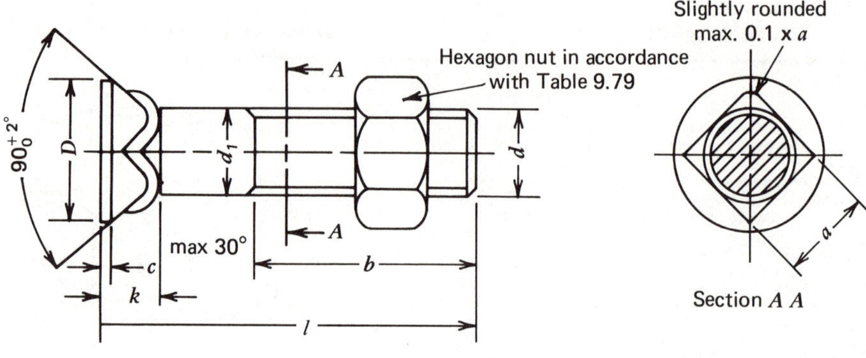

d	M8	M10	M12	M16	M20
Thread Pitch	1.25	1.5	1.75	2	2.5
d_1	Blank rolling diameter				20 h15
a	8 h14	10 h14	12 h14	26 h14	20 h15
b Minimum	From square section 2× pitch max				
b Nominal	34	40	46	58	70
c	1	1.2	1.2	1.4	2.5
D h15	14	18	21	30	36
k h15	5.5	7	8	10.5	13.5
l js17	Standard-lengths are indicated by x				
20	x				
25	x	x			
30	x	x	x		
35	x	x	x		
40		x	x	x	
45		x	x	x	x
50		x	x	x	x
60			x	x	x
70				x	x
80				x	x
90				x	x
100				x	x

1. All dimensions are in millimeters.
2. Screw threads are according to ISO 965. See Section 8 on modifiied ISO fastener screw threads (page 8.20) and thread tolerance 6g.
3. Mechanical properties to ISO 898 grade 9.8. See page 9.18
4. Bolt marking according to ISO 898, see page 9.20.
5. Designation: Example for the designation of a plow bolt, thread sized d = M12, nominal length l = 30 mm. and property class 9.8 (Plow bolt ISO 5 713)

Figure 1.11 ISO standard on plow bolts. Reprinted with permission © Knut Kverneland, World Metric Standards for Engineering, Industrial Press.

HOMEWORK PROBLEMS

1.1 The object of this problem is for you to become familiar with standards that are used as guidelines when designing agricultural and industrial equipment. Note the year in which the standard was last revised and use the most recent yearbook.
Substitute the first digit of your student number or social security number for "X" when answering the following questions.
   a. Determine the maximum and minimum yield strength for a 1X0,000 psi tensile strength steel (SAE J413a).
   b. SAE J414 gives mechanical properties of steel. Determine the difference in cold drawn and hot rolled yield strength for SAE 10X0 steel.
   c. 5X.0 on the Rockwell hardness C scale is what Brinell number? What yield strength would this steel have?

1.2 What is the radius of fillet (R) of a 5/16 diameter hex bolt? What wrench size does this bolt take? How does this radius of fillet (R) compare with a bolt 1 in. in diameter?

1.3 What SAE STD describes the driver's field of view?

1.4 What is the difference in static loaded radius of 18.4-34 R1 and R2 tires?

1.5 SAE J898a recommends operator control locations. Find a tractor and determine if the steering wheel is in the optimum or maximum envelope and sketch. What controls are not in this envelope?

1.6 ASAE S217.10 defines STD–3 point tractor hitches. What is the mast height and how many hitch categories exist? (HINT: Do not forget lawn and garden tractors.)

1.7 Are hydraulic accumulators for tractors covered by ASME Code or NFPA Code? State code number and note why.

1.8 Using ANSI B93 in the NFPA Code, calculate the wall thickness for a 1 in. outside diameter (o.d.) tube used at 5000 psi.

1.9 A steel part has a tensile strength of 120,000 psi. Determine the part's factor of safety for a yield condition when the actual stress was 60,000 psi. (HINT: Refer to SAE STD for material variability.)

1.10 Derive a formula for determining factor of safety when the material properties vary normally and the load is constant with zero standard deviation.

REFERENCES

1. *Agricultural Engineers Yearbook,* American Society of Agricultural Engineers, St. Joseph, Mich., 1981.

2. *ASME Boiler and Pressure Vessel Code,* ASME, United Engineering Center, New York, N.Y., 1977.
3. *Canons of Ethics of Engineers,* National Society of Professional Engineers, Washington, D.C., 1963.
4. *Code of Federal Regulations 37,* Patent, Trademarks and Copyrights, U.S. Government Printing Office, Washington, D.C.
5. Crossley, E. "Make Science a Partner in Your Design," *Machine Design* 52(9):51-55, 1980.
6. *Fluid Power Standards,* National Fluid Power Association, Inc., Milwaukee, Wis., 1981.
7. Gasson, P.C. *Theory of Design,* 1st edition, Harper & Row Publishers, New York, N.Y., 1973.
8. Kverneland, K. O. *World Metric Standards for Engineering,* Industrial Press, New York, N.Y., 1978.
9. Peters, L.C. "Some Engineering Criteria for Determining Design Defects," ASME Paper 80-WA/DE-16, The American Society of Mechanical Engineers, New York, N.Y., 1980.
10. Richey, C. B. Notes, Purdue University, West Lafayette, Ind., 1975.
11. Rohrbach, R.P. "Senior Level Design Courses: A Philosophy of Teaching Creative Design," ASAE Paper No. 78-5512, American Society of Agricultural Engineers, St. Joseph, Mich., 1978.
12. *SAE Handbook,* Society of Automotive Engineers, Inc., Warrendale, Pa., 1980.
13. Sissom, L.E. "Death, Injury and Destruction of Property by Design," ASME Paper 80-WA/DE-13, The American Society of Mechanical Engineers, New York, N.Y., 1980.
14. Smith, S. "Ethical Issues in Design Engineering Practice," National Design Engineering Conference, Chicago, Ill., March 24, 1980.

CHAPTER TWO

MATERIALS AND MANUFACTURING ECONOMICS

2.1 MATERIALS

A primary goal in designing a machine component is that the strength of the part have a factor of safety greater than one. To meet this goal the designer must know the properties of the material. Tests on a particular design can provide useful and precise data. Laboratory tests can indicate if the material of a design had the correct heat treatment, surface finish, size, and shape for reliability under field conditions. The design engineer should be acquainted with heat-treatment techniques and methods of manufacturing or consult with a professional metallurgist and industrial engineer to ensure the best possible selection of manufacturing techniques and materials. Common materials used in the design of agricultural machinery include iron, steel, aluminum, lead, rubber, glass, zinc, and plastics.

The emphasis on energy conservation during the late 1970s greatly affected the composition of vehicles. Figure 2.1 depicts automotive trends of the 1980s that will also affect tractors, combines, and other farm machinery as designers attempt to increase the payload or to save energy. Lighter materials, such as plastics and aluminum, are becoming more popular. Tables 2.1 and 2.2 depict some common materials and their properties.

For specific data on materials prior to designing, refer to American Society for Testing and Materials (ASTM), ISO, SAE, and other standards listed in this chapter's references. Also refer to the appendix for further tables, including metric ISO sizes for common shapes of standard steel stock. Note that this is just a partial list and suppliers should be consulted during design to determine availability of sizes and shapes.

2.2 NUMBERING SYSTEMS FOR CARBON STEELS AND ALUMINUMS

A numbering system devised by the Society of Automotive Engineers and the American Iron and Steel Institute helps engineers specify a particular steel.

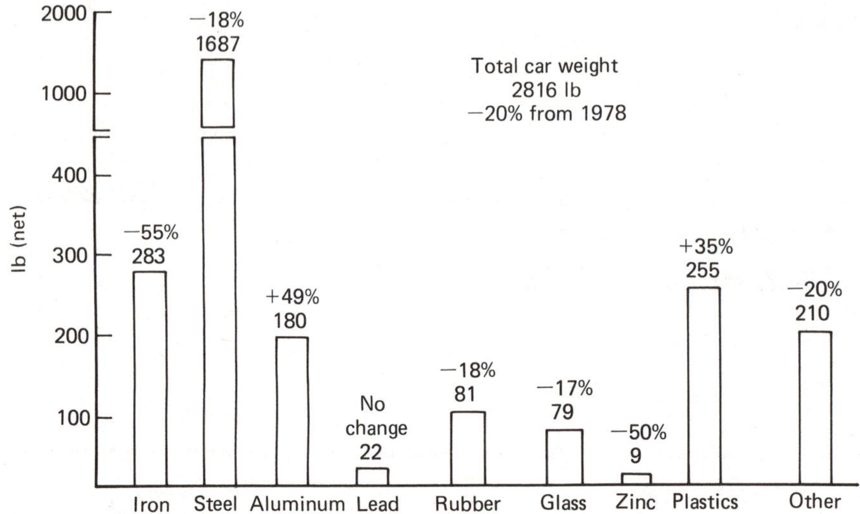

Figure 2.1 Material trends in 1987 composite car.
Reprinted with permission © 1982 Society of Automotive Engineers, Inc.

The system, based on the chemical composition of the material, includes just four digits. The first two digits depict the type of alloy, and the last two digits give the carbon content (0.XX%). In some cases, when the carbon content exceeds 1%, three digits are used. For example, SAE 1010 steel is a plain carbon steel with 0.10% carbon content. The numbering scheme for various types of steels is given in Table 2.3. Common mechanical properties of steels used in farm-equipment design are given in Table 2.4.

The numbering system for aluminum alloys also uses four digits. The first digit specifies the alloy type, the second digit defines the alloy modification, and the last two digits relate to the specific alloy or aluminum purity. The alloys are indicated in Table 2.5.

Cast iron is designated by its tensile strength. For example, a No. 40 cast iron is expected to have a tensile strength greater than 275 MPa (40,000 psi). Again, refer to ASTM and SAE standards for the various types.

2.3 NOMENCLATURE FOR MATERIAL PROPERTIES

Particular material properties are important in the design of a machine component. One design might require more of one property than another, depending on the application and loading.

Tensile strength, sometimes called the ultimate strength, is the greatest stress seen by the material, or the material's resistance to being pulled apart. Values are given for uniaxial tests. Yield strength is usually designated at a 0.2% permanent elongation (see Fig. 2.2).

Table 2.1 **Engineering Material Properties (Metric)**

Material at 20° C	*Yield (mPa)*	*Tensile (mPa)*	*Reduction in Area*	*E(GPa)*	*G(GPa)*	*ν*
1015 steel	315	420	61	211	82	0.29
1050 steel	415	725	40	211	82	0.29
4340 steel (tempered at 200° C)	1675	1876	38	211	82	0.29
2024 aluminum T-3	345	485	18	70	26	0.34
7178 aluminum T-6	540	605	11	70	26	0.34
ABS plastic	—	46	—	—	—	—
Nylon 66	—	59−83	—	1.2−2.9	—	—
Polyethylene	—	1−3	—	1.1−1.6	—	—
Polystyrene	—	6−9	—	2.7−4.2	—	—

Table 2.2 **Engineering Material Properties (English)**

Material at 68° F	*E* (10^6 psi)	*G* (10^6 psi)	*ν*
Metals			
Aluminum	10.2	3.8	0.345
Chromium	40.5	16.7	0.210
Copper	18.8	7.0	0.343
Gold	11.3	3.9	0.44
Iron	30.6	11.8	0.293
Magnesium	6.5	2.5	0.291
Nickel	28.9	11.0	0.312
Silver	12.0	4.4	0.367
Titanium	16.8	6.35	0.321
Tungsten	59.6	23.3	0.280
Vanadium	18.5	6.8	0.365
OtherMaterials			
Diamond	≈ 140	—	—
Glass (heavy flint)	11.6	4.6	0.27
Nylon 66	0.17	—	—
Polyethylene (high density)	0.058−0.19	—	—
Polypropylene	0.16−0.39	—	—
Polystyrene	0.39−0.61	—	—
Quartz (fused)	10.6	4.5	0.170
Tungsten carbide	77.5	31.8	0.22

Reprinted with permission © *Tables of Physical and Chemical Constants,* 14th edition, Longman Group Limited, London, 1973.

Table 2.3 **Nomenclature for AISI and SAE Steels**

AISI or SAE Number	*Composition (%)*
10XX	Plain carbon steels
11XX	Plain carbon (resulfurized for machinability)
13XX	Manganese (1.5–2.0%)
23XX	Nickel (3.25–3.75%)
25XX	Nickel (4.75–5.25%)
31XX	Nickel (1.10–1.40%), chromium (0.55–0.90%)
33XX	Nickel (3.25–3.75%), chromium (1.40–1.75%)
40XX	Molybdenum (0.20–0.30%)
41XX	Chromium (0.40–1.20%), molybdenum (0.08–0.25%)
43XX	Nickel (1.65–2.00%), chromium (0.40–0.90%), molybdenum (0.20–0.30%)
46XX	Nickel (1.40–2.00%), molybdenum (0.15–0.30%)
48XX	Nickel (3.25–3.75%), molybdenum (0.20–0.30%)
51XX	Chromium (0.70–1.20%)
61XX	Chromium (0.70–1.10%), vanadium (0.10%)
81XX	Nickel (0.20–0.40%), chromium (0.30–0.55%), molybdenum (0.08–0.15%)
86XX	Nickel (0.30–0.70%), chromium (0.40–0.85%), molybdenum (0.08–0.25%)
87XX	Nickel (0.40–0.70%), chromium (0.40–0.60%), molybdenum (0.20–0.30%)
92XX	Silicon (1.80–2.20%)

XX, carbon content, 0.XX%.
Mn, All steels contain ≈ 0.50% manganese.
BOF, Prefixed to show Basic oxygen furnace.
B, Prefixed to show Bessemer steel.
C, Prefixed to show open-hearth steel.
E, Prefixed to show electric furnace steel.

Reprinted with permission © *Elements of Material Science,* VanVlack, Addison Wesley Publishing Co.

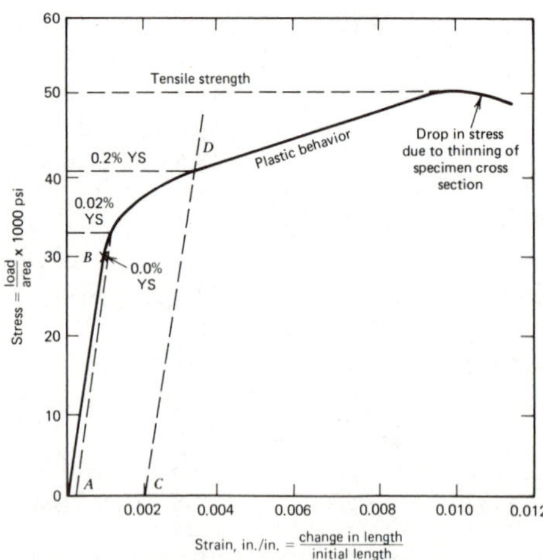

***Figure** 2.2* Methods for determining yield strength (YS). Courtesy of Huntington Alloys, Inc.

Table 2.4 **Estimated Mechanical Properties and Machinability Ratings of Nonresulfurized Carbon Steel Bars, Manganese 1.00% Maximum**

			Estimated Minimum Values							*Average*
UNS No.	*SAE and/or AISI No.*	*Type of Processing*	*Tensile Strength, psi*	*Tensile Strength, MPa*	*Yield Strength, psi*	*Yield Strength, MPa*	*Elongation %*	*Reduction Area, %*	*Brinnell Hardness*	*Machinability (CD 1212 = 100%)*
G10080	1008	Hot rolled	44,000	303	24,500	170	30	55	86	55
		Cold drawn	49,000	340	41,500	290	20	45	95	
G10100	1010	Hot rolled	47,000	320	26,000	180	28	50	95	55
		Cold drawn	53,000	370	44,000	300	20	40	105	
G10150	1015	Hot rolled	50,000	340	27,500	190	28	50	101	60
		Cold drawn	56,000	390	47,000	320	18	40	111	
G10190	1019	Hot rolled	59,000	410	32,500	220	25	50	116	
		Cold drawn	66,000	460	55,000	380	15	40	131	70
G10200	1020	Hot rolled	55,000	380	30,000	210	25	50	111	
		Cold drawn	61,000	420	51,000	350	15	40	121	65
G10300	1030	Hot rolled	68,000	470	37,500	260	20	42	137	
		Cold drawn	76,000	520	64,000	440	12	35	149	70
G10400	1040	Hot rolled	76,000	520	42,000	290	18	40	149	
		Cold drawn	85,000	590	71,000	490	12	35	170	60
G10450	1045	Hot rolled	82,000	570	45,000	310	16	40	163	
		Cold drawn	91,000	630	77,000	530	12	35	179	55
		ACD[a]	85,000	590	73,000	500	12	45	170	65
G10500	1050	Hot rolled	90,000	620	49,500	340	15	35	179	
		Cold drawn	100,000	690	84,000	580	10	30	197	45
		ACD[a]	95,000	660	80,000	550	10	40	189	55

[a]ACD represents annealed cold drawn.

Table 2.5 **Aluminum Alloy Designation System**

Aluminum 99%	1XXX
Copper	2XXX
Manganese	3XXX
Silicon	4XXX
Magnesium	5XXX
Magnesium and silicon	6XXX
Zinc	6XXX
Other elements	7XXX

Ductility indicates how much a material is deformed as it is stretched—the greater the deformation, the more ductile the material. Ductility is reported as elongation, or as reduction in area.

$$\text{elongation } \epsilon \text{ (in./in.)} = \frac{L - L_o}{L_o} \qquad [2.1]$$

where L_o = the initial length
L = the final length

$$\text{reduction } RA = \frac{A_o - A}{A_o} \qquad [2.2]$$

where A_o = the initial cross-sectional area
A = the final cross-sectional area

The more ductile, the greater the part can be deformed without failure and the greater its ability to resist impact loads.

Hardness is the material's resistance to penetration and permanent deformation. Hardness testers create a dent on the material subjected to a standard load. Hardness is then expressed as a function of the dent size. Several hardness schemes are compared in Figs. 2.3 to 2.5.

One way to estimate the Brinell hardness (BH) number for steel is

$$\text{BH} = \frac{\sigma_{ult}}{500} \qquad [2.3]$$

where σ_{ult} = the tensile strength in psi

or

$$\text{BH} = \frac{\sigma_{ult}}{3.45} \qquad [2.4]$$

where σ_{ult} = the tensile strength in MPa

Brittleness is the tendency of a material to break under impact (see Fig. 2.6).

Compression strength is a material's resistance to being compressed.

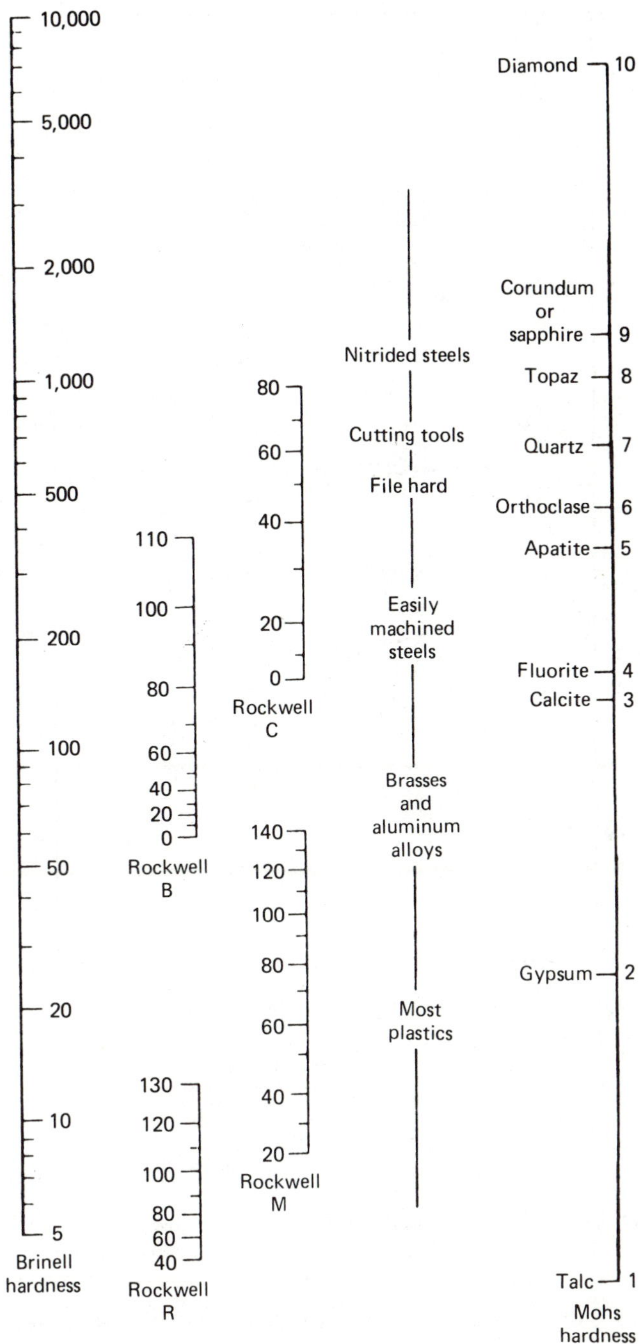

Figure 2.3 Comparison of hardness scales. Reprinted with permission © Faupel and Fisher, Engineering Design, John Wiley and Sons, Inc.

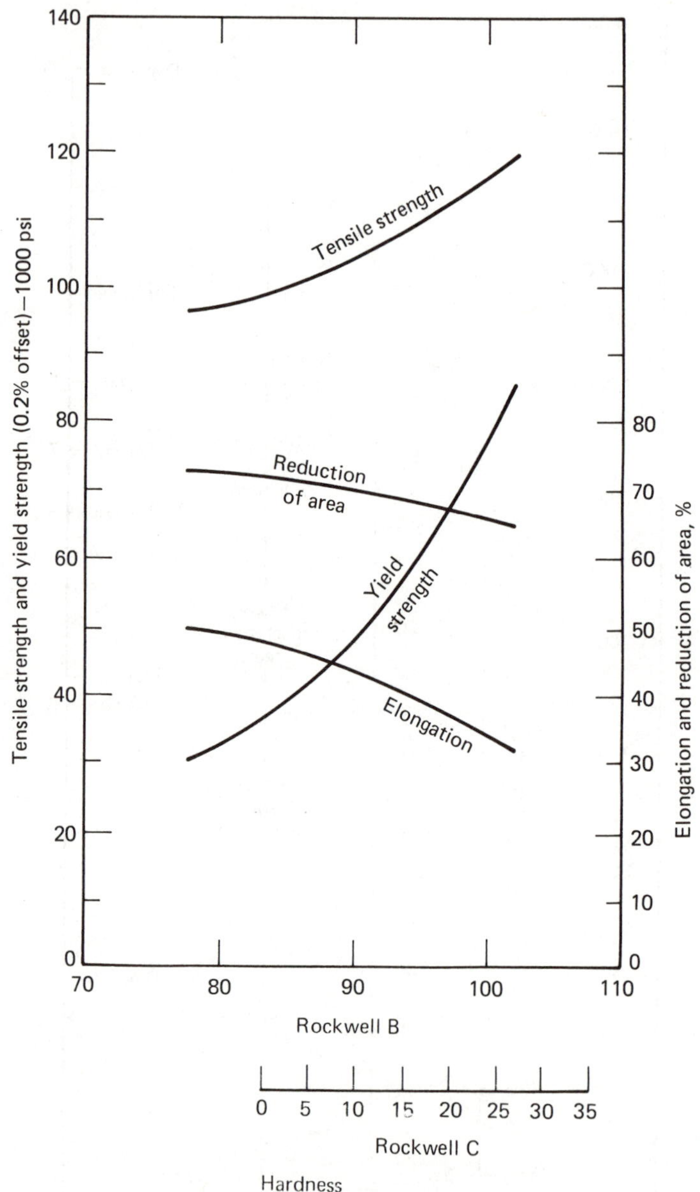

Figure 2.4 Approximate relationship between tensile properties and hardness of cold-drawn duranickel alloy 301 rod. Courtesy of Huntington Alloys, Inc.

Malleability is the capability of a material to be shaped and formed (see Figs. 2.7 and 2.8).

Toughness is the ability of a material to withstand bending and twisting and is a desirable property for farm loader frames, crankshafts, and rockshafts.

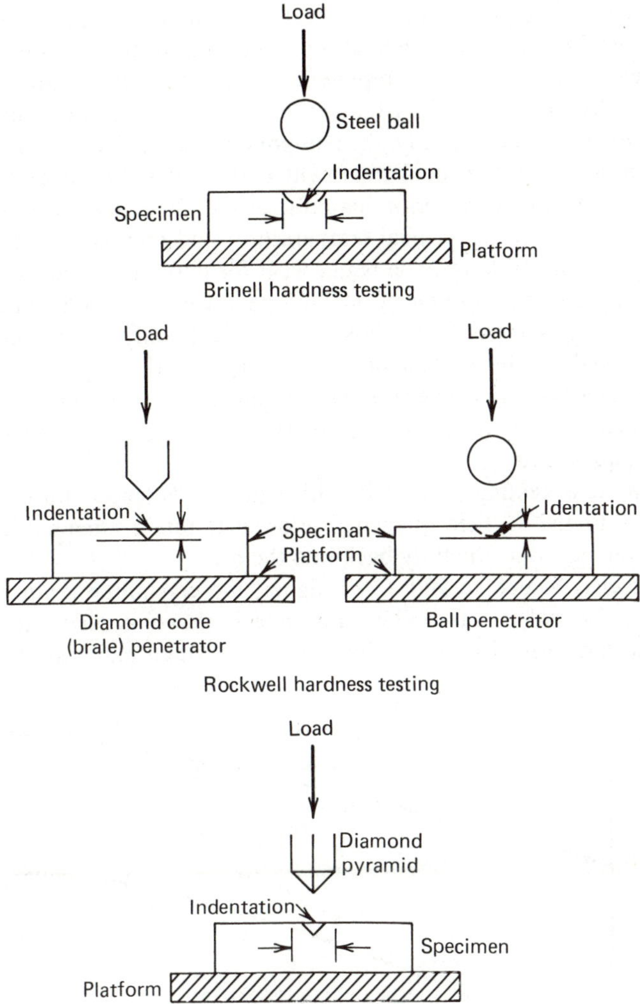

Figure 2.5 Various hardness testing apparatus.
Courtesy of Huntington Alloys, Inc.

Grain size is the microscopic measurement of metals that shows differences in sizes that relate directly to strengths.

Alloys of steel are made by adding various metals to steel to enhance specific characteristics. The overall alloy content is usually less than 1%. A common steel alloy is chromium, which increases the strength of steel and helps prevent rust. Manganese and nickel also increase strength; molybdenum increases toughness; tungsten retains hardness properties during heat treatment; and vanadium retards grain growth (grain growth lowers the material's strength).

Heat treatment is a process that can be used to change the physical properties of steels after they have been shaped into their final form. The process of heat treating begins by heating the component to a specified temperature (which is a function of the material). That temperature is then held for some time in a controlled environment. Finally, the component is cooled at a controlled rate.

Annealing is a common heat treatment that softens the metal and relieves residual stresses caused by prior manufacturing. The steel part is heated to around 800° C (called the critical temperature) and then cooled slowly in air. Hardening is intended to provide better wear for parts subjected to friction and requires bringing the part to the critical temperature, followed by fast cooling.

Case hardening is a specific process in which up to only 0.2 cm depth of hard surface is sought. The part is heated to the critical temperature and then exposed to a carbon atmosphere or carbon particles. After the carbon diffuses into the metal to the required depth, the part is cooled, then reheated to 200 to 500° C, and quenched.

Tempering is a heating process in which the part's temperature is raised to between 200 to 550° C, then quenched in order to retain some hardness and at the same time put some ductility back into the piece.

When steel is heated to 700° C or higher, the crystalline grain structure is called austenite and is soft. When austenite is slowly cooled, the crystalline structure is transformed into pearlite, which is a tough material. Fast cooling

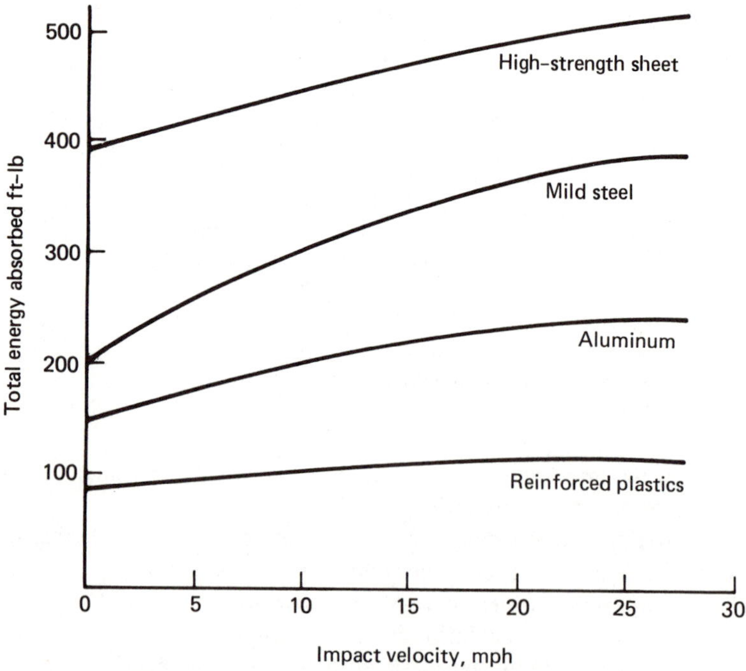

Figure 2.6 Impact strength of steel, aluminum, and plastic.

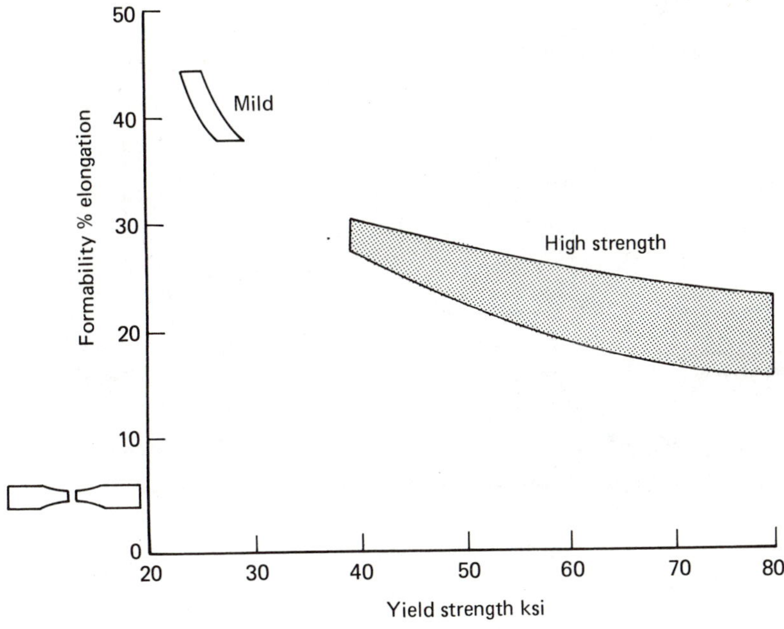

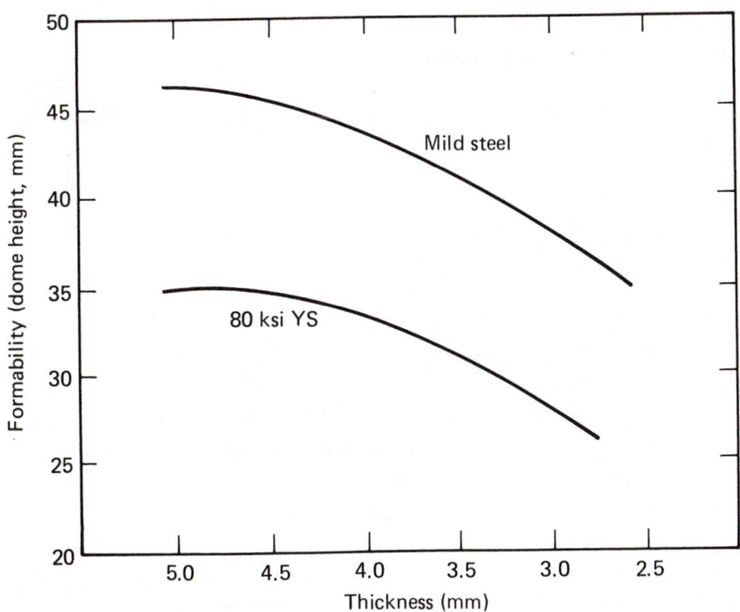

Figure 2.7 Formability compared to strength for steels. Reprinted with permission © Bethlehem Steel.

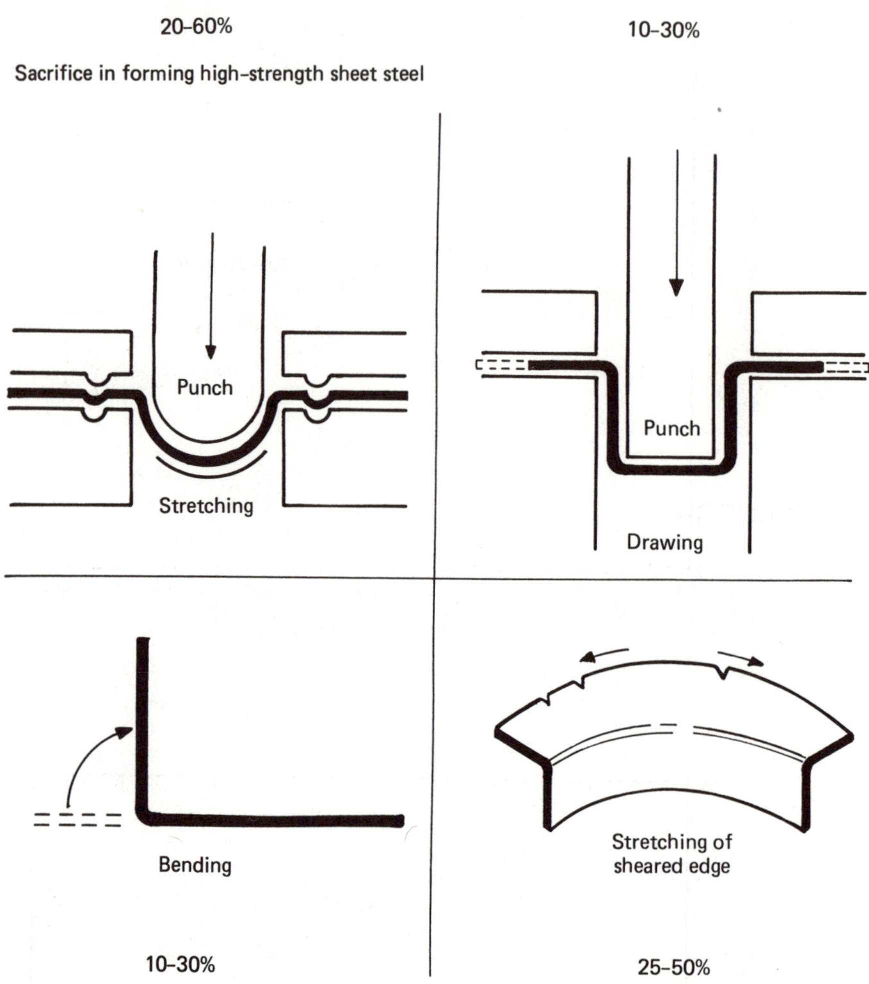

Figure 2.8 Allowable stretch when shaping parts using a punch process. Reprinted with permission © Bethlehem Steel.

(water quenching) yields grains called martensite, which are brittle and have the highest tensile strengths. Oil cooling (which is slower than fast quenching) yields a substance with less hardness than martensite, called troostite, which is used in files. Spring steel, called sorbite, results from cooling in a salt bath more slowly than troostite but faster than pearlite. The endless possible combinations of heat treatments all depend on the original material, its size, its critical temperature, and the relative desired toughness or strength. A single batch of steel from one factory can have many varied properties, depending on when and how it was processed. The designer must be aware of this variability, depicted by the SAE STD J413a, and must specify a minimum strength.

Cast irons contain more carbon (2 to 4%) than do steels. Special forms of cast iron (nodular and malleable), which can withstand much more impact than plain gray cast iron, are made by a special heat-treatment process that holds the material at 900° C for 24 h for every 2.5 cm of thickness. The critical temperature in gray cast iron is 900° C; as the carbon content of steel changes, so does the critical temperature used in heat treatment.

Chemical reactions occur while making steel. The most critical reaction, between oxygen and carbon, continues until the steel solidifies. If small amounts of aluminum are added the oxygen is tied up chemically and does not react with the carbon, resulting in a product called killed steel. Production costs are higher for killed steel (see Fig. 2.9). An option called inclusion control is available in making killed steel. Inclusion control improves the microstructure, which contains impurities that result in minute crack formations. If this crack initiation is reduced, the killed steel will exhibit better forming characteristics. High-strength low-alloy steel (HSLA) is normally obtained by alloying mild steel with columbium or vanadium.

2.4 MANUFACTURING PROCESSES

Experience in engineering helps reduce costs by exposing the designer to many different processes of forming metals. Steel is usually purchased in standard size sheets; it is important to use as much of the sheet as possible to reduce scrap cost. By nesting the shape of a design or many designs, the engineer can reduce scrap. Figure 2.10 shows possible nestings for various shapes. The engineer should be aware that the cross-hatched sections could be utilized on another machine; therefore, size is important.

Familiarity with the available forming processes is necessary before beginning component design. An experienced engineer has the advantage of having seen scores of parts formed (Figs. 2.11 to 2.19).

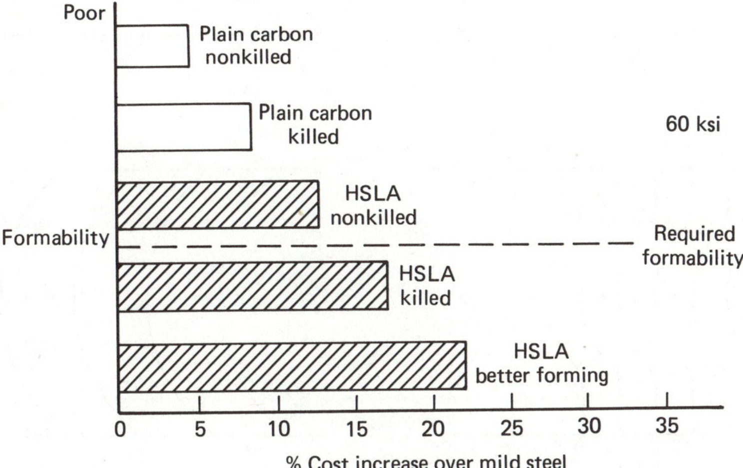

Figure 2.9 Cost of various steels when processed for better formability. Reprinted with permission © Bethlehem Steel.

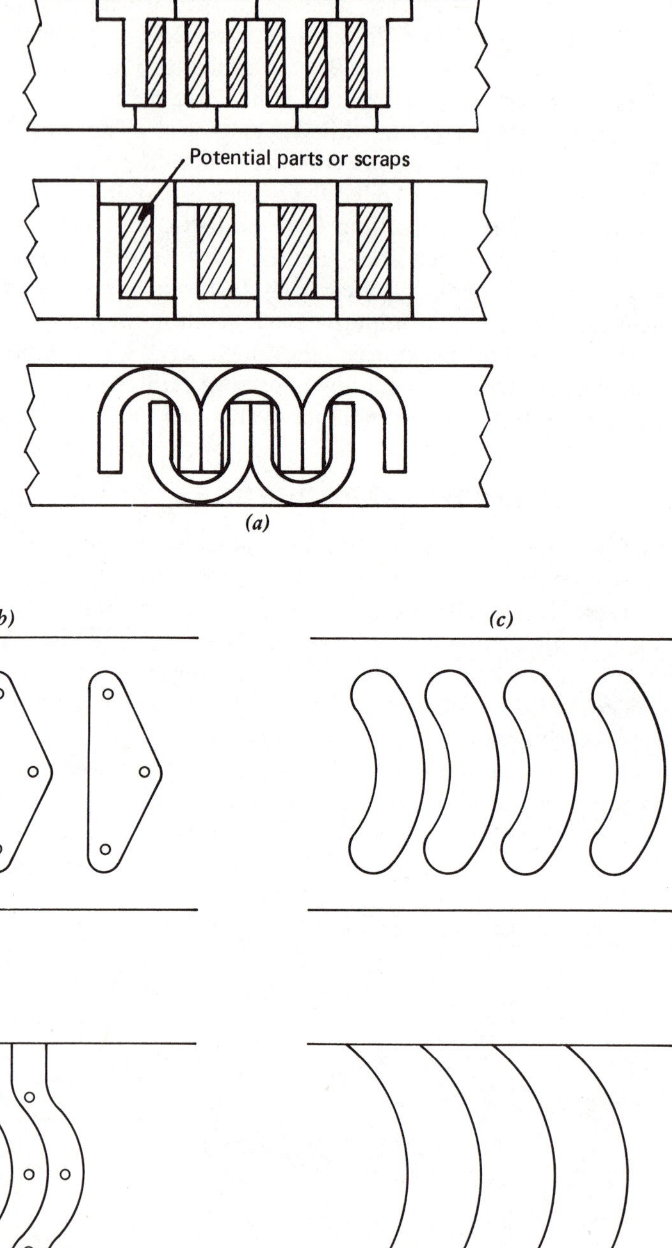

Figure 2.10 Nesting parts and utilizing scrap from common-size stock.
Reprinted with permission © William Wai-Chung Chow, Cost Reduction in Product Design, Von Nostrand Reinhold Co.

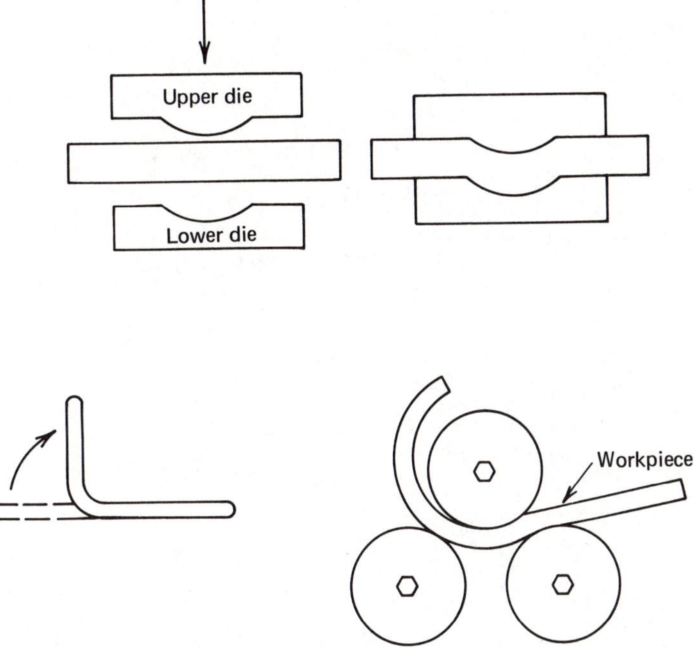

Figure 2.11 Bending operations used to form parts.
Courtesy of Huntington Alloys, Inc.

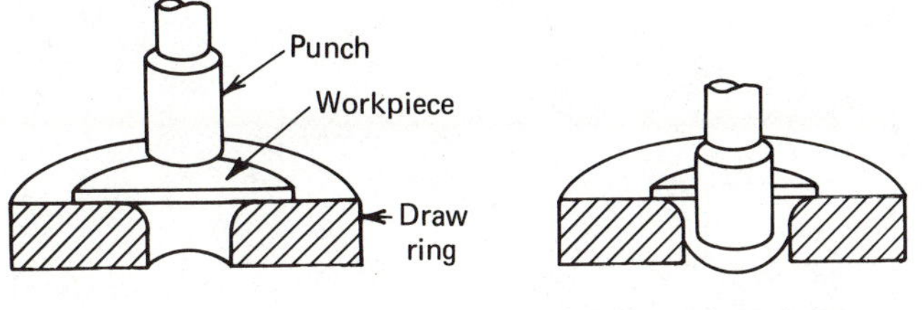

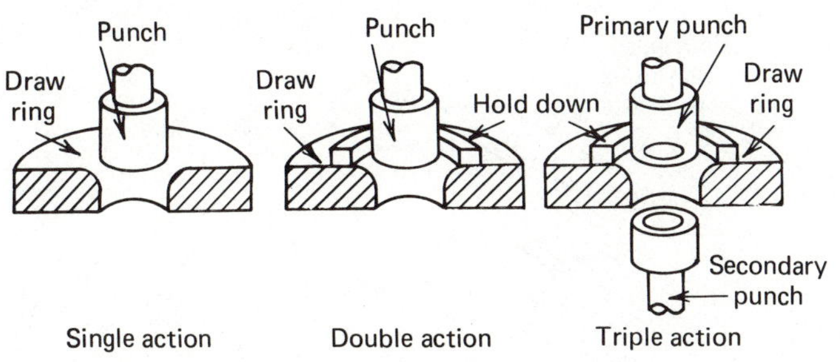

Figure 2.12 Deep drawing operations.
Courtesy of Huntington Alloys, Inc.

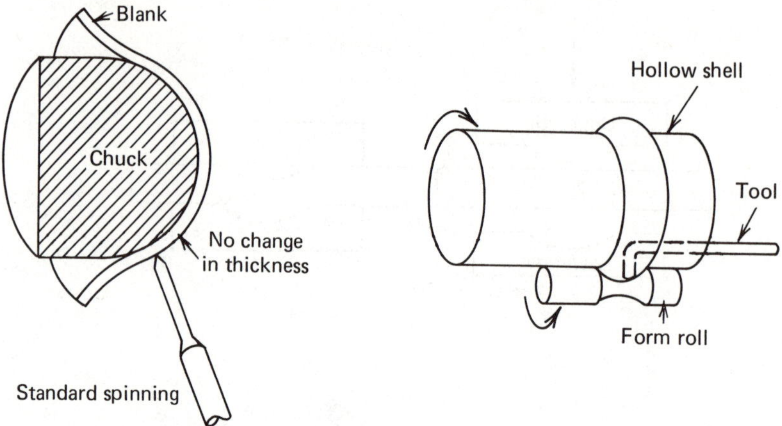

Figure 2.13 Spinning methods that form curved surfaces.
Courtesy of Huntington Alloys, Inc.

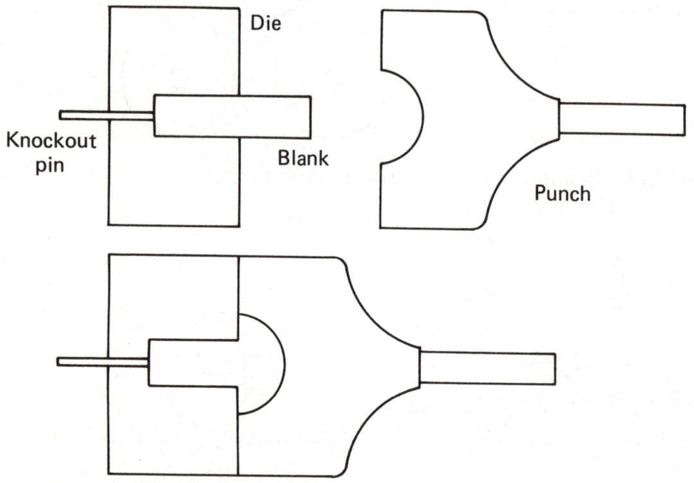

Figure 2.14 Single-blow heading for forming bolts and rivets from bar stock.
Courtesy of Huntington Alloys, Inc.

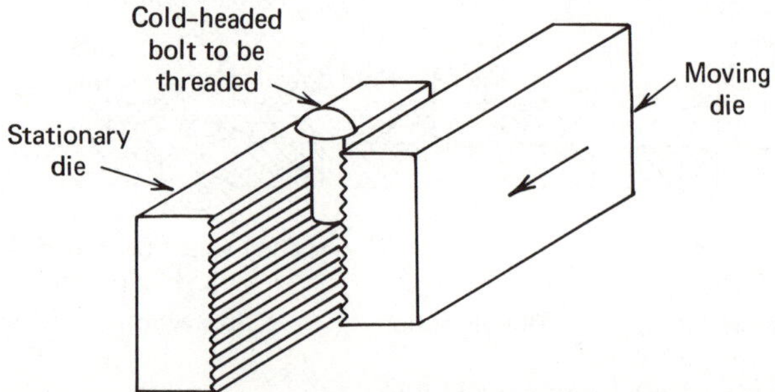

Figure 2.15 Thread-forming process used to Mass Production bolts.
Courtesy of Huntington Alloys, Inc.

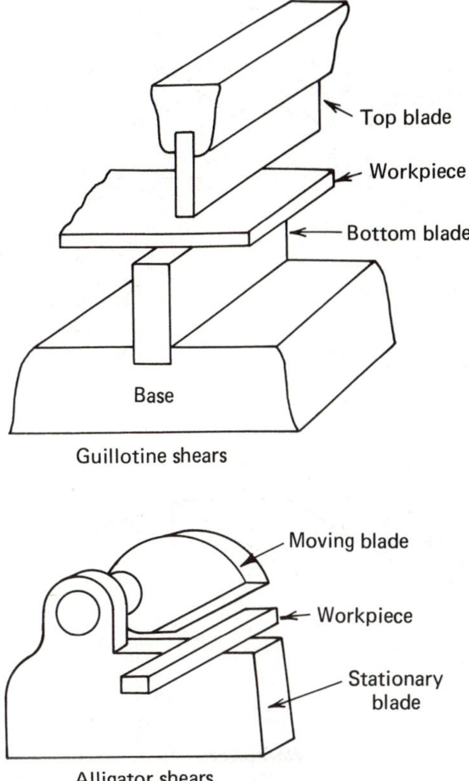

Figure 2.16 Two common types of shears. Courtesy of Huntington Alloys, Inc.

Common sense applied to design will eliminate potential problems. Commonality of parts increases usage but might cause problems during assembly. For example, in the attachment of power-steering hoses to the pump and steering gear on trucks, it is important to use different-size fittings to avoid mix-ups in assembly that could cause a vehicle to turn left when the steering wheel is moved for a right turn. In this application, the use of two different male and two different female fittings, one on each hose, ensures the proper connection of the pump and valve. Another example of common sense is using the same-size bolts for the worst load, which will increase bolt usage and might reduce bolt costs as volume increases over 10,000. Other considerations in design are shown in Fig. 2.20.

2.5 TOOLING

The fixtures, robots, and many machines used in mass production reduce the cost of manufacturing parts (see Fig. 2.21). A part's tooling can range from a few hundred dollars to hundreds of millions of dollars. Tooling requires the purchase of capital equipment to manufacture many parts quickly and allows for part interchangeability (including the use of less-skilled labor). To justify a

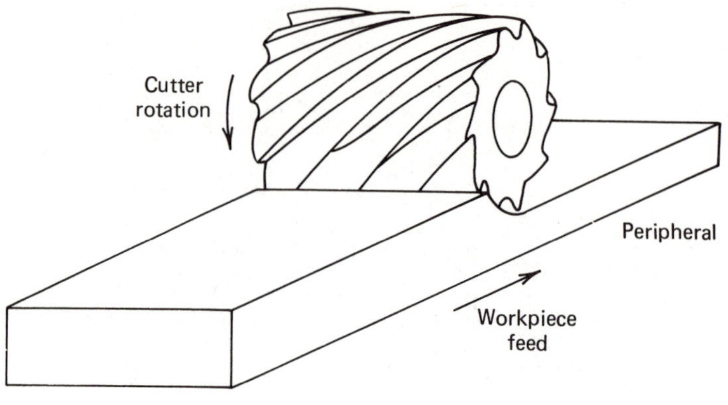

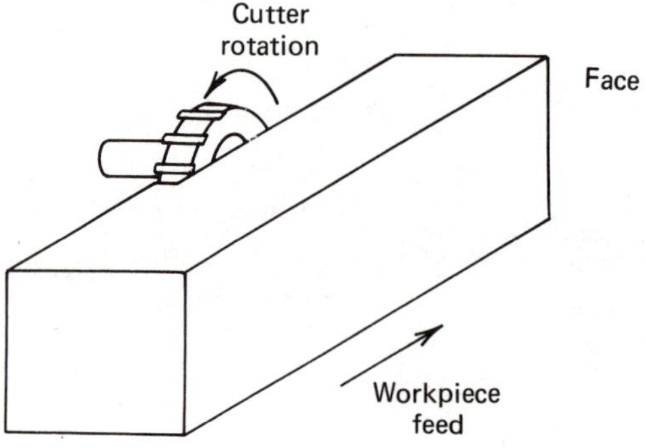

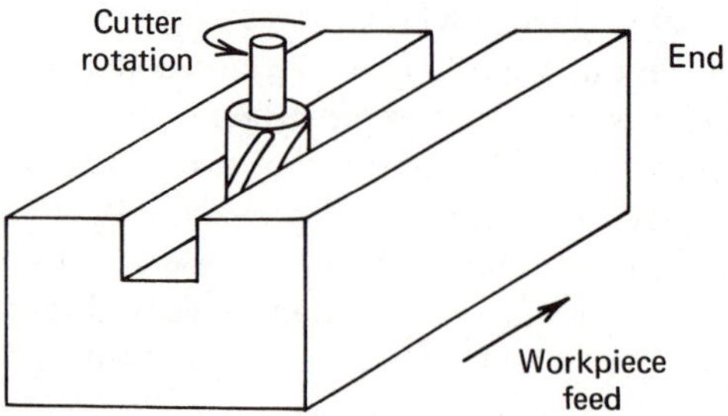

Figure 2.17 Milling.
Courtesy of Huntington Alloys, Inc.

capital expenditure for tooling, the market for a product must be predictable (e.g., tractor sales in the United States have recently varied between 150,000 to 200,000 a year).

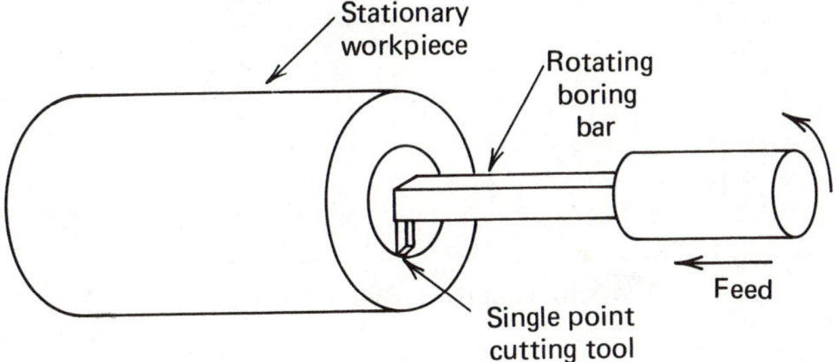

Figure 2.18 Boring is used to refine an inside diameter. Courtesy of Huntington Alloys, Inc.

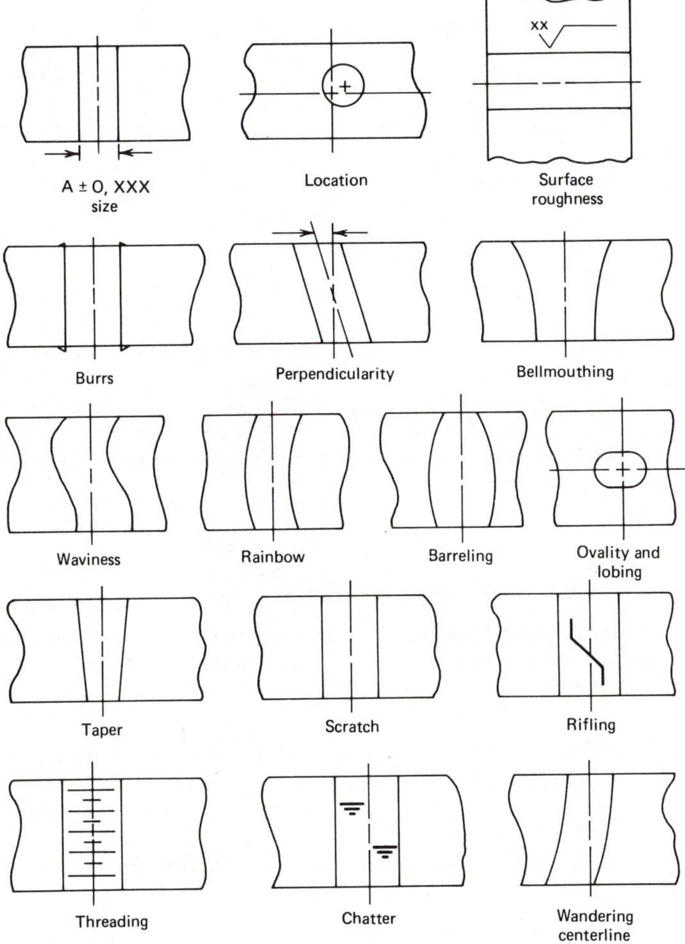

Figure 2.19 Aberrations that occur in small holes.

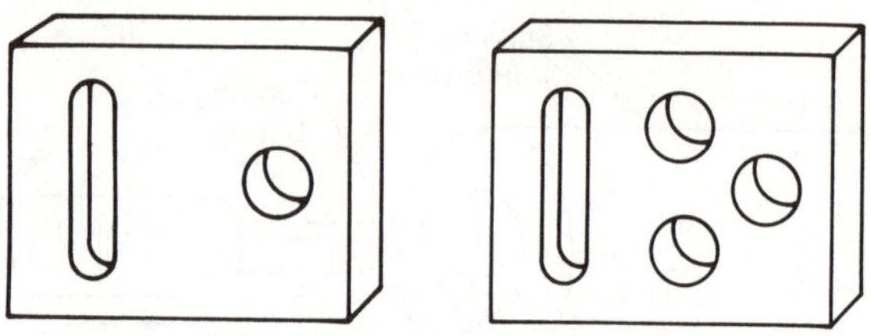

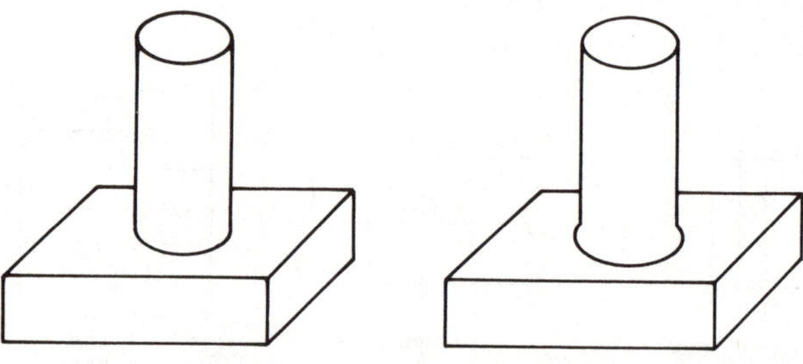

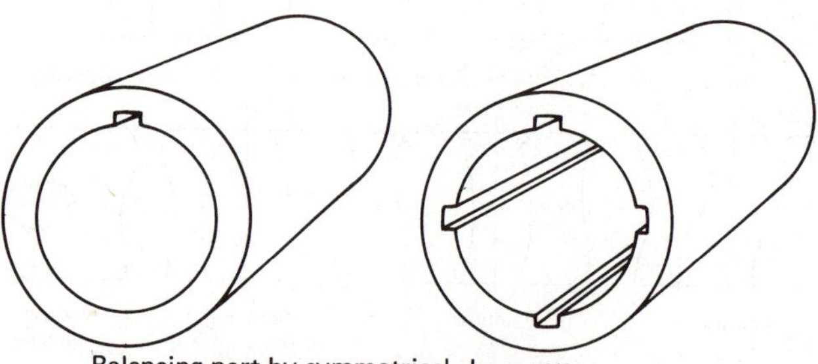

Figure 2.20 Examples of correct design for heat treating. Reprinted with permission © William Wai-Chung Chow, Cost Reduction in Product Design, Von Nostrand Reinhold Co.

An economic analysis for tooling is necessary before spending millions of dollars. At the break-even point, yearly operating savings equal total fixed costs. Let

$$N = \text{number of pieces manufactured per year to break-even} \qquad [2.5]$$

then

$$N = \frac{\text{fixed cost}}{\text{savings per part}} \qquad [2.6]$$

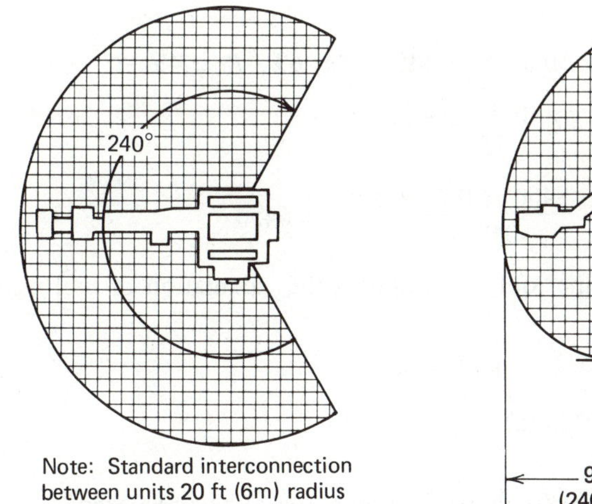

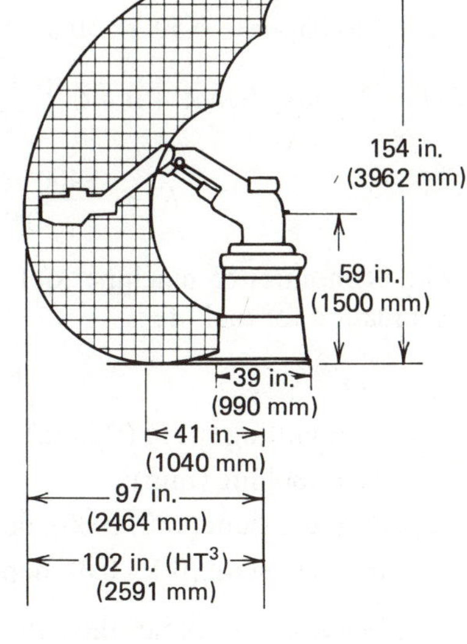

Grid scale: one block = 6 in. (150mm)

Figure 2.21 A Robot's basic range and floor-space drawing. Courtesy of Cincinnati Milacron, Inc.

or

$$N = \frac{I\,(A + B + C + \frac{1}{H}) + Y(N')}{S\,(1 + t)} \qquad [2.7]$$

where I = fixture cost ($)
S = unit savings in direct labor ($)
t = percent overhead on labor saved
A = interest rates
B = fixed charges (%), taxes, insurance
C = upkeep rate as a percent of maintenance
H = equipment life (in years) amortization time
Y = estimated cost on each setup ($)
N' = number of setups

Example 2.1 Tooling Break-even Calculation

If an estimated fixture cost is $400, how many pieces must be run in one lot to break-even on cost in 2 years. Note: Most companies must pay off tooling in the first or second year and estimate their needed return on investing in a new product equal to the current interest rate plus approximately 10%.

Given:

$S = \$0.03;\ t = 0.50, \textit{or } 50\%;\ Y = \$10;\ A = 6\%;\ B = 4\%;\ C = 10\%;$

If $H = 2$ years, then $A + B + C + \frac{1}{H} = 70\%$

$$N = \frac{\$400 \times 0.70 + \$10}{0.03 \times 1.5} = 6450$$

The economics of machine tool cost requires the calculation of tool wear. The equation for tool life is

$$VT^{n} = \mathrm{K} \qquad [2.8]$$

where V = cutting speed (ft/min)
T = tool life (min)
K = constant (180–1000), depending on material being worked
n = constant (0.1 – 0.4), depending on material of tool

Again, by using economics the cutting speed (V_1) can be determined in ft/min at minimum cost.

$$V_1 = \frac{K\, C_1^{n}}{\left[\left(\frac{1}{n} - 1\right)\left(C_2 + rC_1\right)\right]^{n}} \qquad [2.9]$$

where K and n are the same as in Eq. 2.8

C_1 = cost of the machine plus labor per unit time
C_2 = cost of the tool/number of regrinds possible plus cost of regrinding
r = tool change time

If bottlenecks are a problem, then the cutting speed for minimum operation time (V_2 in ft/min) is defined as

$$V_2 = \frac{\mathrm{K}}{\left[\left(\frac{1}{n} - 1\right)r\right]^{n}} \qquad [2.10]$$

Close tolerances also increase costs, as seen in Fig. 2.22.

2.6 USER ECONOMICS

To justify purchase of a tractor or combine the user must be able to foresee a profit. The amount of work that a machine can do is called its field capacity and is usually designated in acres per hour or tons per hour.

$$C = \frac{SWE}{8.25} \qquad [2.11]$$

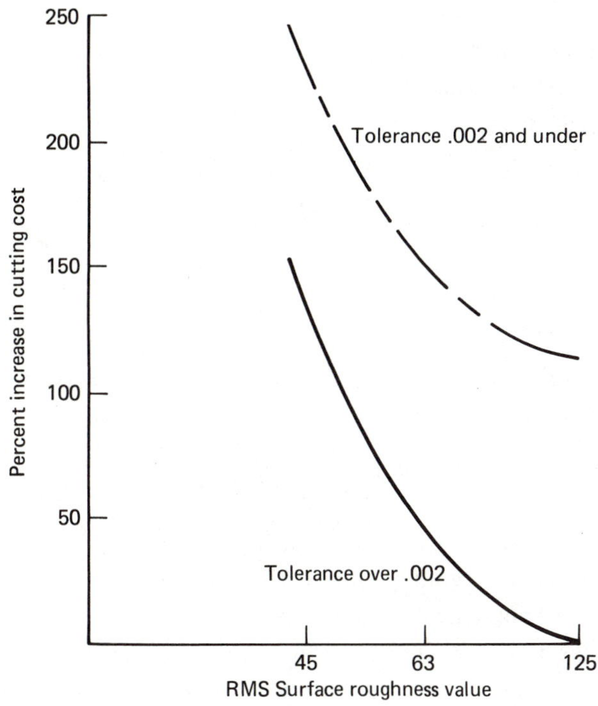

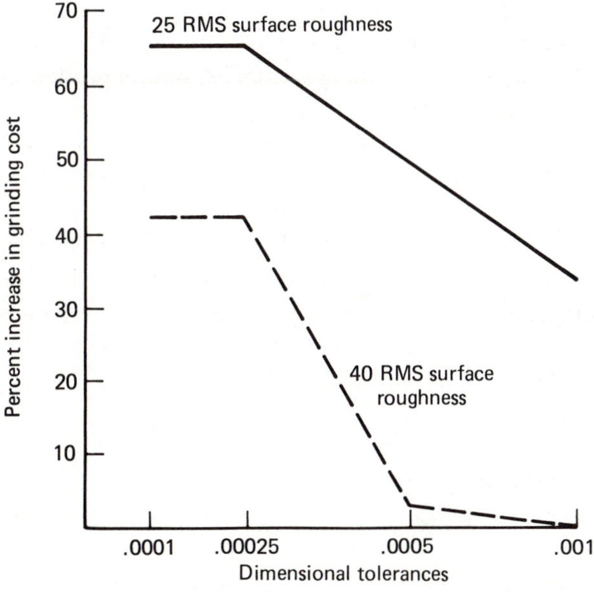

Figure 2.22 Cost-tolerance relationship. Reprinted with permission © William Wai-Chung Chow, Cost Reduction in Product Design, Von Nostrand Reinhold Co.

where C = capacity (a/h)
S = speed (mph)
W = effective width (ft)
E = field efficiency

In metric units

$$C = \frac{SWE}{10} \quad \text{(ha/h)} \qquad [2.12]$$

where S = speed (km/h)
W = width (m)

Field efficiency is reduced when time is lost in turning corners, in unproductive rest periods, such as filling tanks, and in similar situations.

$$E_f = \text{field efficiency} = 100 \times \frac{T_e}{T_o + Th + T_a} \qquad [2.13]$$

where T_o = theoretical time per acre
T_e = effective operating time = $T_o \times 100/K$
K = % of implement width utilized
Th = lost time per acre (not proportional area)–rests, breaks, idle travel
T_a = lost time per acre (proportional area)–filling tanks, yields

Time studies have shown that efficiencies can be as low as 25% and are usually not greater than 90%. Some sample field efficiencies for farming operations are shown in Table 2.6.

The user's cost of machinery consists of fixed and variable costs. Included in fixed costs are taxes, shelter, insurance, interest, and depreciation. Depreciation may be straight line, double declining balance, sum of digits, or accelerated. Variable costs, a function of usage, include fuel, lubrication, maintenance, repairs, labor, rent of tractor for pull-type machines, and a timeliness value used when comparing two machines.

Hunt developed the relationship for determining a machine's accumulated yearly cost.

$$\text{accumulated yearly cost } (AYC) = \text{fixed cost} + \text{variable cost} \qquad [2.14]$$

$$AYC = FC\,\%P + \frac{8.25A}{SWE}(R + L + O + F + T)$$

where A = total acres
R = repair costs ($/h)
L = labor cost ($/h)
O = oil cost ($/h)
F = fuel cost ($/h)
T = timeliness cost, or tractor rental cost ($/h)

S = speed (mph)
W = width (ft)
E = efficiency
$FC\%P$ = fixed cost as a percent of purchase price

In metric units the variable coefficient term (for Eq. 2.14) changes to

$$\frac{10 \times ha}{SWE}$$

where S = km/h
W = m
ha = total hectares

Table 2.6 **Field Efficiencies for Common Operations**

Tillage	
Moldboard plow	75–80%
Disk harrow	77–90%
Field cultivator	75–85%
Spring tooth or Spike tooth harrow	65–76%
Cultivation	
Row crop	68–85%
Rotary hoe	80–88%
Seeding	
Corn planter	
(corn only)	60–75%
(with fertilizer or pesticide attachment)	45–65%
Grain drill	65–80%
Broadcast	65–70%
Harvesting	
Mower	77–85%
Rake	62–89%
Baler	69–80%
Forage harvester	50–70%
Combine	60–75%
Corn picker	55–70%
Cotton picker	60–75%
Swather	70–85%
Miscellaneous	
Sprayer	55–65%

Fixed costs as a percent of purchase price (P) are illustrated by the following:

$$\text{depreciation } (D; \text{straight line}) = \frac{P - S}{L} = 0.09P$$

where salvage (S) = 10% of P
life (L) = 10 yr

$$\text{interest on investment} = \left[\frac{P + S}{2}\right] i = \frac{1.1P}{2}(i)$$

where i = interest rate, such as 0.2
taxes = $0.015P$ (from farm records)
insurance = $0.0025P$ (from farm records)
shelter = $0.01P$ (from farm records)

Total fixed cost is the sum of depreciation, interest on investment, taxes, insurance, and shelter. When i equals 0.2, the total fixed cost can be expressed as a function of P, in this case, total fixed cost = $0.238P$.

Repair, fuel, and oil costs vary with location in the United States and depend on transportation costs and the type of farming enterprise. Taxes, insurance, and shelter are usually constant in each state and are a fixed percentage of purchase price. Figure 2.23 is an example of repair cost for a tractor as a function of purchase price. Combine repair cost is shown in Fig. 2.24. Because care and preventive maintenance vary from user to user, actual recorded expenses over time are preferred to national averages. Repair costs for most machines are a function of usage hours. For tractors, major engine overhaul occurs around 2000 and 5000 hours, and increases repair costs in those years. Refer to the books by Hunt, Richey, and Parsons for repair data. Average fuel usage can be estimated by following:

$$\text{gal/h} = PTO \text{ hp} \times 0.06 \quad \text{(gas)} \tag{2.15}$$

or

$$\text{liters/h} = \text{kW} \times 0.012 \quad \text{(gas)}$$

$$\text{gal/h} = PTO \text{ hp} \times 0.044 \quad \text{(diesel)} \tag{2.16}$$

or

$$\text{liters/h} = \text{kW} \times 0.009 \quad \text{(diesel)}$$

For large tractors, Nebraska tests give fuel-consumption data. Oil, filters, and lubrication cost approximately 15% of fuel costs.

From the user's standpoint, a time to trade machinery can be calculated by using the accumulated yearly cost equation. If exact repair data are available, they can be used to determine accumulated yearly costs. When the yearly cost begins to become greater than the accumulated yearly cost, trading should be considered (see Fig. 2.25). Changes in technology can be incorporated into the

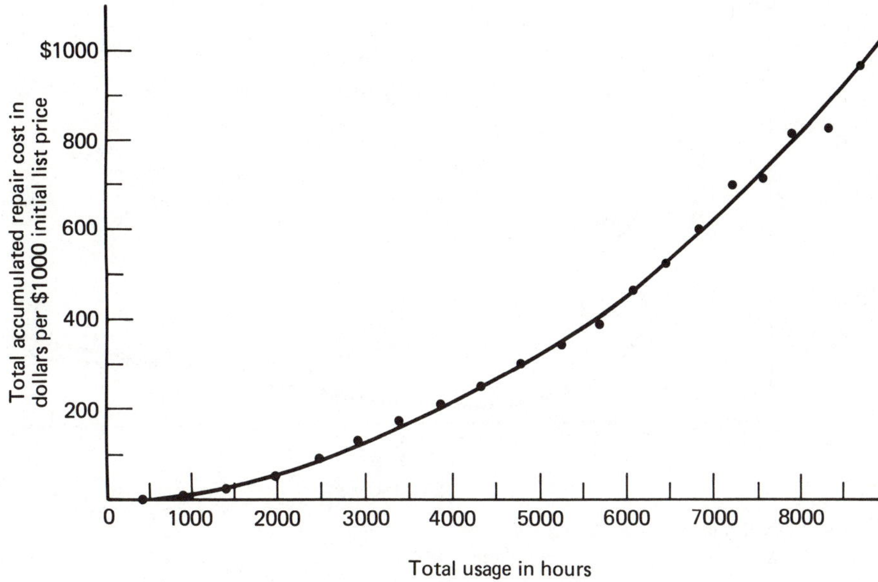

Figure 2.23 Wheel tractor repair cost (1185 tractors from 1 to 20 years old).

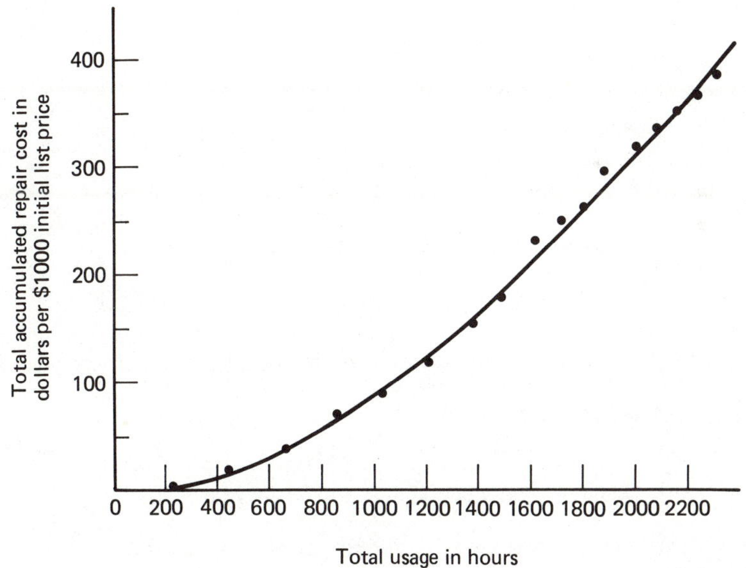

Figure 2.24 Self-propelled combine repair cost (567 combines from 1 to 17 years old).

economic analysis by charging the older machine a timeliness variable cost penalty ($/h). The following example shows how to determine the values for deciding the time to trade.

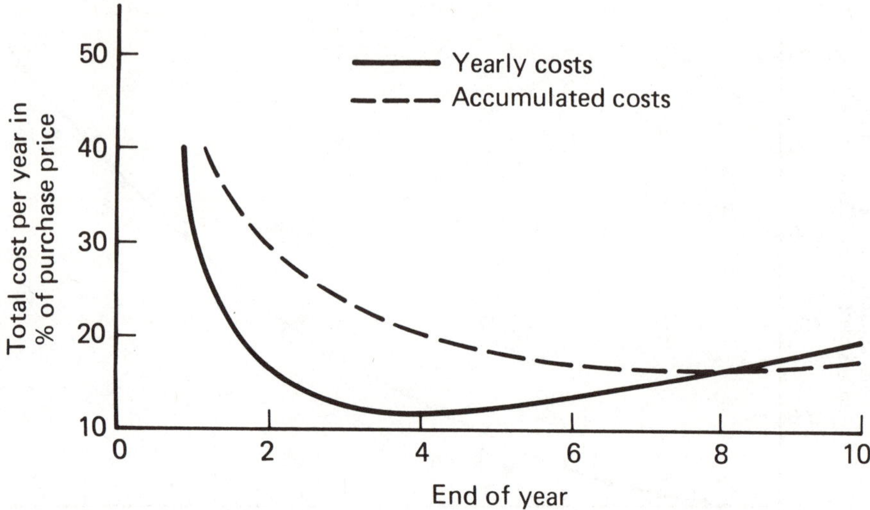

Figure 2.25 Determining break-even time.

Example 2.2 User Machinery Cost Break-even Calculation

Given a $26,000 tractor with the following usage and repair, calculate break-even by constructing an accumulated fixed cost chart.

	Data		
Year	*1*	*2*	*3*
Usage hours	500	1000	1500
Accumulated repairs	0	260	780
Accumulated fixed cost	9,776	12,870	15,782
Total	9,776	13,130	16,562
Average accumulated yearly	9,776	6,565	5,520
Yearly costs		3,354	3,432

Yearly cost value is obtained by subtracting the total between any two years.

$13,130 (year 2 total)
– 9,776 (year 1 total)

$ 3,354 (between year 1 and 2) = yearly cost

Note: Yearly costs increased as repairs rose, and average accumulated yearly cost decreased to a projected break-even around 8 years (see Fig. 2.25).

This 8 years of life is the point at which most machinery should be traded, but interest rates, depreciation charges, and inflation also have a great influence on whether and when to trade.

Another economic analysis beneficial to the design engineer is the determination of the added cost that can be justified to the purchaser of a new machine. The following example depicts this determination for a new tractor to be added to the product line. The old tractor price is known.

Example 2.3 Determine New Machine List Prices
Given:

Current 450-hp tractor sells for $80,000 and pulls a 20-18 in. (30 ft) plow at 5 mph.
Assuming that efficiencies are the same and labor, oil, repair, and fuel are $20/h for both tractors, how much should a new 650-hp tractor cost if it can pull a 26-18 plow at 6 mph? The farm size must be estimated for such a tractor (assume 3600a for a 450-hp tractor). Use Eq. 2.14 and an interest rate of 7%, which results in the fixed cost equal to 0.16P.

$$AYC_{old} = 0.16\,(80{,}000) + \frac{8.25\,(3600)}{5 \times 30 \times e}\,(\$20/\text{h}) \qquad [2.17]$$

$$AYC_{new} = 0.16\,(X) + \frac{8.25\,(3600)}{6 \times 39 \times e}\,(20) \qquad [2.18]$$

Solve for X after setting $AYC_{old} = AYC_{new}$ and canceling e (assumed the same). The new cost $X = \$88{,}884$.

The difference is what the farmer can afford for the engineer and manufacturer to add. Other options for the user include long-term leasing and renting. If annual hour usage is low, then renting or custom hiring is economical. Reliability and timeliness might warrant purchase rather than renting. Timeliness is a cost penalty that changes Example 2.3 by being added to the variable cost side of AYC_{old}. Leasing equipment is for those short on capital or for those who need capital to expand their business.

HOMEWORK PROBLEMS

2.1 Compare the hardness of Rockwell C 30 to Brinell hardness.

2.2 If a material has a yield strength of 30,000 psi, what is its hardness?

2.3 How would you determine the steel makeup of a competitor's front axle. (HINT: Do a hardness test.)

2.4 A design states that the material has a 100,000 psi tensile strength. With normal variation, what is the potential range of yield strength material that a purchasing agent might buy?

2.5 Using the largest constant values (K,n), determine the life of a milling machine head traveling at 10 m/min.

2.6 Determine the minimum cost-cutting speed of a milling machine if the cost of the machine is $100/h and workers are paid $10/h. The tool needs to be changed every 10 hours. The tool cost is $50 and can be ground twice at a cost of $50 each time.

2.7 A farmer is using a 3-18 plow traveling at 6 mph on a 10a field. The farmer started at 7 a.m. Will the job be finished before 5:00 p.m.?

2.8 Study a farmer planting corn on a weekend. Determine the farmer's field efficiency. (HINT: Have the farmer log in and out on the tractor.)

2.9 What should a company charge for a new pull-type combine if the specs are as follows:

	New	Old
Capacity	4 a/h	3 a/h
Variable cost	$16/h	$13/h
Purchase price	?	$26,000
Interest rate	20%	20%
Acres	300	300
Timeliness cost	0	$10/h

NOTE: The new machine is working faster, providing a penalty to the old machine of $10/h.

REFERENCES

1. American Society for Testing and Materials, Philadelphia, Pa.
2. *Automotive Materials in the 80's,* 88(3):56, Society of Automotive Engineers, Inc., Warrendale, Pa., 1980.
3. Bain, E.C. and H.W. Paxton. *Alloying Elements in Steel,* American Society for Metals, Metals Park, Ohio, 1966.
4. *Bethlehem High Strength Steel Sheets Catalog 3498,* Bethlehem Steel Corp., Bethlehem, Pa., 1978.
5. *Bethlehem Structural Shapes, Catalog 3277,* Bethlehem Steel Corp., Bethlehem, Pa., 1978.
6. *Fabricating Huntington Alloys,* International Nickel Co., Inc., Huntington, W.V., 1970.
7. Hunt, D. *Farm Power and Machinery Management,* 6th edition, Iowa State University Press, Ames, Iowa, 1976.
8. Kaye, G.W.C. and T.H. Laby. *Tables of Physical and Chemical Constants,* 14th edition, Longman Group Limited, London, 1973.
9. Krutz, G.W. "Time and Motion Study Used to Determine Tractor Design Parameters," SAE Paper No. 750814, Society of Automotive Engineers, Inc., Warrendale, Pa., 1978. Published in the *ASAE Transactions,* Amer. Soc. of Agric. Eng., St. Joseph, Mich., 1978.
10. Kverneland, K. O. *World Metric Standards for Engineering,* Industrial Press, Inc., New York, N.Y., 1978.

11. Parsons, S.D. *Supplemental Tables for Determining Usage Costs for Farm Tractors and Field Machinery,* Agricultural Engineering Dept., Purdue University, West Lafayette, Ind., 1976.
12. *Principles of Forging Design,* American Iron and Steel Institute, New York, N.Y.
13. Richey, C.B. and D. Hunt. "Determining Usage Costs for Farm Tractors and Field Machinery," Bulletin AE-81, Cooperative Extension Service, Purdue University, West Lafayette, Ind., 1971.

CHAPTER THREE

STATISTICAL TOLERANCE DESIGN

3.1 TOLERANCES AND ALLOWANCES

Because draftsmen frequently determine tolerance specifications on parts, engineers may forget the significance of these specifications. Tighter dimensions can result in high piece costs for parts. One manufacturing firm relates that changing tolerance specifications to a value greater than + 0.032 in. does not reduce the part cost, yet assembly lines have been shut down when tractor doors no longer fit. Almost all quality control problems result from parts that do not conform to the dimensions specified. A good background in tolerance design could mean the difference in cost for a new component (i.e., success or failure), and, therefore, an understanding of the concepts in this chapter is important.

Tolerance is defined as the permissible variation in a dimension and may be unilateral or bilateral. In the United States, the American Standards Association has adopted the unilateral system. In Fig. 3.1a, 3.000 and 2.998 are called the basic dimensions. Machinists aim for the basic dimensions, and the tolerance provides leeway for inaccuracy in their work. Allowance is the difference in dimensions for the tightest fit of the mating parts.

$$\text{allowance}, a = \text{minimum hole diameter} - \text{maximum shaft diameter} \quad [3.1]$$

In Fig. 3.1*a*,

$$a = 3.00 - 2.998 = 0.002$$

Bar diagrams are another common means of displaying tolerance (see Fig. 3.2).

Superfluous dimensioning is a common problem when dimensioning tolerance.

Common drafting regulations make this note: In a given direction, a point should be located by one and only one dimension (see Fig. 3.3). In Fig. 3.3*b*, *AB* and *AC* are within the specified limits, but *BC* is not. In Fig. 3.3*c*, *AC* and

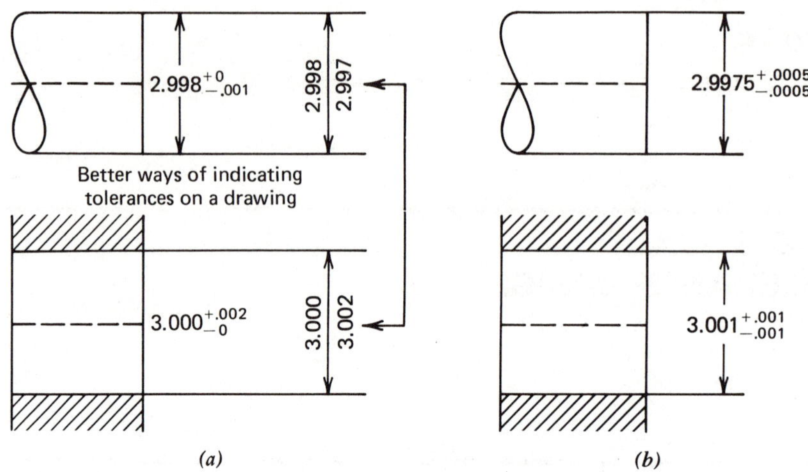

Figure 3.1 Tolerances for a shaft and hole. (*a*) Unilateral tolerances. (*b*) Bilateral tolerances.

h = total tolerance on hole
s = total tolerance on shaft
a = allowance, difference in dimensions for tightest fit
b = difference in dimensions for loosest fit
c = clearance (hole diameter − shaft diameter)
C_{max} = maximum hole diameter − minimum shaft diameter
C_{min} = minimum hole diameter − maximum shaft diameter
$C_{max} - C_{min}$ = clearance range

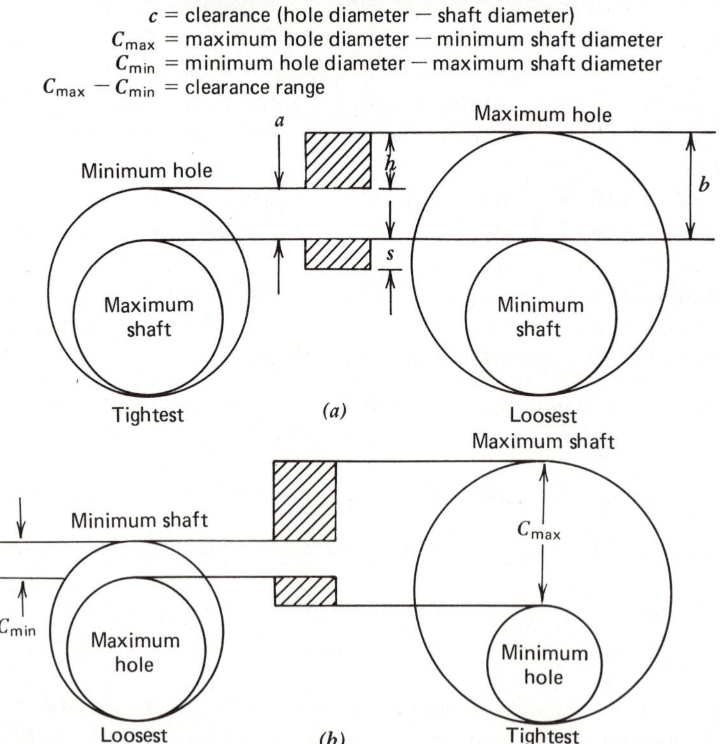

Figure 3.2 Bar diagrams for tolerance stackups. (*a*) Clearance fit. (*b*) Interference fit.

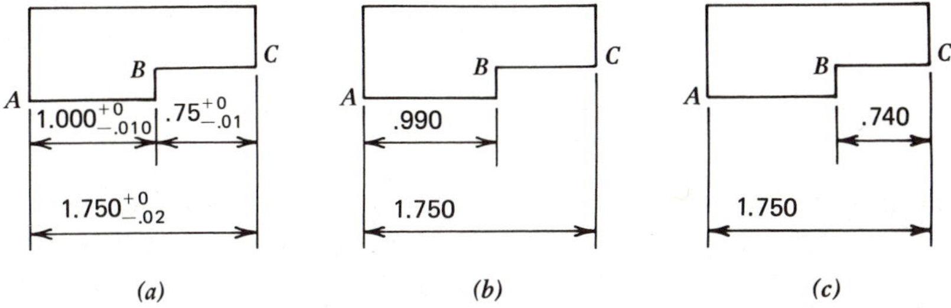

Figure 3.3 Examples of incorrect parts made because of an incorrect dimensioning procedure. (*a*) Incorrect detail drawing. (*b*) and (*c*) Actual parts made.

BC are within the specified limits, but *AB* is not. To eliminate the difficulty one of the dimensions in Fig. 3.3*a* should be eliminated. Which dimensions are retained depends on the functional requirements of the part.

Cumulative and noncumulative tolerances are depicted in Fig. 3.4. Cumulative tolerances are used when lengths *AB, BC, CD,* and *DE* must be held closely. Noncumulative tolerances are used when *B, C, D,* and *E* from datum line must be held closely.

Gauges are commonly used for measuring diameters and holes prior to assembly or when quality problems arise. Go-NoGo plug gauges are used to check hole diameters (see Fig. 3.5). If plug *A* will not go through a hole, the hole is too small; if both *A* and *B* go through a hole, the hole is too big; if *A* goes through the hole but *B* does not, the hole is within an acceptable range. Snap gauges are used to check shaft and bolt diameters.

3.2 APPLICATION OF STATISTICS TO MANUFACTURING

The basic problem in a quantitative experiment is to determine the value of some quantity, for example, the size of a kernel of wheat. Although the true value of anything can never be measured, an approximation to the true value can be found. Refining the methods of measurement can lead to approximations that are closer and closer to the true value, but there is always a limit beyond which refinements have not or cannot be made.

Consider the measurement of angle X with a spectrometer that can be read to $0.001'$. Although readings are taken with equal care, successive measurements will not agree. Suppose that the following values were obtained.

60°	**0.320′**	**60°**	**0.320′**
	0.318		0.319
	0.317		0.322
	0.320		0.320
	0.321		0.319
	0.319		0.321

Of this list of measurements, what value is nearest the true value? It is common to take the arithmetic mean $\bar{X}$ as the best approximation to the true value.

$$\bar{X} = \frac{\Sigma X}{N} \tag{3.2}$$

where X = individual readings
N = number of readings

A histogram, which is a plot of the frequency distribution of these measurements, is shown in Fig. 3.6. The smallest amount by which two measurements can differ and still be detected is the least count. In Fig. 3.6, the least count is 0.001′. The diagram shows that the reading 60° 0.320′, representing the interval 60° 0.3195′ to 60° 0.3205′, was obtained four times. A comparison of the

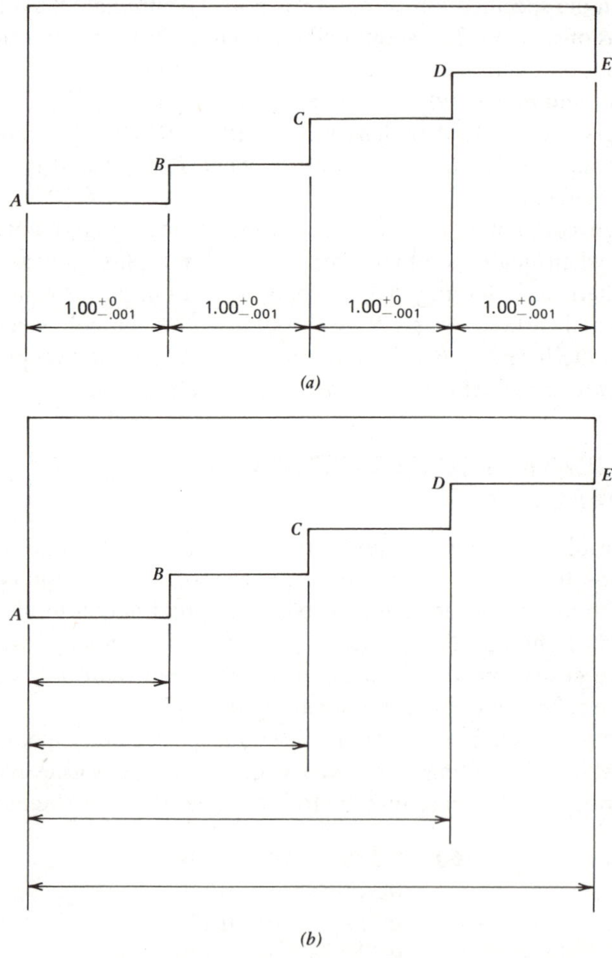

Figure 3.4 Cumulative and noncumulative tolerances for engineering drawings. (*a*) Cumulative tolerances used when length *AB, BC, CD,* and *DE* must be held closely. (*b*) Noncumulative tolerances used when *B, C, D* and *E* from datum line must be held closely.

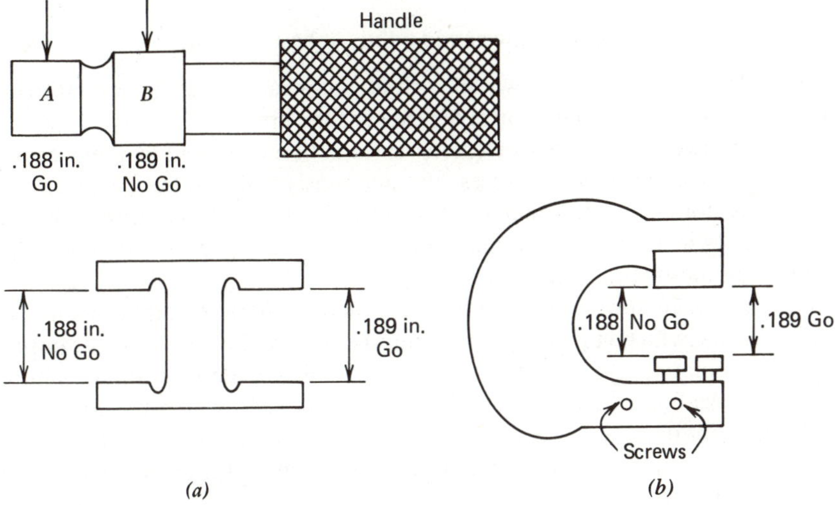

Figure 3.5 Hole plugs and gauges. (*a*) Solid snap gauge. (*b*) Adjustable limit snap gauge.

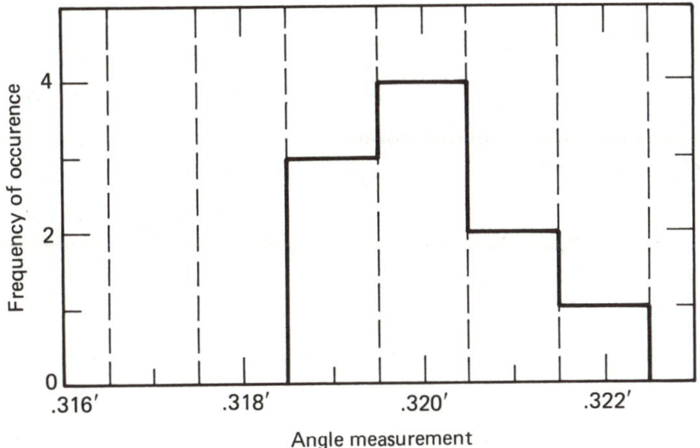

Figure 3.6 Histogram of angle measurements.

rectangular area centering at 60° 0.320′ to the total area under the curve indicates that the relative frequency of occurrence of this reading is one-third. If the number of readings were increased indefinitely and the least count made as small as possible, the distribution curve would become a smooth bell shape. An error can be calculated from measurements.

$$\text{error} = \text{measured value} - \text{true value} \qquad [3.3]$$

If measured value is less than the true value, then error is negative. If measured value is greater than the true value, then error is positive.

Also,

$$\text{deviation} = \text{measured value} - \text{best estimate} \qquad [3.4]$$

Since the true value is not known, the arithmetic mean is used as best estimate. Then, from Eq. 3.4, the deviation is regarded as the error.

There are two broad types of errors: systematic errors and random errors. Systematic errors arise from causes that act consistently under the given circumstances; for example, a metal rule calibrated at one temperature will read systematically incorrectly at another. Such an error can and should be avoided by suitable experimental techniques.

When all systematic errors have been eliminated, there remain accidental or random errors, which consist of a large number of small effects, such as imprecision in estimating the smallest division mark on a scale or a small, natural (random) variation in temperature from the standard at which the equipment has been calibrated. Some of these effects are positive, others negative. Usually, their effect on the value of the observation is in a plus or minus direction with equal probability.

If the systematic errors are eliminated and only the random errors remain, the process is said to be under statistical control, and it can be shown mathematically that the random errors result in a frequency distribution diagram as shown in Fig. 3.7.

Summation of rectangular areas

$$\Sigma y \, \Delta X = 1 \qquad [3.5]$$

where y = the frequency of deviation

ΔX = the increment of measurement plotted

$$\Sigma y \, \Delta X = \frac{\Sigma \text{ of number of occurrences}}{N} \qquad [3.6]$$

where N = number of readings.

Then

$$y \, \Delta X = \frac{\text{number of occurrences}}{N} \qquad [3.7]$$

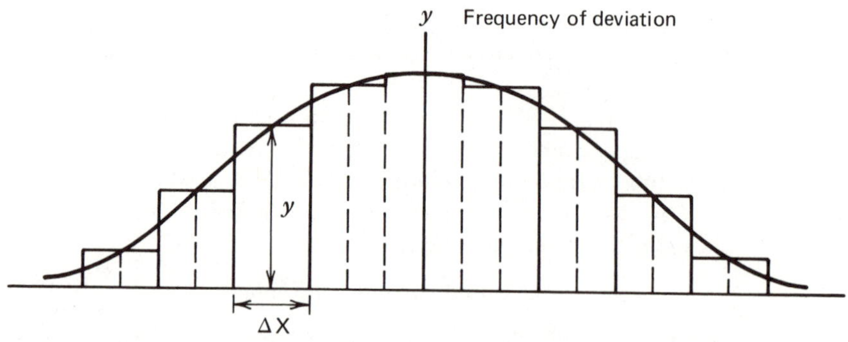

Figure 3.7 Normal frequency distribution curve.

and

$$y = \frac{\text{number of occurrences}}{N(\Delta X)} \qquad [3.8]$$

The width of the rectangles = $\Delta X =$ least count. As ΔX is made to approach zero, the heights of the rectangles become the ordinates of the smooth curve, as shown in Fig. 3.8. The following statements pertain to this normal distribution curve or to any continuous probability distribution.

1. The curve extends from $X = -\text{infinity}$ to $X = +\text{infinity}$ and is asymptotic to the x-axis at both ends.
2. The total area under the curve integrates to 1. That is, the sum of all the probabilities (y values) of all the possible events (X values) that can occur is 1.

$$\int_{-\infty}^{+\infty} y \, dX = 1 \qquad [3.9]$$

3. The area under the curve lying between the mean value (0) and some value X represents the probability that the events with values $\geqslant X$ and $\leqslant 0$ can occur. Since

$$y = \frac{1}{s\sqrt{2\pi}} \exp\left[\frac{-X^2}{2s^2}\right] \qquad [3.10]$$

where s = standard deviation. This probability is

$$F(X) = \int_0^X y \, dX = \frac{1}{s\sqrt{2\pi}} \int_0^X \exp\left[\frac{-X^2}{2s^2}\right] dX \qquad [3.11]$$

$F(X)$ cannot be actually solved as an integral; however, it can be approximated, and these approximations can be found in tables. The distribution found in most tables has a mean of zero and a variance of (s^2). Almost no real data actually have a mean of zero and a variance of 1. Therefore, the data

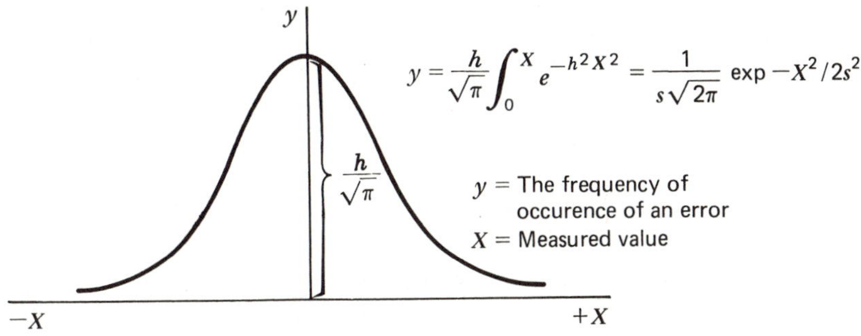

Figure 3.8 Bell-shape frequency distribution.

must be normalized before the tables are used by computing the Z score which is

$$Z = \frac{X - \bar{X}}{s} \qquad [3.12]$$

where $\bar{X}$ = estimated mean and s = estimated standard deviation (see Table 3.1).

Assuming, before normalization, a normal distribution with a mean greater than or less than zero (a noncentral normal distribution), the resulting Z score will represent the number of standard deviations that the observation X is from the mean. By looking in a table for the normal distribution, the probability that an event will fall between X and the mean can be found, or the probability will be given for the case where an event will be greater than X (or less than, since tables usually give only one side of the normal distribution).

The maximum ordinate of the frequency distribution in Fig. 3.8 is $\frac{h}{\sqrt{\pi}}$ where

$$h = \left[\frac{N - 1}{2\,\Sigma\,X^2}\right]^{1/2} \text{ and } (\bar{X} = 0) \qquad [3.13]$$

N = number of observations

Since

$$s = \left[\frac{\Sigma X^2}{N - 1}\right]^{1/2} \text{ (where } \bar{X} = 0\text{) then } h = \frac{1}{(\sqrt{2})\,s} \qquad [3.14]$$

and hence

$$F(X) = \frac{h}{\sqrt{\pi}} \int_{o}^{+X} e^{-h^2X^2} dx \qquad [3.15]$$

Notice that

$$h^2X^2 = \frac{X^2}{2\,s^2} \text{ for } (\bar{X} = 0) \qquad [3.16]$$

That is, suppose that X did not have a mean zero, then

$$h^2(X - \bar{X})^2 = \frac{(X - \bar{X})^2}{2\,s^2} = \tfrac{1}{2}\,Z^2 \qquad [3.17]$$

and

$$h\,(X - \bar{X}) = \frac{1}{(\sqrt{2})}\,Z \qquad [3.18]$$

Since the standard deviation is the square root of the variance and a small standard deviation represents a very pointed curve, a large h value represents the same thing (see Fig. 3.9). A larger value of h or a smaller value of s represents a more accurate set of data.

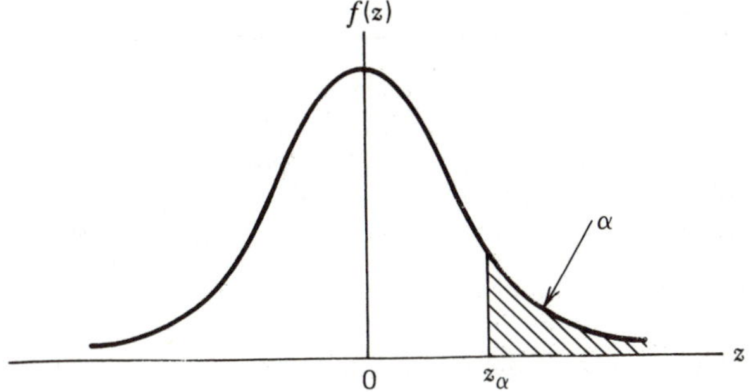

Table 3.1 **Normal distribution (1)** *

z_α	*.00*	*.01*	*.02*	*.03*	*.04*	*.05*	*.06*	*.07*	*.08*	*.09*
0.0	.5000	.4960	.4920	.4880	.4840	.4801	4761	.4681	.4641	
0.1	.4602	.4562	.4522	.4483	.4404	.4364	.4325	.4286	.4247	
0.2	.4207	.4168	.4129	.4090	.4052	.4013	.3974	.3936	.3897	.3859
0.3	.3821	.3783	.3745	.3707	.3669	.3632	.3594	.3557	.3520	.3483
0.4	.3446	.3409	.3372	.3336	.3300	.3264	.3228	.3192	.3156	.3121
0.5	.3085	.3050	.3015	.2981	.2946	.2912	.2877	.2482	.2810	.2776
0.6	.2743	.2709	.2676	.2643	.2611	.2578	.2546	.2514	.2483	.2451
0.7	.2420	.2389	.2358	.2327	.2296	.2266	.2236	.2206	.2177	.2148
0.8	.2119	.2090	.2061	.2033	.2005	.1977	.1949	.1922	.1894	.1867
0.9	.1841	.1814	.1788	.1762	.1736	.1711	.1685	.1660	.1635	.1611
1.0	.1582	.1562	.1529	.1515	.1492	.1469	.1446	.1243	.1401	.1379
1.1	.1357	.1335	.1313	.1292	.1271	.1251	.1230	.1210	.1190	.1170
1.2	.1151	.1131	.1112	.1093	.1075	.1056	.1038	.1020	.1003	.0985
1.3	.0968	.0951	.0934	.0918	.0901	.0885	.0869	.0853	.0838	.0823
1.4	.0808	.0793	.0778	.0764	.0749	.0735	.0721	.0708	.0694	.0681
1.5	.0668	.0655	.0643	.0630	.0618	.0606	.0594	.0582	.0571	.0559
1.6	.0548	.0537	.0526	.0516	.0505	.0495	.0485	.0475	.0465	.0455
1.7	.0446	.0436	.0427	.0418	.0409	.0401	.0392	.0384	.0375	.0367
1.8	.0359	.0351	.0344	.0336	.0329	.0322	.0314	.0307	.0301	.0294
1.9	.0287	.0281	.0274	.0268	.0262	.0256	.0250	.0244	.0239	.0233
2.0	.0228	.0222	.0217	.0212	.0207	.0202	.0197	.0192	.0188	.0183
2.1	.0179	.0174	.0170	.0166	.0162	.0158	.0154	.0150	.0146	.0143
2.2	.0139	.0136	.0132	.0129	.0125	.0122	.0119	.0116	.0113	.0110
2.3	.0107	.0104	.0102	.00990	.00964	.00939	.00944	.00889	.00866	.00842
2.4	.00820	.00798	.00776	.00755	.00734	.00714	.00695	.00676	.00657	.00639
2.5	.00621	.00604	.00587	.00570	.00554	.00539	.00523	.00508	.00494	.00480
2.6	.00466	.00453	.00440	.00427	.00415	.00402	.00391	.00379	.00368	.00357
2.7	.00347	.00336	.00326	.00317	.00307	.00298	.00289	.00280	.00272	.00264
2.8	.00256	.00248	.00240	.00233	.00226	.00219	.00212	.00205	.00199	.00193
2.9	.00187	.00181	.00175	.00169	.00164	.00159	.00154	.00149	.00144	.00139

Note: *Tabulation of the values of α versus z_α for the standardized normal curve.

$$\alpha = P(z > z_\alpha) = \int_{z_\alpha}^{\alpha} \frac{1}{\sqrt{2\pi}} e^{-z^2/2} dz$$

= area under the standardized normal curve from $z = z_\alpha$ to $z = \alpha$.

Table 3.1 (Cont.)

z_α	.00	.01	.02	.03	.04	.05	.06	.07	.08	.09
3	.00135	$.0^{3}988$	$.0^{3}687$	$.0^{3}483$	$.0^{3}337$	$.0^{3}233$	$.0^{3}159$	$.0^{3}108$	$.0^{4}723$	$.0^{4}481$
4	$.0^{4}317$	$.0^{4}207$	$.0^{4}133$	$.0^{5}854$	$.0^{5}541$[a]	$.0^{5}340$	$.0^{5}211$	$.0^{5}130$	$.0^{6}793$	$.0^{6}479$
5	$.0^{6}287$	$.0^{6}170$	$.0^{7}996$	$.0^{7}579$	$.0^{7}333$	$.0^{7}190$	$.0^{7}107$	$.0^{8}599$	$.0^{8}332$	$.0^{8}182$
6	$.0^{9}987$	$.0^{9}530$	$.0^{9}282$	$.0^{9}149$	$.0^{10}777$	$.0^{10}402$	$.0^{10}206$	$.0^{10}104$	$.0^{11}523$	$.0^{11}260$

[a] $.0^{5}541$ means .00000541.

The probable error, *p*, is defined as an error of such magnitude that half the measured values will have an error less than this amount. This definition is illustrated in Fig. 3.10, where the shaded area *F(X)* is equal to one-half.

Integrating between $-p$ to $+p$ in Eq. 3.15 and substituting for h using Eq. 3.14 yields

$$p = \frac{0.477}{h} = (\sqrt{2})(s)(0.477) \qquad [3.19]$$

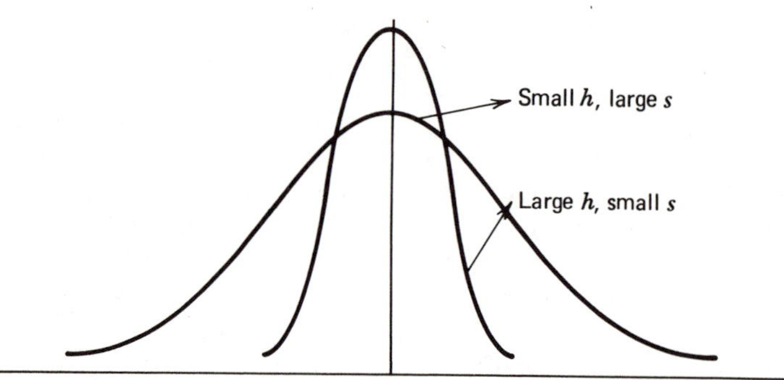

Figure 3.9 Effect of frequency distribution shape caused by small or large standard deviations.

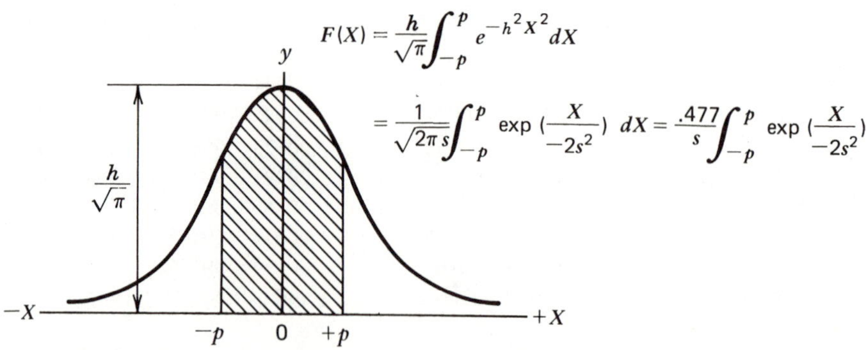

Figure 3.10 Probable error shown on a frequency curve.

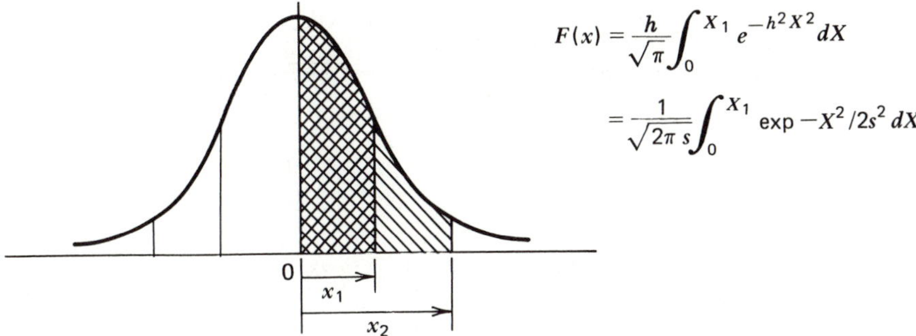

Figure 3.11 Error between mean and some value $(X_1 - 0)$.

The probability of some plus error of magnitude lying between 0 and X_1 but not including 0 and X_1 is represented by the xxxx area in Fig. 3.11 and is:

$$F(X) = \frac{h}{\sqrt{\pi}} \int_o^{X_1} \exp^{-h^2X^2} dX = \frac{1}{\sqrt{2\pi}\, s} \int_o^{X_1} \exp^{\left[\frac{-X^2}{2s^2}\right]} dX \quad [3.20]$$

The probability of a + error of magnitude lying between X_1 and X_2 but not including X_1 and X_2 is represented by the ///// area and is:

$$F(X) = \frac{h}{\sqrt{\pi}} \int_{X_1}^{X_2} \exp^{-h^2X^2} dX = \frac{1}{\sqrt{2\pi}\, s} \int_{X_1}^{X_2} \exp^{\left[\frac{-X^2}{2s^2}\right]} dX \quad [3.21]$$

A table of the normal probability distribution can be consulted for the approximate values of this integral between certain desired limits of X.

$$F(X_2) - F(X_1) = \frac{1}{\sqrt{2\pi}\, s} \int_{X_1}^{X_2} \exp^{\left[\frac{-X^2}{2s^2}\right]} dX \quad [3.22]$$

$$= \frac{1}{\sqrt{2\pi}\, s} \left[\int_o^{X_2} \exp^{\left[\frac{-X^2}{2s^2}\right]} dX - \int_o^{X_1} \exp^{\left[\frac{-X^2}{2s^2}\right]} dx \right]$$

Standard deviation is the square root of the mean squared deviation.

$$s = \left[\frac{\Sigma X^2}{N - 1} \right]^{1/2} \quad [3.23]$$

assuming $\bar{X} = 0$; if $\bar{X} \neq 0$, then

$$s = \left[\frac{\Sigma (X - \bar{X})^2}{N - 1} \right]^{1/2}$$

For a particular manufacturing process, s is obtained by experience. That is, a number of readings of a quantity are made, and the value of s is computed using Eq. 3.23. From Eqs. 3.13 and 3.23,

$$s = \frac{1}{(\sqrt{2})h} \quad [3.24]$$

Table 3.2. **Probability of Area Outside Various Standard Deviations**

Limit	*P*
$\bar{X} \pm s$	0.33
$\bar{X} \pm 2s$	0.05
$\bar{X} \pm 3s$	0.003

The probability of a part falling outside of $\bar{X} \pm ts$, where t is a constant, is given in Table 3.2.

Tolerances on a part of $\pm$ 3s are called natural tolerances. The actual value of the Z score at 0.997 is something like 2.97, but 3s is conveniently used. Therefore, 3s is usually taken as a sufficient approximation. The use of $\pm$ 3s tolerance is an arbitrary convention, used primarily in the United States. In Great Britain, $\pm$ 2s tolerance (0.95) is the usual standard (see Table 3.2). For a part in mass production, if the tolerances are specified equal to $\pm$ 3s, then only 3 parts of 1000 are likely to have dimensions that fall outside the limits specified on the detail drawing.

Fig. 3.12*b* illustrates a production process in which high accuracy is maintained and almost all parts produced fall within the limits specified on the detail drawing in 3.12*a*. In Fig. 3.12*c*, a less accurate process is illustrated, but only 3 parts in 1000 fall outside the specified limits. Figure 3.12*d* illustrates a still less accurate process brought about by a worn tool or bad fixture. Because the natural tolerance is greater than the tolerance specified on the drawing, more than 3 parts of 1000 will fall outside the detail drawing specifications.

A theorem in statistics states: If $X_1, \cdots, X_j$ are independent random variables with mean values $\mu_1, ..., \mu_j$ and variances $s_1^2, ..., s_j^2$, respectively, and if $k_1, ..., k_j$ are constants and y is a linear combination of the X's, then

$$y = k_1 X_1 \pm k_2 X_2 \pm \cdots \pm k_j X_j \qquad [3.25]$$

and y is a random variable with the following properties:

$$\mu_y = k_1 \mu_1 \pm k_2 \mu_2 \pm \cdots \pm k_j \mu_j \qquad [3.26]$$

$$S_y^2 = k_1^2 s_1^2 + k_2^2 s_2 + \cdots + k_j^2 s_j^2 \qquad [3.27]$$

If $X_1, ..., X_k$ are each normally distributed, then y is normally distributed. If the X_j's are not all normally distributed and if the variances (s_j^2) are generally homogeneous, then, from the central limit theorem, as j increases the y distribution rapidly approaches the normal distribution. Even if $X_1, ..., X_j$ are not normally distributed but are independent random variables, y will be normally distributed if the number of X variables, j, is large enough.

Suppose that 5 parts ($X_1, ..., X_5$) fit together to form assembly y, then as long as there is no reason to assume that the characteristics of each part, X_j, depend on the other parts, X_j (i.e., each part is an independent random variable of a distribution), the theorem can be used to make certain assumptions about y, namely, that if

$$y = \bar{X}_1 + \cdots + \bar{X}_5 \quad [3.28]$$

then

$$\mu_y = \mu_1 + \cdots + \mu_5 \quad [3.29]$$

and

$$s_y^2 = s_1^2 + \cdots + s_5^2 \quad [3.30]$$

If each part X_j has a normal distribution of characteristics (not an unreasonable assumption in most cases especially if the part has a variety of quality

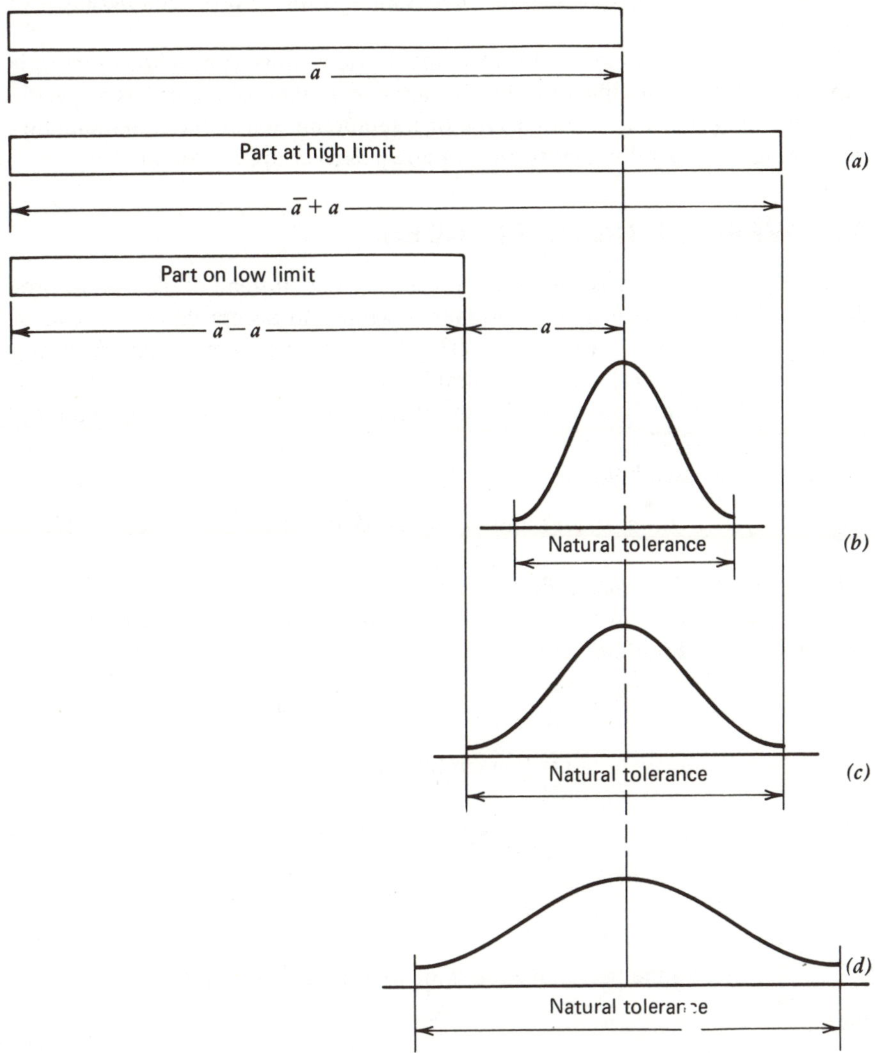

Figure 3.12 Examples of parts with natural tolerance. (*a*) Tight tolerance. (*b*) Average tolerance. (*c*) Wide tolerance.

characteristics, that is, only the most simple kinds of parts can have a nonnormal distribution of characteristics), then, according to the central limit theorem, y will have a normal distribution. If y is the sum of, for example, 4 or more parts, then, by the central limit theorem, y should have a normal distribution regardless of the distribution of the parts X_i.

There are two ways to determine the tolerance of y with the given tolerances of the parts X_i. One way is by additive tolerance. If all five parts, X_i have a tolerance of $\pm$ 0.0005, then y will have an additive tolerance of (5) $\times$ ($\pm$0.0005), or $\pm$ 0.0025, or 0.005 range.

If all five parts have a certain tolerance, an assumption can be made that 100% of the time y will have a tolerance equal to the sum of the tolerances of each part.

However, from a manufacturing point of view, this type of tolerancing is expensive. It is often difficult to be absolutely sure that each part has a specific tolerance of $\pm$ 0.0005. Another type of tolerancing that reduces manufacturing costs uses probability theory and is known as overlapping tolerances.

3.3 OVERLAPPING TOLERANCES

Suppose that a designer is willing to take a 0.3% risk, instead of 0% risk, of producing a bad part; that is, the designer is willing to accept 99.7%, instead of 100%, good assemblies. Taking a 0.003 risk could save costs in manufacturing (e.g., hydraulic valves). Some hydraulic valve parts require tolerances of 0.0001 in., which are hard to hold without cost penalties. From Table 3.2 for p = risk (λ) = 2 (2.97 s_y) $\approx$ $6s_y$.

Then, for an assembly,

$$3s_y = 0.0025 \text{ and } 6s_y = 0.005, \text{ or the range.} \qquad [3.31]$$

Example 3.1 Calculating Part Tolerances

For $6s_y = 0.005$, find the tolerance of each part, assuming a risk of 0.003. From the central limit theorem.

$$s_y^2 = s_{x1}^2 + s_{x2}^2 + s_{x3}^2 + s_{x4}^2 + s_{x5}^2$$

$$s_y^2 = \text{assembly variance}$$

$$s_x^2 = \text{component variance}$$

If

$$s_{x1}^2 = s_{x2}^2 = s_{x3}^2 = s_{x4}^2 = s_{x5}^2$$

(if all the parts have the same tolerance specification)

then

$$s_y^2 = 5s_x^2 \text{ (5 parts)}$$

and, from Eq. 3.31, for an assembly with a normal distributed tolerance of ± 0.0025, then

$$s_y = \frac{0.005}{6} \Longrightarrow s_y^2 = \frac{(0.005)^2}{6^2}$$

Since

$$s_y^2 = 5s_x^2 \Longrightarrow \frac{(0.005)^2}{6^2}$$

then

$$s_x^2 = \frac{(0.005)^2}{(5)(6)^2}$$

or

$$s_x = \frac{0.005}{(\sqrt{5})6}$$

Hence the component tolerances will be normally distributed ($\pm\ 3s_x$) and will have a $6s_x$ range.

$$6\,s_x = \frac{0.005}{\sqrt{5}} = 0.002236$$

$$3\,s_x = \frac{0.005}{(2)(\sqrt{5})} = 0.001118$$

Example 3.2 Finding Tolerances for a Different Risk

Suppose that, in Example 3.1, the risk was $\lambda = .01$; then $\frac{\lambda}{2} = 0.005$. Using Table 3.1 for a normal distribution (which will give only one tail of the curve) and $\frac{\lambda}{2} = 0.005$, then the Z score that corresponds to λ is between 2.5 and 2.6. By interpolation,

$$\frac{\lambda}{2} = 0.005 \Longrightarrow Z \approx 2.58$$

Then

$$2Zs_y = 5.16\,s_y = 2(0.0025) = 0.005$$

or

$$5.16\,s_y = 0.005$$

$$s_y^2 = \frac{(0.005)^2}{(5.16)^2}$$

and knowing

$$s_y^2 = 5s_x^2$$

$$5\,s_x^2 = \frac{(0.005)^2}{(5.16)^2}$$

or

$$s_x^2 = \frac{(0.005)^2}{(5)(5.16)^2}$$

finally

$$s_x = 0.000433$$

Again, for normally distributed tolerances on the parts

$$6s_x = 6(0.000433) = 0.0026$$

or

$$3s_x = 0.0013$$

Example 3.3 Varying Risk at Each Side of an Assembly's Tolerance Specification The risk in Example 3.1 is changed so one tail of the distribution is different from the risk on the other tail of the distribution. For $\lambda_1 = 0.0228$ on one tail and $\lambda_2 = 0.0179$ on the other tail, find the part tolerance. The corresponding Z scores are 2 and 2.1 and hence

$$2s_y + 2.1\,s_y = 2(0.0025) = 0.005$$

then

$$4.1\,s_y = 0.005$$

$$s_y^2 = \frac{(0.005)^2}{(4.1)^2}$$

$$s_y^2 = 5\,s_x^2$$

then

$$5\,s_x^2 = \frac{(0.005)^2}{(4.1)^2}$$

$$s_x = \frac{(0.005)}{(\sqrt{5})(4.1)} = 0.000545$$

and hence

$$6\,s_x = 6(0.000545) = 0.00327$$

or

$$3\,s_x = 0.001636$$

Example 3.4 Relaxing Assembly Tolerances
Relax the tolerances on y instead of on the X_i. Fixed component tolerances will determine the assembly dimensions. Again, five parts with the same tolerance constitute the assembly.

First, assume that

$$\pm zs_x = 0.0005 \text{ for each component}$$

or

$$6s_x = 0.001$$

then

$$s_x^2 = \frac{(0.001)^2}{(6)^2}$$

and since

$$s_y^2 = 5\, s_x^2$$

substituting for s_x^2

$$s_y^2 = 5\frac{(0.001)^2}{(6)^2}$$

$$s_y = \frac{(\sqrt{5})(0.001)}{6}$$

$$s_y = \frac{(2.24)(0.001)}{(6)} = 0.0003732$$

Thus the overlapping tolerance for the normally distributed assembly will be

$$6\, s_y = 6(0.0003732) = 0.0022392$$

or

$$3\, s_y = 0.0011198$$

It should be noted that in all cases the part overlapping tolerance will be larger than the additive tolerance, and manufacturing costs may be reduced.

3.4 ASSEMBLY CONDITION OF CLEARANCE

A common assembly condition that involves cumulative tolerances is clearance between fitting parts, such as a hole and shaft. The tolerance function in this case is

$$Y = X_1 - X_2 \qquad [3.32]$$

where Y = clearance
X_1 = interior component quality characteristic (hole)
X_2 = exterior characteristic (shaft)

Example 3.5 Additive Tolerance Problem

Suppose that a basic hole size is 0.500 and that clearances may vary from between +0.0002 and 0.0012, with a mean clearance value μ_y of +0.0007. The additive specifications are calculated as follows.
Given

$$\mu_y = \frac{+0.0012 + 0.0002}{2} = 0.0007$$

also,

$$C_{\max} = X_{1\max} - X_{2\min} = 0.0012$$

and,

$$C_{\min} = X_{1\min} - X_{2\max} = 0.0002$$

$$X_{1\min} = 0.5000 \text{ nominal}$$

therefore,

$$X_{2\max} = 0.5000 - 0.0002 = 0.4998$$

The clearance range $C_{\max} - C_{\min} = +0.0012 - 0.0002 = 0.001$

$$X_{2\min} = X_{1\min} - \text{half of the clearance range} = 0.4993$$

$$X_{1\max} = X_{1\min} + \text{half of the clearance range} = 0.5005$$

Example 3.6 Reducing Match Parts Cost by Adding Risk Probability

Reduce cost by enlarging the component tolerances in Example 3.5 under some risk λ. Suppose that

$$\lambda = 0.0456$$

From the central limit theorem,

$$s_y^2 = s_1^2 + s_2^2$$

and if the component tolerances are identical,

$$s_y^2 = 2\,s_x^2 \text{ where } s_x = s_1 = s_2$$

Then

$$\frac{\lambda}{2} = 0.0228$$

from Table 3.1

$$Z = 2$$

$$4\, s_y = 0.0012 - 0.0002 = 0.001$$

(clearance range in naturally distributed assembly)

then

$$s_y^2 = \left[\frac{0.001}{4}\right]^2$$

$$s_x = \left[\frac{1}{\sqrt{2}}\right]\left[\frac{0.001}{4}\right] = 0.000177$$

Component tolerance is

$$6\, s_x = 6(.000177) = .001$$

Hence, if X_1 and X_2 are normally distributed independent random variables with mean averages corresponding to their respective nominal specification values, additive tolerances may be converted to overlapping tolerance. For every component the tolerances can be increased from 0.0005 to 0.001 in. Assuming that the mean clearance remains the same, the new clearance, assuming overlapping tolerance, becomes

$$C_{\min} = +\, 0.0007 - 0.001 = -0.0003$$

$$C_{\max} = +\, 0.0007 + 0.001 = +0.0017$$

that is, minimum clearance becomes -0.0003 and maximum clearance becomes $+0.0017$. Thus, the overlapping specifications become

X_1	X_2
0.5000	0.5003
0.5010	0.4993

$X_{1\max} - X_{2\max} = 0.0003$, or an interference fit.

Given these specifications, it can be assumed that 95.5% of the assemblies will be between $+0.0002$ and $+0.0012$ and 100% of the assemblies will be between -0.0003 and 0.0017.

Example 3.7 Computing Risks for an Assembly

Suppose that a clearance less than $+0.0002$ in Example 3.6 is considered to be functionally more critical than a clearance greater than $+0.0012$. Overlapping component tolerance is 0.001, as computed above. If the design engineer arbitrarily changes the value of the mean clearance to $\mu_y = 0.0009$, what risk (λ) are you taking with the change?

Computing the Z scores for the new mean value,

$$Z_{+0.0002} = \frac{(+0.0002)-(0.0009)}{\frac{(0.001)}{4} = \text{estimate of } s_y} = -2.8$$

$$Z_{+0.0012} = \frac{(+0.0012)-(0.0009)}{\left[\frac{0.001}{4}\right]} = +1.2$$

Looking in Table 3.1 at the normal distribution

$$P(y \leqslant 0.0002) = 0.00256$$

Therefore, a 0.25% chance exists that fits are too tight; $P(y) \geqslant 0.0012 = 0.1151$, so an 11.5% chance exists that fits are too loose. The risk of bad assemblies is $11.5 + 0.25$.

$$\lambda = 0.1175$$

3.5 TOLERANCES FOR PARTS ASSEMBLED WITH BOLTS

Problems in assembling parts with bolts, can be costly if the correct relationship between tolerances on hole diameter H, hole center distance M, and maximum bolt diameter are not determined.

For one bolt that connects two parts, each of which has a hole (see Fig. 3.13), a designer needs to calculate hole and bolt tolerances. Two parts, 1 and 2, are to be connected by a bolt and are to fit against part 3. Figure 3.13*b* shows that the minimum opening for passage of the bolt through the parts occurs when the diameters are the minimum permissible and their center distances are at the permissible limits.

$$\text{maximum bolt diameter} = H - 2m \qquad [3.33]$$

Inspection of hole diameter in each part could be made with a Go-NoGo plug gauge. The location of the hole can be checked with a gauge such as that shown in Fig. 3.13*c*, and each part is checked separately. Fig. 3.13*d* shows that such a gauge passes parts whose holes are not of the minimum diameter, resulting in center distances that can exceed the limits on the drawing. In fact, when the hole diameters are a maximum, the limits on the center distance could exceed the limits specified on the drawing by the greatest amount, as shown in Fig. 3.13*d*. Although this gauge is not capable of rejecting parts whose center distance exceeds the limits on the drawing, it will pass all parts that are capable of assembly, which is its primary function. For two bolts that connect two parts, each with two holes, as shown in Fig. 3.14, tolerances can also affect the assembly.

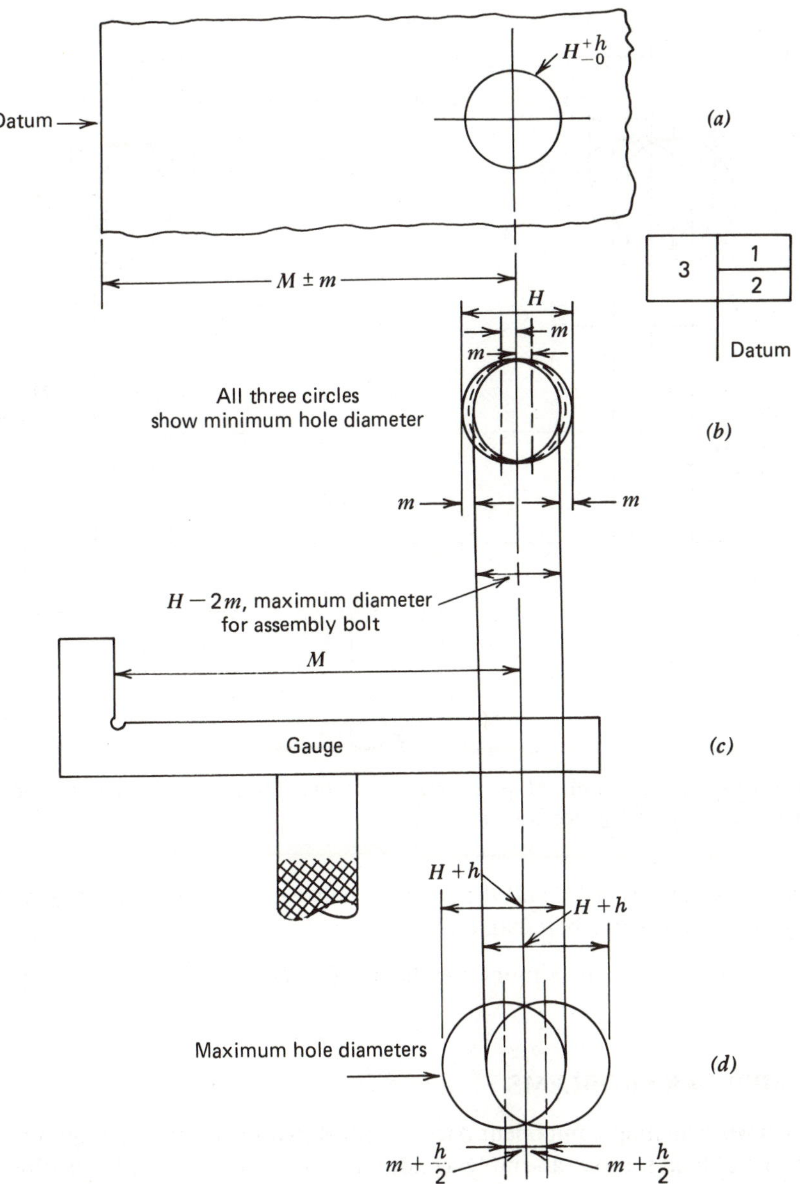

Figure 3.13 One bolt connecting two holes on plates 1 and 2 butted to plate 3. (*a*) Two plates with holes. (*b*) Assembly of parts when hole diameters are a minimum and difference in center distance M is maximum permissible. (*c*) Gauge for holding dimension M. (*d*) Holes with maximum permissible diameter. Gauge passes, but parts have center distances larger than allowable specified on drawing.

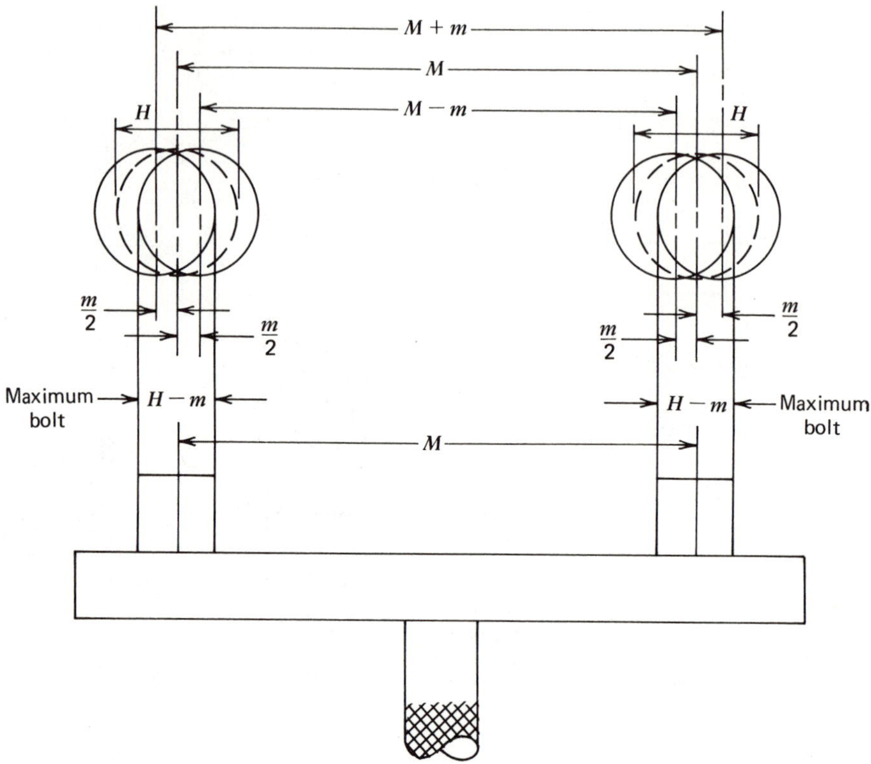

Figure 3.14 The assembly of parts when hole diameters are a minimum and the difference in center distance is a maximum.

If hole diameter is H^{+h}_{-0} and the distance between hole center is $M \pm m$, the maximum permissible bolt diameter is

$$\text{maximum bolt diameter} = H - m \qquad [3.34]$$

HOMEWORK PROBLEMS

3.1 Two matching component parts A and B involve clearance requirements to obtain for the assembly condition which is a normally distributed, independent, random variable. Nominal size is 5/8 in. Clearances may vary between 0.0003 and 0.0019. Suppose that the risk of a bad assembly is $\lambda = 0.0456$. Assume equal tolerance allocation to components.

a. Compute additive specifications.

b. Compute the overlapping tolerance and the overlapping specifications for each component dimension. (HINT: Overlapping tolerance for each component dimension is approximately 0.0017.)

c. Suppose that tight fits are more critical than loose fits and the designer arbitrarily establishes overlapping specifications of 0.6250/0.6267 and 0.6250/0.6233. Compute the risk λ.

3.2 Three components are assembled adjacent to each other. Their respective additive sizes are $X_1 = 0.500 \pm 0.001$ in.; $X_2 = 0.500 \pm 0.001$ in.; and $X_3 = 0.500 \pm 0.001$ in.. Assume that X_1, X_2, and X_3 are normally distributed independent random variables.

a. Prepare an additive tolerance stack up.

b. Considering the y specification to be fixed, assuming a risk value of $\lambda = 0.01242$, and allocating component tolerance equally, determine the overlapping component tolerance specifications.

c. Suppose that economics dictate a risk of $\lambda = 0.03174$. Compute the overlapping tolerances.

d. Suppose that the component tolerances were fixed as 0.500 ± 0.001 in. Compute the probable variation of the assembly value y 99.7% of the time.

3.3 Five parts are bolted together. The assembly cannot vary by more than + 0.1 mm except in 1 out of 10^7 times. Assume normal distribution when manufacturing the part. Determine the tolerance for each part.

3.4 Two identical plates are being bolted together with a 6000 lb clamping load. The hole centers are dimensioned 10 +0.002 and −0.000 in. Specify the bolt size, torque and material for a hole = .500 (HINT: see Table 6.2).

3.5 Measure a sample of 15 pins and their associated holes accurate to 0.001 in. Determine the risk in of these parts not fitting.

REFERENCES

1. Haugen, E. *Probabilistic Mechanical Design,* John Wiley and Sons, Inc., New York, N.Y., 1980.
2. Johnson, H.R. III. "Interference Fits for Mechanical Drives," *Machine Design* 53(12):89-94, 1981.
3. Lipson, C. and Sheth, N.J. *Statistical Design and Analysis of Engineering Experiments*, McGraw-Hill, Inc., New York, N.Y., 1973.
4. Neville, A.M. and J.B. Kennedy. *Basic Statistical Methods for Engineers and Scientists,* International Textbook Co., Scranton, Pa., 1964.

CHAPTER FOUR

SIZING AND USING STRESSES

4.1 UNIAXIAL STRESSES

Relationships between stress and strain are important to the engineer concerned with design and structural analysis. The discussion in this chapter will be limited to solids and isotropic materials. Fortunately, most of the materials engineers use to design agricultural machinery can be considered isotropic. Wood and reinforced concrete are exceptions and thus will not be covered here.

The state of stress is important when sizing a member or choosing its material. The process requires analyzing the state of stress on a series of infinitesimal elements where stress is thought to be maximum. Unfortunately, the maximum stress cannot be determined easily when a combination of loading conditions exists.

Simple loading conditions can be divided into four groups: bending, axial, torsion, and transverse shear (see Fig. 4.1).

For simple bending, the stress σ at point A in Fig. 4.1a is given by

$$\sigma = \frac{MC}{I} \qquad [4.1]$$

From strength of materials,

M = moment (force $\times$ distance)

C = distance from neutral axis to outer fiber

I = moment of inertia

The value of stress at point B in Fig. 4.1a is less than at point A because the moment (i.e., moment arm) is smaller. When calculating stresses on a structure it is important to exercise good judgment in choosing locations where stress might be a maximum. Judgment is more difficult in combined loading cases.

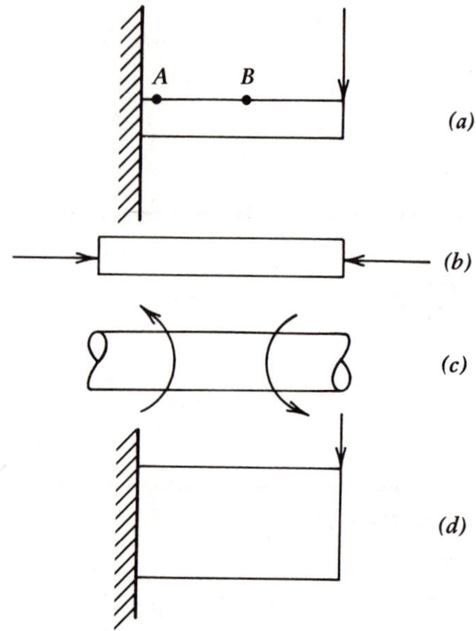

Figure 4.1 Four simple loading conditions. (a) Bending. (b) Axial. (c) Torsion. (d) Shear.

Formulas for calculating stress on the other loading cases are:

$$\text{axial} \qquad \sigma = \frac{P}{A} \tag{4.2}$$

where P = axial force
A = cross-sectional area

$$\text{torsion} \qquad \tau_{xy} = \frac{TC}{J} \tag{4.3}$$

where τ_{xy} = shear stress
T = torque
C = distance from neutral axis to outer fiber
J = polar moment of inertia

$$\text{transverse shear} \qquad \tau_{xy} = \frac{V}{It} \int_{\text{area}} y\, dA \tag{4.4}$$

where V = shear force
t = width if rectangular
y = distance from neutral axis

For rectangular members in transverse shear see Fig. 4.2

$$\tau_{\max} = \frac{V}{Ib} A\bar{y} = \frac{V}{b^2h^3/12} \times \frac{bh}{2} \times \frac{h}{4}$$

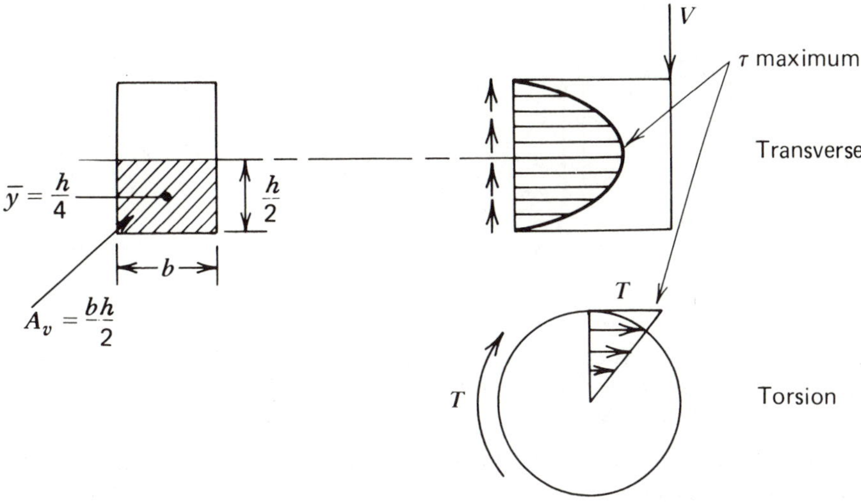

Figure 4.2 Comparing the location of maximum transverse and torsional shear stress.

Therefore,

$$\tau_{\max} = \frac{3V}{2A} \qquad [4.5]$$

at the neutral axis, and for circular cross sections in transverse shear

$$\tau_{\max} = \frac{4V}{3A} \qquad [4.6]$$

at the neutral axis. For circular hollow cross sections

$$\tau_{\max} = \frac{2V}{A} \qquad [4.7]$$

and for I-beams, channels, or wide flange beams, an approximation is given as

$$\tau_{\max} = V/A_w \qquad [4.8]$$

where A_w is area of web only.

It is important to note that transverse shear stress is zero at the outer fiber and maximum at the neutral axis as shown in Fig. 4.2. Shear stress caused by torsion is maximum at the outer fiber and zero at the neutral axis. Transverse shear is usually neglected for parts that have significant bending.

Also, when evaluating a stress state on a member, it is not necessary to use transverse shear stress if the location is at the surface.

Combined loading cases usually have the maximum stress on an outer surface, which reduces the infinitesimal element into two dimensions (Fig. 4.3*a*). This infinitesimal element can be rotated to give maximum principal stresses (normal stresses), as shown in Fig. 4.3*b*.

Maximum normal stresses $\sigma_{n\ \max}$ occur where the shear stresses τ_{xy} are zero. The maximum shear stress $\tau_{\max}$ can be determined if the maximum principal

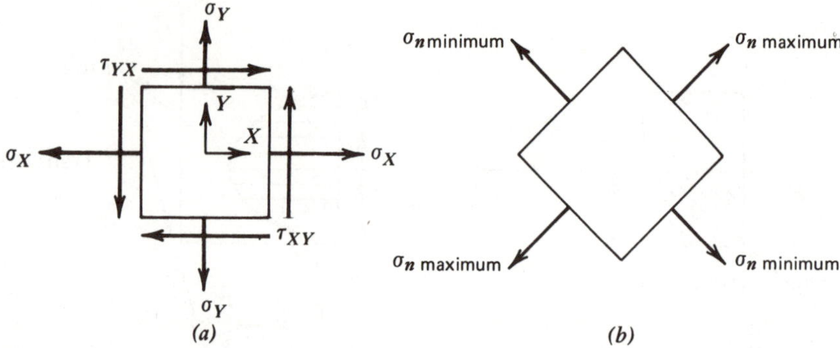

Figure 4.3 State of stress on a two-dimensional infinitesimal element. (a)Actual state of stress on element. (b)Rotated element to result in maximum and minimum normal stresses without shear stress.

stress element is rotated 45° (review Mohr's circle). Maximum shear stress can be found from the principal stress

$$\tau_{\max} = \frac{\sigma_{n\ \max} - \sigma_{n\ \min}}{2} \qquad [4.9]$$

4.2 COMBINED STRESSES

Many machine parts are subjected to loading that develops both normal and shear stress. These parts are loaded in more than one axis and are designated as subjected biaxial or triaxial stress.

Example 4.1

For the bevel gear loaded in Fig. 4.4, determine in both English and metric units the maximum shear stress on the shaft axis at section A-A.

$$M_{xy} = 1334\text{N} \times 0.203\text{m} = 270.8(\text{N}\cdot\text{m}) \qquad (2400\ \text{in.}\cdot\text{lb})$$

where M_{xy} = moment in xy plane

$$M_{xy} = -667\text{N} \times 0.127\text{m} = -84.7(\text{N}\cdot\text{m}) \qquad (-750\ \text{in.}\cdot\text{lb})$$

$$M_{xy\ \text{net}} = 186.1(\text{N}\cdot\text{m}) \qquad (1650\ \text{in.}\cdot\text{lb})$$

$$M_{xz} = 4448\text{N} \times 0.203\text{m} = 902.9(\text{N}\cdot\text{m}) \qquad (8000\ \text{in.}\cdot\text{lb})$$

Calculate Moments (English units)

$$M_x = 300\ \text{lb} \times 8\ \text{in.} = 2400\ \text{in.}\cdot\text{lb}$$

$$M_{xy} = 150\ \text{lb} \times 5\ \text{in.} = -750\ \text{in.}\cdot\text{lb}$$

$$M_{xy}\ \text{net} = 1650\ \text{in.}\cdot\text{lb}$$

$$M_{xz} = 1000\ \text{lb} \times 8\ \text{in.} = 8000\ \text{in.}\cdot\text{lb}$$

Resultant Moment

$$M_{\max} = [(902.4)^2 + (186.1)^2]^{1/2} = 921.9(\text{N}\cdot\text{m}) \qquad (8168\ \text{in.}\cdot\text{lb})$$

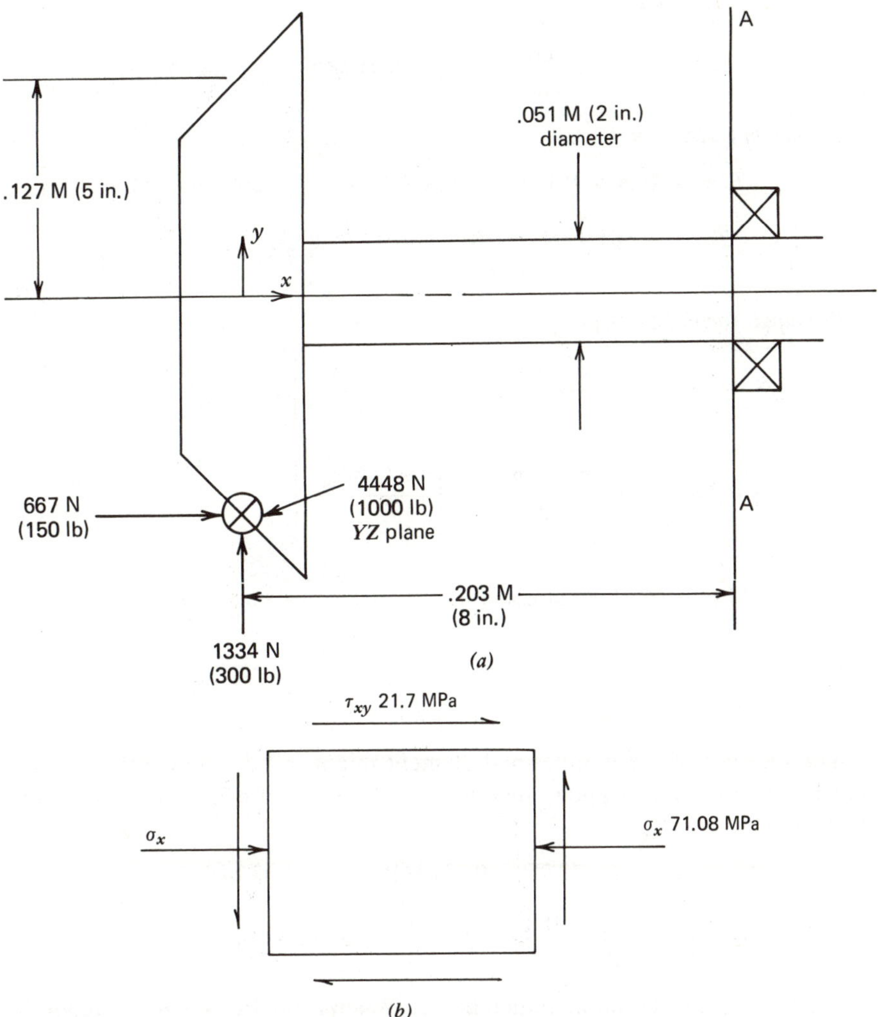

Figure 4.4 Combined stress calculation at section A-A for example 2.1. (a) Gear under load. (b) Stress on element at section A-A.

At outer fiber

$$\sigma_{x\ \max} = \frac{M_{\max} C}{I} = \frac{(921.9)(0.0255)4}{\pi(0.0255)^4} \cong 70.75 \text{ MPa} \qquad (10{,}400 \text{ psi})$$

Resultant Moment in a New Plane (English units)

$$M_{\max} = [(8000)^2 + (1650)^2]^{1/2} = 8168 \text{ in.} \cdot \text{lb}$$

At outer fiber

$$\sigma_{x\ \max} = \frac{M_{\max} C}{I} = \frac{8168 \text{ in.} \cdot \text{lb} \times 1 \text{ in.}}{0.7854 \text{ in.}^4} = 10{,}400 \text{ psi}$$

Axial Load of 667 N

$$\sigma_x = -\frac{P}{A} = -\frac{667 \text{ N}}{0.002 \text{ m}^2} = -0.33 \text{ MPa} \qquad (-47.7 \text{ psi})$$

Torque on Outer Fibers

$$T = 4448\text{N} \times 0.127\text{m} = 564.9(\text{N} \cdot \text{m}) \qquad (5000 \text{ in.} \cdot \text{lb})$$

$$T_{zy} = \frac{TC}{J} = \frac{(564.9\ (\text{N} \cdot \text{m}))(0.255)(2)}{\pi(0.0255)^4} = 21.7 \text{ MPa} \qquad (3183 \text{ psi})$$

Principal Normal Stresses

$$\sigma_{n,\,\max} = \frac{\sigma_x + \sigma_y}{2} + \left[\left(\frac{\sigma_x - \sigma_y}{2}\right)^2 + \tau_{xy}^2\right]^{½}$$

$$\sigma_{n,\,\min} = \frac{\sigma_x + \sigma_y}{2} - \left[\left(\frac{\sigma_x - \sigma_y}{2}\right)^2 + \tau_{xy}^2\right]^{½}$$

$$\text{Maximum shear stress} = \tau_{\max} = \left[\left(\frac{\sigma_x - \sigma_y}{2}\right)^2 + \tau_{xy}^2\right]^{½}$$

For this example

$$\sigma_y = 0$$

The location of the infinitesimal element might fall on the neutral axis for bending in a particular plane; therefore no stress (σ_y or σ_x) would result from that moment and

$$\sigma_x = (-70.75 - 0.33) \text{ MPa} = -71.08 \text{ MPa}$$

$$\tau_{\max} = \left[\left(\frac{-71.08 - 0}{2}\right)^2 + 21.7^2\right]^{½} = 41.64 \text{ MPa} \qquad (6120 \text{ psi})$$

A few key rules should be remembered. When using Eq. 4.9, $\sigma_{n\ \min}$ might be zero, which should not be overlooked in calculating the maximum shear stress. As seen in the example, whenever a force creates a torque, that same force will create a moment in a different plane which adds to the combined stress state. That is, moments are usually created by a force times the distance from the neutral axis, which creates either another moment or torque in a different plane (i.e., combined stresses).

4.3 STRESS FAILURE THEORIES

Engineering materials are only tested for uniaxial loading (tension or torsion) to provide designers with yield and ultimate stress levels. If a component is subjected to biaxial stress, a criterion for failure must be established. Failure theories have been postulated by researchers for static loading and isotropic materials; they do not take into account such loading conditions as impact, fatigue, creep, and buckling. Refer to advanced strength of material text for impact and buckling theories. Stress concentrations, considered later in this chapter, are not included in the failure theories used here. If the material is brittle (i.e., low temperature), stress concentration can be significant, with the exception of nonhomogeneous materials such as gray cast iron.

Maximum-Principal-Stress Theory (Rankine theory)

Concept. Failure occurs when maximum principal tensile or compressive stress is equal to the ultimate tensile or compressive stress in the axial direction. According to this theory, shear strength in torsion is equal to tensile strength. This theory is true only for brittle materials, such as gray cast iron.

For biaxial tension and shear stress

$$\sigma' = \frac{\sigma_x + \sigma_y}{2} \pm \left[\left(\frac{\sigma_x - \sigma_y}{2}\right)^2 + \tau_{xy}^2\right]^{1/2} \qquad [4.10]$$

where σ' = maximum principal tensile stress
σ_x and σ_y = stress in x and y directions
τ_{xy} = shear stress on an infinitesimal element

Maximum-Principal-Strain Theory (Saint Venant)

Concept. Maximum principal strain is equal to strain at failure as determined from a tensile test. Current uses are limited to the design of gun barrels, porcelain, and concrete.

For uniaxial tension and shear with ν (Poisson's ratio) = 0.25

$$\sigma' = \frac{3\sigma_x}{8} + \frac{5}{8}(\sigma_x^2 + 4\tau_{xy}^2)^{1/2} \qquad [4.11]$$

Maximum-Shear-Stress Theory (Coulomb)

Concept. Failure occurs when maximum shear stress is equal to the shear stress at failure when loaded in pure shear. This theory yields conservative results for brittle and ductile materials because it predicts shear properties equal to one-half the comparable tensile properties.

For biaxial torsion and shear

$$\tau_{max} = \left[\left(\frac{\sigma_x - \sigma_y}{2}\right)^2 + \tau_{xy}^2\right]^{1/2} \qquad [4.12]$$

or

$$\sigma' = [(\sigma_x - \sigma_y)^2 + 4\,\tau_{xy}^2]^{1/2} \qquad [4.13]$$

Maximum-Strain-Energy Theory (Beltrami)

Concept. Failure occurs when strain energy per unit volume at a point is equal to maximum strain energy per unit volume that a material can absorb by uniaxial loading.

For uniaxial tension with shear stress

$$\sigma' = [\sigma_x^2 + 2(1 + \nu)\, \tau_{xy}^2]^{1/2} \qquad [4.14]$$

The development of the distortion energy theory resulted from the maximum-strain-energy theory.

Distortion-Energy Theory (Hencky-von Mises)

Concept. Failure occurs when energy of distortion is equal to energy of distortion under uniaxial stress.

For biaxial stresses and shear stress

$$\sigma' = (\sigma_x^2 - \sigma_x \sigma_y + \sigma_y^2 + 3\, \tau_{xy}^2)^{1/2} \qquad [4.15]$$

where shear properties are 57.7% of tensile.

Maximum-Octahedral-Shear-Stress Theory

Concept. Three principal stresses are resolved into two states: 1. equal tensile or compressive in all directions, which may cause fracture (not yield); 2. eight octahedral shear stresses responsible for inelastic action. This theory yields the same results as those given by the distortion-energy theory.

Example 4.2 Comparing failure-theory stress levels

Compare the maximum principal stresses calculated by the the principal-stress theory, maximum distortion-energy theory, and the maximum-shear-stress theory.

Given: 0.2 m diameter shaft loaded in bending (M = 8000 (N · m) and torsion (12,000 N · m). See Fig. 4.5.

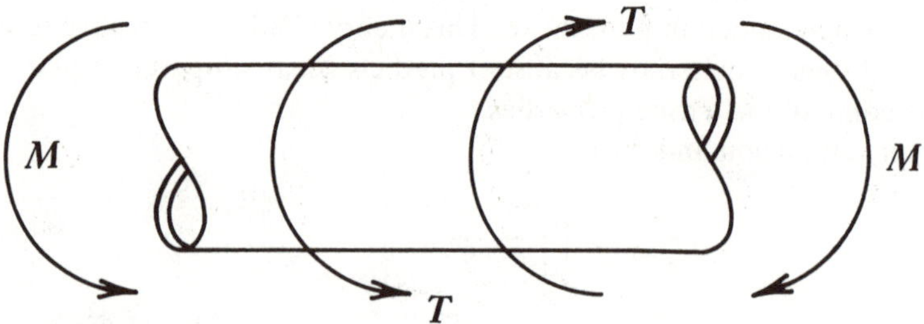

Figure 4.5 Shaft subjected to bending and torsion.

Maximum Bending Stresses (at outer fiber)

Section modulus for a circular section is

$$Z_M = \frac{C}{I} = \frac{\pi D^3}{32} = 0.007854 \text{ m}^3$$

$$\sigma_{\max} = \frac{M}{Z_M} = \frac{8000 \text{ (N} \cdot \text{m)}}{.007854 \text{ m}^3} = 1.02 \text{ MPa}$$

Maximum Torsion Stresses (at outer fiber)

Section modulus for a circular section is

$$Z_t = \frac{C}{J} = \frac{\pi D^3}{16} = 0.00157 \text{ m}^3$$

$$\tau_{xy \max} = \frac{T}{Z_t} = 0.76 \text{ MPa}$$

Using Maximum-Principal-Stress Theory

$$\sigma' \text{ (max. prin. stress)} = \frac{\sigma_x}{2} \pm \left[\left(\frac{\sigma_x}{2} \right)^2 + \tau_{xy}^2 \right]^{1/2} = 1.4 \text{ MPa or } -0.4 \text{ MPa}$$

Note: $\sigma_y = 0$

Using Distortion-Energy Theory

$$\sigma' = (\sigma_x^2 + 3\tau_{xy}^2)^{1/2} = 1.67 \text{ MPa}$$

The distortion-energy theory gives a more conservative answer of 17% greater stress.

Using Maximum Shear Stress Theory

$$\sigma' = (\sigma_x^2 + 4\tau_{xy})^{1/2} = 1.8 \text{ MPa}$$

It can be seen from the previous example that each theory predicts a different maximum level of stress. The designer must decide which answer to use. Experimental work has been done to correlate these theoretical calculations with measured values for different materials.

The following guideline is suggested for use in design.

- A general conservative approach, widely used in designing for all types of materials, is the maximum-shear-stress theory. This theory can overestimate the stress by approximately 15% in some materials.
- For brittle materials, the maximum-principal-stress theory closely approximates reality, and for ductile materials the distortion-energy theory is the best aproximation.

4.4 STRESS CONCENTRATIONS

The theory of elasticity shows that a high stress region occurs at locations that have changes in geometries, such as holes, fillets and notches. This localized high stress is called stress concentration. The theoretical stress concentration factor K_t is defined as follows.

$$K_t = \frac{\sigma_{\text{max}}}{\sigma_{\text{nominal}}} \quad \text{or} \quad \frac{\tau_{\text{max}}}{\tau_{\text{nominal}}} \qquad [4.16]$$

where K_t is always greater than 1.

Stress concentrations can also be obtained experimentally by techniques such as photoelasticity or strain gage. Figure 4.6*a* depicts the development of K_t in an axial loaded plate that has a hole. Figure 4.6*b* depicts noncircular-hole stress concentrations. For $a < W/10$ it has been determined that $\sigma_{\text{max}} = \sigma_{\text{nominal}}\,(1 + 2\,a/b)$, where $(1 + 2\,a/b) = K_t$. σ_{max} will be greater if a is larger than b than if a is less than b. If $a = 4b$, the stress-concentration factor can be determined as follows.

$$\sigma_{\text{max}} = \sigma_{\text{nominal}}\,[1 + 2(4)] \quad \text{or} \quad K_t = 9$$

If $b = 4a$, then

$$\sigma_{\text{max}} = \sigma_{\text{nominal}} \left[1 + 2\left[\frac{1}{4}\right]\right] \qquad \text{or} \qquad K_t = 1.5$$

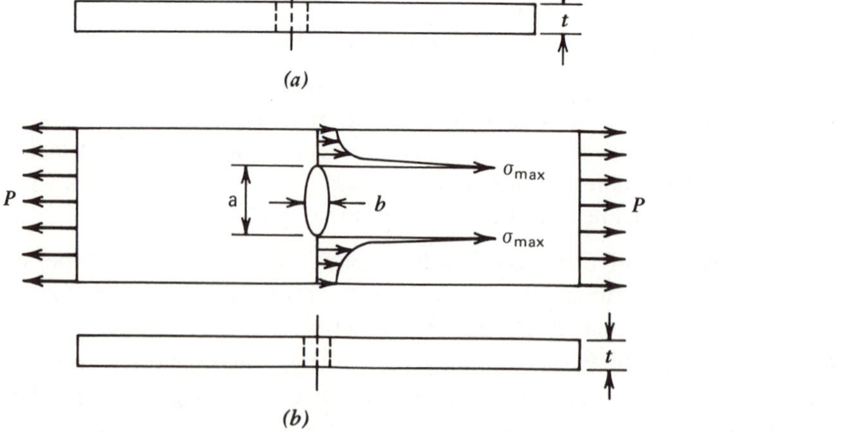

Figure 4.6 Plate or bar with hole.

These equations show why cracks that run perpendicular to load P are more likely to spread than those which run parallel to load P.

Figure 4.7 shows the stress intensities for stepped shafts loaded in tension, bending, and torsion. *Note*: The nominal stress (σ_{nom}) is calculated using the smaller section.

Stress concentrations can be caused by nonhomogeneity in the material, such as cracks at the surface or within the material. Air holes in castings, concrete, or welds also create internal stress concentrations. If loading exists on a small area, stress concentrations must be taken into account during design. Typical high stress contact problems exist between ball and race in bearings, cam and cam follower, wheel and railroad track, and gear teeth.

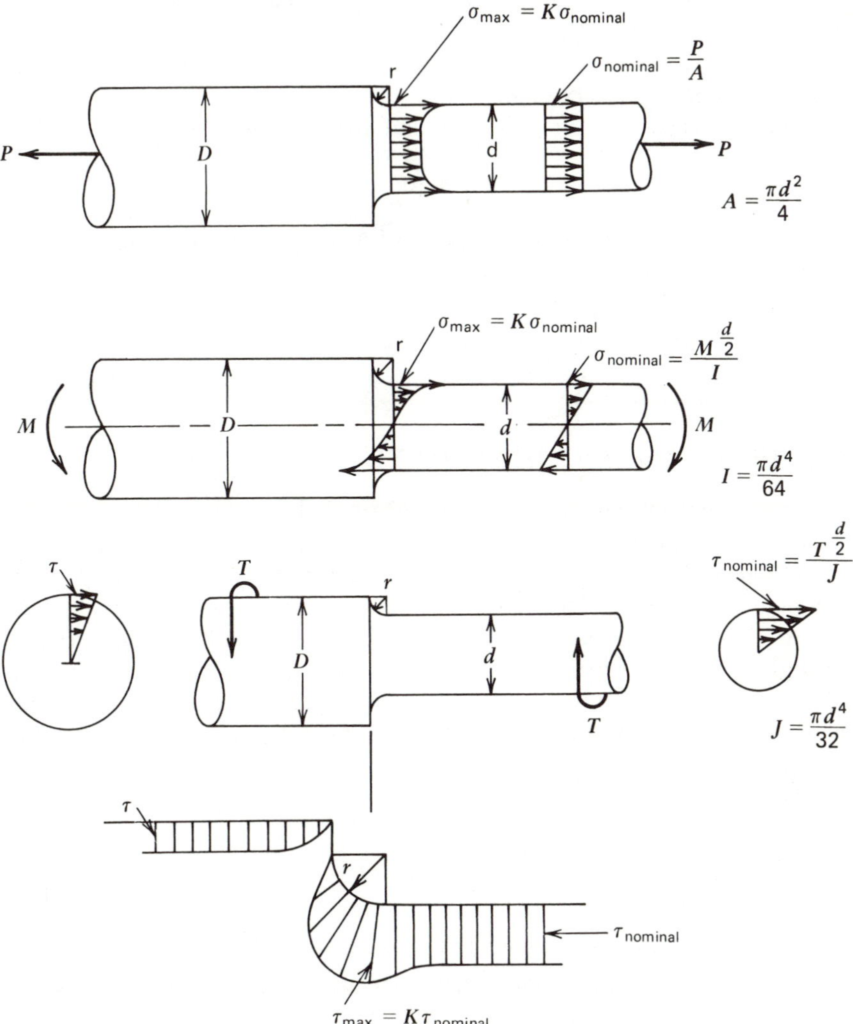

Figure 4.7 Stress intensities for stepped shafts.

Because of local yielding effects stress concentrations can be neglected for ductile materials when making static design calculations. Brittle materials do not exhibit local yielding, and the stress will build up to a rupture point if loading is increased. Therefore, for brittle materials subjected to static loading, stress concentration must be included in design calculations.

For members subjected to varying loads, the stress-concentration factor should always be included in design computations because the maximum stress must not exceed the endurance limit, which is less than the ultimate stress and the yield stress. If the maximum stress were to exceed the endurance limit, then a fatigue crack would eventually occur and propagate under repeated loading of the member, thus reducing the cross-sectional area of the member until it is insufficient. The member fails by breaking apart and is said to have failed in fatigue.

Peterson's book provides K factors on all shapes and loading conditions as demonstrated in Figs. 4.8 through 4.11.

Another stress-concentration factor found in the literature is K_{ty}.

$$K_{ty} = \frac{\sigma_{\max}}{\sigma} \qquad [4.17]$$

where K_{ty} = stress concentration factor based on gross stress

$\sigma_{\max}$ = maximum stress (true stress)

σ = applied stress away from any geometry change

K_{ty} takes into account stress increases that result from decreased section and geometry. For fatigue calculations use only K_t found in Figs. 4.8 through 4.11.

4.5 BEAM BENDING AND SHAFTS

Various components, such as a hook, chain link, or clamps, are cast (or formed) into a curved shape. When loaded with a moment the bending stress doesn't follow the linear pattern that it follows in a straight beam. Figure 4.12 shows the higher order of stress distribution for a curved beam. The difference between the stress distribution of a curved beam and that of a straight beam is the stress at the center of gravity for the curved beam is not zero and the neutral axis is closer to the radius of curvature origin. The stresses at the outer fiber are usually most important. Inside fiber stress is composed of bending stress plus normal axial stress.

For bending stress only on inside fiber

$$\sigma_{\text{bending}} = \frac{F\,a\,h_i}{A\,eR_i} \qquad [4.18]$$

where Fa = bending moment

h_i = the distance from neutral axis to the inside fiber ($h_i = R - e - R_i$)

e = eccentricity distance between neutral and center-of-gravity axis

R_i = radius of curvature of inside fiber

A = area

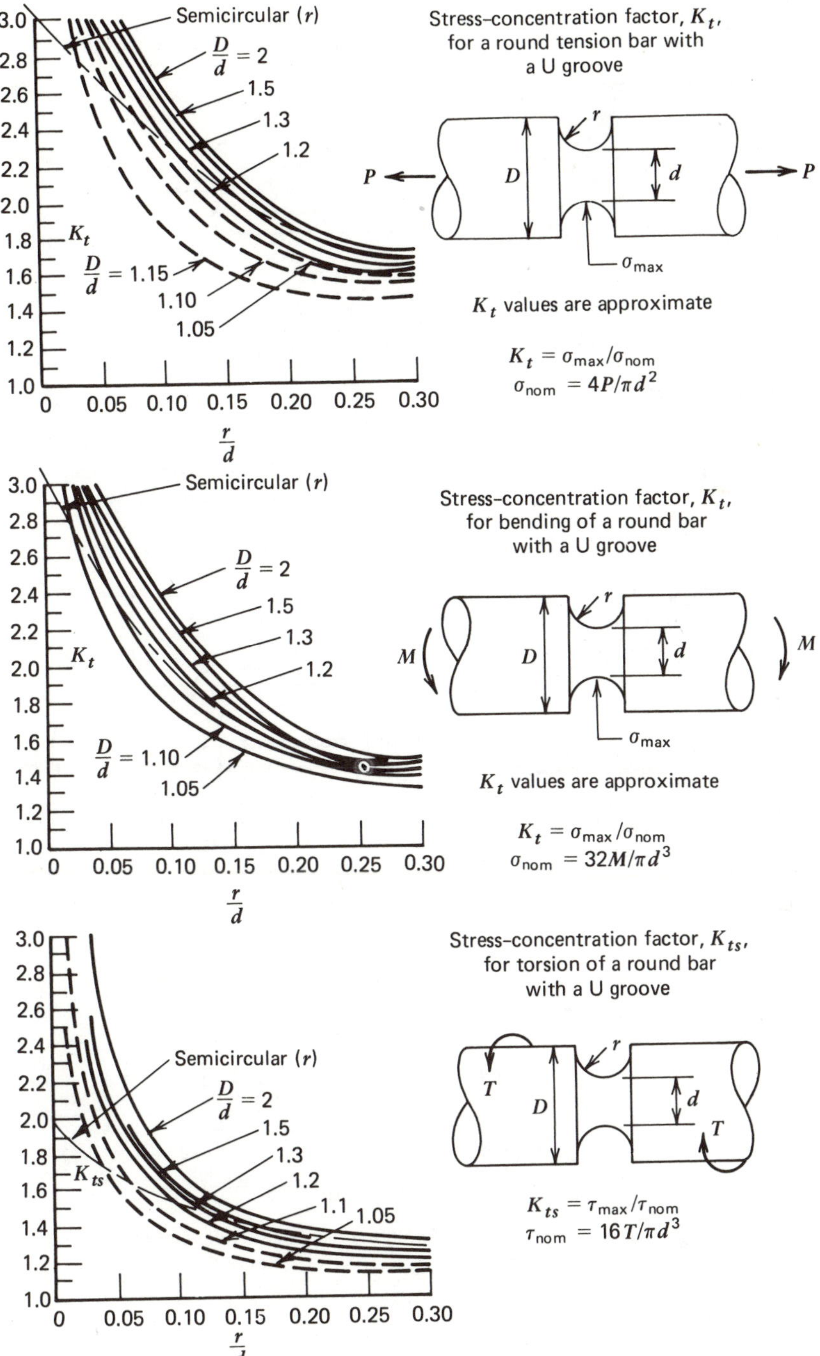

Figure 4.8 Peterson's stress concentrations for shafts. Reprinted with permission © Peterson, R.E., Stress Concentration Factors, John Wiley and Sons, Inc.

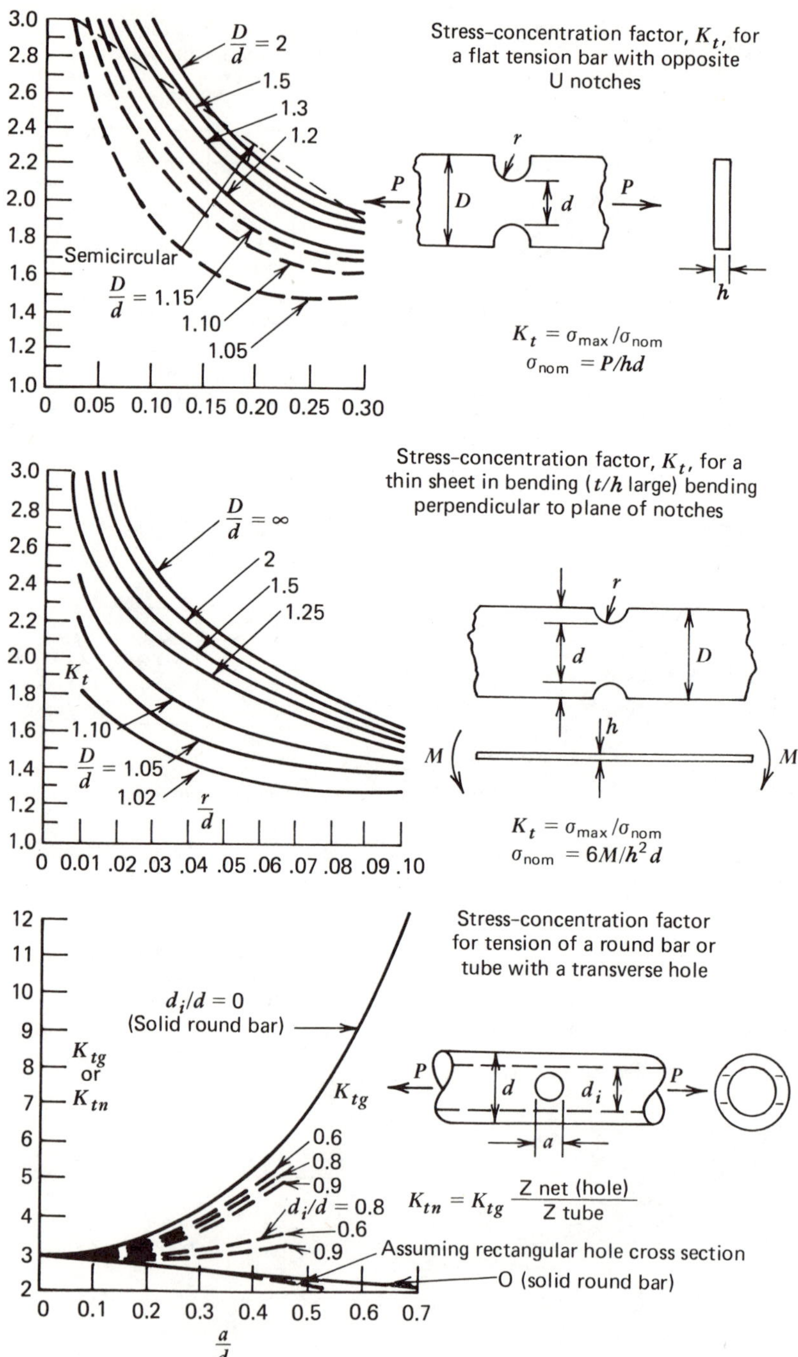

Figure 4.9 Peterson's stress concentrations for bars and hollow tubes. Reprinted with permission © Peterson, R.E., Stress Concentratrion Factors, John Wiley and Sons, Inc.

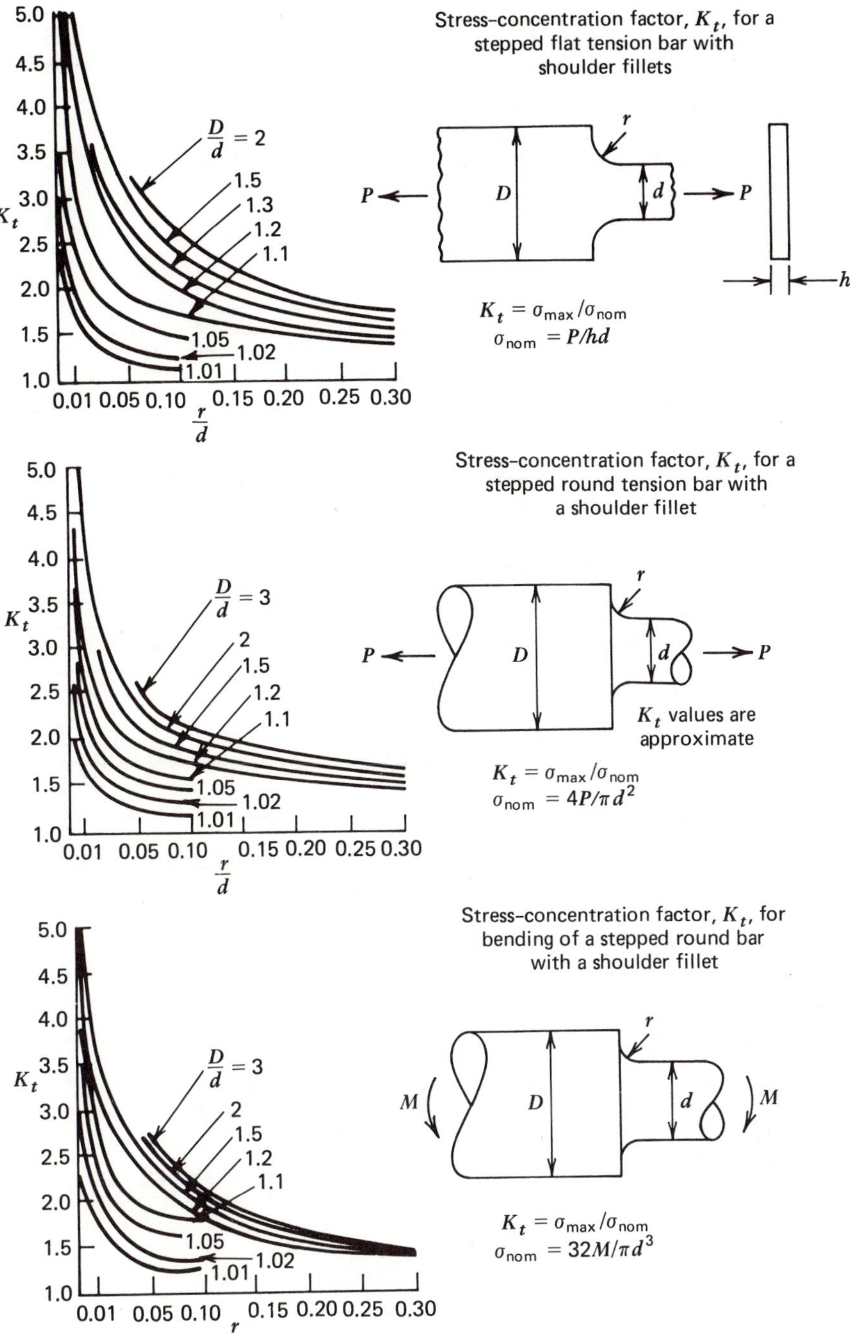

Figure 4.10 Peterson's stress concentrations for various sections. Reprinted with permission © Peterson, R.W., Stress Concentration Factors, John Wiley and Sons, Inc.

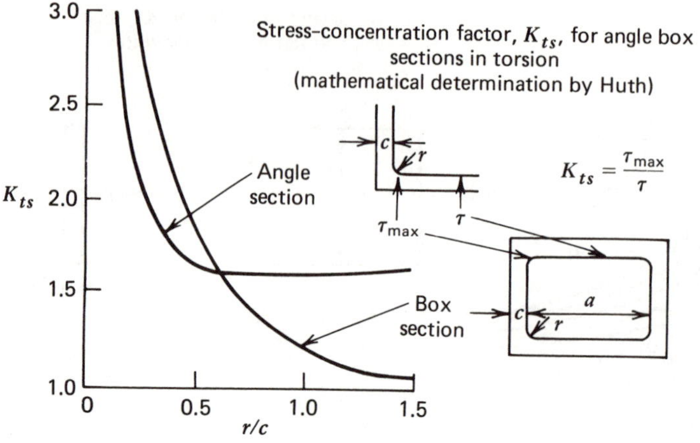

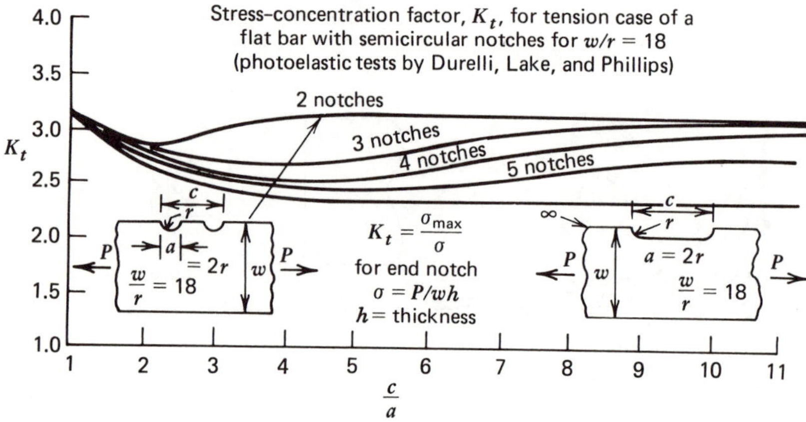

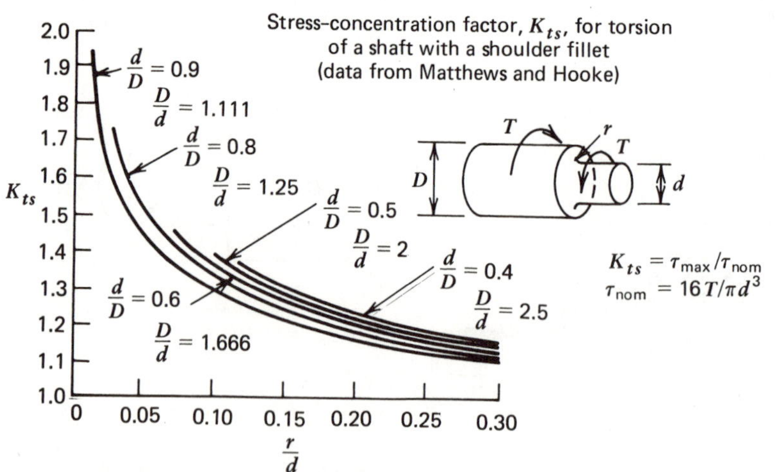

Figure 4.10 (*continued*)

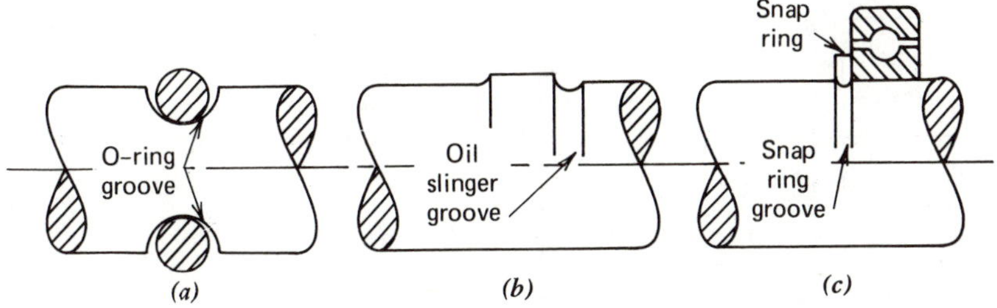

Figure 4.11 Examples of stress concentrations. (a) O'ring groove. (b) Oil slinger groove. (c) Snap ring groove.

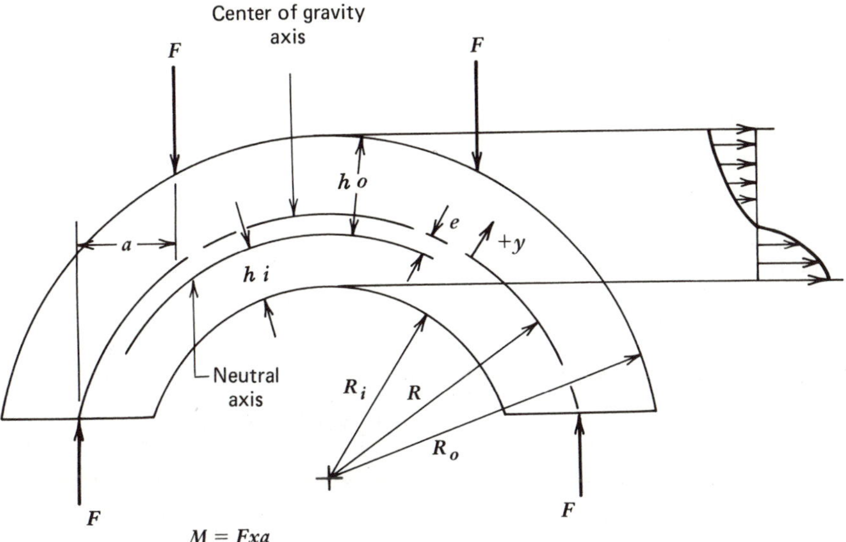

Figure 4.12 Curved beam stress distribution.

Outside fiber stress consists of the bending stress plus axial stress.

For bending stress only on outside fiber

$$\sigma_{\text{bending}} = \frac{F\ a\ h_o}{A\ e\ R_o} \qquad [4.19]$$

where h_o = the distance from neutral axis to the outside fiber ($h_o = R_o + e - R$)

R_o = the radius of curvature to the outside fiber

Figure 4.13 depicts symmetrical cross sections where the neutral axis location may be found.

Note: For symmetrical sections the maximum stress occurs at the inside fiber, and the axial stress must be added or subtracted depending on the sign convention.

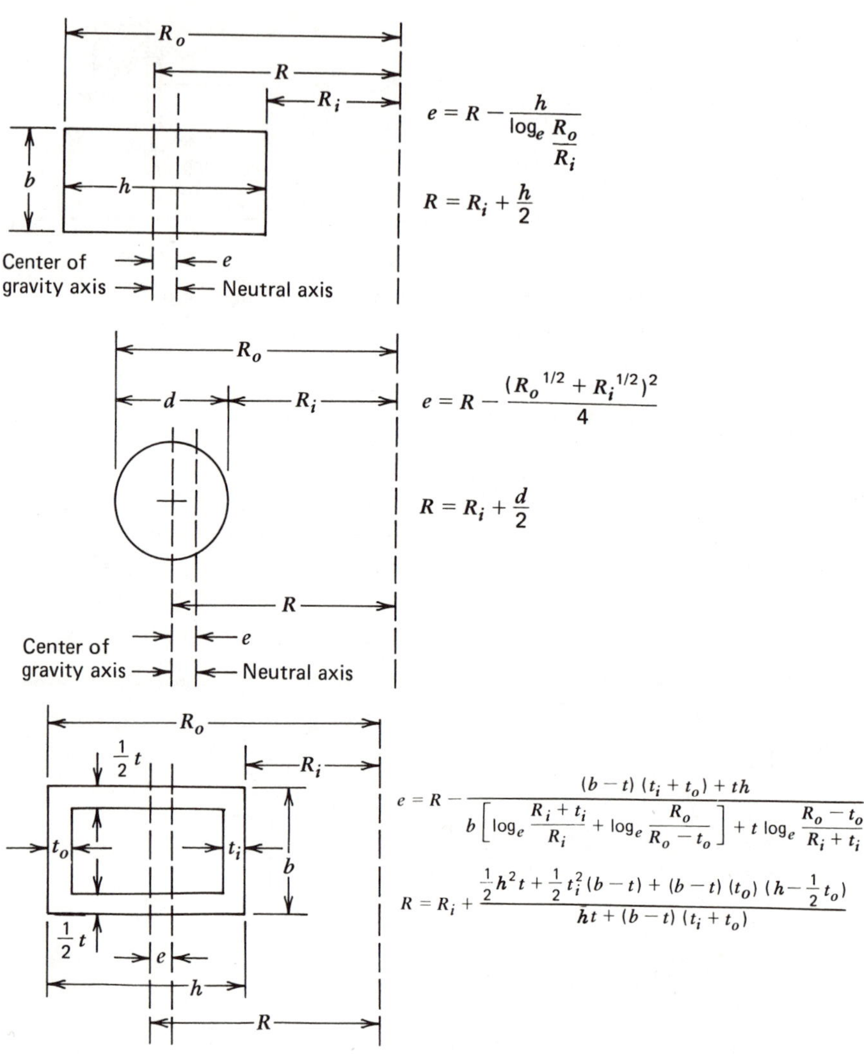

Figure 4.13 Sample cross sections for curved beams.

Example 4.3 Calculation of stresses on a hook at Section A-A. (Figure 4.14)

$$\sigma_{\max} = \sigma_b + \frac{P}{A}$$

$$\text{moment } M = 1000\ \text{N} \times 0.15\ m = 150.0\ (\text{N}\cdot m)$$

For a curved beam

$$\sigma_{b_{\max}} = \frac{M\ h_i}{A\ e\ R_i}$$

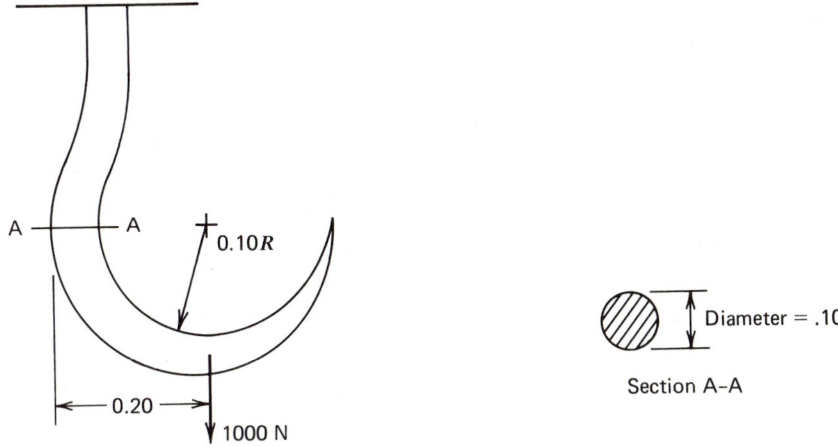

Figure 4.14 A hook under load for Example 4.3.

Determining e, h_i for a circular section

$$e = R_i + \frac{d}{2} - \frac{(R_o^{1/2} + R_i^{1/2})^2}{4}$$

$$e = 0.1 + \frac{0.1}{2} - \frac{(0.2^{1/2} + 0.1^{1/2})^2}{4} = 0.0043 \text{ m}$$

$$h_i = R_i + \frac{d}{2} - e = .1 + 0.1/2 - 0.0043 = 0.1457 \text{ m}$$

$$\sigma_{b_{\max}} = \frac{1000 \times 0.15 \times 0.1457}{\pi \times (0.1/2)^2 \times 0.0043 \times .1} = 6.47\text{MPa}$$

Therefore,

$$\sigma_{\max} = \sigma_{b_{\max}} + \frac{P}{A} = 6.47 + \frac{1000}{\pi(0.5)^2} = 7.7\text{MPa}$$

Note: Because e is small, its value can cause considerable variation in the stress value and care must be taken in its calculation.

4.5.1 Shafts

Shafts are used on machinery to transmit torque and speed. The type of loading can fluctuate, and stress concentrations are present at keyways or changes in sections. The speed of a shaft should be separate from its critical rotating speed or large vibrations can result.

The Dunkerley equation is usually used to determine the critical speed of a shaft that tends to cause whirling because it has on it a mass such as a pulley or gear. (See Figure 4.15.)

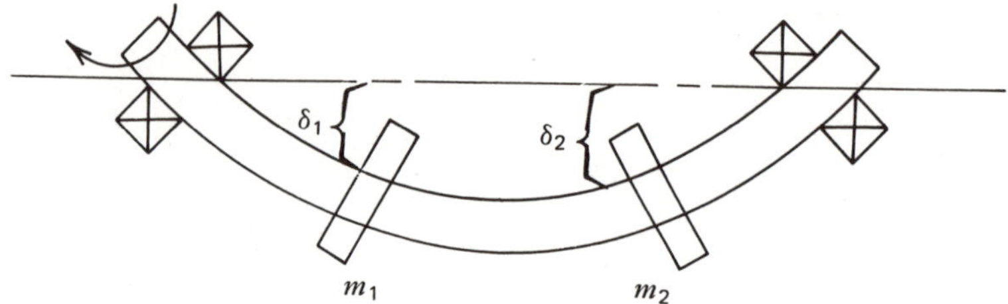

Figure 4.15 Whirling shaft with two concentrated weights.

The critical speed is estimated as

$$\frac{1}{\omega_c^2} = \frac{1}{\omega_s^2} + \frac{1}{\omega_1^2} + \frac{1}{\omega_2^2} + \ldots \qquad [4.20]$$

where ω_c = shaft critical speed (radian/time)

ω_1 = critical speed with one mass $m_1 = (g/\delta_1)^{1/2}$

g = gravitational constant 9.81 m/sec^2

δ_1 = static deflection caused by force at the m_1 location ($F = m_1 g$)

$\omega_s \cong (5/4g/\delta_s)^{1/2}$ = critical speed of shaft without any weights, determined as a result of a deflection by its own weight

where

δ_s = maximum shaft deflection caused by its own weight

For eccentric masses (set screws and keys) on a shaft, the deflection at that mass is equal to

$$\delta_e = \frac{m_e\ e\ \omega^2}{K - m_t w^2}$$

where δ_e = eccentric shaft deflection

ω = shaft speed

e = eccentricity

m_e = eccentric mass

m_t = total mass

K = shaft-spring constant

Current manufacturing processes based on low cost do not allow for precise concentric masses on a shaft (for example, car tires and rims and combine threshing cylinders). The eccentricity is therefore reduced by balancing: weights are added to nullify the centrifugal force caused by eccentricity, or metal is removed (a common practice on crankshafts). Friction also reduces the effects of critical speeds to some lower value.

The horsepower transmitted by the shaft is

$$\text{hp} = \frac{T \times \text{rpm}}{63{,}025} \qquad [4.21]$$

where T = torque (in · lb)

or

$$hp = \frac{F \times V}{33{,}000} \qquad [4.22]$$

where F = force (lb)

V = velocity at the force in ft/min

Watts (W) or joules (J) per second are the common metric power units.

$$1\ kW = \frac{T \times rpm}{9550} = \text{kilowatts} \qquad [4.23]$$

where T is in N · m

Note: 1000 N · m/sec = k W. The conversion relationship is thus 1 kW = 1.341 hp.

Because shafts carry both bending and torsion there is a combined stress. The maximum static shear stress, from the maximum-shear-stress theory (Mohr's circle), results in

$$\tau_{max} = \frac{0.5\,\sigma_y}{FS} \left[\left[\frac{\sigma_x}{2} \right]^2 + \tau_{xy}^2 \right]^{1/2} \qquad [4.24]$$

where

$$\sigma_x = \frac{32\,M}{\pi\,d^3} \quad \text{and} \quad \tau_{xy} = \frac{16T}{\pi\,d^3} \qquad [4.25]$$

where M = moment

T = torque

Therefore

$$\tau_{max} = \frac{0.5\,\sigma_y}{FS} = \frac{16}{\pi\,d^3}\sqrt{M^2 + T^2} \qquad [4.26]$$

or

$$d = \left[\left[\frac{32\,FS}{\pi\,\sigma_y} \right] (M^2 + T^2)^{1/2} \right]^{1/3}$$

ASME has developed a code for transmission shaft design, B17c-1927. The code allows for shock loading in the design of rotating shafts.

$$d^3 = \frac{32\,FS}{\sigma_y\,\pi}\sqrt{(K_m\,M)^2 + (K_T\,T)^2} \qquad [4.27]$$

	K_m	K_T
Steady load	1.5	1.0
Minor shocks	2.0	1.5
Heavy shocks	3.0	3.0

Equation 4.27 can be modified to account for the endurance limit of materials that will be discussed in the chapter on fatigue.

If the shaft is hollow the factor $16/\pi d^3$ can be replaced by r/J, where r is the outside radius and J is the polar moment of inertia computed on the net area.

There are two additional important stress considerations in designing shafts. The stress concentrations caused by keyways in a shaft must be added to the stress; and, if a gear or hub is press fit on a shaft, additional compressive stresses are present. One way to handle each of these problems is to increase the factor of safety in Eq. 4.27.

4.5.2 Keyways

Shafts, pulleys, and hubs of gears are usually fastened together by keys. Several common types are shown in Fig. 4.16 and Table 4.1.

The most common keyway is the square keyway. To keep the keyway from sliding axially down the shaft and becoming loose, a setscrew is employed to reduce potential axial shifting. If axial movement is required, splines are used (i.e., pto shafts). Assuming that all the torque is carried by the shaft's outer radius, the force on the keyway can be determined.

$$F = \frac{T}{r} \qquad [4.28]$$

Table 4.1 **American National Standard Square Key**

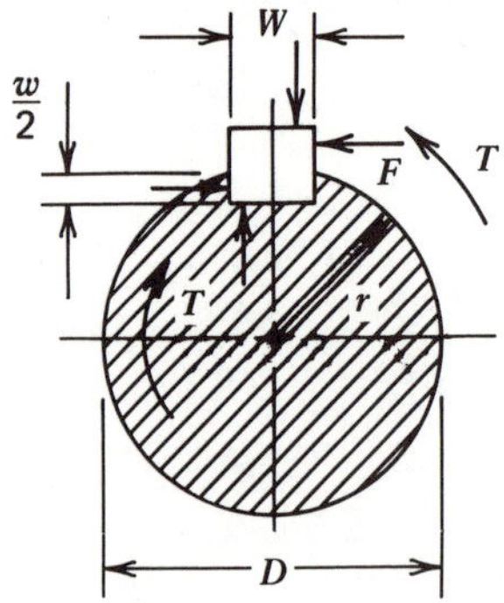

Shaft Diameter—nominal		*Square Stock,*	*Shaft Diameter—nominal*		*Square Stock,*
Over	*To (inclusive)*	*W*	Over	To (inclusive)	*W*
5/16	7/16	3/32	1 3/4	2 1/4	1/2
7/16	9/16	1/8	2 1/4	2 3/4	5/8
9/16	7/8	3/16	2 3/4	3 1/4	3/4
7/8	1 1/4	1/4	3 1/4	3 3/4	7/8
1 1/4	1 3/8	5/16	3 3/4	4 1/2	1
1 3/8	1 3/4	3/8	4 1/2	5 1/2	1 1/4

Notes: All dimensions in inches.
See appendix for other keys.

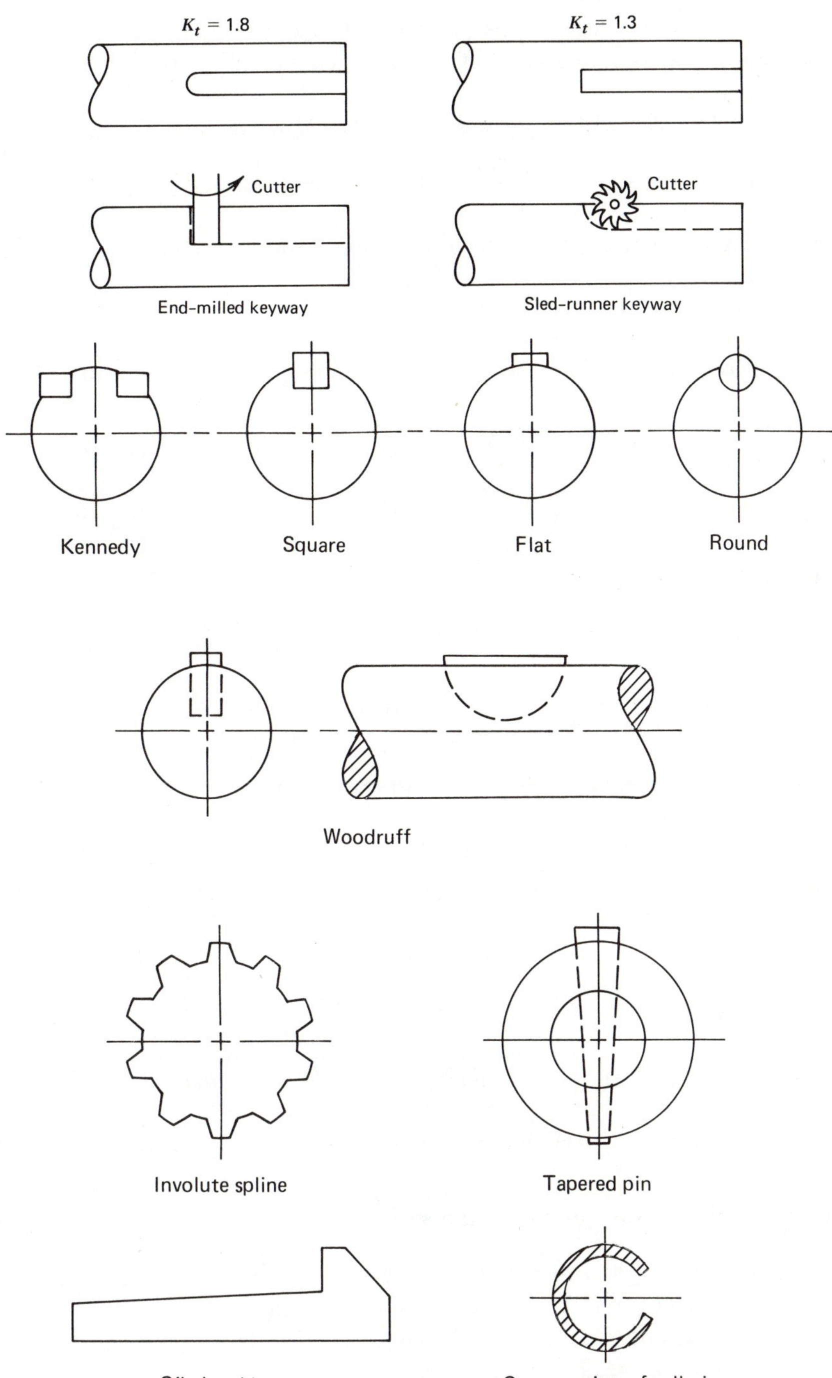

Figure 4.16 Types of keyways and stress concentrations for keyways. NOTE: Use $K_t = 3$ when shaft is in bending and torsion.

The key is usually designed as the weakest link (shaft and gear are made from stronger material) and should fail first if an overload occurs.

Example 4.4 Key Design for a Torque of 1000 in. · lb on a 1-in. shaft

$$F = \frac{T}{r} = \frac{1000}{1/2} = 2000 \text{ lb at shaft surface}$$

The key length can be calculated from the bearing stress at yield (σ_{all})

$$\sigma_{allowable} = \frac{F}{A}$$

Assume 25,000 psi material yield strength (includes FS) and 1/4 in. square key

$$25{,}000 = \frac{F}{1/4 \times L}$$

$$L\,(\text{in.}) = \frac{8000 \text{ lb/in.}}{25{,}000 \text{ lb/in.}^2} = 0.32 \text{ in.}$$

If the shaft material is 32,000 psi

$$L = \frac{8000}{32{,}000} = 0.25 \text{ in.}$$

If the key is in shear, design using allowable shear stress (τ_{all}). Again, using 25,000 psi for σ_{all}, find the length needed to resist shear.

$$\tau_{all} = 1/2\,\sigma_{all} = 12{,}500 = \frac{3F}{2A}$$

$$12{,}500 \times 2A = 3F$$

and the shear area A is

$$A = \frac{1}{4} \times L$$

with $F = 2000$ lb

$$12{,}500 \times \frac{L}{2} = 3(2000) \qquad L = \frac{6000}{6250} = 0.96 \text{ in.}$$

Note: Because shear is critical in this case the longer key length is required.

4.6 DESIGNING FOR DEFLECTION

A good rule in design is to pattern the new component after one of similar shape, size, and material, with loads in the same magnitude. If this practice results in a high value (i.e., 100), for the factor of safety the designer must question whether the part has been overdesigned.

A different failure criterion was used in the original design process for many agricultural and construction machines, these devices are not usually overdesigned. The difference could have been that a certain mass (weight) was needed for reduced vibration, increased tractor traction, or increased protection.

One of the most common criteria on high-calculated factor-of-safety parts is the need for low deflections. Once the load is determined, the cross section can be changed to minimize deflections.

Deflection is important in the design of transmission shafts. If these shafts deflect too much, the teeth will not mesh properly and impact will be more violent, creating wear, higher stresses, and noise. The deflection of cylinders in forage choppers and combines must also be considered during the design, because machine performance is related to the clearance setting.

The fundamental equation of beam theory, Eq. 4.29, provides the basics for finding deflections of various structures.

$$\frac{d^2 y}{d\,x^2} = \frac{M}{EI} \qquad [4.29]$$

For the cantilever beam shown in Fig. 4.17, the reaction forces and moments can be solved for

$$R_1 = P \qquad \text{and} \qquad M_1 = -PL$$

Subtitute for M into Eq. 4.29 integrating twice, and evaluating the constants of integration will provide an equation for deflection as a function of load, material properties, cross section, and distance from $x = 0$. At any x location, $M = P\,(L - x)$. Remember that the sign convention for moments is negative.

$$EI\,\frac{dy^2}{dx^2} = Px - PL \qquad [4.30]$$

Integrating once yields

$$EI\,\frac{dy}{dx} = \frac{Px^2}{2} + PLx + c_1 \qquad [4.31]$$

Since $dy/dx = 0$ at $x = 0$, then $C_1 = 0$. Integrating twice yields

$$EI\,y = \frac{Px^3}{6} + \frac{PLx^2}{2} + C_2 \qquad [4.32]$$

at the left end $x = 0$, and $y = 0$; therefore, $C_2 = 0$.

$$y = \frac{P}{EI}\left[\frac{x^3}{6} + \frac{Lx^2}{2}\right] \qquad [4.33]$$

and y_{max} at $x = L$ equals

$$y_{max} = \frac{PL^3}{3\,EI} \qquad [4.34]$$

Fig. 4.18 shows maximum deflections for beams with different loading and support conditions.

L
P
M₁
x
R₁
y

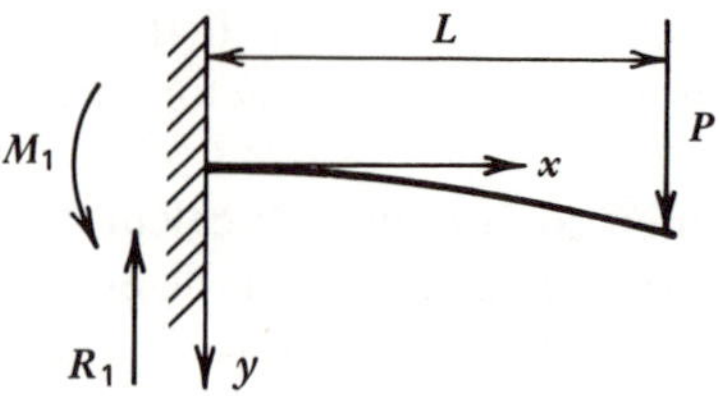

Figure 4.17 Forces on a cantilever beam.

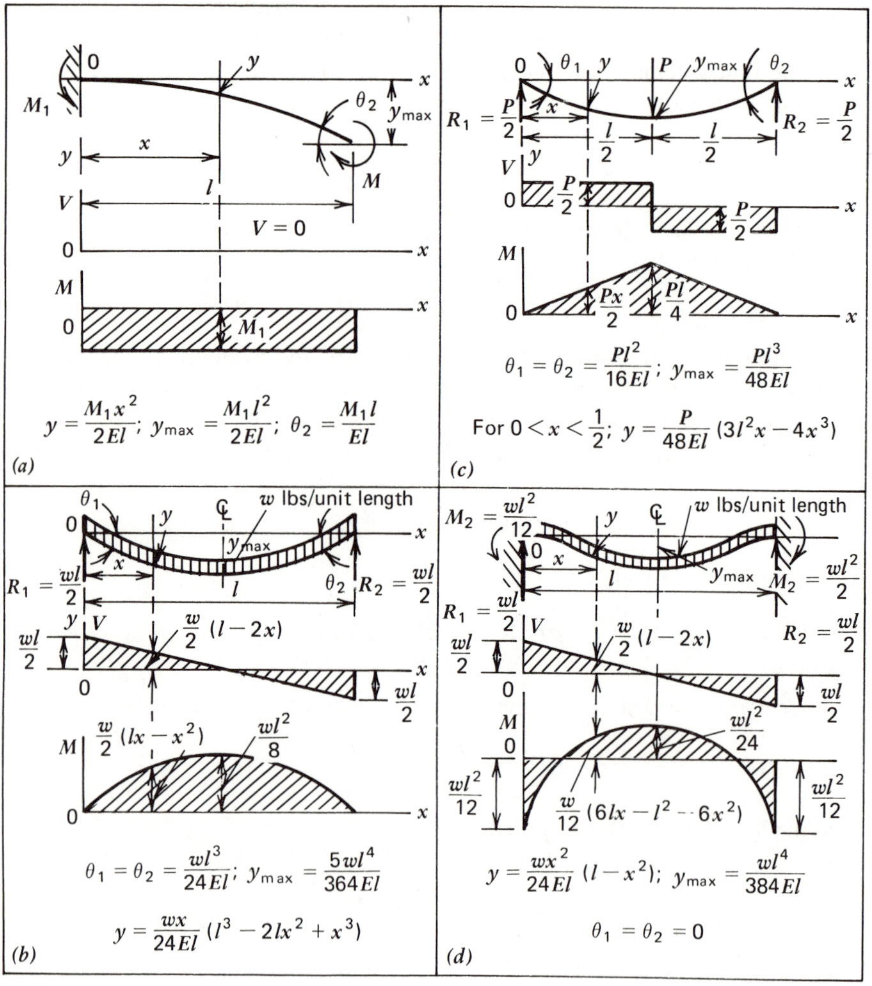

(a)
$y = \frac{M_1 x^2}{2EI}$; $y_{max} = \frac{M_1 l^2}{2EI}$; $\theta_2 = \frac{M_1 l}{EI}$
V = 0
(b)
w lbs/unit length
$R_1 = \frac{wl}{2}$
$R_2 = \frac{wl}{2}$
$\frac{w}{2}(l-2x)$
$\frac{w}{2}(lx-x^2)$
$\frac{wl^2}{8}$
$\theta_1 = \theta_2 = \frac{wl^3}{24EI}$; $y_{max} = \frac{5wl^4}{364EI}$
$y = \frac{wx}{24EI}(l^3 - 2lx^2 + x^3)$
(c)
$R_1 = \frac{P}{2}$
$R_2 = \frac{P}{2}$
$\frac{Px}{2}$
$\frac{Pl}{4}$
$\theta_1 = \theta_2 = \frac{Pl^2}{16EI}$; $y_{max} = \frac{Pl^3}{48EI}$
For $0 < x < \frac{1}{2}$; $y = \frac{P}{48EI}(3l^2x - 4x^3)$
(d)
w lbs/unit length
$M_2 = \frac{wl^2}{12}$
$R_1 = \frac{wl}{2}$
$R_2 = \frac{wl}{2}$
$\frac{w}{2}(l-2x)$
$\frac{wl^2}{24}$
$\frac{wl^2}{12}$
$\frac{w}{12}(6lx - l^2 - 6x^2)$
$y = \frac{wx^2}{24EI}(l-x^2)$; $y_{max} = \frac{wl^4}{384EI}$
$\theta_1 = \theta_2 = 0$

Figure 4.18 Deflections, shear, and moment diagrams for uniform cross-section beams.

Shear stress can also cause additional deflections and warping to a minor degree, and therefore neither should be overlooked if deflection is critical to the machine function. Also, for beams with multiple types of loading, the principle of superposition (i.e., using individual loads) to find deflection might be the best approach. With this method, the general beam equation and Castigliano's theorem are commonly used to solve indeterminate structure problems (i.e., a loading condition where $\sum F$ and $\sum M$ will not provide all the forces, deflections, and stresses).

Another method for determining deflections is Castigliano's theorem, which states that displacement from any force on an elastic system is obtained by taking the partial derivative of the total strain energy with respect to that force.

For a member under tension, the strain energy (U) or potential is the average force times the deflection, $U = F/2\ \delta$. Substituting for $\delta = FL/AE$, then:

$$U = \frac{F^2 L}{2AE} \quad [4.35]$$

Also, for straight beams the strain energy for moment M is

$$U = \int \frac{M^2\,dx}{2EI} \quad [4.36]$$

and for transverse shear, V, is

$$U = \int \frac{K\,V^2\,dx}{2AG}$$

Using Castigliano's theorem on the beam in Fig. 4.17,

$$U = \int \frac{M^2\,dx}{2EI} = \int_0^L \frac{(Px)^2\,dx}{2EI} = \frac{P^2L^3}{6EI}$$

The deflection is $\partial U/\partial P = y = PL^3/3EI$, obtained by integrating the beam equations.

The weight of the member itself can result in sizable deflections and is important in designing rollers that are in contact with each other while rotating.

One method of reducing deflection and increasing strength at a weight and cost savings is by adding stiffening ribs. Because incorrect proportions of rib stiffness can increase rather than decrease stress, the change of I and c should be checked, as discussed by Orthwein.

4.7 STRAIN DETERMINATION

Many agricultural machine parts have complicated shapes that can be analyzed only at critical sections by theoretical beam formulas and the engineers must rely on past experience in choosing material or sizing sections.

To verify strength, engineers must rely on testing. The most common approach is to measure the strain experimentally for a given load and to interpret that strain for field condition loading. A proportionate relationship from zero to the yield strain is used.

The most common experimental method used to determine strain is to estimate the level of stress at different load conditions using theoretical formulas, use stress coat to determine maximum strain locations, and finally apply foil-type gages at those locations to obtain specific strain readings.

Two types of motion can cause a body or component to displace: rigid-body (as a whole) movements, and movements of body points relative to one another. Strain is related to movements within a body because rigid-body displacements do not produce strains. Normal strain is defined as the change in length of a given distance on the body divided by the original length; shearing strain is defined as the angular change between two line segments on a body that originally were perpendicular to each other. Common relationships are

$$\epsilon_x = \frac{\partial u}{\delta x}, \epsilon_y = \frac{\partial v}{\partial y}, \epsilon_z = \frac{\partial w}{\partial z} \qquad [4.37]$$

$$\gamma_{xy} = \frac{\partial v}{\partial x} + \frac{\partial u}{\partial y}, \gamma_{yz} = \frac{\partial w}{\partial y} + \frac{\partial v}{\partial z}, \gamma_{zx} = \frac{\partial u}{\partial z} + \frac{\partial w}{\partial x}$$

where *u, v,* and *w* are displacements in the *x, y,* and *z* directions, respectively.

4.7.1 Stress Coat

Stress coat is a combination of approximately one-third zinc resinate dissolved in carbon disulfide with a small amount of plasticizer added to change the point at which cracking occurs. The manufacturer denotes the amount of plasticizer by the coating number (the greater the number, the higher the level of threshold strain). Complete details for application of stress coat and strain gages are given in many references.

The part is usually tested inside rather than outdoors because of the moisture-absorbing quality of the coating. The part and six or more calibration bars are sprayed with a uniform thickness of the stress coat. The most commonly used coating is 1205. The part and bars are cured at 80 to 85°F in controlled humidity for about 15 hours or more. The calibration fixture provides a given deflection for the calibration bar and is used throughout the test. A calibration bar is loaded to its constant deflection, and a given strain can be read from a gauge bar where the crack closest to the load occurs.

This crack is designated as the threshold strain ϵ_t^*. The component is then loaded in reasonable increments from zero until the first crack appears. A flashlight is used to look for a crack at each loading increment. When the first crack is found, it is circled and its direction and corresponding load are marked. The stress then is equal to

$$\sigma_1 = E\,\epsilon_t^* \qquad [4.39]$$

where E = modulus of elasticity of part

σ_1 = principal stress at the applied load perpendicular to crack (*Note*: strain is assumed negligible parallel to crack.)

The first crack will be the location of highest strain. Additional load increments are applied to determine other high-strain areas. The direction of each crack is marked for later strain gage application.

Assuming a proportionate relationship between maximum load and load at first crack, the first crack stress (σ_A) can be estimated.

$$\sigma_A = \frac{\text{maximum load}}{\text{load at first crack}} \times \sigma_1 \qquad [4.40]$$

It is important to check calibration during the entire test. If the maximum estimated stress is above the yield point of the material or the endurance limit, further analysis or redesign is needed. Another form of stress coat is colored paint applied by a manufacturer, which will crack locally if high strains are applied.

4.7.2 Strain Gages

Strain gages can be used to provide test information on any of the following,

1. Functional performance.
2. Power requirements.
3. Stresses (strains).
4. Durability.
5. Wear rates.
6. External forces.

Strain gages are usually used to provide a more accurate measure of strain on a specific location, whereas stress coat determines the location of high strains (i.e., electrical gage location) for a surface. Foil-wire gages, as shown in Fig. 4.19, are common.

The basic characteristics of gages include length, gage sensitivity, loading or strain range, and accuracy. Kelvin noted that the resistance in a wire increased with increasing strain. This change, for wire-type gages can be depicted as

$$\frac{dR}{R} = \frac{\partial \rho}{\rho} + \frac{\partial L}{L} - \frac{dA}{A} \qquad [4.41]$$

where R = resistance of the uniform conductor
ρ = the specific resistance $\rho \approx 50 \times 10^{-6}\ \Omega \cdot \text{cm}$
L = length
A = cross-sectional area (where $dA = -\nu\, dL/L$ is its change as a result of transverse shear strain).

Substitution for A gives

$$\frac{\Delta R}{R} = \frac{\partial \rho}{\rho} + \frac{dL}{L}(1 + 2\nu) \qquad [4.42]$$

where $1 + 2\nu$ is equal to approximately 1.6 for metals. If the direction of normal universal strain is known, then the strain sensitivity is defined as

$$S = \frac{\Delta R/R}{\epsilon_x} \qquad [4.43]$$

Figure 4.19 Applying a foil strain gage.

which is the gage factor when $\epsilon_y = -0.285\,\epsilon_x$. Gages must be compensated for temperature changes, and the test temperature must be held nearly constant. Dummy gages are used to balance out the effects of pressure, temperature, and humidity. Other factors that affect strain gages are: a. adhesive (2% elongation); b. number of strain cycles; c. amplitude of strain; d. gage current; e. frequency of loading; f. time (aging); g. humidity and moisture; h. hydrostatic pressure and; i. magnetic fields.

Care is required in applying gages that normally work in a -55°C to 85°C range. A procedure is detailed as follows:

1. Clean surface of paint, scale pits, rust, and so on.
2. Apply a neutralizer and abrade with silicon carbide paper.
3. Wipe abrasive particles off with a sponge.
4. Neutralize again.
5. Draw reference lines to orient gage.
6. Put cellophane tape on top surface of gage to facilitate handling.
7. Apply adhesive to one surface (gage, usually) and let dry for 1 minute at 45°C.
8. Apply glue (adhesive) to surface where gage is being attached.
9. Slide gage onto glued surface starting at a slight angle.
10. Hold finger pressure on gage for 30 seconds.
11. Remove cellophane tape by pulling it at 90° from the gage's principal direction.

Electrical devices commonly used to measure $\Delta R/R$ and to convert this measured resistance change into strain are the potentiometer and the Wheatstone bridge (Fig. 4.20). The Wheatstone bridge can determine both dynamic and static strain gage readings. The change in output voltage is related to strain as follows:

$$\Delta E_{\text{out}} = V_{\text{in}} \frac{R_1 R_2}{(R_1 + R_2)^2} \left[\frac{\Delta R_1}{R_1} - \frac{\Delta R_2}{R_2} + \frac{\Delta R_3}{R_3} - \frac{\Delta R_4}{R_4} \right] \quad [4.44]$$

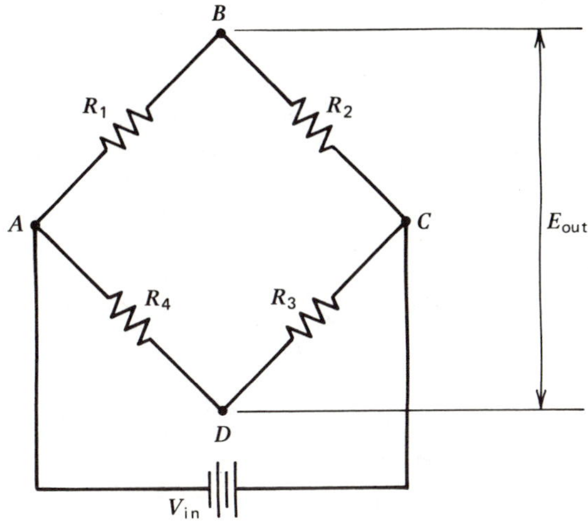

Figure 4.20 Wheatstone bridge—V_{IN} is the input voltage and E_{OUT} is the output voltage proportionate to strain.

In most cases, only one resistor is an active gage, therefore, the change in voltage ΔE_{out} determines $\Delta R/R$ for the active gage. If the gage factor S is known, the strain can be directly calculated. *Note:* Loads are applied in the range of strain for which the adhesive will not fail.

Example 4.5 Strain gage reading and stress determination

Two strain gages are attached on principal stress lines for the steel shaft in Fig. 4.21. Calculate the maximum principal stress by using the graphed readings in Fig. 4.21.

Principal Stresses Can Be Determined from Strain Gages

For biaxial (surface loading)

$$\sigma_1 = \frac{E}{1 - \nu^2} (\epsilon_1 + \nu\, \epsilon_2) \qquad [4.45]$$

$$\sigma_2 = \frac{E}{1 - \nu^2} (\epsilon_2 + \nu\, \epsilon_1) \qquad [4.46]$$

where ν = Poisson's ratio

E = Young's modulus of elasticity

ϵ_1, ϵ_2 = strain measurements

For triaxial loading

$$\sigma_1 = \frac{E}{(1+\nu)(1-2\nu)} [(1-\nu)\epsilon_1 + \nu(\epsilon_2 + \epsilon_3)] \qquad [4.47]$$

$$\sigma_2 = \frac{E}{(1+\nu)(1-2\nu)} [(1-\nu)\epsilon_2 + \nu(\epsilon_1 + \epsilon_3)] \qquad [4.48]$$

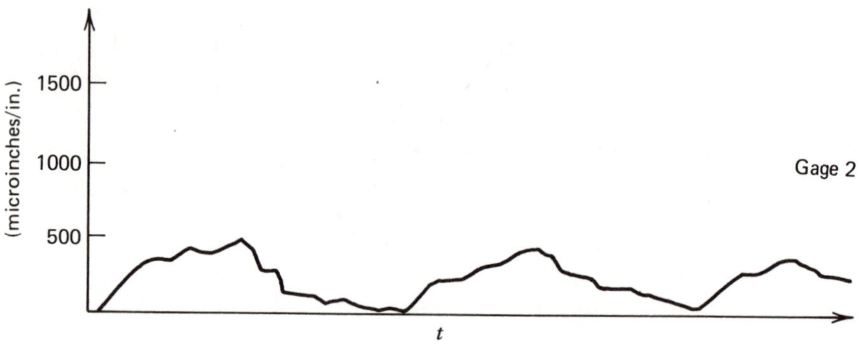

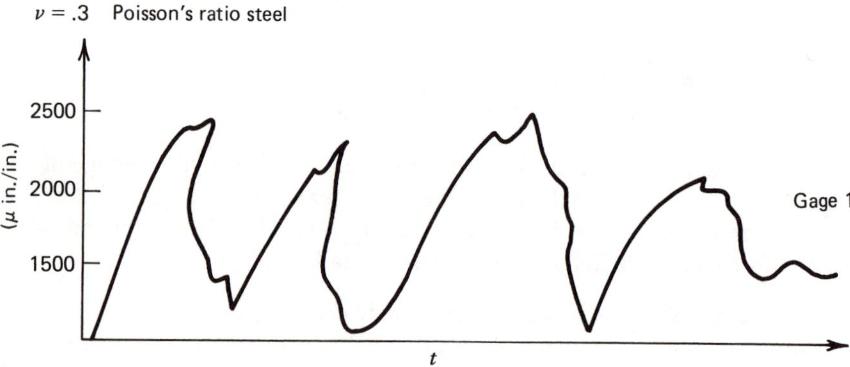

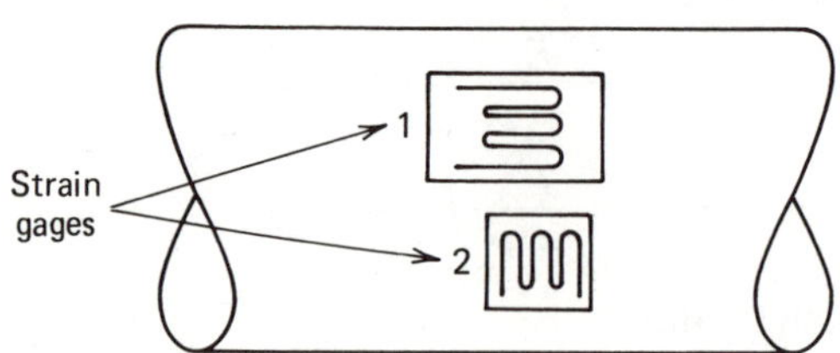

Figure 4.21 Steel shaft strain readings μin./in.

$$\sigma_3 = \frac{E}{(1+\nu)(1-2\nu)}\,[(1-\nu)\epsilon_3 + \nu(\epsilon_2 + \epsilon_1)] \qquad [4.49]$$

For the shaft (Fig. 4.21)

$$\sigma_1 = \frac{30 \times 10^6 \text{ psi}}{1-(0.3)^2}\,(2500 \times 10^{-6} + 0.3(500 \times 10^{-6}) = 87{,}363 \text{ psi}$$

In metric units

$$\sigma_1 = \frac{200 \times 10^6 \text{ kPa}}{1 - (0.3)^2}(2500\,\mu + 0.3(500 \times \mu) = 582{,}420 \text{ kPa}$$

$$\sigma_2 = \frac{30 \times 10^6 \text{ psi}}{1 - (0.3)^2}\;(500 \times 10^{-6} + 0.3(2500 \times 10^{-6})$$

$$= 41{,}209 \text{ psi or } 274{,}727 \text{ kPa}$$

If the material is steel SAE 1040CD, would it have yielded? The yield point of SAE 1040 cold drawn steel is 380,000 kPa (55,000 psi) and

$$\text{FS} = \frac{\sigma_y}{\sigma_{\text{cal}}} = \frac{380{,}000}{582{,}420} = \frac{55{,}000 \text{ psi}}{87{,}363 \text{ psi}} = 0.65$$

Yes, it would have yielded.

Other types of strain, such as photoelastic stress analysis, are not covered by this book.

4.8 PLASTIC DESIGN

Yielding is a surprisingly common occurrence. Many processes, such as bending, rolling, drawing, extruding, and forging, create plastic deformation in the part. Figures 4.22 and 4.23 depict the elastic and plastic state in a part. The

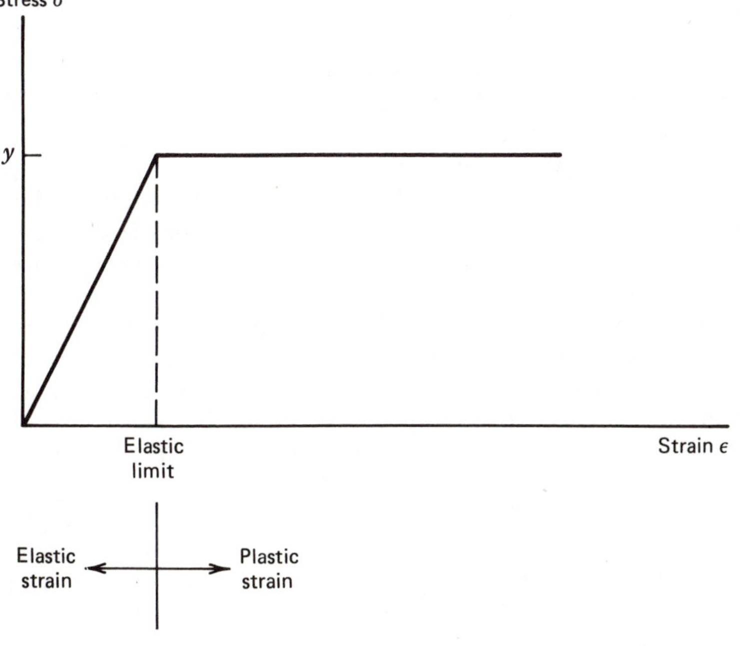

Figure 4.22 Stress strain curve with a zero plastic slope.

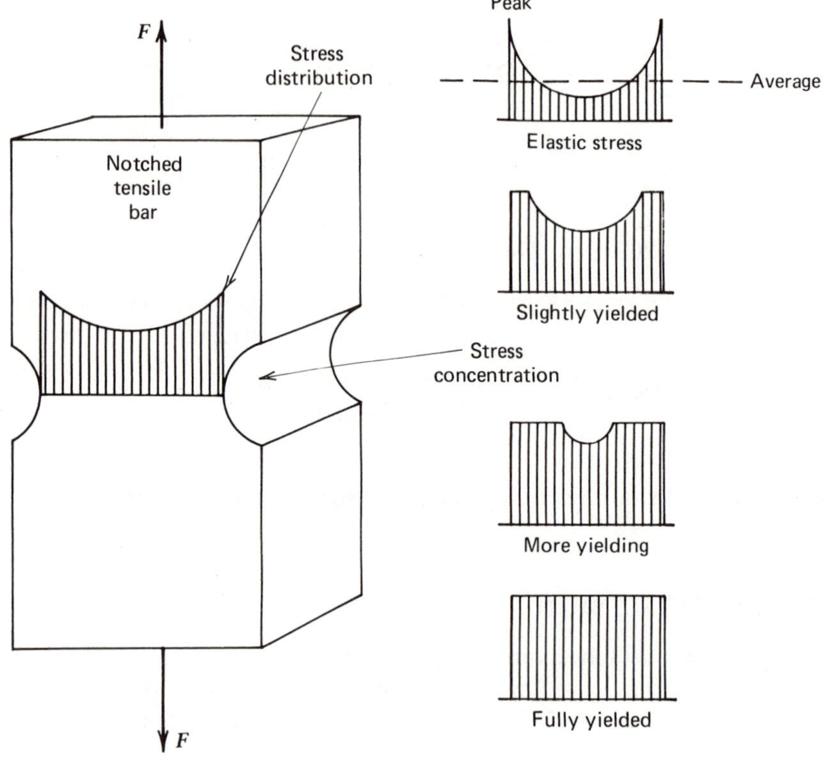

Figure 4.23 Plastically stressed part loaded to yield.

slope of the plastic stress strain relationship can be assumed parallel, as depicted by Fig. 4.22, or it can have an increasing slope. The degree to which a part can accept plastic deformation is related to elongation and reduction in area (e.g., low carbon steels exhibit this property).

A rectangular solid beam can withstand a 50% greater moment above the yield point before it fails. This deformation results in a considerable amount of absorbed energy. Utilizing this energy can help engineers design rollover protection devices and crashworthy vehicles.

Plastic collapse at a point defines a phenomenon called hinges. For complex loading cases, these hinge locations are usually determined experimentally. Engineers then utilize this data in determining the energy under a force-deflection curve when designing rollover bars and the like.

Shape also affects the size of forces that a part can tolerate before separating. Table 4.2 includes several examples.

Considerable care must be taken when designing in the plastic range. Most designs are for one-time occurrences (crashes, etc.), and fatigue loading is not considered. Testing usually is imperative to verify the feasibility of the plastic design. In addition, plastic design saves energy and cost by using the material to its fullest, but parts could have intolerably large deflections.

Table 4.2 Comparing Plastic and Elastic Moments

Beams	Strength ratio $\frac{\text{plastic moment}}{\text{elastic moment}}$
Rectangular (b)	1.5
Circular (R)	1.7
I-section	1.17
T-section (b, b; t is small)	1.8
Hollow rectangle (h, b; t is small)	$\frac{bh+h^2/2}{bh+h^2/3}$
Hollow circle (t is small)	1.27

4.9 STRESS CAUSED BY IMPACT

Components may yield when an impact load is applied to a part. The tractor drawbar provides an example of this situation. When a tractor is put in drive, it achieves some finite velocity while the drawbar pin and other clearances are taken up. The pin and drawbar pinhole experience loads in excess of the drawbar power developed, and there is local yielding. The load on a rectangular bar that is subjected to a weight W can be calculated as

$$F = W + W\left[\frac{1 + 2hAE}{L}\right]^{1/2} \tag{4.50}$$

where h = the height the weight is dropped
A = cross-sectional area of the bar
E = dynamic Young's modulus, which is usually higher than for a static test
L = bar length

If $h = 0$, then the stress is found from Eq. 4.50 as

$$\sigma = \frac{2W}{A} \tag{4.51}$$

or twice the static level, not considering the dynamic Young's modulus effects.

In most agricultural vehicles impact is measured by accelerations. Field vibration tests show that these accelerations seldom exceed three. When parts are subjected to impact, the factor of safety should be increased or the calculated stress should be determined using the dynamic load.

Example 4.6 Impact load calculation on a tractor drawbar

Determine the factor of safety on a drawbar given that static drawbar pull = 20,000 N; area = 0.0004 m^2; and test impact load = 2.5 g. The P_D (dynamic load) = 2.25 × 20,000 = 50,000 N

$$\sigma = \frac{P_D}{A} = \frac{50{,}000 \text{ N}}{0.0004 \text{ m}^2} = 125{,}000 \text{ kPa}$$

for SAE 1020 steel (yield = 262,000 kPa)

$$\text{FS} = \frac{\sigma_y}{\sigma_{cal}} = \frac{262{,}000 \text{ kPa}}{125{,}000 \text{ KPa}} = 2.1$$

SUMMARY: A COMMON MISTAKE IN STRESS ANALYSIS

The location of the infinitesimal element might fall on the neutral axis for bending in a particular plane and therefore no stress would result (i.e., σ_y or σ_x) from that moment!

HOMEWORK PROBLEMS

4.1 Find the maximum shear stress and normal stress on the 4-in. diameter tractor axle in Fig. 4.24. Assume that $K_t = 1$.

4.2 A ladder rung on a combine is made from a 1-in. bar as shown in Fig. 4.25. Neglecting curvature, calculate the maximum shear stress for a 200-lb person standing in the middle of the rung.

4.3 Determine the maximum shear stress in the member loaded as shown in Fig. 4.26. Assume that $K_t = 3$.

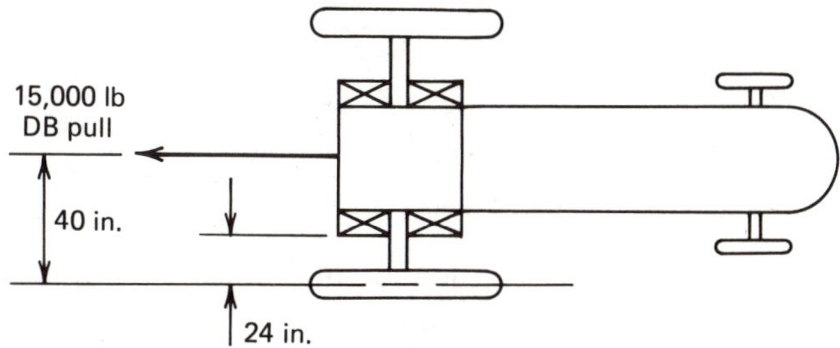

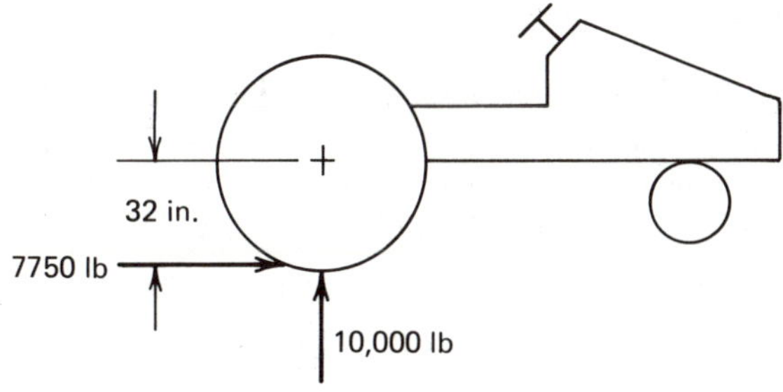

Figure 4.24 A tractor rear-axle force diagram.

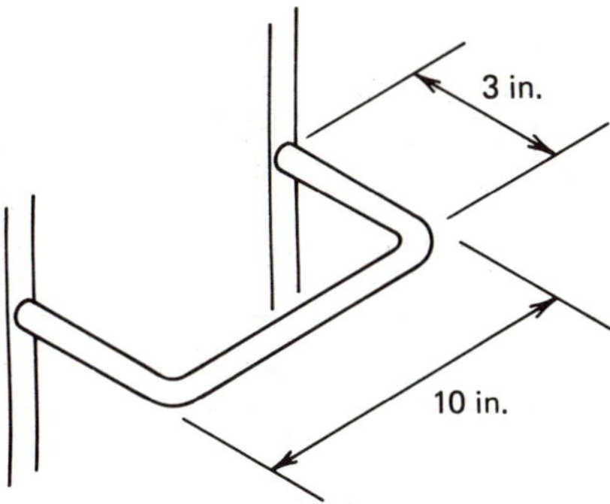

Figure 4.25 A ladder rung.

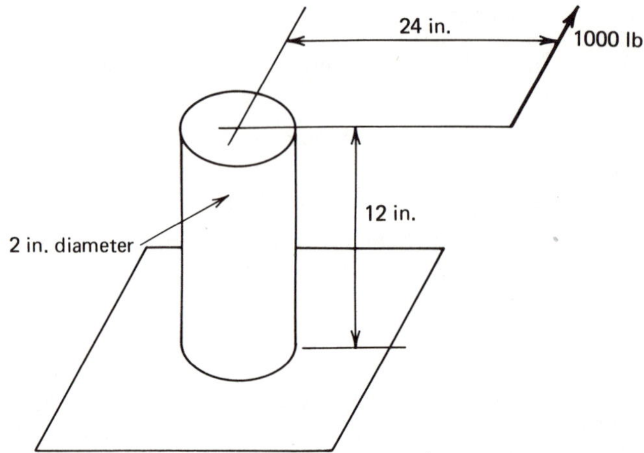

Figure 4.26 Bar under combined stress.

4.4 Determine the stress concentration for an O-ring groove as shown in Fig. 4.11, if the part has an outside diameter of 4 in. and the O-ring groove is a semicircle with a radius of 0.125 in., for the loads in a) bending only, and b) torque only.

4.5 A bar beam 1 cm by 1 cm is bent around a 30-cm inside radius. Calculate the bending stress if the moment applied is 100 N · m.

4.6 What is the critical speed (rpm) of the shaft shown in Fig. 4.27?

4.7 Calculate the shaft size needed in a heavy shock load case where the moment is negligible and the shaft transmits 30 hp at 200 rpm. Assume that FS is 2 and the material yield strength σ_y is 30,000 psi.

4.8 In Fig. 4.18 derive part a using Castigliano's theorem.

4.9 For Problem 4.7, what size keyway should be specified?

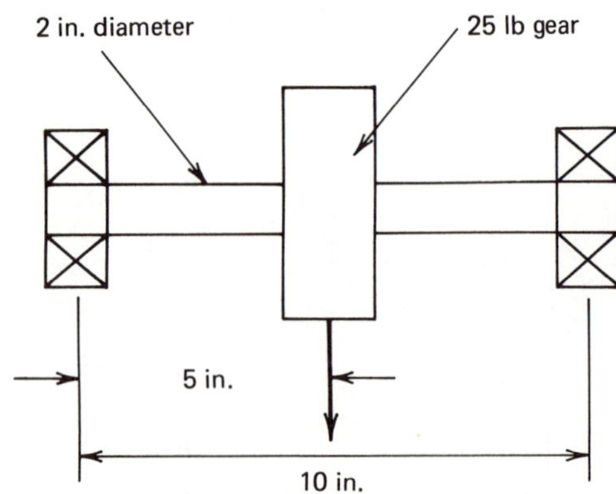

Figure 4.27 Shaft with rotating gear.

4.10 A steel part is loaded while stress coated. The calibration bar gives a threshold strain of 500 μ/in. If the first crack occurs at 1000 lb of load and the maximum expected is 5000 lb, what is the maximum stress at the location of the first crack?

4.11 A tractor axle is subjected to dynamic loads that are twice the static load and has a $K_t = 1.5$ at the bearing. Repeat Problem 4.1 for this new loading case.

4.12 If the person in Problem 4.2 jumps on the rung with 2 g, what is the new maximum stress?

4.13 Calculate the factor of safety for Problem 4.12 if SAE 1020 material is used.

4.14 What is the critical speed in Problem 4.6 if the shaft's diameter is increased by 1 in.?

4.15 Two strain gages located on a part's surface after stress give maximum readings of 300 μ/m (microns per meter) and 2900 μ/m. What are the principal stresses?

REFERENCES

1. Burton, P. "A Modification of the Coulomb-Mohr Theory of Fracture," *Journal of Applied Mechanics* 28(2): 259-262, 1961.
2. Dally, J.W. and W.F. Riley. *Experimental Stress Analysis,* McGraw-Hill Book Co., New York, N.Y., 1965.
3. Faupel, J.H. and F.E. Fisher. *Engineering Design,* 2nd edition, John Wiley and Sons, Inc., New York, N.Y., 1981.
4. Hall, A.S. Jr., A.R. Holowenko, and H.G. Laughlin. *Theory and Problems of Machine Design,* Schaum's Outline Series, McGraw Hill Book Co., New York, N.Y., 1961.
5. Johnson, R. C. *Optimum Design of Mechanical Elements,* 2nd edition, John Wiley and Sons, Inc., New York, N.Y., 1980.
6. Juvinall, R. *Engineering Considerations of Stress, Strain and Strength,* McGraw-Hill Series in Mechanical Engineering, McGraw-Hill Book Co., New York, N.Y., 1967.
7. Lukasik, J.J. "Field Stress Analysis of Industrial Construction and Forestry Vehicles," Society for Experimental Stress Analysis, Ames, Iowa, May 11-16, 1975.
8. Marin, T.J. "Theories of Strength for Combined Stresses and Non-Isotropic Materials," *Journal of Aeron. Science,* 24(4):265-269, 1957.
9. Orthwein, W.C. "Adding Ribs for Maximum Strength," *Machine Design,* 53(4):113-116, 1981.
10. Perry, C.C. *Shear Signs Revisited,* Vishay Measurements Group, Raleigh, N.C., February 1981.
11. Peterson, R.E. *Stress Concentration Factors,* 1st edition, John Wiley and Sons, Inc., New York, N.Y., 1974.
12. Roark, R.J. and W.C. Young. *Formulas for Stress and Strain,* McGraw-Hill Book Co., New York, N.Y., 1975.

13. Smith, R. "Strain Concentration Factor as a Function of Strain in a Design Application," ASME Paper #81-DE-5, The American Society of Mechanical Engineers, New York, N.Y., 1981.
14. Spotts, M.F. *Design of Machine Elements,* 5th edition, Prentice-Hall, Inc., Englewood Cliffs, N.J., 1978.

CHAPTER FIVE

FATIGUE LOADING

5.1 INTRODUCTION OF FATIGUE

Any analysis of machine components should take into account the fact that loads are not constant, but fluctuate and create a variable stress state. Engineers first made this observation as they were looking at railroad-train axle failures, and they coined the term *fatigue* to describe the situation. As a railroad axle shaft rotates, an infinitesimal stress element at the outer fiber is subjected to a maximum compressive stress. As the shaft rotates 90° the stress on the element is zero (except torsional stresses), and after another 90° rotation the stress at the element location results in a maximum tensile stress. This stress variation causes the part to fail at a stress level below the yield point. The failure is catastrophic in nature, usually starting at a small crack and progressing to the point where not enough section is left to meet loading conditions. Stress concentration caused by change in sections, stress risers, material fault, or inclusions in the material are the predominant mechanisms that initiate fatigue failure.

The ability to design and test for fatigue is important to an experienced engineer. Fatigue testing is done with a machine that puts bending stress on rotating shafts or cantilever beams (see Fig. 5.1). Some machines put torsional or combined stresses on the parts. Many parts must be tested at different stress levels to determine the fatigue strength (see Fig. 5.2). These tests are temperature, material, and shape dependent. If one of these variables changes, so will the fatigue plot curve, called the S-N curve. S-N stands for stress versus the number of cycles. The tensile strength of a part is loaded to one-half cycle at failure. S-N curves may be plotted on either semilog or log-log paper. The difference between these two types of plots depicts where the curve levels off to a horizontal slope. Steels have a stress level to which the part may be loaded without failing. This stress level, called the endurance limit, σ_e, is a horizontal line on the S-N curve. Some nonferrous metals, such as pure aluminum, do not exhibit an endurance limit and will fail at some finite number of stress cycles.

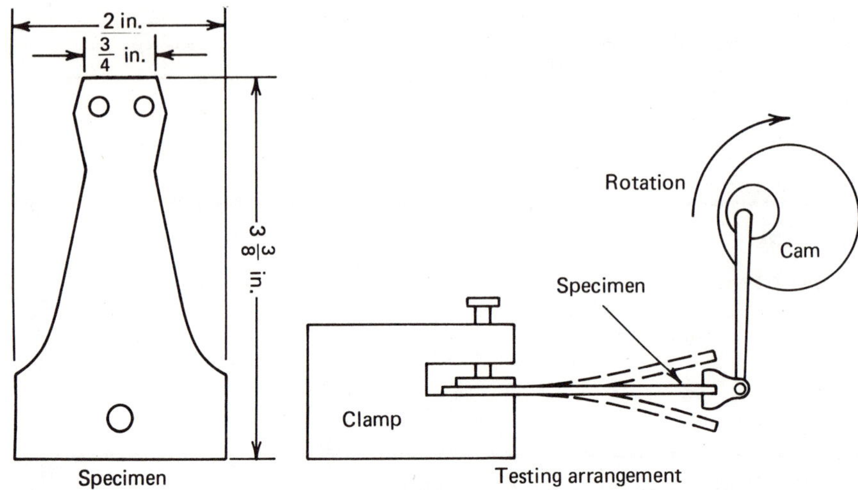

Figure 5.1 Cantilever beam fatigue testing. Courtesy Huntington Alloys, Inc.

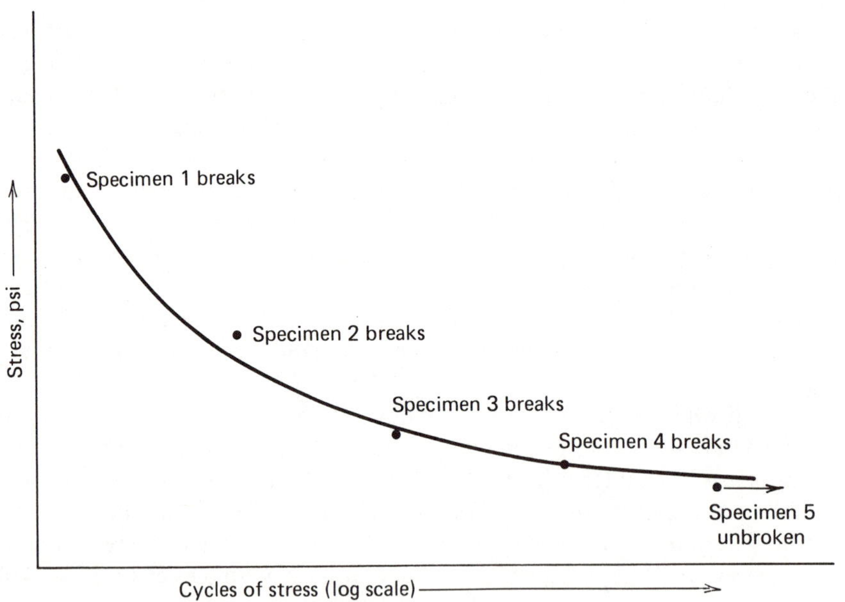

Figure 5.2 Typical plot of fatigue-test results. Courtesy Huntington Alloys, Inc.

Failures commonly occur in rotating shafts or flexing-beam-type components. Engineers should use test data to design a part for fail-safe, reliable designs on such components as airplane wings, car steering, tractor brakes, roofs, or other parts, for which human safety is an issue.

A cycle of stress is the variation from the maximum value to the minimum value and back to the maximum value, as shown in Fig. 5.3. The average stress

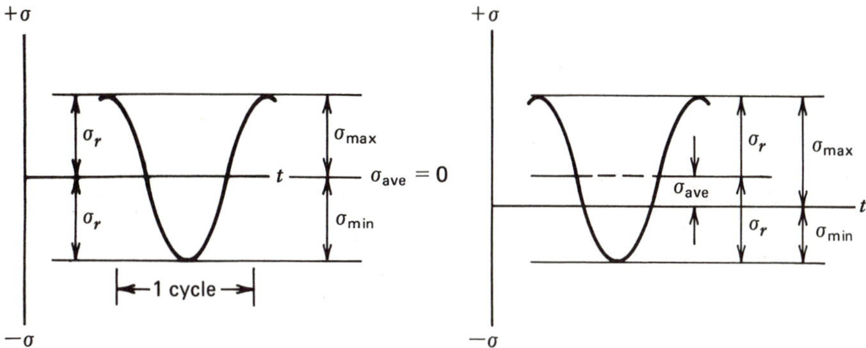

Figure 5.3 Fatigue cycle of a part.

is defined as

$$\sigma_{\text{ave}} = \frac{\sigma_{\text{max}} + \sigma_{\text{min}}}{2} \qquad [5.1]$$

where σ_{max} = maximum algebraic value of stress
σ_{min} = minimum algebraic value of stress
σ_r = stress range $(\sigma_{\text{max}} - \sigma_{\text{ave}})$

Stresses can be normal, principal, or shear. Fig. 5.3 shows that the average value of stress can be greater than zero or less than zero. For example, a tractor axle has one level of stress when stationary but has another higher value as a result of torsion and weight transfer when pulling a plow. This latter is the variable portion. A building truss has a weight-related stress but after a snowfall or during a windstorm, the stress increases to a new level. That stress might be compressive all the time.

Wöhler first studied fatigue systematically between 1860 and 1870. He showed for parts in either tension or compression that, as σ_{ave} is increased, the value of σ_r required to produce fatigue failure is reduced (see Fig. 5.4). Gerber pointed out that Wöhler's results suggested a parabolic curve relating the permissible σ_r for a given σ_{ave}.

Whöler first studied fatigue systematically between 1860 and 1870. He showed for parts in either tension or compression that, as σ_{ave} is increased, the value of σ_r required to produce fatigue failure is reduced (see Fig. 5.4). Gerber pointed out that Whöler's results suggested a parabolic curve relating the permissible σ_r for a given σ_{ave}.

5.2 SODERBERG EQUATIONS

In 1935, Soderberg proposed a straight line variation to Gerber's relationship, for which the yield point stress σ_y was considered the failure criterion. This important assumption depends upon the function of a machine component. If a part yields and is deformed, its new permanent location may or may not

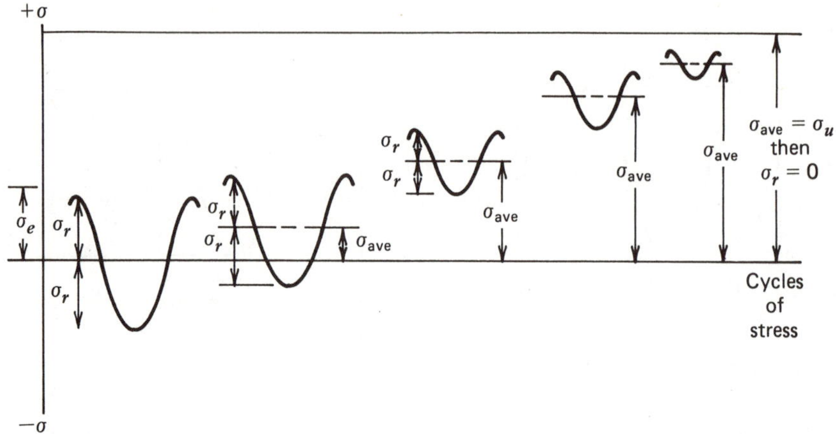

Figure 5.4 Effect of the average stress on the stress range for fatigue.

interfere with another part's function. Some parts can be allowed to deform permanently, whereas small deflections of others are critical to overall system performance.

Using the law of similar triangles to compare triangle *BDC* and *OAE* in Fig. 5.5, the ratios of the similar triangles' sides can be expressed as

$$\frac{BD}{BC} = \frac{OA}{OE} \qquad [5.2]$$

Thus Soderberg's equation is derived by substituting stress values into Eq. 5.2.

$$\frac{\sigma_r}{(\sigma_y/FS) - \sigma_{ave}} = \frac{\sigma_e}{\sigma_y} \qquad [5.3]$$

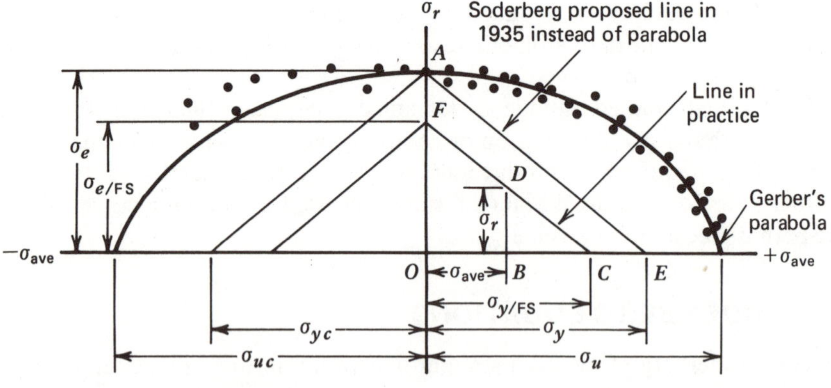

Figure 5.5 Gerber's parabola and the Soderberg line.

For members in which stress concentration exists, experience has shown that the stress-concentration factor K_f must be applied to σ_r in Eq. 5.3. Thus

$$\sigma_{\text{ave}} + K_f\ \sigma_r\ \frac{\sigma_y}{\sigma_e} = \frac{\sigma_y}{\text{FS}} \qquad [5.4]$$

where K_f = fatigue stress concentration
In Fig. 5.5, if a factor of safety is used in a design practice, the line FDC represents Soderberg's method.

Where σ_{ave} is compressive, the yield stress should be used in compression for σ_y. In all instances all quantities are to be substituted into this equation as positive numbers, otherwise the equation will not represent the Soderberg diagram from which it was derived. The endurance limit σ_e is related to the ultimate tensile stress σ_u and the surface condition of the part. Note that as more manufacturing cost is put into the part its endurance limit is improved (see Fig. 5.6). Fig. 5.6 is used to determine values of σ_e .

Example 5.1 Designing with Soderberg Equation

Determine the factor of safety for a cyclically loaded part of low-carbon, hot-rolled steel. $K_f = 2$. Maximum stress is 40 MPa, and minimum stress is -10 MPa.

$$\sigma_u = 1100 \text{ MPa (ultimate stress); } \sigma_y = 620 \text{ MPa (yield point).}$$

Values of σ_u and σ_y are usually found in the SAE Handbook if the type of steel is known. The stress concentration K_t, even though given in this example, must be obtained from Peterson diagrams and will be defined in Eq. 5.6.

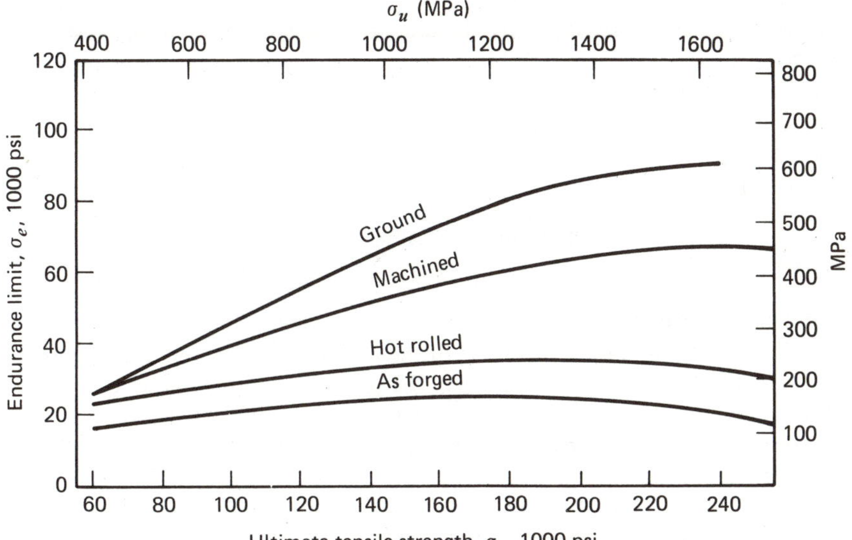

Figure 5.6 Surface conditions effects on endurance limit for steel.

The endurance limit σ_e is obtained from Fig. 5.6 (surface condition versus σ_u).

$$\sigma_e = 230\text{MPa}$$

$$\sigma_{ave} = \frac{\sigma_{max} + \sigma_{min}}{2} = \frac{40\text{ MPa} + (-10\text{ MPa})}{2} = 15\text{ MPa}$$

$$\sigma_r = \sigma_{max} - \sigma_{ave} = 40\text{ MPa} - 15\text{ MPa} = 25\text{ MPa}$$

Using Soderberg's equation

$$\sigma_{ave} + K_f\ \sigma_r\ \frac{\sigma_y}{\sigma_e} = \frac{\sigma_y}{\text{FS}}$$

$$15\text{ MPa} + 2\ (25\text{ MPa})\left[\frac{620\text{ MPa}}{230\text{ MPa}}\right] = \frac{620\text{ MPa}}{\text{FS}}$$

$$\text{FS} = 4.1$$

The stress concentration K_f in variable stress conditions is reduced from the static geometric stress concentration factor K_t by the material's ductility and the severity of the notch. For use in the Soderberg equation,

$$K_f = q(K_t - 1) + 1 \qquad [5.5]$$

where K_f = fatigue stress concentration factor for direct tension or bending
K_t = geometric stress concentration factor for direct tension or modified geometric stress concentration factor for bending
q = notch sensitivity

The notch sensitivity q is a function of the notch radius and the material (see Fig. 5.7). The notch-sensitivity factors are based on bending tests with $t/r > 4$. In the absence of additional information for sharper notches and other types of loading, concentration factors must be approximated for situations in which they are not strictly applicable.

The fatigue stress concentration factor is the extent to which a notch can be expected to actually reduce the fatigue limit of a part. It is defined as

$$\text{K}_\text{f} = \frac{\text{fatigue limit of specimen without the notch}}{\text{fatigue limit of specimen with the notch}} \qquad [5.6]$$

Specimens must have the same effective section when K_f is evaluated experimentally in order that only the effect of the notch is determined, that is, that an additional effect caused by reduction in the section is not included.

Gray cast iron is considerably less notch sensitive than the steels because of the presence of large graphite flakes. Higher-strength irons tend to be more notch sensitive.

The Soderberg equation is versatile and can be used in many ways by the designer. Assuming a factor of safety, the strength of material that should be used can be found (solve for σ_y).

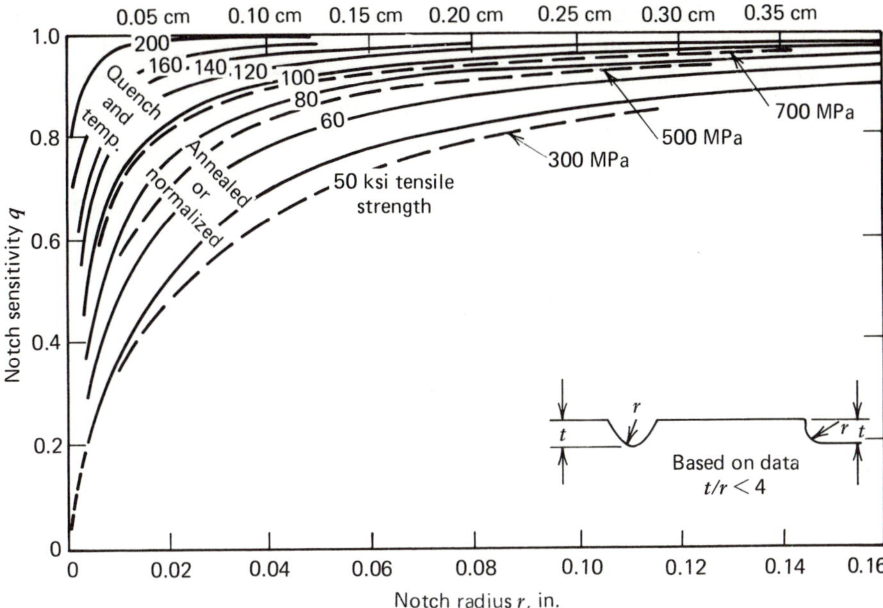

Figure 5.7 Notch-sensitivity curves for steel. Reprinted with permission © Peterson, R.E., Stress Concentration Factors, John Wiley and Sons, Inc.

If the material is known, an acceptable stress concentration can be determined and then the necessary surface finish can be chosen. The part can also be sized with the Soderberg equation by increasing the moment of inertia or area. This process reduces both average stress and stress range. Both σ_r and σ_{ave} are related to the load values and can be replaced in the Soderberg equation (i.e., $\sigma_{ave} = P_{ave}/A$, the axial loading P).

This fatigue equation is a powerful tool because it allows great flexibility in design to achieve a least-cost functional goal. Because the Soderberg equation is intended for infinite life, it is applicable only in the S-N curve area to the right of the bend (see Fig. 5.8). This bend has been found to occur at 10^6 cycles for most parts. It is recommended that all fatigue testing extend past 10^8

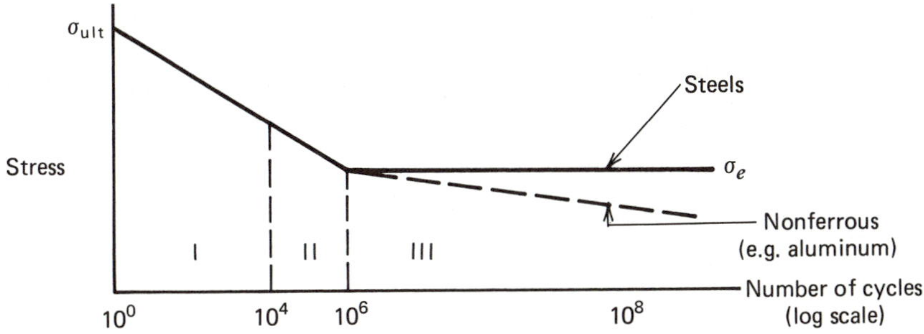

Figure 5.8 Different regions of fatigue failure on an S-N curve.

cycles to ensure that the curve becomes horizontal. *Note*: Some aluminum alloys do not have an endurance limit even after 10^8 cycles. In area I called low-cycle fatigue, internal flaws dominate the causes of failure; in area II microscopic slip-plane failures are predominant; and in area III, in the Soderberg region, the part gets its strength from the atomic lattice structure.

If the tensile strength and endurance limit of a part are known, cycles to failure at a given stress can be estimated from a pseudo S-N curve depicted as a straight line. The S-N plot's abscissa (number of cycles) is graphed as a log scale. Other factors that affect fatigue life and endurance strength that must be included in a design calculation are listed below.

1. Stress concentrations caused by changes in section, such as keyways, holes, grooves, and fillets. These are functions of the geometric stress concentration factor and the sensitivity of the material to notches.
2. Stress concentration as a result of surface roughness.
3. Improper grinding during manufacture, which leaves high residual tensile stresses.
4. Cold forming, which introduces tensile or compressive residual stresses.
5. Heat treatment, which causes the surface to decarbonize.
6. Plating, which usually reduces σ_e.
7. Corrosion caused by the presence of moisture and liquids (e.g., salt water).
8. Clamped, riveted, or press-fitted joints, which result in fretting or galling.
9. Tensile stresses in assembled packages.
10. Larger sections, which usually exhibit a reduced life.
11. Nonmetallic inclusions as a result of a fabrication process.
12. Very high or very low loading frequencies, which usually reduce the σ_e.
13. Shape of the cross section.

Consider the effect of a large section. Two shafts of different diameters loaded to the same stress level and made from the same material with exactly the same surface conditions might be expected to fail at the same number of cycles. Experiments have shown that this is not the case. The large shaft fails first because its greater cross-sectional area increases its probability of its having more inclusions, which results in earlier crack initiation.

The process of designing requires first choosing three of the unrelated variables in the Soderberg equation and then determining if the fourth is within the constraints of cost, manufacturing, and so on.

When designing with ductile steels Eq. 5.4 can be used for loading in tensile, compression, bending, or any combination thereof. For torsion loading, a shear stress version is recommended.

$$\tau_{\text{ave}} + K_f\ \tau_r\ \frac{\tau_y}{\tau_e} = \frac{\tau_y}{\text{FS}} \qquad [5.7]$$

where τ_{ave} = average shear stress
τ_r = range in shear stress
τ_y = yield shear stress of the material
τ_e = endurance shear stress limit ($\approx$ 1/2 σ_e)

If combined stress conditions exist, the maximum-shear-stress theory can be used with Eq. 5.7.

$$\tau_{\max}^2 = \left[\frac{\sigma_x - \sigma_y}{2}\right]^2 + \tau_{xy}^2 \qquad [5.8]$$

Eq. 5.8 should be used to determine τ_r and τ_{ave}, noting that this theory is conservative for ductile and brittle materials. If the material is brittle, stress concentration also affects the σ_{ave} term.

$$K_f\,\sigma_{\text{ave}} + K_f\,\sigma_r\,\frac{\sigma_y}{\sigma_e} = \frac{\sigma_y}{\text{FS}} \qquad [5.9]$$

The factor of safety (index of safety) commonly used in most cases is 2, but it depends on the particular industry and the reliability of load data. Quick approximate estimates for endurance stress follow.

For ductile steels (>5% elongation)

$$\sigma_e = \frac{\sigma_u}{2} \qquad [5.10]$$

For brittle steels

$$\sigma_e = 0.4\sigma_u \qquad [5.11]$$

For nonferrous metals

$$\sigma_e = 0.25\sigma_u \qquad [5.12]$$

5.3 MINER'S RULES

If loading results in stresses above the endurance limit (Area I and II of Fig. 5.8) but below the S-N curve, the part will have a finite life. This analysis has been approached by an approximation technique called Miner's rule, which assumes that when a part is loaded above the endurance stress, some permanent damage results. Assuming linear damage ratios that add up to 1, Miner's rule is expressed as

$$\frac{n_1}{N_1} + \frac{n_2}{N_2} + \frac{n_3}{N_3} + \ldots = 1 \qquad [5.13]$$

or

$$\sum_{i=1}^{x}\left[\frac{n_i}{N_i}\right] = 1 \qquad [5.14]$$

where x = number of stress levels endured by a component
n_i = actual cycles at stress σ_i
N_i = cycles failure at stress σ_i

Experiments have shown that this cumulative summation damage theory does come close to 1 (0.7 – 1.3) in most cases. The variations resulted because the experiments did not take into account all the factors that affect fatigue.

Example 5.2 Miners' Rule Used in Low-Cycle Fatigue

Estimate the life to failure (yield) at the endurance limit if a part is loaded cyclically 9 cycles at 90,000 kPa and 100 cycles at 60,000 kPa given that $\sigma_y =$ 100,000 kPa minimum and $\sigma_e = 40{,}000$ kPa minimum.

Log paper can be used for this estimation as shown in Fig. 5.9. Assume that semilog was used instead of log-log for the S-N curve. Use Miner's rule only as an approximation. Actual tests are preferred for stress levels recorded above σ_e.

Then the slope

$$\frac{\Delta_y}{\Delta_x} = \frac{10{,}000 \text{ kPa}}{\text{each power of 10}}$$

and Miner's rule is

$$\sum_{i=1}^{x} \left[\frac{n_i}{N_i} \right] = 1$$

where $n_1 = 9$
$n_2 = 100$
$n_3 = ?$
$N_1 = 10$
$N_2 = 10^4$
$N_3 = 10^6$

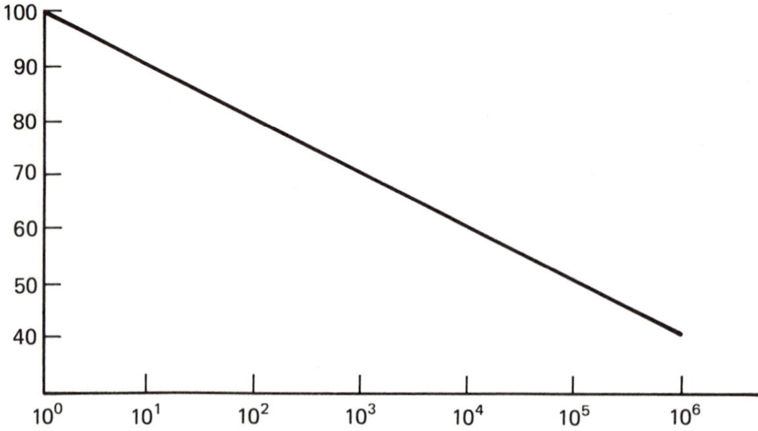

Figure 5.9 Semilog plot of stress versus cycles (S-N curve).

Substituting into Eq. 5.14

$$\frac{9}{10} + \frac{100}{10{,}000} + \frac{n_3}{10^6} = 1$$

Therefore, $n_3 = 0.90 \times 10^5$ cycles (the life if the remainder of the stress levels are at the endurance limit).

5.4 STRAIN-RANGE METHOD

There exists a need to design parts for lower weight and cost, thus allowing a larger stress level (i.e., a lower factor of safety). Some parts are even designed to plastically deform.

A load history is necessary to design at lower factors of safety, especially in the plastic range. Test engineers compile this load history and report the cycles at different levels of strain from strain gaged components. One of the most common load history methods is the rain-flow count. The rain-flow method counts closed stress-strain hysteresis loops as shown in Fig. 5.10. The closed

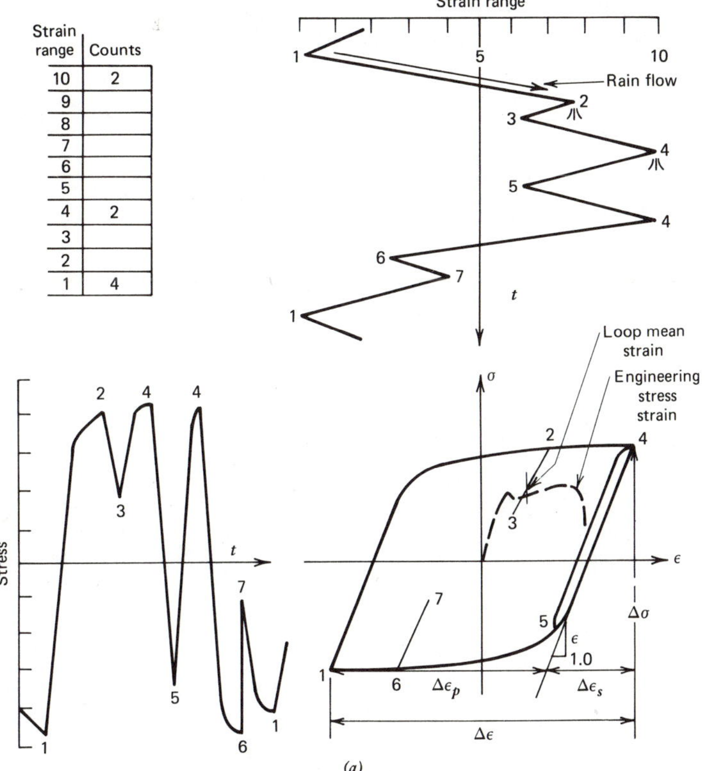

Figure 5.10 Stress-Strain response of metals used in strain range rainflow counting.

loops in Fig. 5.10*a* include 1-4-1, 3-2-3, 4-5-4, and 6-7-6. The number of cycles at the loop mean strain and strain range are recorded. Sample increment of strain range and mean strain is 50 μ/m. If the strain-time history is reoriented so that the largest strain magnitudes occur first and last and time is plotted on the vertical axis, the connecting lines depict a series of pagoda roofs from which rain could flow down on increasing strains to the right (see Fig.5.10*b*)—thus the term rain flow. Following the supposed flow of water and observing where it intersects other flows establishes the closed hysteresis loops of the stress-strain plot. A sample strain-range count is given in Fig. 5.10*b*, and its rules discussed. Rainflow is initiated at each peak and allowed to drip down until it no longer has a larger peak to drip to, thus closing the loop (i.e., 1-4). Also note that rainflow must be initiated at valleys 3, 5, and 6 in order to determine if a hysteresis loop has occurred. Rainflow must stop (i.e., closed loop) if it encounters flow from another, such as roof 1-2 onto 3-4, yielding loop 3-2-3.

When a part is loaded into the plastic region, microscopic inclusions could cause cracks to develop or cold working the part could occur (e.g., hardening it). Fatigue failure in the low cycle range depends on which mechanism is predominant: (a) strain hardening, or (b) mean crack propagation. The monotonic, or one-half cycle, tensile stress-strain curve, compared with the cyclic stress-strain behavior of metals, depicts hardening or softening (see Fig. 5.11).

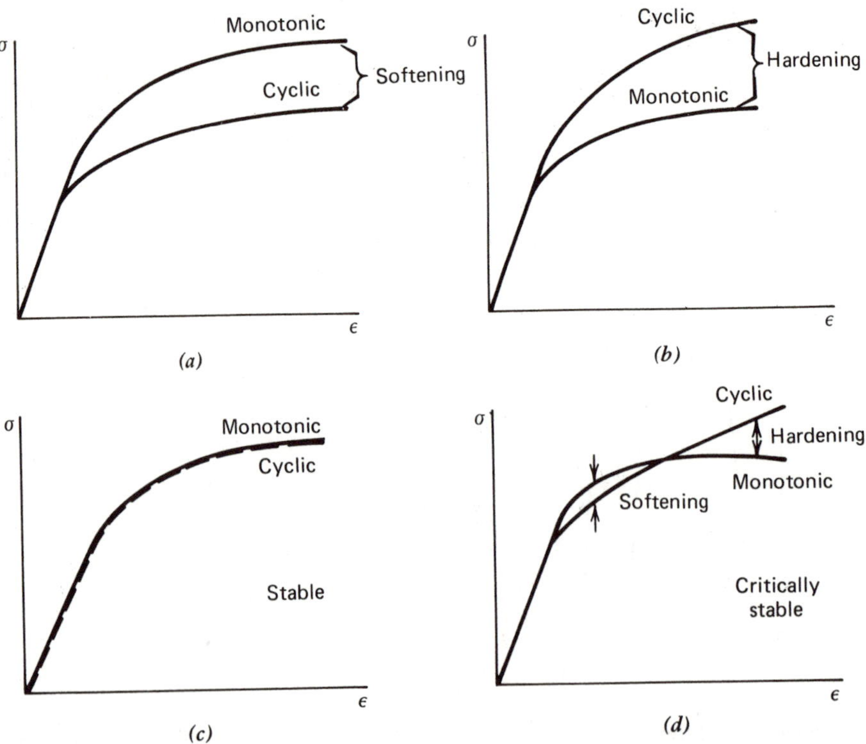

Figure 5.11 Cyclic induced changes in metal depicted by stress-strain curves.

The true stress at fracture, σ_f, differs from the alternate, σ_u, because it is calculated using the minimum cross sectional area at failure, not the original cross sectional area.

$$\sigma_f = \frac{P_f}{A_f}$$

where σ_f = true stress at fracture
A_f = minimum cross section at failure
P_f = load at failure

In Fig. 5.10, point 4 would indicate σ_f if the curve were a fracture test. True plastic strain is

$$\epsilon = \ln(1 + e) \qquad [5.15]$$

where e = engineer's strain calculated for an original length. A plot of true stress versus strain on log scales depicts the strain hardening coefficient (slope) n_f' (see Fig. 5.12). This relationship can then be expressed as

$$\sigma = K'(\epsilon_p)^{n_f'} \qquad [5.16]$$

where n_f' = strain hardening coefficient
ϵ_p = plastic strain
K' = stress at plastic strain of unity

Since the total strain ϵ is equal to the elastic strain ϵ_e plus the plastic strain ϵ_p, then the following relationship can be written.

$$\Delta\epsilon = \Delta\epsilon_e + \Delta\epsilon_p \qquad [5.17]$$

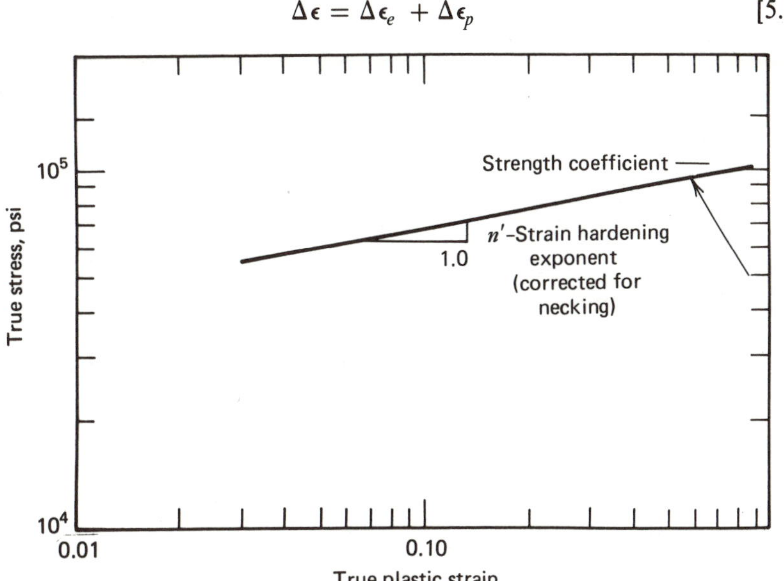

Figure 5.12 True stress-plastic strain plot for determining strain hardening coefficient.

and, substituting Eq. 5.16 and the relationship for uniaxial stress-strain,

$$\Delta\epsilon = \frac{\Delta\sigma}{E} + \left[\frac{\Delta\sigma}{K'}\right]^{1/n_f'} \qquad [5.18]$$

where $\Delta\sigma$ is the stress range ($\Delta\sigma/2$ is the amplitude). Researchers have found in controlled tests that the number of cycle reversals to failure ($2N_f$) correlates well with the plastic strain amplitude $\Delta\epsilon_p/2$ (see Fig. 5.13). A similar linearization of true stress versus cycles to failure is seen in Fig. 5.14. The slope of ϵ_p versus life is called the fatigue ductility exponent, c, and the slope of true stress versus life is called the fatigue strength exponent, b.

$$\frac{\Delta\epsilon_p}{2} = \epsilon_f'(2N_f)^c \qquad [5.19]$$

and

$$\frac{\Delta\sigma}{2} = \sigma_f'(2N_f)^b \qquad [5.20]$$

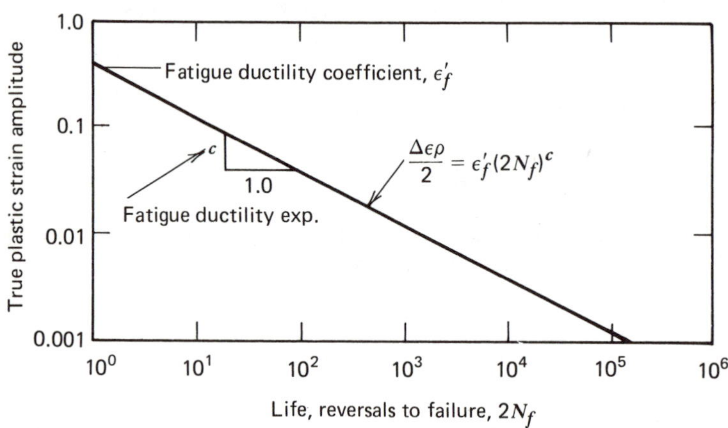

Figure 5.13 Plastic strain amplitude versus reversals to failure, 1020 HR steel.

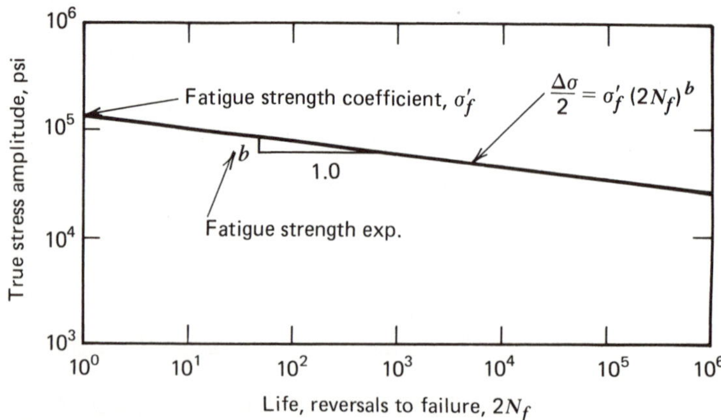

Figure 5.14 Stress amplitude versus reversals to failure, 1020 HR steel.

where ϵ_f' = fatigue ductility coefficient (failure strain at one cycle)
σ_f' = fatigue stress coefficient (stress level for failure at one cycle)

Substituting 5.18, 5.19, and 5.20 into 5.17, the strain-range equation results (see Fig. 5.15).

$$\frac{\Delta\epsilon}{2} = \frac{\sigma_f'}{E}(2N_f)^b + \epsilon_f'(2N_f)^c \qquad [5.21]$$

typical values for b, c, ϵ_f', and $(\sigma_f')^c$ are given in SAE STD J1099. For example, for SAE 1020 HR., $c = -0.51$, $b = -0.12$, $\epsilon_f' = 0.41$, $\sigma_f' = 860$ MPa (130 ksi), with $n_f' \approx b/c$. This experimental cyclical data can be used to design for a finite life under the strain-life curve shown in Fig. 5.15. The fatigue ductility coefficient is approximated from the monotonic tension test and is related to the reduction in area (RA) as

$$\epsilon_f' = \ln\left[\frac{100}{100 - \%RA}\right] \qquad [5.22]$$

A part's life may be extended if the material strengthens itself, for example, during work-hardening fatigue.

Example 5.3 Strain Range Applied to an Agricultural Wagon

A grain wagon running gear is subjected to impact as it is pulled over rough ground that can yield the bolster. Past test data show that over a period of 10 years only 10,000 occurrences are expected (see Fig. 5.16). When full, the wagon carries 350 bushels of wheat. Strain data on the bolster can be calculated using combined stress or determined from strain-gage readings. The bolster is constructed of SAE 1020 HR, 10 cm × 10 cm × 0.6 cm, tubular steel. The strain history is shown in Fig. 5.17. At section A-A,

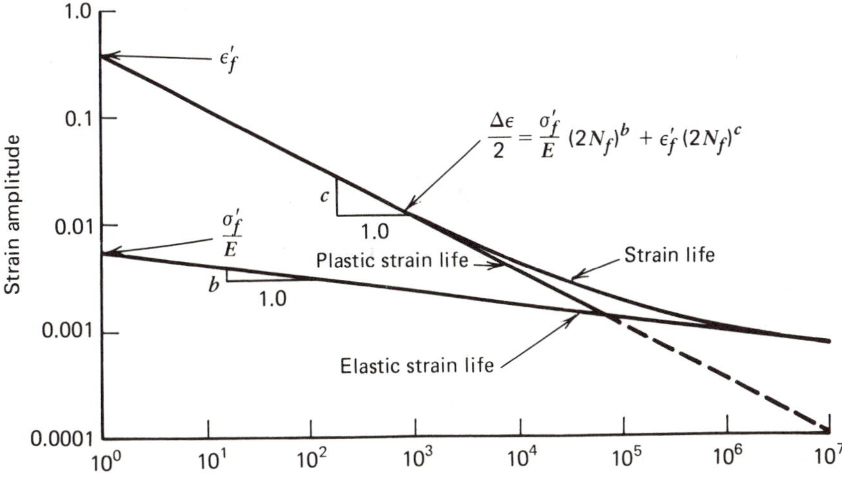

Figure 5.15 Strain-range life, reversals to failure, $2N_f$.

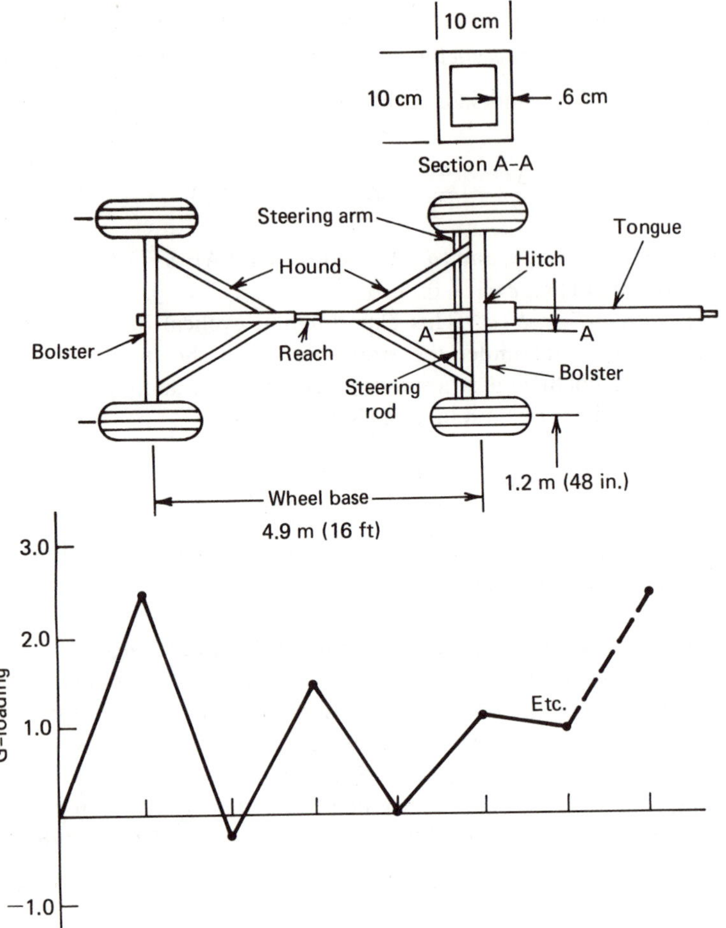

Figure 5.16 Load time history for a farm wagon.

$$\sigma_x = \frac{MC}{I} = \frac{\text{load}}{4 \text{ tires}} \times \frac{1.2\text{m} \times 0.1m}{0.000008 \text{ m}^4} = 3750 \times \text{load} \qquad [5.23]$$

$$\tau_{xy} = \frac{TC}{J} = \text{offset load} \times \text{moment arm } \frac{C}{J}$$

$$\tau_{xy} = \frac{0.1L \times 0.5\text{m} \times 0.1\text{m}}{0.000016 \text{ m}^4} = 312 \times \text{load} \qquad [5.24]$$

where the offset torque load was 10% of the weight per tire impacted at 0.5m off the axis. Therefore,

$$\sigma_{\text{max}} = 3763 \times \text{load} \qquad [5.25]$$

The first step is to check yielding

$$\text{maximum stress} = \epsilon_{\text{max}}\text{E} = 0.0063 \text{ m/m} \times 195{,}000 \text{ MPa} = 1230 \text{ MPa}$$

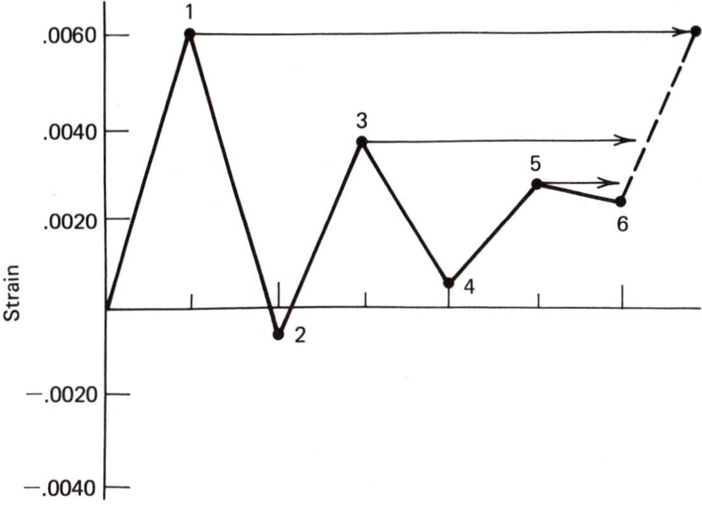

Figure 5.17 Strain history (Example 5.3).

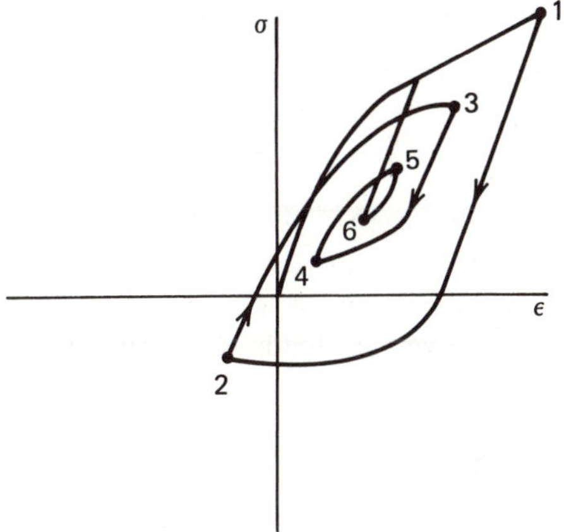

Figure 5.18 Stable strain loops (Example 5.3).

The material is plastically yielding under maximum g load, but deflections do not hinder the wagon's function. Using the rain-flow method, hysteresis loops can be determined from Fig. 5.17, such as 1-2, 3-4, and 5-6 (see Fig. 5.18). For this material $n_f' \approx n$; therefore, no strain hardening is expected. The material is loaded less than the ultimate strain $\epsilon_u = 0.9$ m/m for SAE 1020; thus the strain-range approach can be used.

For loop 1-2, calculate the strain range.

$$\Delta\epsilon = (0.0063 + 0.0005) = 0.0068 \text{ m/m} \tag{5.27}$$

$$\frac{\Delta\epsilon_{1-2}}{2} = 0.0034 \text{m/m} \tag{5.28}$$

For loop 3-4

$$\Delta\epsilon = 0.0038 - 0.0005 = 0.0033 \text{ m/m} \qquad [5.29]$$

$$\frac{\Delta\epsilon_{3-4}}{2} = 0.0016 \text{ m/m} \qquad [5.30]$$

Using Eq. 5.21 and using a Newton-Raphson method to converge the solution, $2N_f$ can be found.

For

$\sigma_f{}' = 860$ MPa, $E = 195{,}000$ MPa, $b = -0.12$, $\epsilon_t{}' = 0.41$, and $C = -0.51$, loop 1-2 has 15,173 cycles and loop 3-4 has 160,444 cycles. Loop 5-6 has such a low strain range that it is behaving elastically. Applying Miner's rule, the blocks to failure F_{LB} can be obtained.

$$F_{LB} = \frac{1}{\Sigma \frac{n}{N_f}} = \frac{1}{\frac{1}{15{,}173} + \frac{1}{160{,}444}} = 13{,}862 \qquad [5.31]$$

where n = number of cycles counted in a sequence

N_f = cycles to failure

The factor of safety is then calculated as F_{LB} divided by the actual cycles

$$\text{FS} = \frac{13{,}862}{10{,}000} = 1.38 \qquad [5.32]$$

depicting that the part will last 10 years for the loading initially assumed. Fig. 5.19 shows how the life of a particular part was increased by a factor of approximately 5 when the material was changed as a result of a strain-range analysis after data for the given test loads and stresses were acquired.

5.5 WEIBULL MODEL

Some materials, such as ceramics, do not behave like metals when cyclically loaded. Fatigue tests show variable data. Similarly, component assemblies, such as truck spring systems have different failure patterns under fatigue loading. This type of failure requires an analysis different from standard fatigue design.

Weibull developed a set of curves that fit this type of scattered-fatigue test data. The results do not provide a factor of safety but do provide life estimates with some reliability associated with that life. The randomness of failure for component assemblies is shown in Fig. 5.20; $f(t)$ is defined as the probability density of the time to failure for a given component. *Note*: $f(t) = 1/2$ for $t = 0{,}1$, the heads and tails of a coin.

$$F(t) = \int_0^t f(t)\, dt \qquad [5.33]$$

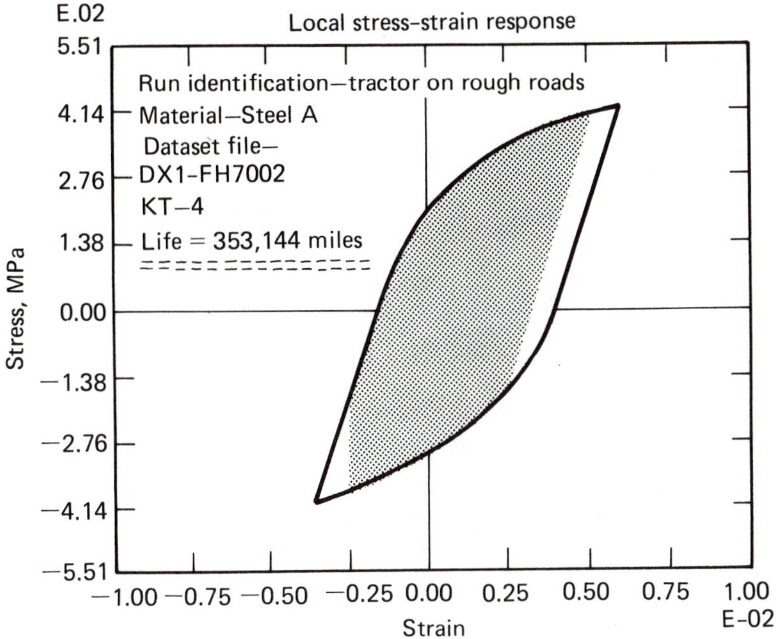

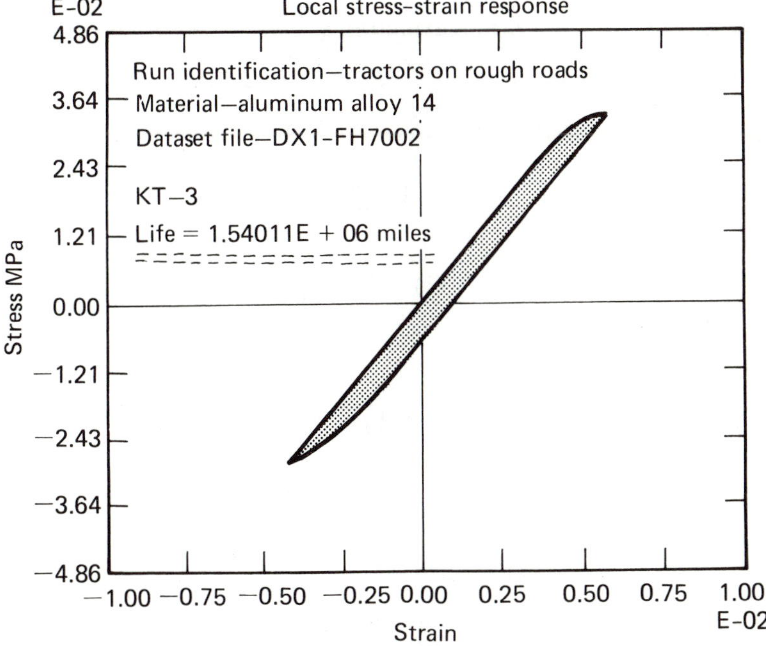

Figure 5.19 Sample strain cycles. Courtesy MTS Systems Corp.

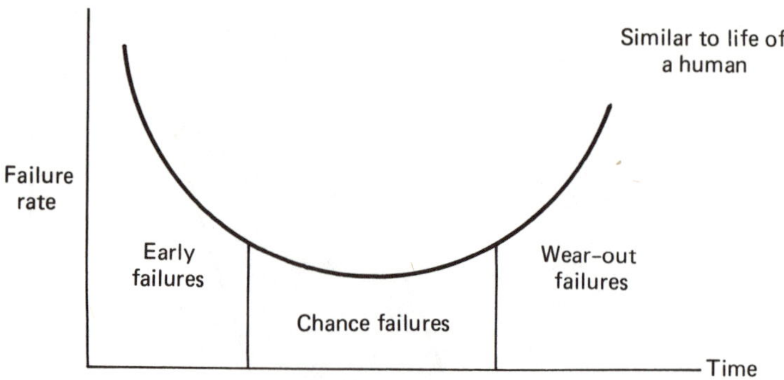

Figure 5.20 Typical random failure curve for a system of components.

Where $F(t)$ is the probability that the component will fail in a given time span. $R(t)$ is the reliability function, expressing probability that the part survives to a time t.

$$R(t) = 1 - F(t) \tag{5.34}$$

The probability that the part will fail from t to $t + \Delta t = F(t + \Delta t) - F(t)$, assuming that the part lasted to time t. The average rate of failure in the interval Δt to $t + \Delta t$ is

$$\frac{F(t + \Delta t) - F(t)}{\Delta t} \times \frac{1}{R(t)} \tag{5.35}$$

as $\Delta t \Rightarrow 0$ the failure rate $Z(t)$ is expressed as

$$Z(t) = \frac{F'(t)}{R(t)} = \frac{f(t)}{1 - F(t)} \tag{5.36}$$

thus

$$R(t) = 1 - F(t) \tag{5.37}$$

also $F'(t) = -R'(t)$ and then

$$Z(t) = -\frac{R'(t)}{R(t)} = \frac{-d[\ln R(t)]}{dt} \tag{5.38}$$

Solving for $R(t)$

$$R(t) = e^{-\int_0^t Z(x)dx} \tag{5.39}$$

knowing that $f(t) = Z(t)R(t)$

$$f(t) = Z(t)\, e^{-\int_0^t Z(x)dx} \tag{5.40}$$

Assuming that failure rate is constant and greater than zero, α can be substituted for $Z(t)$

$$f(t) = \alpha\, e^{\alpha t} \qquad \text{(exponential distribution)} \tag{5.41}$$

where

$$\frac{1}{\alpha} = \text{MTBF} = \text{mean time between failures} \quad [5.42]$$

A constant failure rate is not realistic for Fig. 5.20. The failure rate might increase or decrease smoothly with time. Thus Weibull created a distribution that can fit any random failure by varying β.

$$f(t) = \text{failure times} = \alpha\beta t^{\beta-1}e^{-\alpha t^{\beta}} \quad [5.43]$$

for $t > 0, \beta > 0, \alpha > 0$, when $\beta < 1$, the failure rate decreases with time (see Fig. 5.21); when $\beta = 1$, the failure rate is constant (an exponential distribution); when $\beta = 3$ to 4 the distribution is close to a normal distribution; and when $\beta > 1$, the failure rate increases with time. Association fuctions in Wiebull's distribution are:

$$\text{reliability } R(t) = e^{-\alpha t^{\beta}} \quad [5.44]$$

$$\text{failure rate } Z(t) = \alpha\beta t^{\beta-1} \quad [5.45]$$

The mean of the Weibull distribution having α and β parameters may be obtained by evaluating the integral.

$$\mu = \int_0^{\infty} t\ \alpha\beta^{\beta-1}e^{-\alpha t^{\beta}}\,dt \qquad \text{when} \qquad \mu = \int_0^{\infty} t\, f(t)\,dt \quad [5.46]$$

With further manipulations, mean time to failure for Weibull's model is

$$\mu = \alpha^{-1/\beta}\ \Gamma(1 + \frac{1}{\beta}) \quad [5.47]$$

where, in a table of integrals,

$$\Gamma(1 + 1/\beta) = \int_0^{\infty} \alpha t e^{-\alpha t^{\beta}} d(\alpha t^{\beta}) \quad [5.48]$$

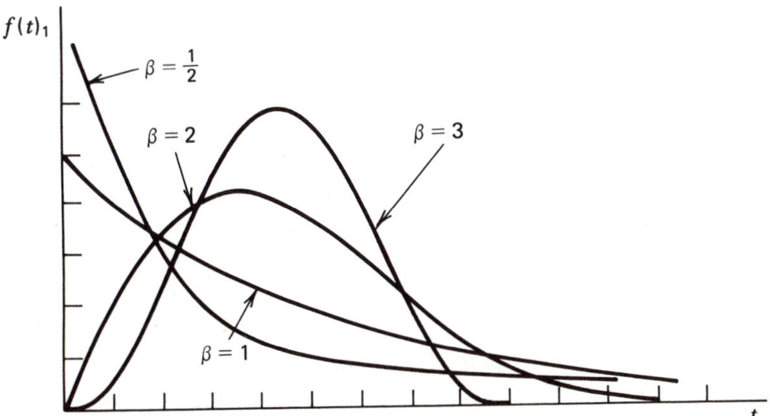

Figure 5.21 Weibull density curves at different B levels.

and the variance is

$$\sigma^2 = \alpha^{-2}/\beta \left\{ \Gamma\left(+ \frac{2}{\beta} \right) - \left[\Gamma\left(+ \frac{1}{\beta} \right) \right]^2 \right\} \qquad [5.49]$$

The gamma function is defined as

$$\Gamma(h) = \int_0^\infty \chi^{h-1} e^{-x}\, dx \qquad \text{for } h > 0 \qquad [5.50]$$

It is difficult to estimate α and β by analytical methods, but they can be estimated by solving a system of transcendental equations. A graphical approach is more commonly used. This method is based on the fact that the reliability function of the Weibull distribution can be transformed into a linear function of $\ln(t)$ by means of a double logarithmic transformation. Take the natural log of $R(t)$

$$\ln \left[\frac{1}{R(t)} \right] = -\alpha t^\beta$$

Again take the ln

$$\ln \ln \left[\frac{1}{R(t)} \right] = \ln \alpha + \beta \ln t$$

To estimate α and β, estimates of $R(t)$ at various t are needed. The usual procedure is to place n units on a life test and observe their failure times. A check is then made to determine if Weibull is a good model. That is, if the values of time of failure versus percent of failure plot as a straight line on log-log paper, then the Weibull method can be used.

The Weibull method would be used if a sample of four failures of axle springs occurred at 77,400, 198,000, 37,000, and 120,000 cycles. The Weibull method used on this sample would determine the cumulative probability to failure.

Median ranks provide the best answer to determining the question of cumulative probability to failure because the smallest measurement in a sample size of four could have a wide range of values. Statistically, sets of samples have been tested and the median (50% rank) has been determined. The values are then given in median-rank tables. For a sample of four, 50% of the measurements will be above and below the median. The first measurement is at the 15.91 percentile of the entire population. Other ranks, 5%, 95%, and so on are used to determine confidence limits. Median ranks for small sample sizes are plotted on vertical-scale Weibull paper. See Table 5.1 and the table of median ranks in the appendix.

Example 5.4 Weibull Graphic Procedure

1. Plot cycles to failure against median rank. If tables of median rank are not available, close estimates can be obtained by

$$r_j = \frac{j - 0.3}{n + 0.4} \qquad [5.51]$$

where j = rank order number
n = sample size
r_j = median rank

2. Draw a straight line fit (use least squares) (see part 5 if a curved line forms).
3. Also, plot 5% and 95% ranks (draw smooth curves). Note that these can be curved.
4. Draw a parallel line to the best fit line (of part 2) through the zero point and read the β parameter. Draw conclusions as follows: $\beta < 1$ decreasing failure rate; $\beta = 1$ constant failure rate; $\beta > 1$ increasing failure rate.
5. If a minimum life is expected subtract it from all values and draw a new straight line (it must be a straight line).
6. Draw other conclusions (e.g., β_{50} life = 50% of population will fail at this time). Characteristic life (Θ) is when 63.2% parts have failed.

Example 5.5 Application of Weibull
Ten shafts have failed at the following cycles.

Cycles to Failure	*% Median Rank*	*5% Rank*	*95%Rank*
53,000	6.7	0.5	25.9
66,000	16.2	3.6	39.4
80,000	25.86	8.7	50.7
82,000	35.51	15.0	60.7
89,000	45.17	22.2	69.7
91,000	54.83	30.4	77.8
94,000	64.49	39.3	85.0
100,000	74.14	49.3	91.3
109,000	83.77	60.6	96.3
118,000	93.3	74.1	99.5

The straight-line plot is shown in Fig. 5.22. The following conclusions can be drawn:

1. $\beta = 4.5$, which is an increasing failure rate (a note of concern should arise).
2. 90% confident that 10% will fail from 36,000 to 78,000 cycles.
3. 95% confident that no more than 10% will fail at 36,000 cycles of fatigue life.
4. β_{50} life is 90,000 cycles.

Some researchers report that a sample size of 15 or more is needed for the accurate implementation of Weibull graphing, otherwise too much human error is introduced.

Table 5.1 **Median, 5, and 95% Ranks***

Median Ranks

j Rank Order	*Sample size, n* 1	2	3	4	5	6	7	8	9	10
1	50.000	29.289	20.630	15.910	12.945	10.910	9.428	8.300	7.412	6.697
2		70.711	50.000	38.573	31.381	26.445	22.849	20.113	17.962	16.226
3			79.370	61.427	50.000	42.141	36.412	32.052	28.624	25.857
4				84.090	68.671	57.859	50.000	44.015	39.308	35.510
5					87.055	73.555	63.588	55.984	50.000	45.169
6						89.090	77.151	67.948	60.691	54.83
7							90.572	79.887	71.376	64.490
8								91.700	82.038	74.142
9									92.587	83.774
10										93.303

Median Ranks

j Rank Order	*Sample Size, n* 11	12	13	14	15	16	17	18	19	20
1	6.107	5.613	5.192	4.830	4.516	4.240	3.995	3.778	3.582	3.406
2	14.796	13.598	12.579	11.702	10.940	10.270	9.678	9.151	08.677	8.251
3	23.578	21.669	20.045	18.647	17.432	16.365	15.422	14.581	13.827	13.147
4	32.380	29.758	27.528	25.608	23.939	22.474	21.178	20.024	18.988	18.055
5	41.189	37.853	35.016	32.575	30.452	28.589	26.940	25.471	24.154	22.967
6	50.000	45.951	42.508	39.544	39.967	34.705	32.704	30.921	29.322	27.880
7	58.811	54.049	50.000	46.515	43.483	40.823	38.469	36.371	34.491	32.795
8	67.620	62.147	57.492	53.485	50.000	46.941	44.234	41.823	39.660	37.710

5% Ranks

j Rank Order	*Sample Size, n* 1	2	3	4	5	6	7	8	9	10
1	5.000	2.532	1.695	1.274	1.021	0.851	0.730	0.639	0.568	0.512
2		22.361	13.535	9.761	7.644	6.285	5.337	4.639	4.102	3.677
3			36.840	24.860	18.925	15.316	12.876	11.111	9.775	8.726
4				47.287	34.259	27.134	22.532	19.290	16.875	15.003
5					54.928	41.820	34.126	28.924	25.137	22.244
6						60.696	47.930	40.031	34.494	30.354
7							65.184	52.932	45.036	39.338
8								68.766	57.086	49.310
9									71.687	60.584
10										74.113

* Reprinted with Permission© LIPSON, C. and Sheth, *N.J. Statistical Design and Analysis of Engineering Experiments,* McGraw Hill Book Co.

Table 5.1 (Cont.)

95% Ranks

j Rank Order	*Sample size, n* 1	2	3	4	5	6	7	8	9	10
1	95.000	77.639	63.160	52.713	45.072	39.304	34.816	31.234	28.313	25.887
2		97.468	86.465	75.139	65.741	58.180	52.070	47.068	42.914	39.416
3			98.305	90.239	81.075	72.866	65.874	59.969	54.964	50.690
4				98.726	92.356	84.684	77.468	71.076	65.506	60.62
5					98.979	93.715	87.124	80.710	74.863	69.646
6						99.149	94.662	88.889	83.125	77.756
7							99.270	95.361	90.225	84.997
8								99.361	95.898	91.274
9									99.432	96.323
10										99.488

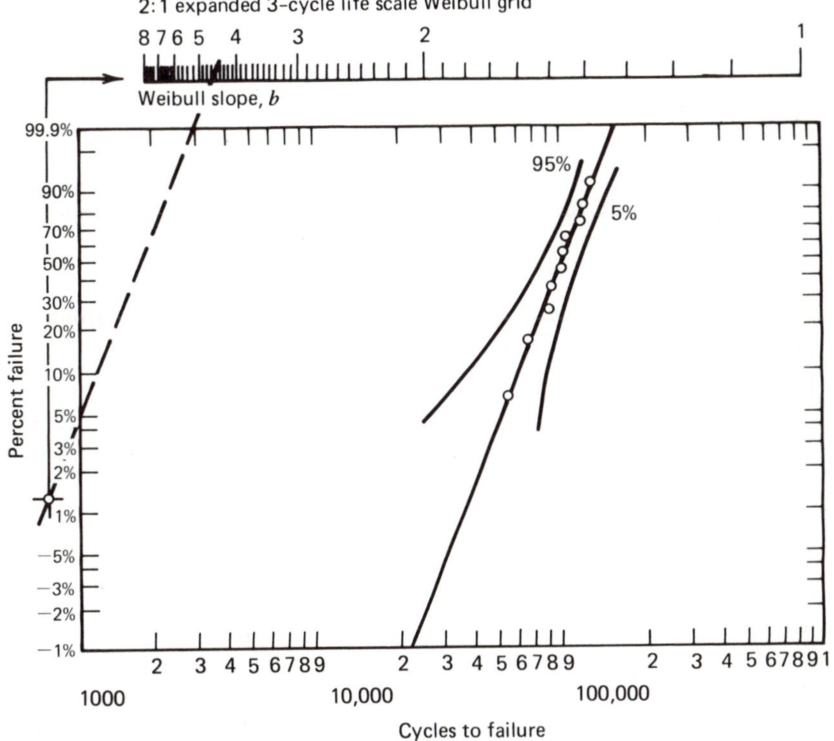

***Figure** 5.22* Confidence bands on shaft life (Example 5.4).

Note: Fatigue failures are said to account for over 90% of design errors because they are not properly analyzed or because they are simply overlooked.

HOMEWORK PROBLEMS

5.1 A 2-in. diameter cast iron bar is cyclically loaded in tension from 0 to 10,000 lb. Determine the maximum fatigue stress concentration that can be machined into it if it retains an infinite life and a factor of safety of 1.5. $\sigma_u = 100$ ksi, the part's surface area is machined, and the reduction in area less than 1%.

5.2 A pump rod of SAE 1045 annealed steel is subjected to a load that varies from 10,000 lb in compression to 12,000 lb in tension. Find the required diameter for the rod, using FS = 2. The surface of the rod is machined. Use the Soderberg equation.

5.3 Determine the life to failure at σ_e for a cyclically loaded part, given $\sigma_y =$ 100,000 psi, $\sigma_e =$ 50,000 psi, 7 cycles at 90,000 psi, and 100 cycles at 60,000 psi.

5.4 The following figures represent the scatter in fatigue strengths of a component. Determine the characteristic strength. Use Weibull.

432 MPa

400 MPa

501 MPa

601 MPa

602 MPa

5.5 Determine the life to failure for a part made from SAE 1025 HR steel having a plastic strain range $\Delta\,\epsilon_p = 400\ \mu$in./in. subjected to completely reversed bending. What is strain at yield with no fatigue?

5.6 A bar of 1025 HR steel is loaded as shown in Fig. 5.23. Determine the factor of safety for an infinite life if the load varies between +20,000 lb to −6000 lb.

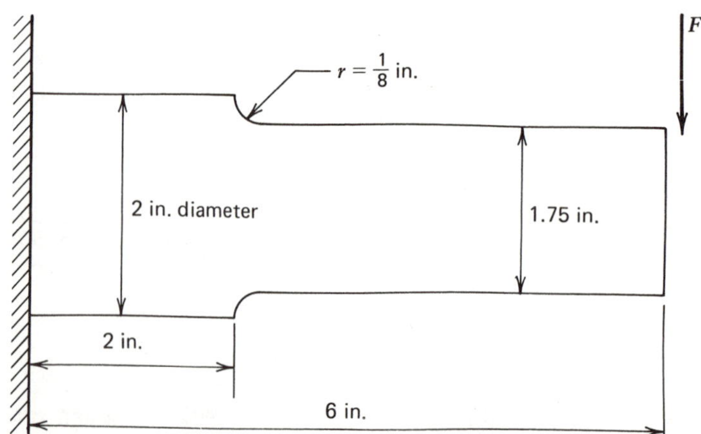

Figure 5.23 Problem 5.6

5.7 A shaft's torque varies from zero to a maximum in a cyclical fashion. Determine this maximum torque (seen in Fig. 5.24).

5.8 A modified Goodman diagram plots stress on the vertical axis and mean stress on the horizontal axis. A particular steel has $\sigma_u = 100{,}000$ psi, a yield of 60,000 psi and an endurance limit of 40,000 psi for reversed bending. Sketch the diagram, shading the area used in a Soderberg analysis.

5.9 In Example 5.3 change the material to SAE 1030 and determine the difference in life.

5.10 Reduce the factor of safety in Problem 5.6 by changing the radius.

5.11 The following figures represent the scatter in fatigue strengths of a component. Determine the characteristic strength. Use Weibull.

150,000 psi
23,874 psi
109,322 psi
95,861 psi
32,157 psi
129,635 psi
37,521 psi

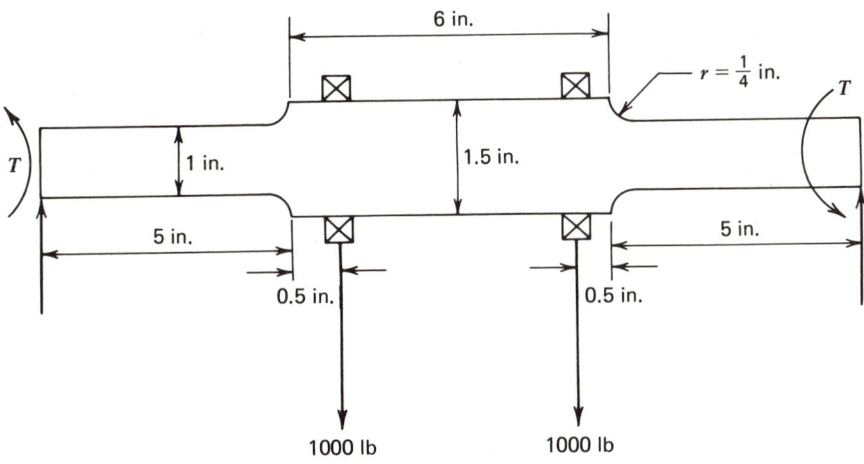

Figure 5.24 Problem 5.7

REFERENCES

1. Aamoth, R.W. "Analysis of Life Characteristics With Weibull," General Motors Institute Report.
2. Bagci, C. "Computer Aided Fatigue Design of Power Transmission Shafts with Strength Constraints Using a Finite Line Element Technique and Proposed Fatigue Failure Criterion," ASME Paper 79-DET-103, The American Society of Mechanical Engineers, New York, N.Y., 1979.
3. Bittence, J.C. "Specifying Materials Statistically," *Machine Design*, 50(2):79–83, 1978.
4. Crooker, T.W. "Fracture Mechanics Fatigue Design." *Mechanical Engineering*, pp. 40–45, June 1977.
5. Fuchs, H.O. and R.I. Stephens. *Metal Fatigue in Engineering* , John Wiley & Sons, Inc., New York, N.Y., 1980.
6. Galliart, D.R. and L.E. Tucker. "Data Acquisition and Fatigue Life Predictions," ASAE Paper 72-622, American Society of Agricultural Engineers, St. Joseph, Mich., 1972.
7. Galliart, D.R., S.D. Downing, and H.D. Bern. "Computer Based Material Properties . . . An Effective Look to Reliable Products," *Closed Loop*, MTS System Corp., Minneapolis, Minn., August 1979.
8. Hillberry, B.M. "Modern Fatigue Analysis," *Contemporary Developments for Design and Project Engineers*, Seminar, School of Mechanical Engineers, Purdue University, West Lafayette, Ind., 1980.
9. Lamberson, L.R. "An Assessment of the Accuracy of Weibull Graph Paper as a Prediction Tool," SAE Paper 800173, Society of Automotive Engineers, Inc., Warrendale, Pa.., 1980.
10. Leever, R.C. "A Users View of Fatigue Life Predictions," SAE Paper 780105, Society of Automotive Engineers, Inc., Warrendale, Pa., 1978.
11. Luk, Y.W., and L.D. Mitchell. "Development of an Interactive Computer Program for Fatigue Analysis," ASME Paper 79-DE-E-4, The American Society of Mechanical Engineers, New York, N.Y., 1979.
12. Martz, J. and V. Wiess. "Mean Stress and Environmental Effects on Mean Threshold Fatigue Lack Growth," ASTM Publication 601, Philadelphia, Pa., 1976.
13. Mattos, R.J. and F.V. Lawrence. "Estimation of the Fatigue Crack Initiation Life in Yields Using Low Cycle Fatigue Concrete," Report FCPN019, College of Engineering, University of Illinois, Urbana, Il., October 1979.
14. Nagao, M. and V. Weiss. "X-Ray Diffraction Study of Low Cycle Fatigue Damage in Plain Carbon Steel," ASME Paper 76-WA/MAT-10, The American Society of Mechanical Engineers, New York, N.Y., 1976.
15. Sherman, A.M., R.G. Davies, and A.R. Krause. "Fatigue Life Predictions for High Strength Steels in Automotive Applications," SAE Paper 810435, Society of Automotive Engineers, Inc., Warrendale, Pa., 1981.
16. Skena, C.C. "Strain-Rate Sensitivity and Energy Absorption of Hot-Rolled Sheets of DQSK, USS EX-TEN FSO, and USS Dual Phase 80 Steels," SAE Paper 810235, Society of Automotive Engineers, Inc., Warrendale, Pa., 1981.
17. Socie, D.F. "Fatigue Life Estimation Techniques, Electro General Report No. 145," Urbana, Il.
18. Socie, D., G. Shifflet, and H. Berns. "A Field/Recording System with Applications to Fatigue Analysis," ASTM Symposium, Atlanta, Ga., Nov. 14–15, 1977.

19. "Technical Report on Fatigue Properties, SAE STD J1099," *Society of Automotive Engineers Handbook*, Society of Automotive Engineers, Inc., Warrendale, Pa., 1981.
20. Topper, T.H., R.M. Wetzel, and J. Morrow. "Neuber's Rule Applies to Fatigue of Notched Specimen," *Journal of Materials*, 4(1), 200–209, March 1969.
21. Weiss, V., Y. Kasai, and K. Sieradzka. "Microstructural Aspects of Fracture Toughness," ASTM Publication 605, Philadelphia, Pa., 1976.
22. Wirsching, P.H. and J.E. Kempert. "Models that Predict Fatigue Failure: The Strain Range Approach," *Machine Design* 48(16):67–68, 1976.

CHAPTER SIX

JOINING PARTS TOGETHER

6.1 BOLTED JOINTS

The two most prevalent means of joining parts together on farm equipment are bolting and welding. For low to medium volume assemblies economics dictate the use of less tooling to join and attach components. Either low-cost welding or air-torque wrenches can be used to fasten one piece to another. This equipment can be easily adapted and reused as designs change, but the mechanism of fastening always stays the same.

The key to a long-lasting, reliable bolted-joint design is the analysis of joint forces. There are two common bolted joints, one used to clamp two parts tightly together and the other used as a shear coupling. When an initial torque (preload) is applied to a bolt clamping two plates together and both the bolt and the plates are assumed to deflect axially, the preload W_1, which results in bolt tension, is given as

$$W_1 = K_b \delta_b = K_j \delta_j \qquad [6.1]$$

where j = joint
b = bolt
δ = deflection (in.)
K = stiffness (lb/in.) = force/deflection = AE / L
K_j = joint stiffness for one member of joint
K_b = stiffness for one bolt
A = area affected (bolt or plate)
E = modulus of elasticity
L = length of bolt or plate not to exceed distance between head and nut (see Fig. 6.1).

The clamping force creates friction between the members joined, providing resistance to shear. The level of clamping is limited to the point at which the bolt either yields or breaks. This level is a function of the joint's design

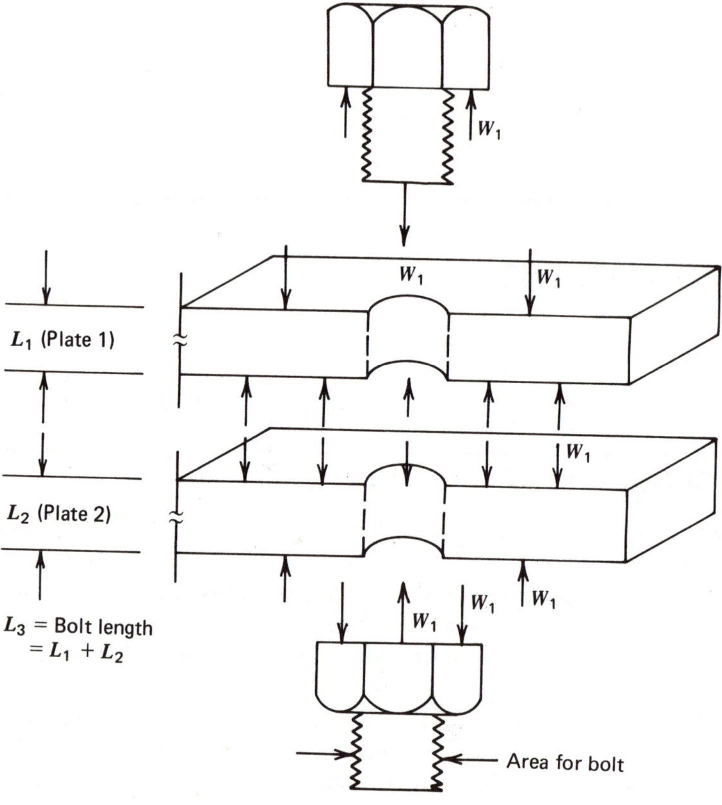

Figure 6.1 Free-body diagram of a bolt with preload

requirements. During initial bolt torque, if the joint stiffness is greater than the bolt stiffness, the deflection of the joint is less than the deflection of the bolt. This situation is common when both the bolt and the plate are made from steel.

When an external load W_2 is applied perpendicular to a bolted plate, the original preload reflects greater percentages of the load onto the joint plates, which is an advantage in fatigue situations (see Fig. 6.2). Osgood warns that higher preloads might lower bolt fatigue life because of high mean stress.

If it is assumed that the change in bolt length equals the change in joint deflection, then the load shifting to the plates can be calculated. If the plate stiffness is twice the bolt stiffness, the bolt endures only one-third of the external load. Stiffness ratios are actually nearer to 10:1.

Therefore,

$$\Delta \delta_b = \frac{W - W_1}{K_b} \qquad [6.2]$$

where W = resultant load on bolt (N)

W_1 = initial preload (N)

K_b = bolt stiffness (N/m)
$\Delta\delta_b$ = change in bolt length

and

$$\Delta\delta_j = \frac{[W_1 - (W - W_2)]}{K_j} \qquad [6.3]$$

where $W - W_2$ = joint resultant load free-body diagram in Fig. 6.2a
K_j = joint stiffness

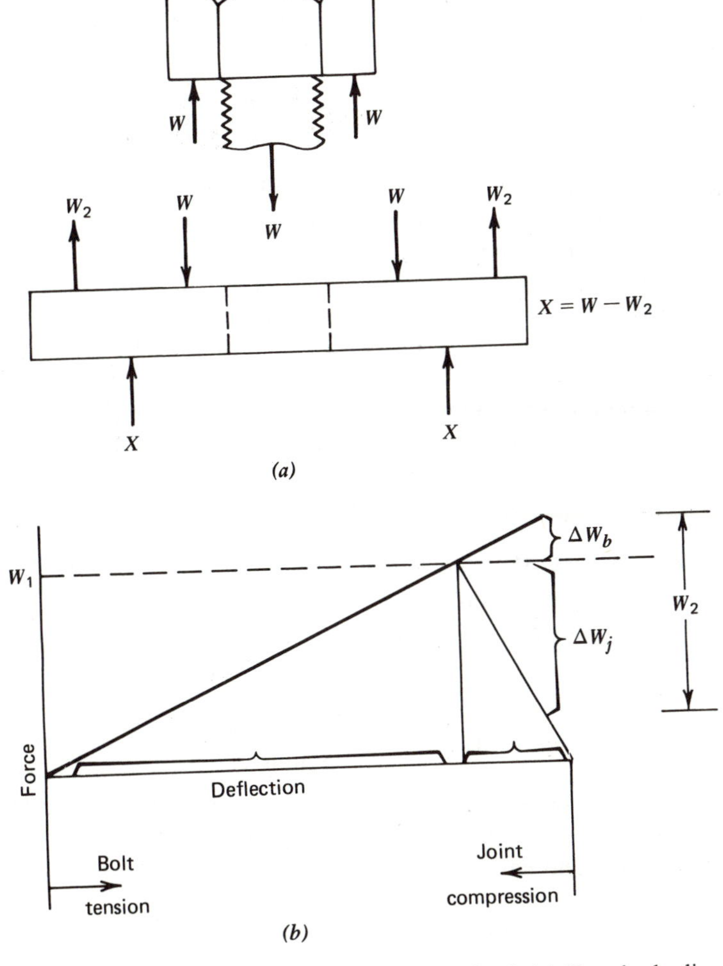

Figure 6.2 Preloaded bolt joint with external load. (*a*) Free-body diagram. (*b*) Force-deflection curve. X = Compression force between plates. W_2 = External load. W = Resultant bolt load. ΔW_b = Change in bolt load (absorbed by the bolt). ΔW_j = Change in joint load (flange relaxation).

$\Delta\delta_j$ = change in joint deflection

When $\Delta\delta_b = \Delta\delta_j$ and combining Eqs. 6.2 and 6.3,

$$\frac{W}{K_b} - \frac{W_1}{K_b} = \frac{W_1}{K_j} - \frac{W}{K_j} + \frac{W_2}{K_j} \qquad [6.4]$$

for $K_b = 1/b$ and $K_j = 1/m$ and rearranging into the more common form for resultant bolt load, W.

$$W = W_1 + W_2\left[\frac{m}{m + b}\right] \qquad [6.5]$$

If more than one bolt or plates of different materials are used, then Eq. 6.5 is modified as follows.

$$W = W_1 + W_2\left[\frac{\Sigma m}{\Sigma m + \Sigma b}\right] \qquad [6.6]$$

Finding proper preload for a bolt is a challenging problem. Not only are preload measurements inaccurate, but theories are almost as numerous as the many measurement techniques.

One theory beginning to gain wide acceptance recommends preloading every bolt to 100% of proof load (the highest load that a bolt can sustain without incurring permanent set). An older and more widely used theory restricts preload to 75% of proof load. When using 100% proof, smaller bolts can be used, resulting in cost savings because smaller bolts are less expensive and drilling smaller holes is less costly. The common equation for preload is

$$W_1 = \frac{T}{0.2d} \qquad [6.7]$$

where W_1 = initial axial preload (lb)
T = torque (in. · lb)
d = nominal diameter of bolt (in.)

In metric units

$$W_1(\mathrm{N}) = \frac{5000\ T(\mathrm{N \cdot m})}{d\,(\mathrm{mm})}$$

The torque coefficient (0.2) varies with different finishes, platings and lubricating coatings on standard fasteners. Platings are used to provide base-metal protection in different environments. Zinc and cadmium are the most popular platings.

A more accurate preload equation is

$$W_1 = T\left[\frac{\dfrac{1}{r_p(\tan\alpha + \mu/\cos\Phi_n)}}{1 - \mu\tan\alpha/(\cos\Phi_n)} + \mu_c r_c\right] \qquad [6.8]$$

where r_c = radius of midpoint rubbing surface on nut or bolt head
r_p = pitch radius of thread
T = tightening torque
W_1 = axial load
α = helix angle of thread at the pitch radius
Φ_n = angle between the tangent to the tooth profile (on loaded side) and a radial line measured in a plane normal to the thread helix, measured at the pitch radius
μ = coefficient of friction between mating threads
μ_c = coefficient of friction between plate and nut or bolt head

The friction coefficients, μ, for nonlubricated surfaces range from 0.12 to 0.25 and for lubricated surfaces from 0.07 to 0.17. Because new coatings are being developed, the fastener manufacturer's product specifications should be consulted. Low-friction surfaces reduce the need for higher torques, but this advantage may be affected by poor vibration resistance. Also, low-friction fasteners can be easily overtorqued, resulting in difficulty in controlling torque for assembly operations.

A number of factors must be evaluated to determine the practicality of using automatic tightening systems. These factors include magnitude and direction of load and mechanical characteristics of the joint and fastener materials.

Fatigue load, in particular, should be carefully evaluated. One rule of thumb is that fatigue load should be no more than 50% of the preload value and should not cause a fastener elongation greater than that caused by the initial tightening. Also, experiments show that in most common bolts only 10% of applied torque goes into fastener tension; the rest is friction between threads and bearing surfaces. Some special elliptical vibration nuts or nylon inserts have a higher percentage of torque going into bolt shear.

Part of the preload clamping force when tightening a bolted joint might be lost when the bolt is embedded in the plate. Embedding has been related to surface finish, which is independent of clamping load. Embedding tends to reduce the load on the bolt and joint. It is estimated that 1.3 μ of embedding are included in all joints with machine-ground surfaces.

Continued use of the same bolt and nut combination under repetitive torque to the former levels will change the friction coefficient value, resulting in a decreased clamping load. After the tenth installation, this decrease can be approximately 50% of the original value.

Using the maximum-shear-stress theory, the shear stress in a bolt axially loaded can be found as

$$\tau_{\max} = \left[\left(\frac{\sigma_x}{2}\right)^2 + \tau_{xy}^2\right]^{1/2} = \left[\left(\frac{W}{2A}\right)^2 + \left(\frac{1.6T_f}{\pi d^3}\right)^2\right]^{1/2} \qquad [6.9]$$

where A = root area

T_f = torque

W = resultant bolt load

τ_{xy} = 10% of the preload torque transmitted to the bolt, resulting in a 1.6 coefficient

d = root diameter

For a static loading case, τ_{max} should be less than the shear yield strength for a given material. For steels, τ_{yield} is approximately one-half the proof load (yield strength). A factor of safety (safety index) can be expressed for the bolt as

$$FS = \frac{\tau_y}{\tau_{max}} \qquad [6.10]$$

In English units bolts are specified by the shank diameter and pitch is specified in threads per inch (e.g., 3/8–16). In metric units the pitch changes to millimeters per thread. Tables 6.1 and 6.2 give common sizes for bolts, along with associated materials. In the English system, material is designated as stress-to-fracture (tensile strength) or as the proof load (stress at yield). Proof-load values from Tables 6.1 and 6.2 can be used in calculating the bolt's factor of safety. Metric bolt-class designation has the tensile strength in 10^8 Pa left of the decimal point and the yield point is the number to the right of the decimal point times the tensile strength. For example, class 10.9 has a proof load 0.9 times 10×10^8. To convert to psi, divide by 6895. When using these tables for clamp loads less than shown, it is recommended that a lower tightening torque be used.

A step-by-step procedure for bolt design follows.

1. Find W_2 (external load on bolt).
2. Assume that W_1 is approximately the same value as W_2 to size the bolt conservatively.
3. Use a chart to pick bolt size by conservatively setting the clamp load equal to W_1, obtained from part 2 above.
4. Calculate joint stiffness $(m/m + b)$.
5. Calculate W (resultant load on bolt).
6. Calculate shear stress and FS on bolt by picking material from Tables 6.1 and 6.2.
7. Adjust torque if FS is too high or low. (FS should be between 1.05 and 1.33 if 75% proof is used to set clamp load.)
8. Check bearing strength of bolted part.
9. Specify torque range on nut or bolt.
10. Use serrated washers or other devices to reduce vibrational loosening.
11. Check for part separation, $W_2 \geqslant W$, which could be a functional joint failure (e.g., a hydraulic cylinder leaks fluid when the joint separates).

Table 6.1 **Torques and Proof Loads for Course-Thread Metric Bolts**

			Class 4.6 Minimum Tensile Strength 400 MPa		Class 8.8 Minimum Tensile Strength 830 MPa		Class 9.8 Minimum Tensile Strength 900 MPa		Class 10.9 Minimum Tensile Strength 1040 MPa		Class 12.9 Minimum Tensile Strength 1220 MPa	
Size	Nearest English Equivalence	Root Area (mm^2)	Torque[a] N·M	Proof Load KN	Torque N·M	Proof Load KN	Torque N·M	Proof Load KN	Torque N·M	Proof Load KN	Torque N·M	Proof Load KN
2.0		2.1	.14	.473	0.37	1.26	0.40	1.35	.52	1.74	.61	2.01
2.5	3	3.4	.28	.763	0.76	2.04	0.82	2.20	1.06	2.82	1.24	3.29
3.0		5.0	.51	1.13	1.35	3.00	1.47	3.27	1.88	4.15	2.21	4.88
3.5	6	6.8	.80	1.53	2.13	4.08	2.31	4.41	2.96	5.64	3.47	6.58
4.0		8.8	1.18	1.98	3.15	5.28	3.41	5.71	4.38	7.30	5.14	8.52
5.0		14.2	2.39	3.20	6.36	8.52	6.90	9.23	8.86	11.8	10.99	13.8
6.0		20.1	4.05	4.52	10.81	12.1	11.72	13.1	15.05	16.7	17.66	19.5
6.3	1/4	22.6	4.78	5.09	12.76	13.6	13.84	14.7	17.77	18.8	20.84	21.9
8.0		36.6	9.84	26.25	22.0	28.46	23.8	36.54	30.4	42.87	35.5	
10.0		58.0	19.49	13.1	51.99	34.8	56.38	37.7	72.38	48.7	84.91	56.3
12.0		84.3	33.99	19.0	90.68	50.6	98.33	54.8	126.25	70.0	148.10	81.8
14.0		115.0	54.10	25.9	144.32	69.0	156.49	74.8	200.93	95.4	235.70	112
16.0	5/8	157.0	84.40	36.3	225.18	94.2	244.17	102	313.50	130	367.76	152
20.0		245.0	164.64	55.1	439.24	147	476.28	159	611.52	203	717.36	238
24.0	1-3/16	353.0	284.66	79.4	759.43	212	823.48	229	1057.31	293	1240.30	342
30.0	1-5/16	561.0	565.49	126	1508.64	337	1635.86	365	2100.38	466	2463.91	544
36.0		817.0	988.24	184	2636.49	490	2858.85	531	3670.62	678	4305.92	792
42.0		1120.0	1580.54	252	4216.66	672	4572.29	728	5870.59	930	6886.66	1086
		U.S. EQUIVALENT SAE 2			SAE 5		SAE 6 & 7		SAE 8		SAE 9	

[a] Torque for zinc-coated fasteners.

Table 6.2 **Suggested Tightening Torque Values**

Size	Tensile Stress Area A(in).²	Bolt Shank Diameter D(in.)	SAE GRADE 2 BOLTS Clamp Load P(lb)	Tensile Strength (min psi)	Proof (Yield) Load (psi)	Tightening Dry K=0.20 lb in.	Torque Lub. K=0.15 lb in.	SAE GRADE 5 BOLTS Clamp Load P(lb.)	Tensile Strength (min psi)	Proof (Yield) Load (psi)	Tightening Dry K=0.20 lb in.	Torque Lub. K=0.15 lb in.	SAE GRADE 8 BOLTS Clamp Load P(lb)	Tightening Dry K=0.20 lb in.	Torque Lub. K=0.15 lb in.
4-40	0.00604	0.1120	240	69,000	55,000	5	4	380	120,000	85,000	8	6	540	12	9
4-48	0.00661	0.1120	280			6	5	420			9	7	600	13	10
6-32	0.00909	0.1380	380			10	8	580			16	12	820	23	17
6-40	0.01015	0.1380	420			12	9	640			18	13	8920	25	19
8-32	0.01400	0.1640	580			19	14	900			30	22	1,260	41	31
8-36	0.01474	0.1640	600			20	15	940			31	23	1,320	43	32
10-24	0.01750	0.1900	720			27	21	1,120			43	32	1,580	60	45
10-32	0.02000	0.1900	820			31	23	1,285			49	36	1,800	68	61
1/4-20	0.0318	0.2500	1,320			66	49	2,020			96	75	2,860	144	106
1/4-28	0.0364	0.2500	1,500			76	56	2,320			120	96	3,280	168	120
5/16-18	0.0524	0.3125	2,160			11	8	3,340			17	13	4,720	25	18
5/16-24	0.0580	0.3125	2,400			12	9	3,700			19	14	5,220	25	20
3/4-16	0.0775	0.3750	3,200			20	15	4,940			30	23	7,000	45	35
3/4-24	0.0878	0.3750	3,620			23	17	5,600			35	25	7,900	50	35
7/8-14	0.1063	0.4375	4,380			30	24	6,800			50	35	9,550	70	55
7/16-20	0.1187	0 4375	4,900			35	25	7,550			55	40	10,700	80	80
1/2-20	0.1419	0.5000	5,840			50	35	9,050			75	56	12,750	110	80
1/2-20	0.1599	0.5000	6,600			55	40	10,700			90	65	14,400	120	90
9/16-12	0.1820	0.5625	7,100	64000	52000	65	50	11,600			110	80	16,400	150	110
9/16-18	0.2050	0.5625	7,900			75	55	12,950			120	90	18,250	170	130
5/8-11	0.2260	0.6250	6,800			90	70	14,400			150	110	20,350	220	170
5/8-18	0.2560	0.6250	10,000			100	80	16,950			180	130	23,000	240	180
3/4-10	0.3340	0.7500	13,000			160	120	21,300			260	200	30,100	380	280
3/4-16	0.3730	0.7500	14,550			180	140	23,800			300	220	33,600	420	320
7/8-9	0.4620	0.8750	9,700	55000	28000	140	110	27,000	115000	78000	400	300	41,600	600	460
7/8-14	0.5090	0.8750	10,700			155	120	29,800			440	320	45,800	640	500
1-8	0.6060	1.0000	12,700			220	160	35,500			580	440	54,500	900	680
1-12	0.6630	1.0000	13,900			240	170	36,800			640	480	59,700	1000	740
1 1/8-7	0.7630	1.1250	16,000			300	220	42,300	105000	74000	800	600	68,700	1280	960
1 1/8-12	0.8560	1.1250	18,000			340	260	47,500			880	660	77,000	1440	1080
1 1/4-7	0.9690	1.2500	20,350			420	320	53,800			1120	840	87,200	1820	1360
1 1/4-12	1.0730	1.2500	22,550			460	360	59,600			1240	920	96,600	2000	1500
1 3/8-6	1.1550	1.3750	24,300			560	420	64,100			1460	1100	104,000	2380	1780
1 3/8-12	1.3150	1.3750	27,600			640	460	73,000			1680	1260	118,400	2720	2040
1 1/2-6	1.4050	1.5000	29,800			740	560	78,000			1940	1460	126,500	3160	2360
1 1/2-12	1.5800	1.5000	33,200			840	620	87,700			2200	1640	142,200	3560	2660

Notes: $T = KDP$
where T = Tightening Torque, lb-in.
K = Torque-friction coefficient
D = Nominal bolt diameter, in.

Clamp load, also refered to as preload or initial load in tension on bolt.
Clamp load in table is calculated at 75% of proof load × tensile stress area
Grade 8 bolts:
Tensile strength (min psi) is 150,000psi; proof load is 120,000psi.

Research has shown that only two or three threads are in contact between the nut and bolt. Again, proper specification for the nut material should be the same as the bolt. Table 6.3 provides strengths of common metric nuts. Table 6.4 gives suggested torque values for nuts that should be specified on all assembly drawings.

Some control over torque is needed. If the fastener is torqued too high, there is danger of 1.) stripping the threads, 2.) breaking the bolt, or 3.) yielding the threads. On the other hand, if the torque is too low, clamp load might not be high enough in a vibration condition. For every bolted joint there exists an optimum preload, which is obtained by proper torque and fastener choice. For vibrational problems, self-locking nuts, cotter pins, castle nuts, and adhesives or plastic locking systems are used (see appendix and reference). Figure 6.3 shows loosening can be combated with special-toothed lock washers or adhesives. The endurance limit can be generally approximated for grade 8 bolts by

$$F_e = 4500d^{1.59} \tag{6.11}$$

where F_e = endurance bolt (lb)
d = nominal diameter (in.)

The effective area, A_{eff}, is the portion of the joint held together by clamp load (see Fig. 6.4).

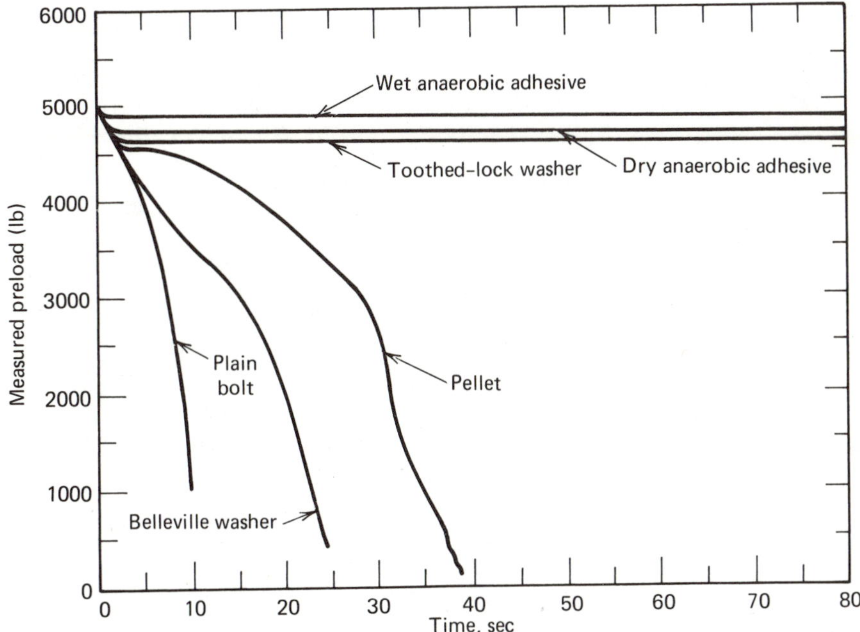

Figure 6.3 Bolted joint subjected to vibration, adhesives and tooth- locked washers reduce preload loss.

Table 6.3 **Hexagon Nuts Coarse Threads**

Nominal Size	*Nearest English*	*Class 5* *Use Class 4.6 Bolts*	*Class 9* *Use Class 8.8 and 9.8 Bolts*	*Class 10* *Use Class 8.8, 9.8, 10.9 Bolts*
mm		*Proof Load kN*	*Proof Load kN*	*Proof Load kN*
2.0		1.18	1.86	2.15
2.5	3	1.93	3.05	3.52
3.0		2.87	4.53	5.23
3.5	6	3.86	6.10	7.05
4.0		5.00	7.90	9.13
5.0		8.09	14.1	14.8
6.0		11.4	19.9	20.9
6.3	1/4	12.9	22.4	23.5
8.0		20.9	36.2	38.1
10.0		33.1	57.4	60.3
12.0		48.1	83.5	87.7
17.0		65.6	114	120
16.0	5/8	89.5	155	163
20.0		170	223	255
24.0	13/16	201	321	367
30.0	1-13/16	320	511	583
36.0		466	743	850
42.0		638	1020	1165

Table 6.4 **Suggested Torque Values (ft lb)**

Bolt or Nut Size	*Grade "5" Regular Nut*		*Grade "5" Prevailing Torque Nut*	
	Nominal	*Specified*	*Nominal*	*Specified*
1/4—20	6	5-7	9	7-11
5/16—18	13	11-15	18	15-21
3/8—16	23	20-26	30	26-34
7/16—114	35	30-40	43	37-49
1/2—13	55	47-63	68	58-78
9/16—12	80	68-92	97	82-112
5/8—11	110	99-121	135	122-148
3/4—10	200	180-220	233	210-256
7/8—9	300	270-330	350	315-385
1—8	440	396-484	507	456-558

Tolerance of specific torque: 0-10 ft lb ±20% of nominal; 10-100 ft lb ±15% of nominal; 100 ft lb ±10% of nominal.

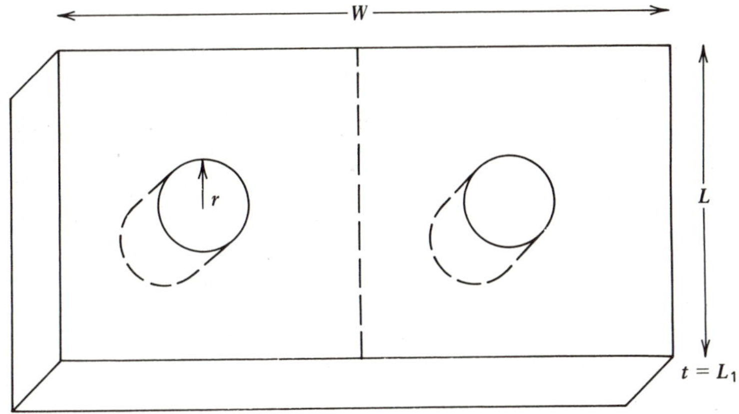

Figure 6.4 Determining effective area per bolt on a joint.

$$A_{\frac{\text{eff}}{\text{bolt}}} = \frac{W \times L - 2\pi r^2}{2} \qquad [6.12]$$

Osgood analyzed the effects of washers in distributing the load per bolt, because compressive stresses between clamped parts decrease nonlinearly as the distance from the fastener head increases. His study showed that a standard flat washer provided the following effective area:

$$A_{\text{eff}} = \frac{\pi}{4}\left[\left(d_w + \frac{L_1}{2}\right)^2 - d_d^2\right] \qquad [6.13]$$

where d_w = washer diameter
L_1 = one flange thickness
d_d = hole diameter

Example 6.1 Analysis of a Bolted Joint (see Fig. 6.5)
Three steel parts are bolted together. Check the bolt design.

Given: Bolt area = 1 cm², A_{eff} = 10 cm², E = 207,000 MPa, for steel

$$b = \frac{L}{AE} = \frac{2.0 \text{ cm}}{1 \times 207{,}000 \text{ MPa}} = \frac{1 \text{ cm}}{103{,}500 \text{ MPa}} \qquad [6.14]$$

$$m = \frac{L}{AE} = \frac{2 \text{ cm}}{10 \times 207{,}000 \text{ MPa}} = \frac{1 \text{ cm}}{1{,}035{,}000 \text{ MPa}}$$

$$\frac{m}{m + b} = \frac{\dfrac{1}{1{,}035{,}000}}{\dfrac{1}{1{,}035{,}000} + \dfrac{1}{103{,}500}} = 0.09 \qquad [6.15]$$

If a gasket material was used for the middle section, m becomes much larger and changes the joint stiffness.

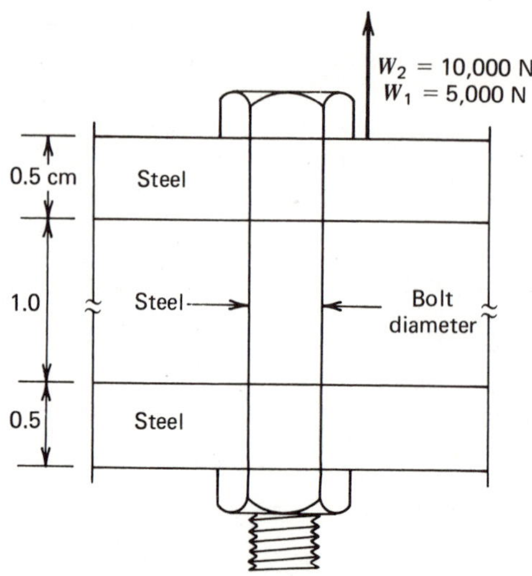

Figure 6.5 Three-plate bolted joint (Example 6.1).

$$W = W_1 + W_2\left(\frac{m}{m+b}\right) = 5000 + 10000(0.09) = 5900\text{N} \qquad [6.16]$$

W_2 is larger than W, and the joint has separated, indicating a possible functional failure!

Example 6.2 Sizing a Bolt

Two bolts holding a joint together as shown in Fig. 6.6 are subjected to an external moment M.

Given: 16.0 mm diameter class 8.8 bolts; $M = 220{,}000$ N·m; torque = 200 N·m ± 20 N·m; $m/(m+b) = 0.21$.

From Table 6.1, proof load = 94.2 kN. Assume that plates 1 and 2 are very rigid and that moment M causes them to displace about point A. Also, small elongations in the bolts δ_1 and δ_2 only create tension in the bolts (negligible bending).
Then $\Sigma M_A = 0$

$$10\,R_2 + 2\,R_1 = M \qquad [6.17]$$

From similar triangles

$$\frac{\delta_2}{\delta_1} = \frac{\dfrac{R_2 L_2}{A_2 E}}{\dfrac{R_1 L_1}{A_1 E}} \qquad [6.18]$$

Therefore, $5\,R_1 = R_2$. Using Eq. 6.17, R_2 (the larger load) can be found.

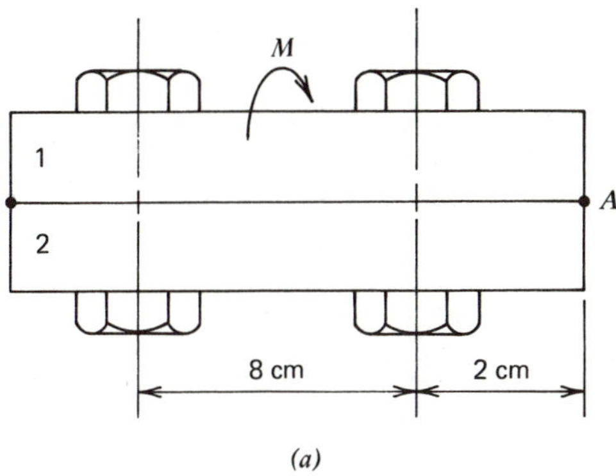

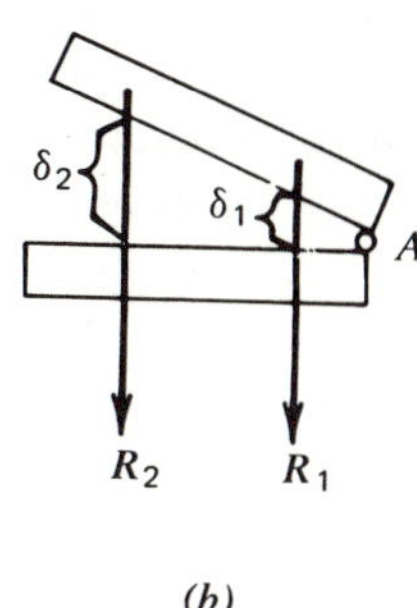

Figure 6.6 Moment applied to a bolted joint. (*a*) Subjected to moment *M*. (*b*) Pivoted about point *A*.

Choosing equal-size bolts,

$$10.4\, R_2 = M(220{,}000\ \text{N}\cdot\text{m}) \qquad [6.19]$$

Therefore,

$$R_2 = 21{,}150\ \text{N} = W_2$$

$$W = W_1 + W_2\left(\frac{m}{m+b}\right) = 62{,}500 + 21{,}150(0.21) = 66{,}900\ \text{N} \qquad [6.20]$$

Checking for point, separations, $W_2 < W$. Therefore, the joint doesn't separate.
Maximum shear stress can be calculated as

$$\tau_{\max} = \left[\left(\frac{W}{2A}\right)^2 + \left(\frac{1.6T_f}{\pi d^3}\right)^2\right]^{1/2} \qquad [6.21]$$

If transverse shear is greater, it should be used in place of torsional shear. Assume that only 1/10 of torque is transmitted into shear stress.

$$\tau_{\max} = \left[\left(\frac{66900}{2 \times 157}\right)^2 + \left(1.6 * \frac{200{,}000}{\pi 16^3}\right)^2\right]^{1/2} \qquad [6.22]$$

$$\tau_{\max} = 214\ \text{MPa} \qquad \tau_y \text{ for a class 8.8} = \frac{\sigma_y}{2} = 332\ \text{MPa} \qquad [6.23]$$

$$\text{FS} = \frac{\tau_y}{\tau_{\max}} = \frac{332}{214} = 1.55 \qquad [6.24]$$

Using bolts that shear in order to protect machine components at couplings is another common application on farm machines. Replacing bolts when they shear can cause downtime costs and added labor costs. For small shafts, 4.5 in. in diameter or less, custom-made off-the-shelf torque limiters are suggested for most applications. If diameters are larger than 4.5 in., a shear bolt design is suggested, especially for stone crushers, heavy earth moving equipment, and balers. This topic is covered in most strength-of-materials courses, and the reader is encouraged to refer to those references.

The equation for special shear bolts is given as

$$d = P\left[\frac{KT}{RNY}\right]^{1/2} \qquad [6.25]$$

where d = neck diameter chosen less than shank diameter D
P = 1.49 for bolts with little tensile preload
P = 1.4 for bolts with axial preload one-half of yield
K = stress concentration factor, usually designed to be 1.05
T = initial shear torque
R = radius at which the bolts are located
Y = yield point in shear for bolt material
N = number of bolts

To ensure that shear bolts fail according to Eq. 6.25 the design shouldn't allow bending moments in the bolt or pretensioning that creates resistive friction between coupled surfaces.

Bolt-hole sizing is related to the need for easy component assembly. Refer to Chapter 3 for determining hole sizes and tolerances. Tables 6.5 and 6.6 provide medium-clearance fits for drilling and punching holes. Usually, the drill or punch is oversized by 1/32 in., as an initial choice. Tolerances and part variation or part function might require different clearances for bolts and their respective holes.

The added torque capability and clamp load achieved when going from a coarse thread to a fine thread is usually about 10% and does not warrant the expense in most applications. Bolt costs usually are reduced substantially when bolts are ordered in quantities that require mass-production tooling (a 2-week run). Usually, 1 million pieces provide the lowest cost parts. Bolted joint design on moving vehicles has quite a few safety implications, such as holding the wheel rim to the axle. Thus this design requires more in-depth design analysis than merely picking a bolt from a chart or table (see Fig. 6.7).

6.2 WELDED JOINTS

Welding is one of the most common methods of joining farm machinery together. Some of the advantages that have made welding popular are its low manufacturing capital costs for small-volume machines (less than 1000 units per year) and its versatility, especially in assembling complicated joints. Because it is used on nonsymmetrical geometries, welding does not lend itself well to mathematical treatments and thus adds uncertainty and difficulty for the engineer who must specify processes and weld joint size. To adjust for

Table 6.5 **Specifying Drilled Holes**

Nominal Drill	Decimal Equivalent	Specify Limits M/M	Nominal Drill	Decimal Equivalent	Specify Limits M/M
#37	.1040	.108 (2.769) .102 (2.591)	19/32	.5938	.602 (15.291) .592 (15.037)
1/8	.1250	.130 (3.302) .123 (3.124)	39/64	.6094	.617 (15.672) .607 (15.418)
9/64	.1406	.146 (3.708) .139 (3.531)	5/8	.6250	.633 (16.078) .623 (15.824)
5/32	.1562	.161 (4.089) .154 (3.912)	41/64	.6406	.649 (16.485) .639 (16.231)
11/64	.1719	.144 (4.496) .170 (4.318)	21/32	.6562	.664 (16.866) .654 (16.612)
3/16	.1875	.132 (4.877) .186 (4.724)	43/64	.6719	.680 (17.272) .670 (17.018)
13/64	.2031	.208 (5.283) .201 (5.105)	11/16	.6875	.656 (17.678) .686 (17.424)
7/32	.2188	.224 (5.690) .217 (5.512)	45/64	.7031	.711 (18.059) .701 (17.805)
15/64	.2344	.235 (6.071) .232 (5.893)	23/32	.7188	.727 (18.466) .717 (18.212)
1/4	.2500	.255 (6.477) .248 (6.299)	47/64	.7344	.742 (18.847) .732 (18.593)
17/64	.2656	.274 (6.960) .264 (6.706)	3/4	.7500	.758 (19.253) .748 (18.999)
9/32	.2812	.289 (7.341) .279 (7.087)	49/64	.7656	.776 (19.710) .763 (19.380)
19/64	.2969	.305 (7.747) .295 (7.493)	25/32	.7812	.791 (20.091) .778 (19.761)
5/16	.3125	.320 (8.128) .310 (7.874)	51/64	.7969	.807 (20.498) .794 (20.268)
21/64	.3281	.336 (8.534) .326 (8.280)	13/16	.8125	.822 (20.879) .810 (20.574)
11/32	.3438	.352 (8.541) .342 (8.687)	53/64	.8281	.838 (21.285) .825 (20.955)
23/64	.3594	.367 (9.322) .357 (9.068)	27/32	.8438	.854 (21.692) .841 (21.361)
3/8	.3750	.383 (9.728) .373 (9.474)	55/64	.8594	.865 (22.073) .856 (21.742)
25/64	.3906	.399 (10.135) .389 (9.881)	7/8	.8750	.883 (22.475) .872 (22.145)
13/32	.4062	.414 (10.516) .404 (10.262)	57/64	.8906	.900 (22.860) .886 (22.504)
27/64	.4219	.430 (10.922) .420 (10.668)	29/32	.9062	.916 (23.266) .902 (22.911)
7/16	.4375	.446 (11.328) .436 (11.074)	59/64	.9219	.932 (23.673) .917 (23.292)
29/64	.4531	.461 (11.709) .451 (11.455)	15/16	.9395	.948 (24.079) .933 (23.698)
15/32	.4688	.477 (12.116) .467 (11.862)	61/64	.9531	.963 (24.260) .949 (24.105)
31/64	.4844	.492 (12.497) .482 (12.243)	31/32	.9688	.975 (24.867) .964 (24.486)
1/2	.5000	.508 (12.903) .498 (12.643)	63/64	.9844	.994 (25.248) .980 (24.892)
33/64	.5156	.524 (13.310) .514 (13.056)	1	1.000	1.010 (25.654) .996 (25.298)
17/32	.5312	.539 (13.691) .529 (13.437)	1-1/64	1.0156	1.026 (26.060) 1.011 (25.679)
35/64	.5469	.555 (14.097) .545 (13.843)	1-1/32	1.0312	1.041 (26.441) 1.027 (26.086)
9/16	.5625	.570 (14.478) .560 (14.224)	1-3/64	1.0469	1.057 (26.848) 1.042 (26.476)
37/64	.5781	.586 (14.884) .576 (14.630)	1-1/16	1.0625	1.072 (27.229) 1.058 (26.873)

Table 6.6 **Specifying Punched Holes**

Nominal Drill	Decimal Equivalent	Specify Limits M/M	Nominal Drill	Decimal Equivalent	Specify Limits M/M
7/64	.1094	.124 (3.150) .104 (2.642)	19/32	.5938	.602 (15.469) .589 (14.961)
1/8	.1250	.140 (3.556) .120 (3.048)	39/64	.6094	.624 (15.850) .604 (15.342)
9/64	.1406	.156 (3.962) .136 (3.454)	5/8	.6250	.640 (16.256) .620 (15.748)
5/32	.1562	.171 (4.343) .151 (3.835)	41/64	.6406	.656 (16.662) .636 (16.154)
11/64	.1719	.187 (4.750) .167 (4.242)	21/32	.6562	.671 (17.043) .651 (16.535)
3/16	.1875	.202 (5.131) .182 (4.623)	43/64	.6719	.687 (17.450) .667 (16.942)
13/64	.2031	.218 (5.537) .198 (5.029)	11/16	.6875	.702 (17.831) .682 (17.325)
7/32	.2188	.234 (5.944) .214 (5.436)	45/64	.7031	.718 (18.237) .698 (17.729)
15/64	.2344	.249 (6.325) .229 (5.817)	23/32	.7188	.734 (18.644) .714 (18.136)
1/4	.2500	.265 (6.731) .245 (6.223)	47/64	.7344	.749 (19.025) .729 (18.517)
17/64	.2656	.281 (7.137) .261 (6.629)	3/4	.7500	.765 (19.431) .745 (18.923)
9/32	.2812	.296 (7.518) .276 (7.010)	49/64	.7656	.781 (19.837) .761 (19.329)
19/64	.2969	.312 (7.925) .292 (7.417)	25/32	.7812	.796 (20.218) .776 (19.710)
15/16	.3125	.328 (8.331) .308 (7.823)	51/64	.7969	.812 (20.625) .792 (20.117)
21/64	.3281	.343 (8.712) .323 (8.204)	13/16	.8125	.828 (21.031) .808 (20.523)
11/32	.3438	.359 (9.119) .339 (8.611)	53/64	.8281	.843 (21.412) .823 (20.904)
23/64	.3594	.374 (9.500) .354 (8.992)	27/32	.8438	.859 (21.815) .839 (21.311)
3/8	.3750	.390 (9.906) .370 (9.398)	55/64	.8594	.874 (22.200) .854 (21.652)
25/64	.3906	.406 (10.312) .386 (9.804)	7/8	.8750	.850 (22.606) .870 (22.098)
13/32	.4062	.421 (10.693) .401 (10.185)	57/64	.8906	.906 (23.012) .886 (22.504)
27/64	.4219	.437 (11.100) .417 (10.592)	29/32	.9062	.921 (23.393) .901 (22.885)
7/16	.4375	.452 (11.481) .432 (10.973)	59/64	.9219	.937 (23.800) .917 (23.292)
29/64	.4531	.468 (11.887) .458 (11.379)	15/16	.9395	.952 (24.181) .932 (23.673)
15/32	.4688	.484 (12.294) .464 (11.786)	61/64	.9531	.968 (24.587) .948 (24.079)
31/64	.4844	.499 (12.675) .489 (12.167)	31/32	.9688	.984 (24.994) .964 (24.486)
1/2	.5000	.515 (13.081) .495 (12.573)	63/64	.9844	.999 (25.375) .979 (24.867)
33/64	.5156	.531 (13.487) .511 (12.979)	1	1.000	1.015 (25.781) .995 (25.273)
17/32	.5312	.546 (13.868) .526 (13.360)	1-1/64	1.0156	1.031 (26.187) 1.011 (25.679)
35/64	.5469	.562 (14.275) .542 (13.767)	1-1/32	1.0312	1.046 (26.568) 1.026 (26.060)
9/16	.5625	.578 (14.681) .558 (14.173)	1-3/64	1.0469	1.062 (26.975) 1.042 (26.476)
37/64	.5781	.593 (15.062) .573 (14.554)	1-1/16	1.0625	1.078 (27.381) 1.058 (26.873)

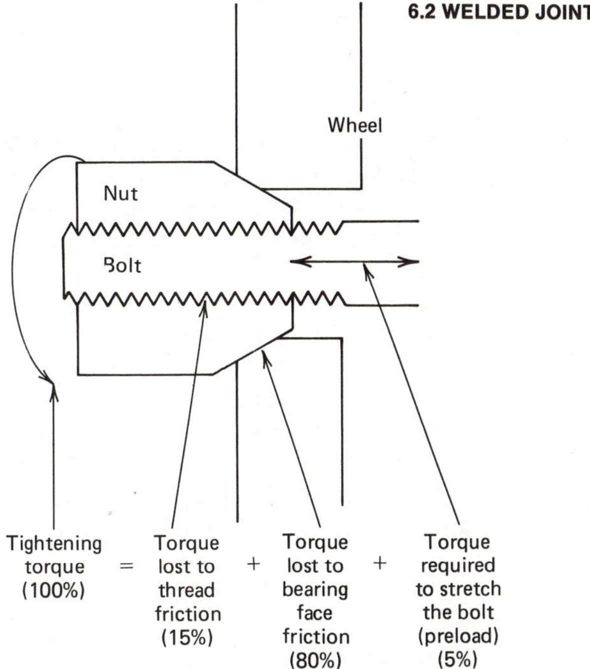

Figure 6.7 Appropriate wheel torque distributions.

hard-to-control manual-welding quality and to ensure safe and reliable joints on machines large safety factors are assumed. This assures safe and reliable joints on machines.

Weldments are usually fabricated from low carbon steels that have been cut and shaped from flat steel stock. This process allows the designer to buy common steel plates in large quantities and to design many components, that reduces the purchase cost of the raw material.

The most common types of welding include oxyacetylene, shielded metal arc, submerged arc, gas metal arc (MIG), gas tungsten arc (TIG), electroslag, election beam, plasma arc, spot, seam, projection, and friction. All but the first two can be mechanized and controlled to eliminate operator skill and provide a uniform product. As manufacturing processes change and robots are used and adapted to mechanized processes, product quality will increase and designers will be able to reduce safety factors. In addition, it will be possible to remove operators from the welding area to safer environments. The gases and ultraviolet light emitted during welding are common hazards. A 1-second exposure of an observer to a welding arc can cause discomfort, and longer periods of exposure can cause permanent eye damage.

Welding is a high-temperature process that creates a metallurgical bond between two parts and converts them into one continuous piece of metal. Bonding of metals by welding occurs as a result of using heat to melt two surfaces and then allowing them to cool. High-temperature gradients create metallurgical changes in the parent metal in the vicinity of the weld, known as the heat-affected zone (usually abbreviated HAZ). Fast cooling can cause residual stresses in welds that can be relieved by heat treatment or mechanical cold forming, such as peening. Another problem is the effect of air on molten steel. Gases formed during welding can be entrapped in the weld and cause porous conditions. To combat this problem, a flux or inert-gas shield protects the molten steel during welding.

Figure 6.8 shows how the electrode coating on the manual arc process provides a protective atmosphere. The slag material helps protect, while the molten bath solidifies the weld. The slag is removed after welding because it does not provide any strength and, with multiple-pass welds it could become embedded between passes and reduce joint strength. The electrode's function is to produce a gas shield around the arc, promote electrical conduction across the arc, add slag elements, add alloying elements, and control the bead shape. Welding electrodes are classified by the American Welding Society (AWS) as

$$EXXYZ$$

where E = electrode

XX = minimum tensile strength of electrode filler material (ksi)

Y = position of electrode: 1 = all (including overhead); 2 = horizontal and fillet; 3 = fillet

Z = current and rod covering

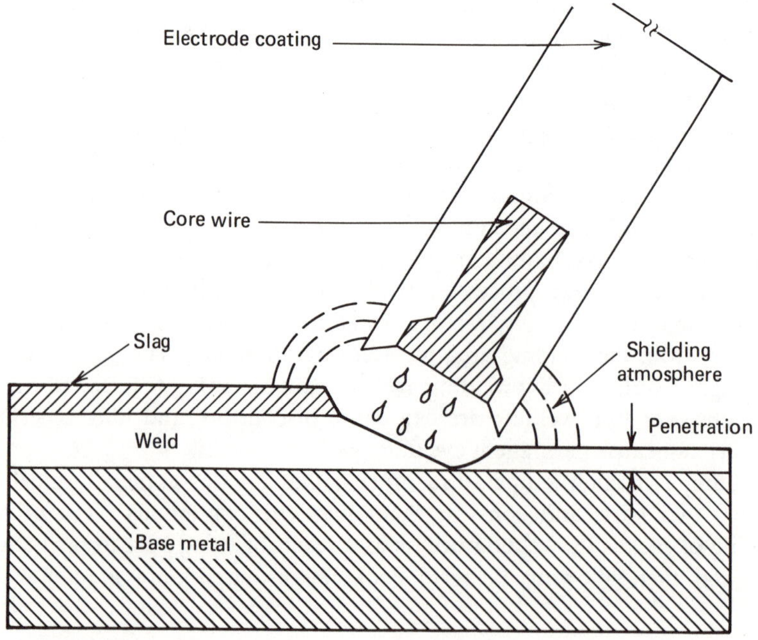

Figure 6.8 Shielded metal arc.

For example, an E6010 electrode has a minimum tensile strength of 60,000 psi and can be used in all positions; Z = 0 refers to a high-cellulose coating that contains high quantities of H_2. Hydrogen can become entrapped in welds and reduce their strength.

When the current density at the electrode tip heats up (because of I^2R resistance), it causes the tip to melt. The current flow in the electrode creates a magnetic field, which pinches off globules of melted electrodes and transfers them across the arc. Molten temperature reaches an estimated 9000°F in some materials being welded. The basic types of joints and welds are shown in Figs. 6.9 and 6.10.

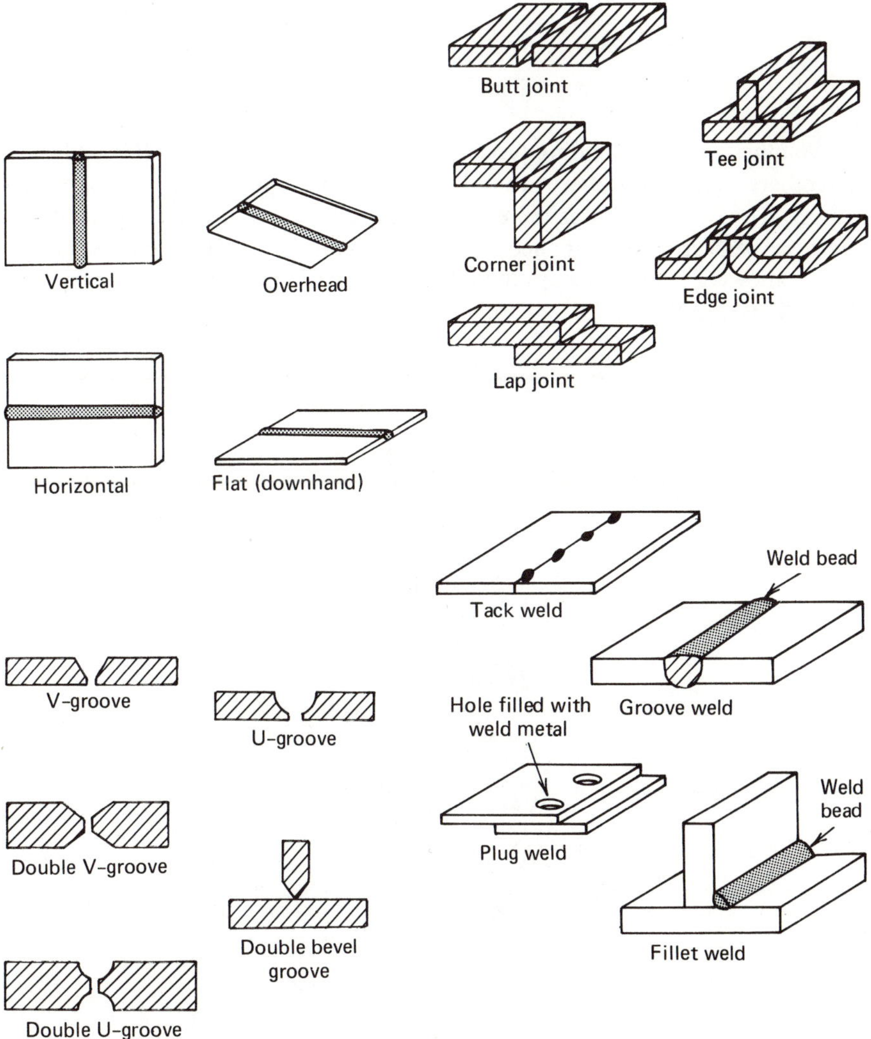

Figure 6.9 Common weld joints and types.

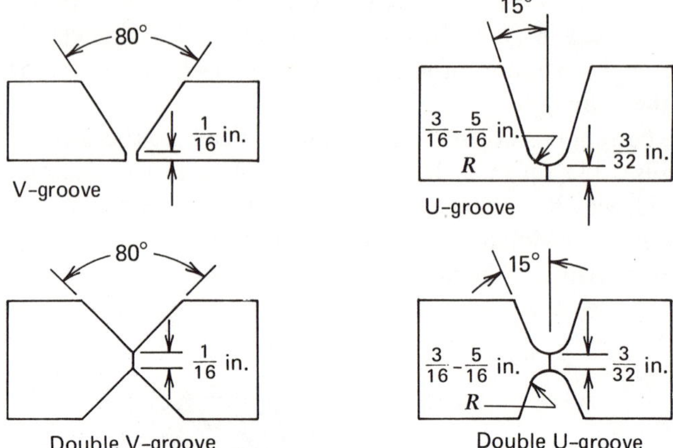

Figure 6.10 Common joint designs.

Two other common welding processes are tungsten inert gas (TIG) and submerged arc. The atmosphere around the tungsten arc is usually inert argon or helium. Tungsten is used as the electrode because of its high resistance to extreme temperatures and its nonconsumable (nonmelting). The arc is extremely steady, allowing it to be accurately directed into a joint, and it can be used in all positions (see Fig. 6.11).

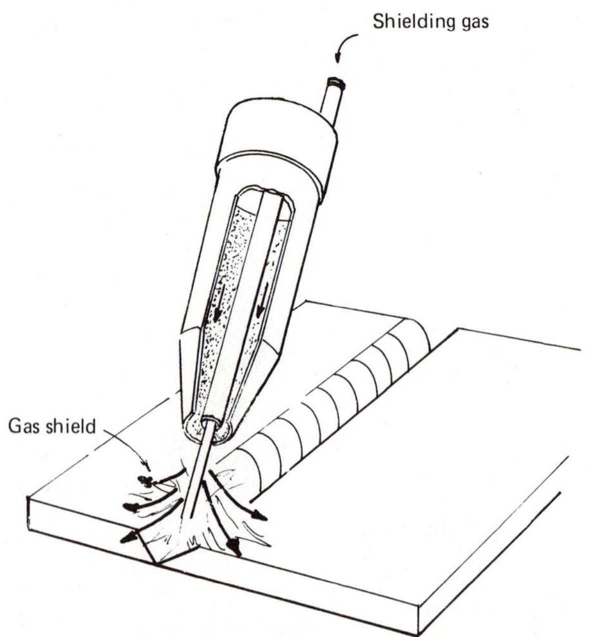

Figure 6.11 TIG welding process.

Welding can be done with or without a filler rod and has been popular with stainless steel. Extensive applications of TIG include dairy and food-processing containers made from sheets of stainless steel.

The submerged arc process, which uses a consumable electrode, is employed for welding thicker metals (Fig. 6.12). In this process a thick blanket of flux covers the joint. There is no visible arc during welding, and the process is usually automated for flat applications. Some of the flux is consumed and protects the cooling metal, and the rest is recycled into a hopper. Current settings for these common processes are given in Table 6.7.

The two most common types of welds are the butt and fillet, seen in Fig. 6.13 and Fig. 6.14. Two types of fillet welds are transverse (e.g., perpendicular to the load) and parallel to the load. The side of a fillet weld is the leg designated as *b*, and the throat *t* is the minimum section used in strength calculations. Tests have shown that maximum strains have been found at the throat of welds. If a butt weld is subjected to force *P*, then the stress is

$$\sigma_w = \frac{P}{A} = \frac{P}{Lt} \qquad [6.26]$$

where A = cross-sectional area
P = load
L = length of weld
t = depth of weld (maximum possible is the thickness of the metal)
σ_w = allowable weld stress

It is generally assumed that when fillet welds fail, bending caused by the slight eccentricity is negligible. The shear stress can be calculated as

$$\tau_w = \frac{P}{0.707\, tL} \qquad [6.27]$$

where τ_w = allowable weld stress for fillets (other terms are the same as in Eq. 6.26)

Different length of welds alter the *L* value in these equations, as shown in Fig. 6.15. Table 6.8 depicts allowable stress σ_w and τ_w, which already contains a factor of safety of 3. For fatigue, K_f must be included in the calculation, which accounts for the surface condition.

For compressive stress on butt welds, AWS suggests that static load values are acceptable in cyclically loading cases because the welds have been tested to

Table 6.7 **Common Welding-Process Settings**

Type (voltage 10–40)	*Current Range (amperes)*
Metal arc	70–300
Submerged arc	350–1000
TIG and MIG	60–500

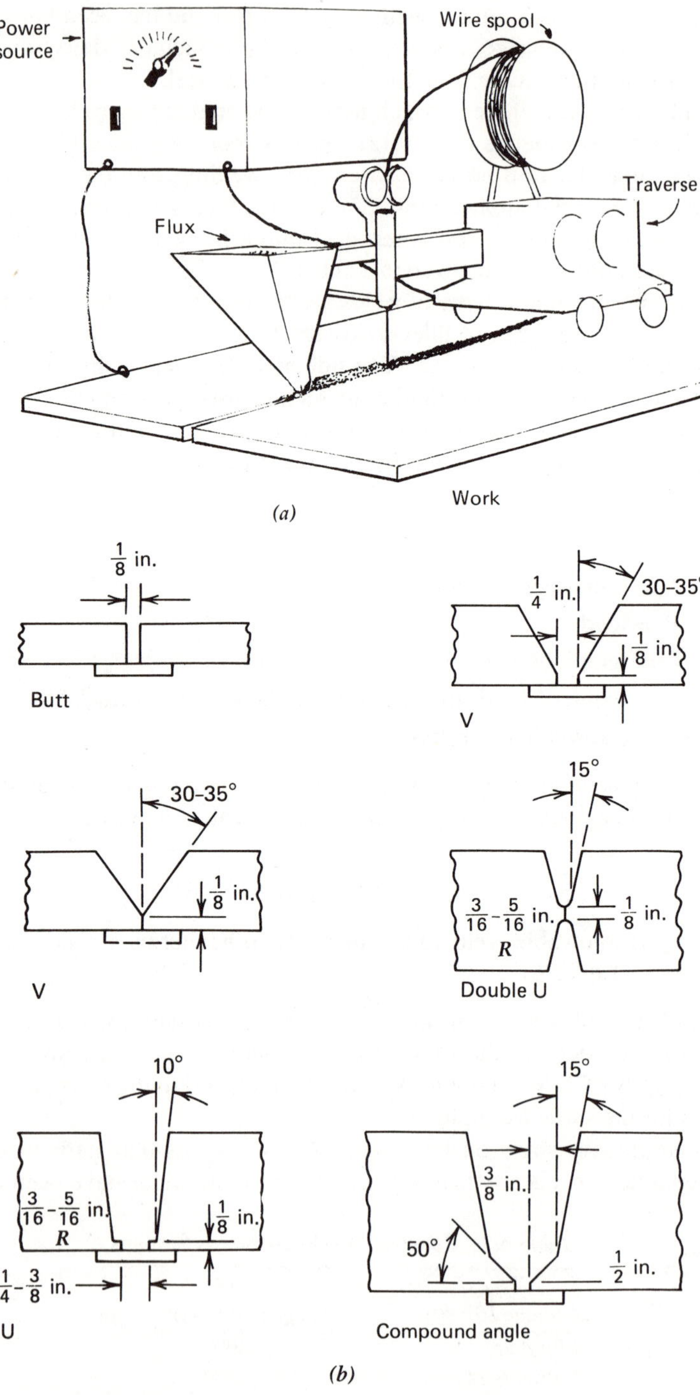

Figure 6.12 Submerged arc process.

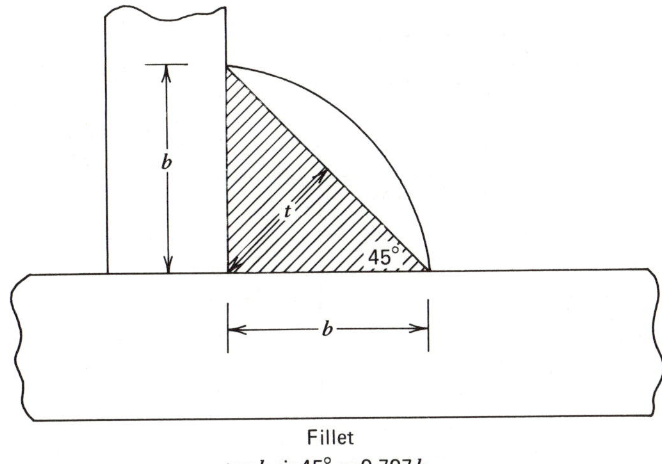

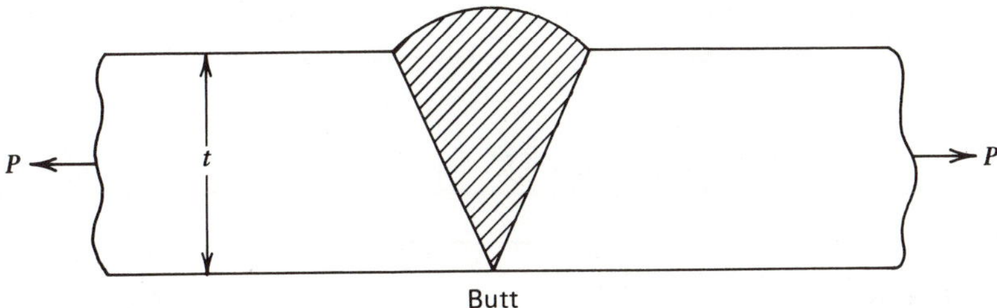

Figure 6.13 Butt and fillet weld (t = throat, b = leg size).

Table 6.8 **Allowable Stress for Electrode Welds on Low-Carbon Steels**

Type of Weld	*Static Loads, $MN/m^2 (lb/in.^2)$*	*Reverse Fatigue Loads, $MN/m^2 (lb/in.^2)$*
Butt welds		
Tension	110 (16,000)	55 (8000)
Compression	124 (18,000)	55 (8000)
Shear	67 (10,000)	34.5 (5000)
Fillet welds		
Transverse and parallel welds	96.5 (14,000)	34.5 (5000)

last 2×10^6 cycles. If a single fillet weld is subjected to loading as shown in Fig. 6.16, the bending stress, σ_b, is calculated as

$$\sigma_b = \frac{3P}{hL} \tag{6.28}$$

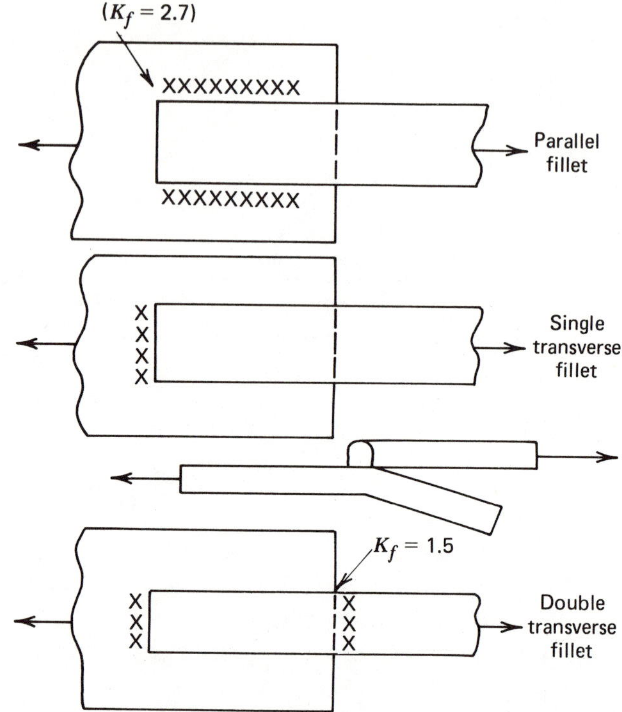

Figure 6.14 Lapping fillet welds.

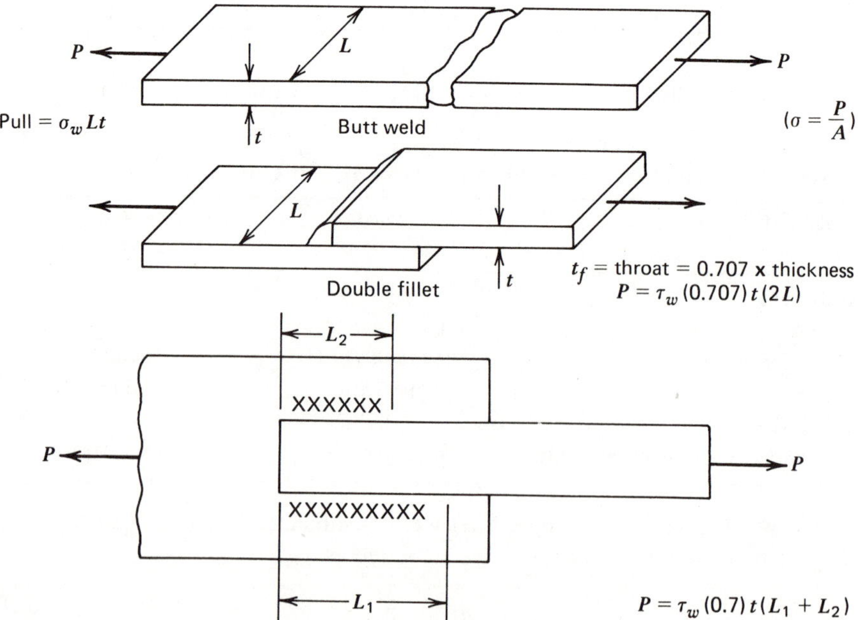

Figure 6.15 Length of weld affecting allowable load *P*.

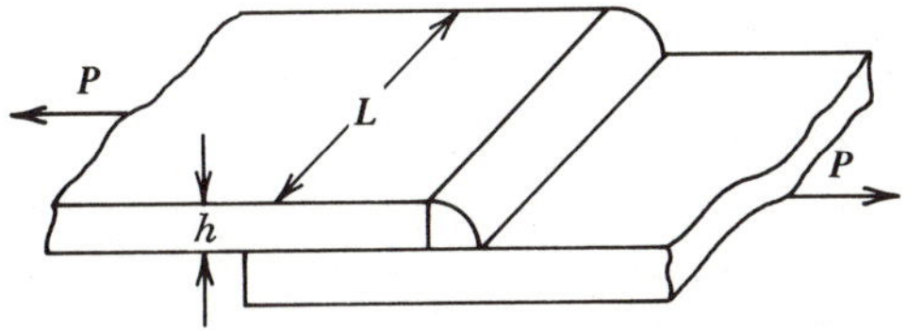

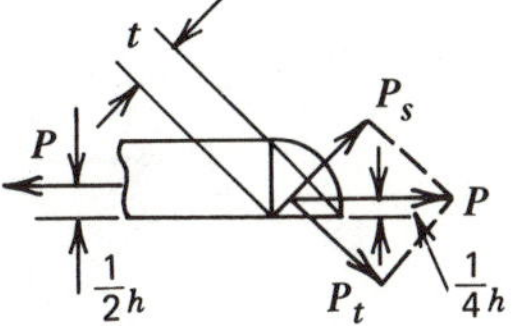

Single fillet lap joint in tension

Figure 6.16 Free-body diagram of a single fillet lap joint.

where h = metal thickness
L = length of weld

This stress adds to the tensile stress. Using the maximum-principal-stress theory, the maximum stress is

$$\sigma = \frac{4.2\,P}{hL} \tag{6.29}$$

which is a significant increase. To alleviate this bending stress, double parallel fillets are recommended.

Table 6.9 provides a starting place for estimating the weld size, given plate thickness. The American Institute of Steel Construction (AISC) in New York provides another code for designers to use in sizing welds on buildings.

Example 6.3 Weld Size Determination (Fig. 6.17)

Determine the length of weld required for two 1-cm plates cyclically loaded to ±40,000 N for 1.) Parallel fillet and 2.) Transverse fillet.

For parallel fillet

$$P = \frac{\sigma_W}{K_f}A = \frac{\tau_w}{K_f}(L_1 + L_2) \times \text{throat } W$$

$$40{,}000 = \frac{34.5\text{ N }(2L_1)(0.7)(0.01\text{ m})}{1.5}$$

$L_1 = 0.22$ m

Table 6.9 **Minimum Weld Size Rule of Thumb***

Plate Thickness (in.)	*Weld size (Throat)*
1/8–3/10	1/8
1/4–5/16	3/16
3/8–5/8	1/4
3/4–1	3/8
1-1/8–1-3/8	1/2
1-1/2	3/4

Notes: 1. Cheaper to increase length than size
2. Allow +½ in. for starting and stop.
3. Check strength for larger size.

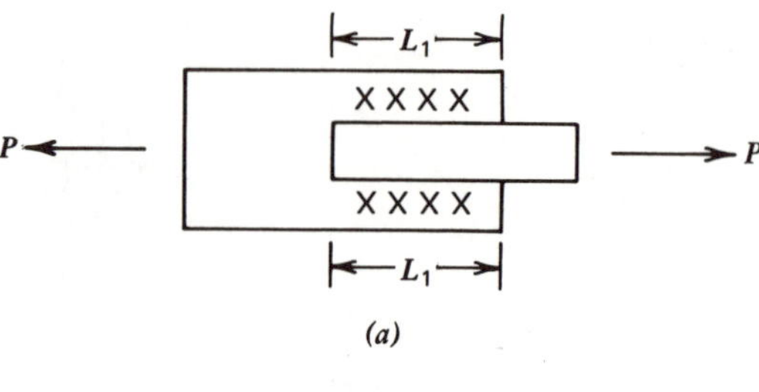

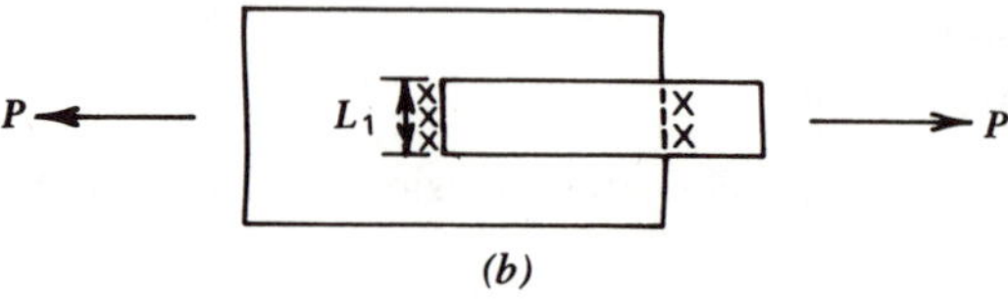

Figure 6.17 Weld design for Example 6.3.

For transverse fillet

$$40{,}000 = \frac{34.5\ \text{N}\ (2L_1)(0.7)(0.01\ \text{m})}{1.5}$$

$L_1 = 0.12$ m

When a weldment is subjected to bending or torsional loading, another approach in determining its strength, called treating the weld as a line, is used. It can be shown that the section modulus of an area is almost equivalent to the section treated as a line, multiplied by its thickness. For bending moment M

$$\sigma_W = \frac{M}{Z_w W} \qquad [6.30]$$

where Z_w = section modulus of weld (mm^2)
W = weld throat (mm)

Let f represent $\sigma_w \cdot W$ then Eq. 6.30 becomes

$$f = \frac{M}{Z_w} \qquad \text{N/mm (lb/in.)} \qquad [6.31]$$

This equation permits finding the size of the weld directly from suggested strength tables and section moduli.

For torsion

$$f = \frac{Tc}{J_w} \qquad \text{(N/mm)} \qquad [6.32]$$

where J_w = unit polar moment of inertia for welds

Values for J_w and Z_w are given in Tables 6.10 and 6.11. Analysis becomes more difficult for combined loading cases. Experimental stress techniques should be used to verify sizes for difficult design cases.

Example 6.4 Treating a Weld as a Line (see Fig. 6.18)
A circular bar is welded to a steel plate. Find the weld size.

Given: Bending moment = 10,000 N × 10 cm = 1000 N · m
Shear force = 10,000 N

Z_W (see Table 6.10) $= \frac{1}{4}\pi d^2 = 19.6 \text{ cm}^2$

Force per centimeter of weld at points A and B

$$f_A = \frac{M}{Z_w} = \frac{100{,}000 \text{ N}\cdot\text{cm}}{19.6 \text{ cm}^2} = 5102 \text{ N/cm} \qquad [6.33]$$

For vertical shear assume uniform distribution (this is only an assumption!).

$$f_\tau = \frac{\text{force}}{\text{weld length}} = \frac{10{,}000 \text{ N}}{5\pi} = 637 \text{ N/cm} \qquad [6.34]$$

Because this is combined loading, the resultant load is determined vectorally.

$$f_R = (5102^2 + 637^2)^{1/2} = 5142 \text{ N/cm} \qquad [6.35]$$

Size of weld b is $= \frac{f_R}{f_{\text{all}}} = \frac{5142}{9{,}650 \times 0.707} = 0.8$ cm (leg size of fillet weld)
where $f_{\text{allowable}} = 0.707b\ \sigma_{\text{allowable}}$ for a static load. For transverse fillets f_{all} is about 20% higher than for parallel fillets, as shown in Example 6.3. If a combined load is applied, however, AWS recommends the use of the more conservative f_{all} for parallel welds.

Specifying the weld on the drawing is an important parameter in controlling the design. AWS has a standard symbol for welds, shown in Fig. 6.19. It is important to remember that welds on drawings are only shown as lines with the arrowhead pointing to the line. Also, for fillet welds the right-triangle right angle is always first (on left side), no matter in which direction the arrowhead points.

Distortion caused by high heat gradients creates problems because steel changes during heating from a body-center cubic atomic structure to a face-center cubic. During cooling the solid part restrains the plastic portion as the

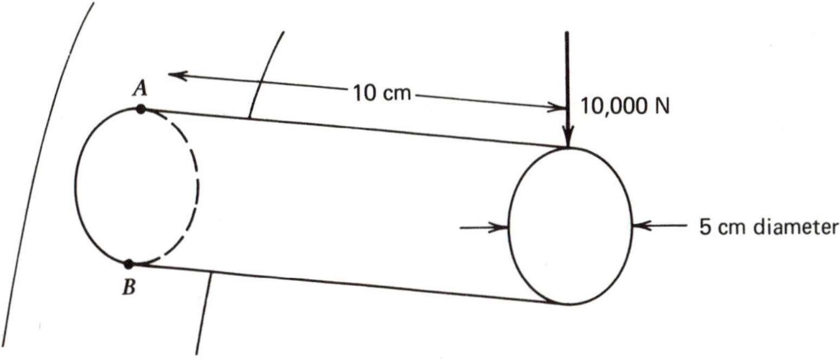

Figure 6.18 Weld as a line (Example 6.4).

Table 6.10 **Bending Properties of Fillet Welds**

Weld	Location of CG	Unit Moment of Inertia	Section Modulus About $x-x$ Axis
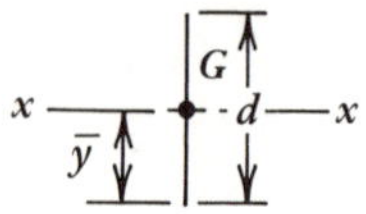	$\bar{x}=0$ $\bar{y}=d/2$	$I=\frac{d^3}{12}$	$Z_w=\frac{d^2}{6}$
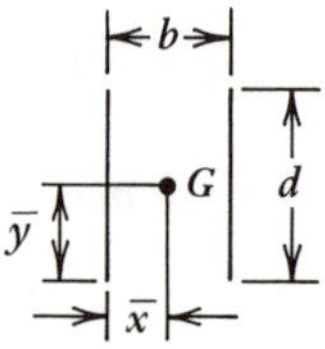	$\bar{x}=b/2$ $\bar{y}=d/2$	$I=\frac{d^3}{6}$	$Z_w=\frac{d^2}{3}$
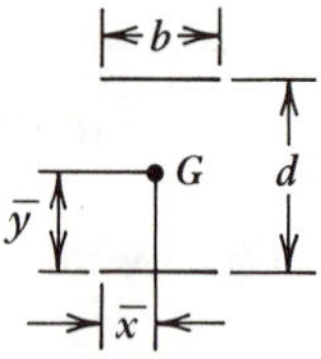	$\bar{x}=b/2$ $\bar{y}=d/2$	$I=\frac{bd^2}{2}$	$Z_w=bd$
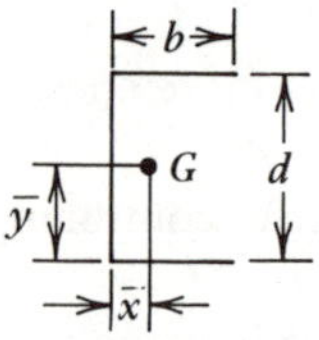	$\bar{x}=\frac{b^2}{2b+d}$ $\bar{y}=d/2$	$I=\frac{d^2}{12}(6b+d)$	$Z_w=bd^2+\frac{d^2}{6}$
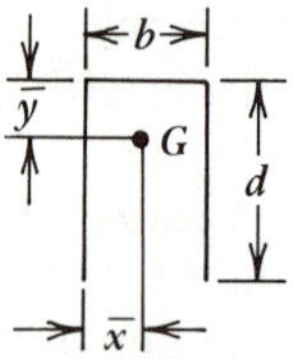	$\bar{x}=b/2$ $\bar{y}=\frac{d^2}{b+2d}$	$I=\frac{2d^3}{3}-2d^2\bar{y}+(b+2d)\bar{y}^2$	$Z_w=\frac{2bd+d^2}{3}$ Top
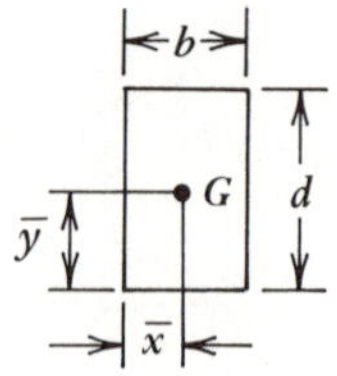	$\bar{x}=b/2$ $\bar{y}=d/2$	$I=\frac{d^2}{6}(3b+d)$	$Z_w=bd+\frac{d^2}{3}$
		$I=\frac{\pi r^3}{2}$	$Z_w=\frac{\pi d^2}{4}$

Note: The unit moment of inertia, I is taken about a horizontal axis through the centroid of the weld group *CG*.

Table 6.11 **Torsional Properties of Fillet Welds**

Weld	*Location of CG*	*Unit Polar Moment of Inertia*
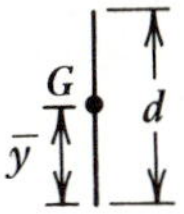	$\bar{x}=0$ $\bar{y}=d/2$	$J_w=d^3/12$
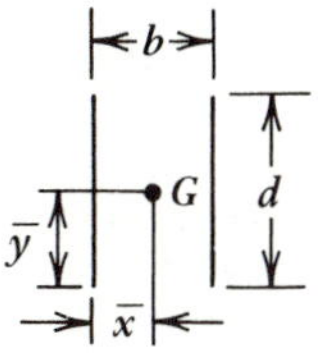	$\bar{x}=b/2$ $\bar{y}=d/2$	$J_w=\dfrac{d(3b^2+d^2)}{6}$
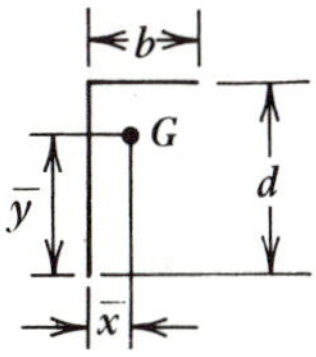	$\bar{x}=\dfrac{b^2}{2(b+d)}$ $\bar{y}=\dfrac{d^2}{2(b+d)}$	$J_w=\dfrac{(b+d)^4-6b^2d^2}{12(b+d)}$
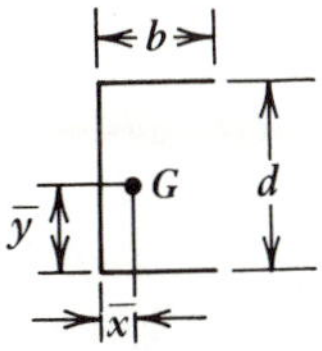	$\bar{x}=\dfrac{b^2}{2b+d}$ $\bar{y}=d/2$	$J_w=\dfrac{8b^3+6bd^2+d^3}{12}-\dfrac{b^4}{2b+d}$
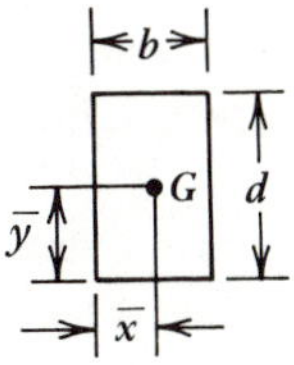	$\bar{x}=b/2$ $\bar{y}=d/2$	$J_w=\dfrac{(b+d)^3}{6}$
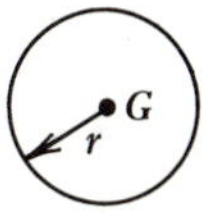		$J_w=2\pi r^3$

Note: *CG* is the centroid of the weld group.

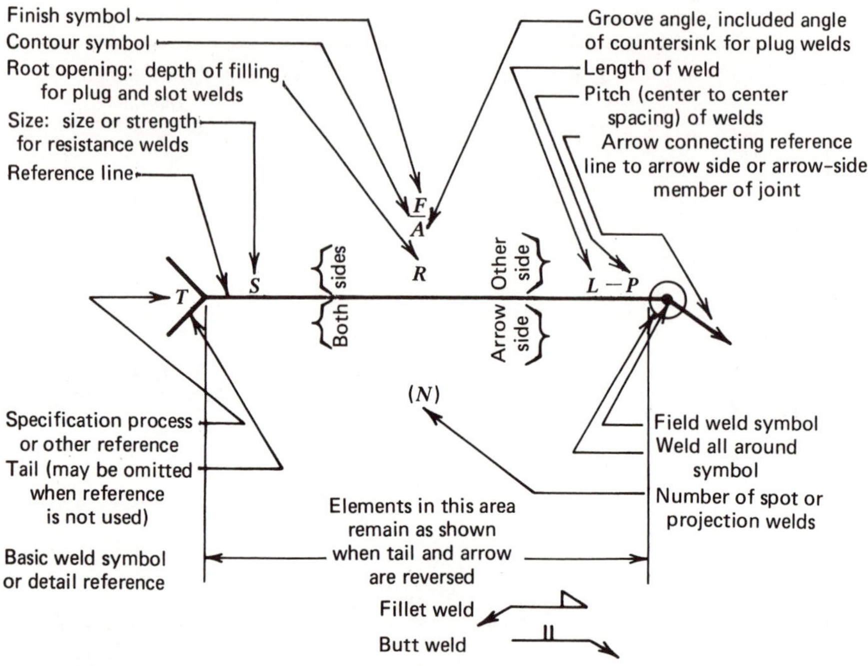

Figure 6.19 Standard AWS symbol for welds.

part cools towards the weld center, thus causing distortion. Proper positioning in a fixture can help reduce distortion problems, as shown in Fig. 6.20. The location of welds and machine settings are also important, as depicted in Fig. 6.21.

6.3 SOLDERED JOINTS

Soft solders secure attachment of tin parts by a metal solvent or intermetallic solution action that takes place at relatively low temperatures. Soft solders, therefore, are not to be confused with hard solders, or brazing alloys, whose action involves the formulation of a fusion of the metal that is joined, nor are they to be confused with welding alloys, whose action again involves actual fusion of the respective metals.

Because the soldering process involves a metallurgical or metal solvent action between the solder and the metal being joined, it is obvious that a soldered joint is chemical in character rather than purely physical; that is, the attachment is formed in part by chemical action rather than by mere physical adhesion. The properties of the soldered joint therefore differ from those of the original solder because the solder is partly converted to a new and different alloy by a solvent action between the respective metals, with the formulation of a completely new metallic contact.

The solder alloy lends itself well to torsional stress and other strains, such as temperature change, without rupture of the joint.

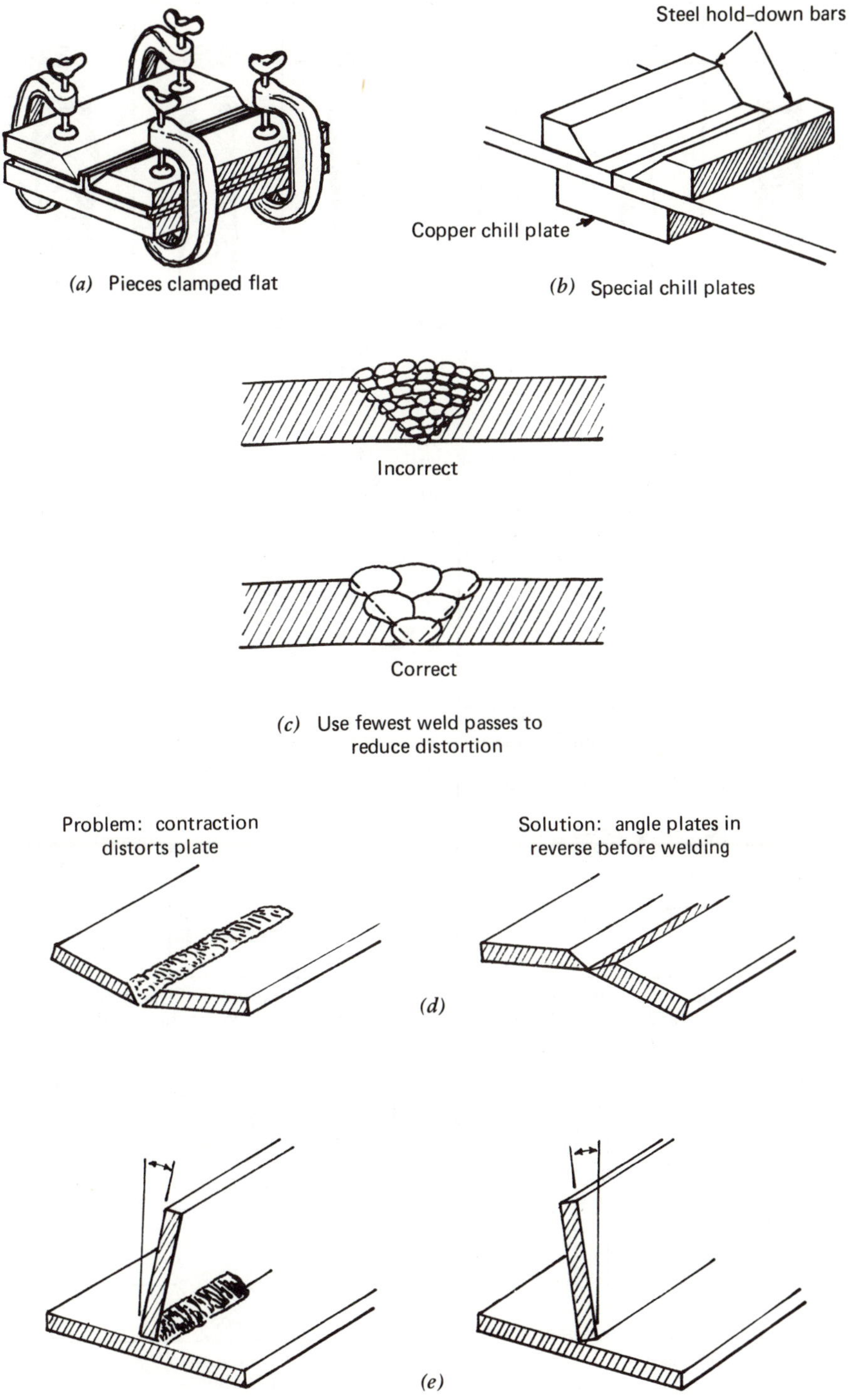

Steel hold-down bars

Copper chill plate

(*a*) Pieces clamped flat

(*b*) Special chill plates

Incorrect

Correct

(*c*) Use fewest weld passes to reduce distortion

Problem: contraction distorts plate

Solution: angle plates in reverse before welding

(*d*)

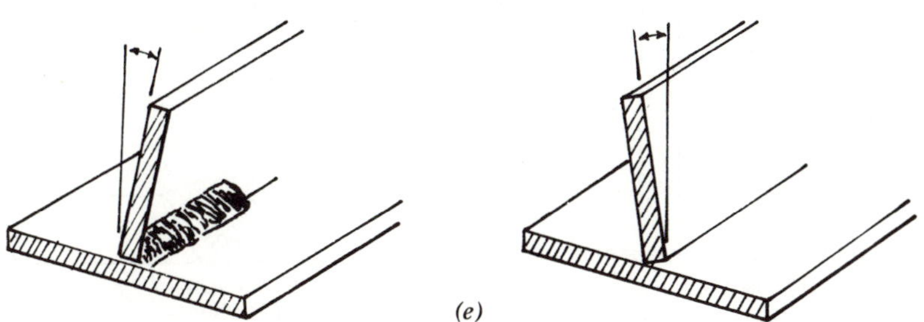

(*e*)

Figure 6.20 Proper positioning reduces weld distortion.

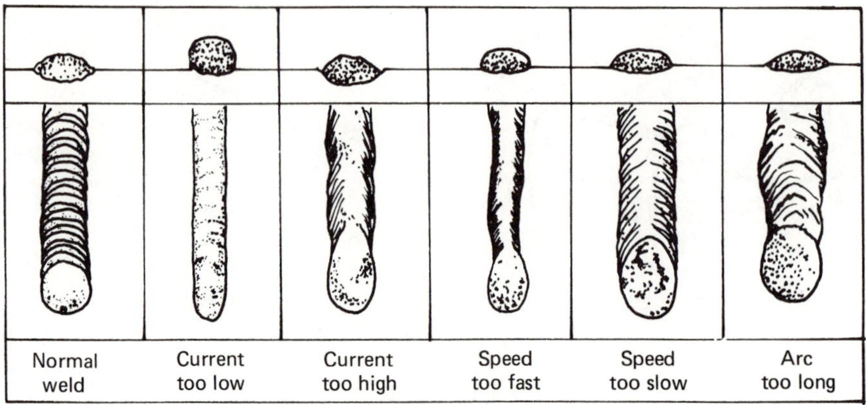

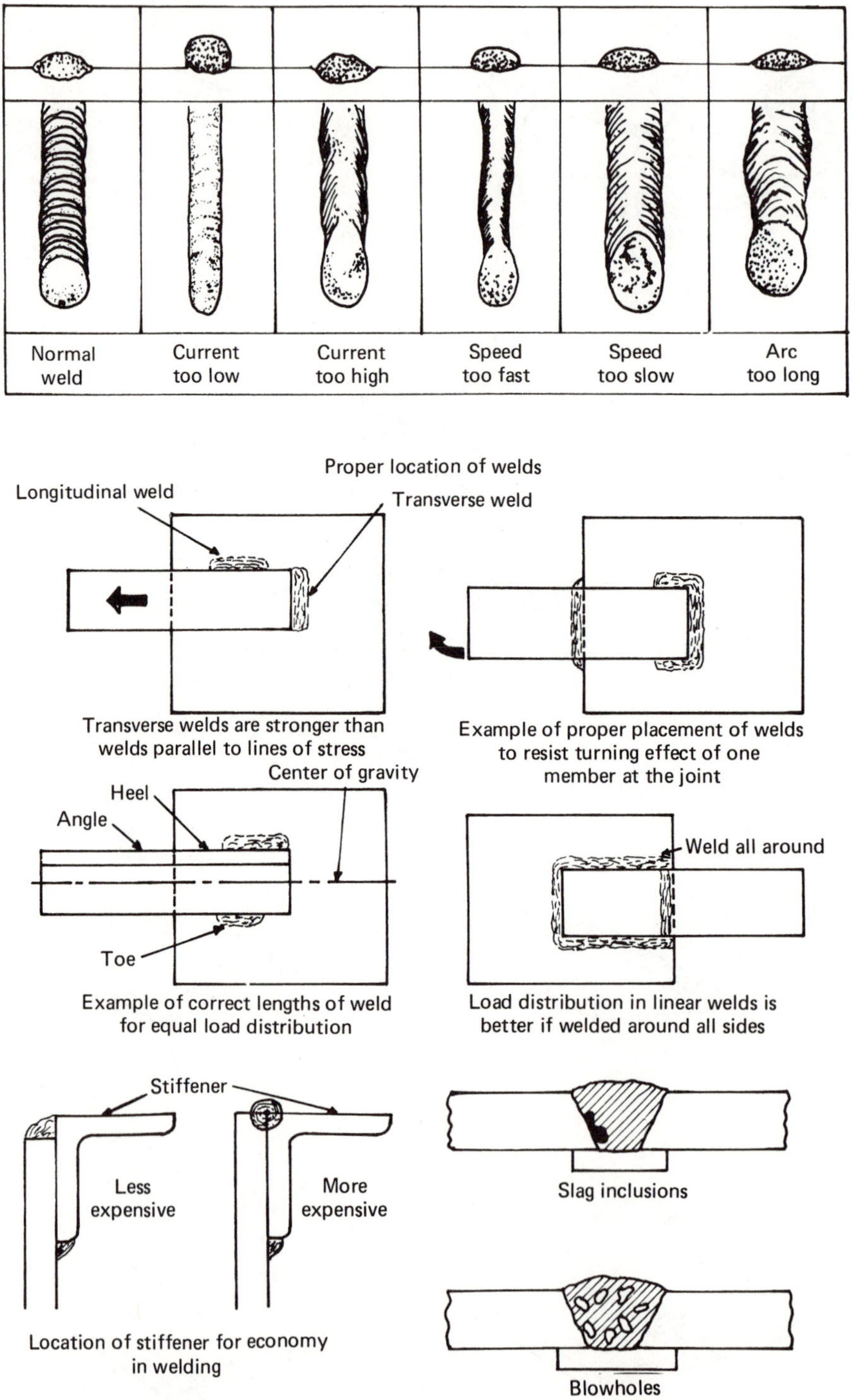

Normal weld
Current too low
Current too high
Speed too fast
Speed too slow
Arc too long
Proper location of welds
Longitudinal weld
Transverse weld
Transverse welds are stronger than welds parallel to lines of stress
Example of proper placement of welds to resist turning effect of one member at the joint
Center of gravity
Heel
Angle
Toe
Example of correct lengths of weld for equal load distribution
Weld all around
Load distribution in linear welds is better if welded around all sides
Stiffener
Less expensive
More expensive
Location of stiffener for economy in welding
Slag inclusions
Blowholes

Figure 6.21 Common problems in welding.

Because the soldering operation involves the partial creation of a new alloy between the solder and the metal that is soldered, it follows that the physical properties of this new alloy are not necessarily the same as those of the original solder. The tensile strength, shear strength, creep strength, and similar physical properties of a soldered connection depend on the extent to which alloy formation has taken place during soldering and are subject to wide variation as a result of the inherent variables in soldering technique.

The quality of a solder joint is governed by several factors, such as speed of alloy formation, flow and spread of solder, chemical stability of the soldered connection, soundness and porosity of the joint, and physical resistance to shock, strain, and stress. It would seem these apparently diverse factors could hardly be checked in any one single test. However, it should be remembered that the primary purpose of solder is to unite two or more metals in continuous metallic contact. Naturally, the tendency for this union to persist and to resist bending or torsional strain and stress reflects the joint quality of the connection. Figures 6.22 and 6.23 show the variation in content that affects a soldered joint's strength and melting point.

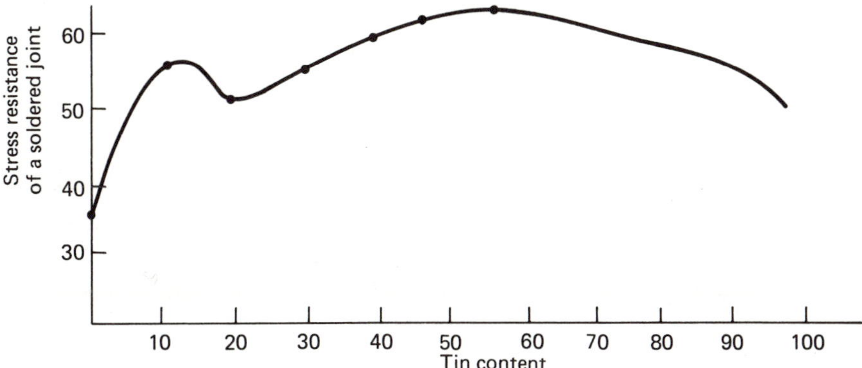

Figure 6.22 Fractional tin content of solder increases strength required to rupture a connection.

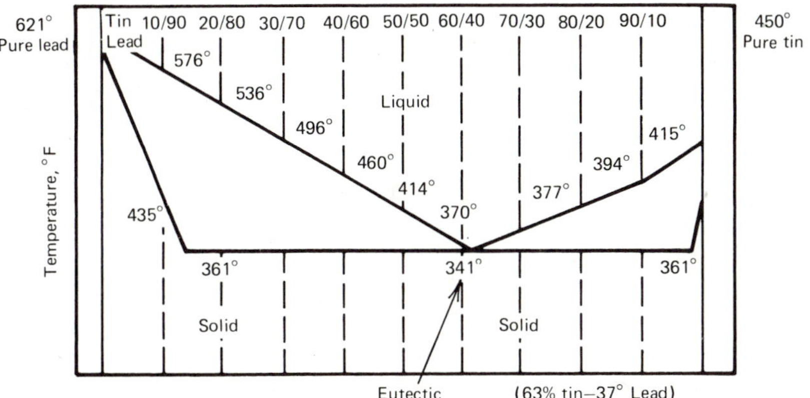

Figure 6.23 Tin-lead fusion diagram.

HOMEWORK PROBLEMS

6.1 Specify the bolt size and torque needed to attach a rollover bar to a tractor's rear axle. Use the moment diagrams supplied and assume a maximum load of 5000 lb. at the top of the rollover bar, as shown in Fig. 6.24. Show assumptions and calculations for credit.

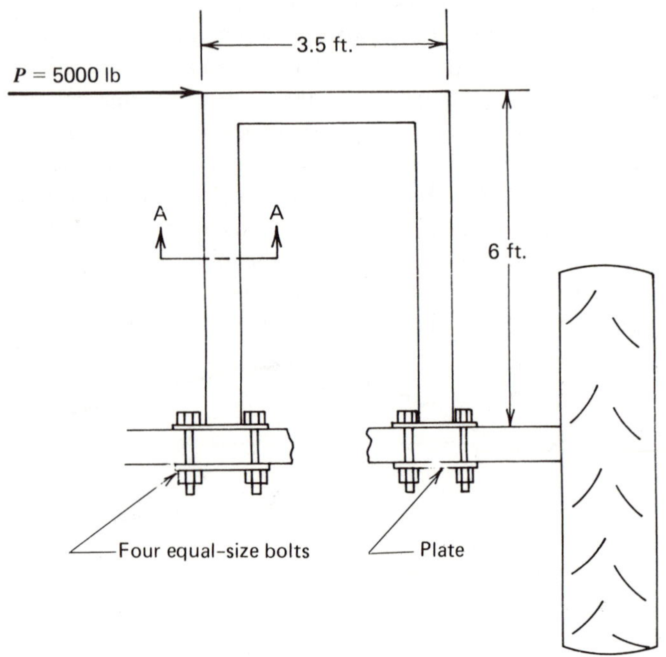

Section A-A

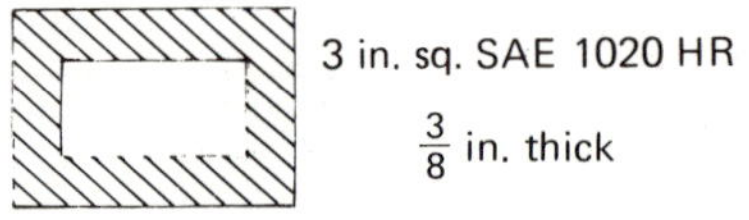

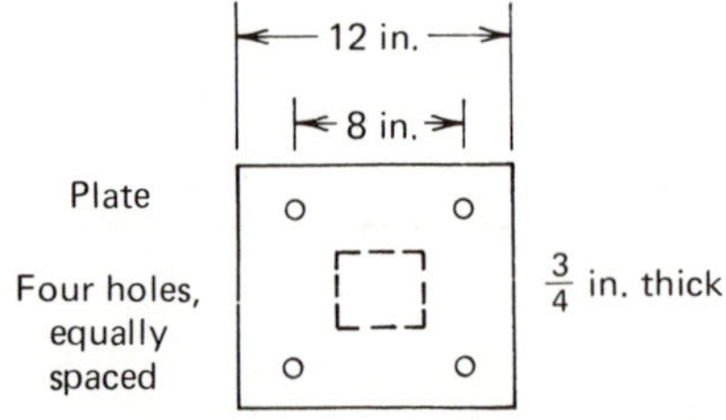

Figure 6.24 Problem 6.1.

3-member space frame

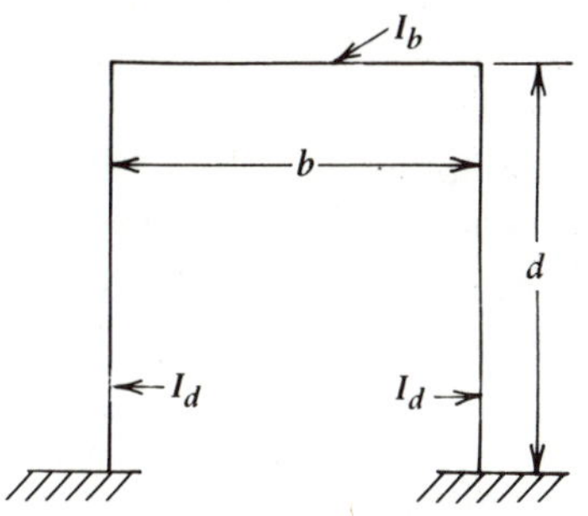

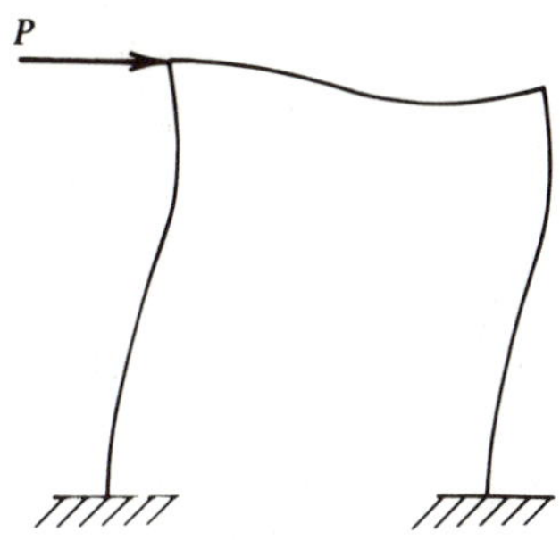

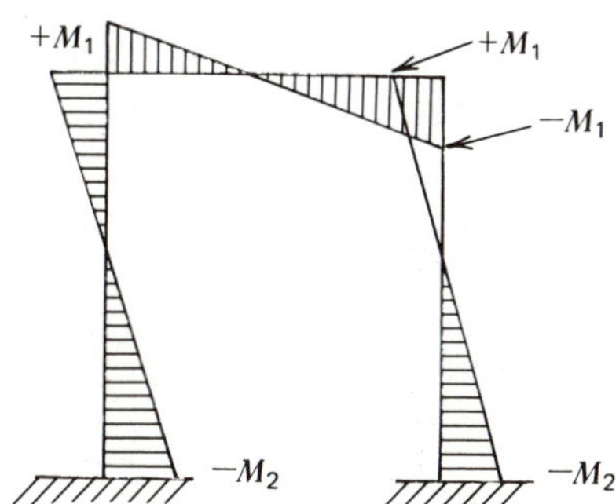

$$M_1 = \frac{Pd}{2}\left[\frac{\dfrac{3d}{I_d}}{\dfrac{b}{I_b} + \dfrac{6d}{I_d}}\right]$$

$$M_2 = \frac{Pd}{2} - M_1$$

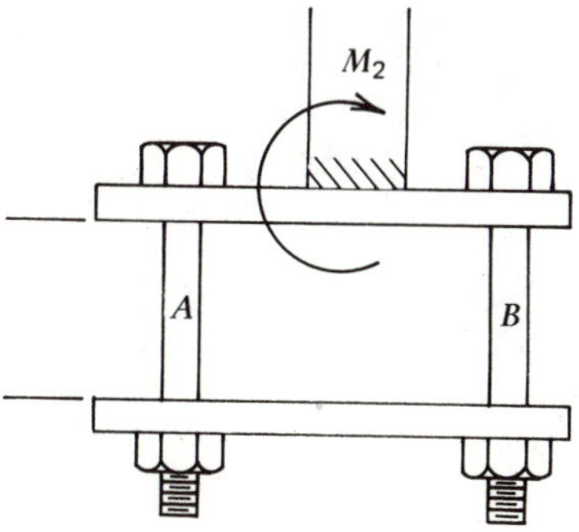

Figure 6.24 (Cont.)

6.2 Design a set of bolts on a bracket that will attach a 200-lb air conditioner to the side of a rollover bar. The tractor is traveling in plowed fields that cause dynamic loads of 3 g's. Specify the bolts needed and torque required (see Fig. 6.25).

6.3 A bolted assembly only puts axial load on a bolt preloaded to 1500 lb. If the ratio of the deflection per pound of load for the joint members to deflection per pound of load for the bolt is 1/4, what is the magnitude of the bolt load when an external 2000-lb load is applied, creating axial bolt tension.

6.4 If a clamp load of 5000 lb is needed for each bolt, specify a choice of bolt sizes and material. Assume that 10% of the external load is applied to the bolt because of a good joint design.

6.5 Redo Problem 6.4 and put the torque specifications on the bolt chosen for the load application.

6.6 Calculate the allowable load for a fillet connection as shown in Fig. 6.26. The throat depth is 0.60 in.

6.7 Determine the section modulus for a weld treated as a line for a 4 × 4 tube welded all around. (HINT: Refer to statics.)

6.8 A 2-in. diameter shaft is welded to a plate and is subjected to an axial load of 10,000 lb, 5 in. from the plate. Sketch the problem and determine the weld size.

6.9 Redo Problem 6.8 for a cyclical load of ± = 10,000 lb. Assume an infinite life. Refer to AWS standards if life is 500,000 cycles or less.

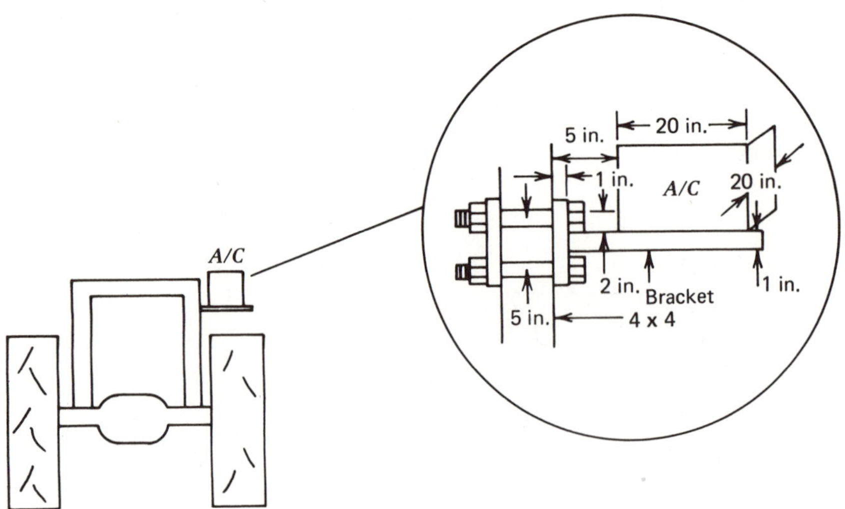

Figure 6.25 Problem 6.2.

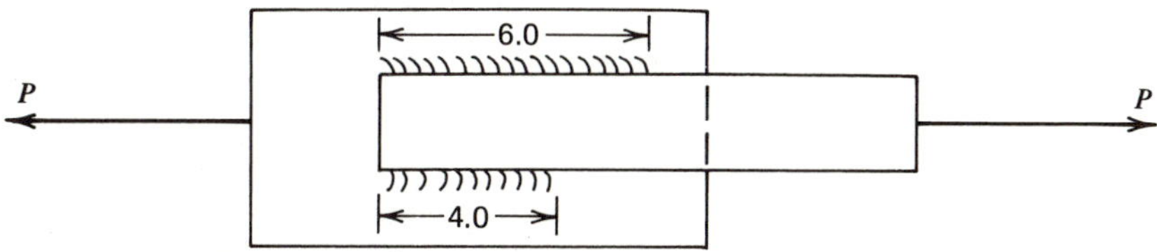

Figure 6.26 Problem 6.6.

REFERENCES

1. Aaronson, Stephen F., "Analyzing Critical Joints," *Machine Design* 54(14):95–101, 1982.
2. Aronson, R. "Solving the Fastener Preload Puzzle," *Machine Design* 50(4):100–103, 1978.
3. Batson, R. and J. Tokarski. "Keeping Fasteners Tight," *Machine Design* 47 (23):86–89, 1975.
4. Dann, R.T. "How Much Preload for Fasteners?" *Machine Design* 47(20):66–69, 1975.
5. Dewalt, W.J. and R.E. Mack. "Design Factors for Threaded Aluminum Fasteners," *Machine Design* 52(15):109–110, 1980.
6. ESNA Industrial Catalog 777, Division Amerance Corp., Union, N.J., 1980.
7. Fazekas, G. "On Optimal Bolt Preload," ASME Paper 75-WA/DE-14, *Transactions of the ASME Journal of Engineering for Industry*, The American Society of Mechanical Engineers, New York, N.Y., 1975.
8. Malcomb, R. Notes, Cummins Engine, Columbus, Ind., 1980.
9. Metric and Multistandard Components Corp., Catalog 3010, Burridge, Ill., 1980.
10. Osgood, C.C. "Saving Weight in Bolted Joints," *Machine Design* 51(24):128–133, 1979.
11. Spotts, M.F. *Design of Machine Elements*, Prentice Hall, Inc., Englewood Cliffs, N.J., 1978.
12. Stone, W. "Predicting Initial Bolt Load," *Machine Design* 54(3):111–112, 1982.
13. The Technology of Threaded Fasteners, Bulletin 729, Loctite Corp., Newington, Conn., 1974.
14. *Welding Handbook*, 4th edition, American Welding Society, New York, N.Y., 1960.
15. Zelenda Catalog, No. 1010, Metric, Forest Hills, N.Y., 1975.

CHAPTER SEVEN

FINITE ELEMENT ANALYSIS

7.1 INTRODUCTION

The finite-element method is fast becoming the most-used engineering design tool. All major farm equipment manufacturers make finite-element programs available to their engineers. Because these programs have been written by engineers who have advanced technical knowledge in their areas of expertise, they sometimes pose a problem to engineers who have little background in finite elements and other dynamic-simulation computer programs. The design engineer must rely on a program written by others and use the analysis technique as a black box–entering input and receiving output.

This chapter provides some background on how the finite-element technique works. Readers are also advised to review the references and consult with experts. The degree of success in using a finite-element program is related to the assumptions and models developed by the program originator. Understanding the technical aspects of the method requires a background in calculus of variations as well as graduate-level courses in partial differential equations and plasticity.

The finite element method is an approximation of true results. One application would be estimating deflections of a machine component given some field loading conditions. This method predicts the theoretical solution to a certain degree of accuracy. If the model predicts to 10% of what actually happens, the user should feel confident in applying the method. If the degree of error is greater than 10%, models must be refined considerably and higher costs will result. There is a break-even point at which an engineering judgment is needed (see Fig. 7.1). With this method, components that formerly required many years of engineering analysis can be modeled and analyzed in hours, and engineers can redesign and reduce the costs of components with a higher degree of reliability. Recent advances include interactively designing a part on a CRT graphics terminal, analyzing the same part with finite elements, and making the tooling from the same computer data file. This process is called computer-aided design/computer-aided manufacturing (CAD/CAM) or com-

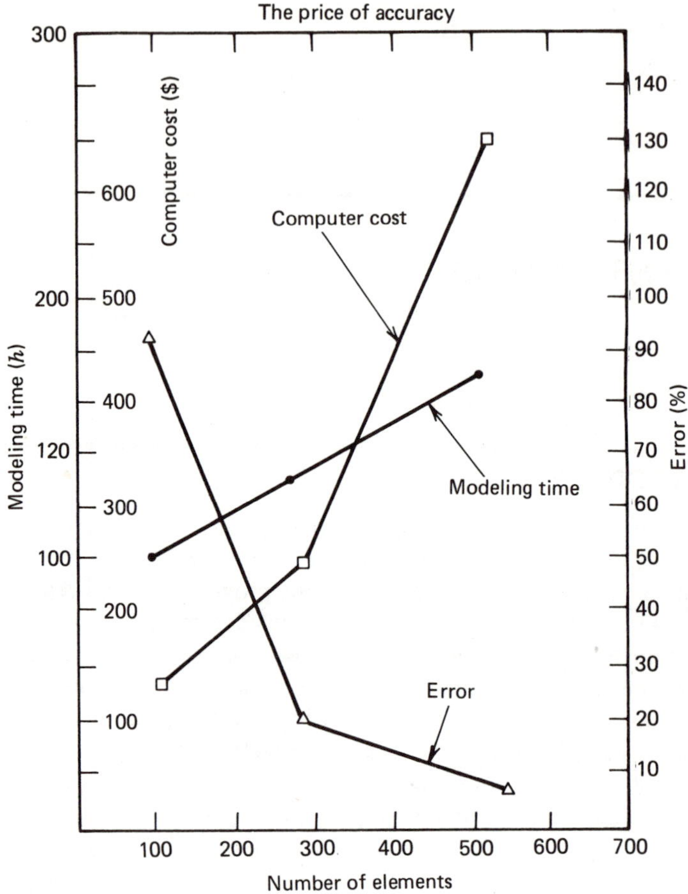

Figure 7.1 Trade-offs for accuracy using finite elements.

puter aided engineering (CAE). Interactive graphics and CRT terminals help reduce potential input problems before the finite-element simulation and reduce the need for a large number of output pages (see Figs. 7.2 and 7.3).

7.2 BASIC CONCEPTS

The finite-element method is a technique by which a continuous quantity, such as a partial differential equation, can be approximated by a discrete model composed of a set of piecewise continuous functions. This discretization is made on a small but finite portion of a large component. Figure 7.4 shows a fin that has been modeled by four elements. On each end of a one-dimensional element is an arbitrarily numbered node. The overall model is composed of a finite number of nodes. At each node the value of the variable (i.e., pressure, flow, temperature, deflections) from the differential equation is either known or

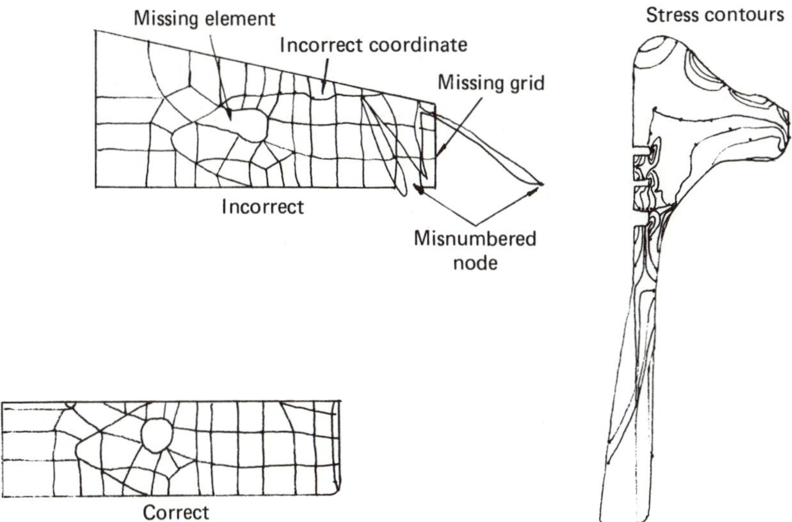

Figure 7.2 Visual review of finite-element output reduced solution errors.

Figure 7.3 Finite-element output on CRT shows areas of high stress concentrations.

unknown. Between the nodes are subdomains, called elements. Mathematics converts the partial differential equation into matrix equations that describe each element. The individual elements (in matrix form) are joined together to form the overall model. The results of a finite element simulation output are given in terms of the variable at each nodal point. After the node variables are obtained, internal quantities can be estimated. A sample structural problem would solve for the deflection on a loaded component and then evaluate the stresses inside each element.

The main advantage of the finite-element method is that a computer program can be written to create the overall model matrix; for different applications (e.g., shape geometry) input is easily changed and a new solution obtained with the same computer program. Other important advantages include:

1. Each element can have different material properties (see Fig. 7.5).
2. Irregular-shape geometrics can be modeled.
3. The size of elements can be varied.
4. Mixed boundary conditions can be handled (e.g., loads and temperatures).

The number of nodes in an element can vary (see Fig. 7.6) with the addition of internal nodes that provide a means to curve the element or specify a nonlinear variable. Many other types of elements are available (see Fig. 7.7). Again, the user must realize that the accuracy of the results from a finite-element model are a function of engineering judgment in selecting the element type and size and of the computer program and its associated mathematical assumptions.

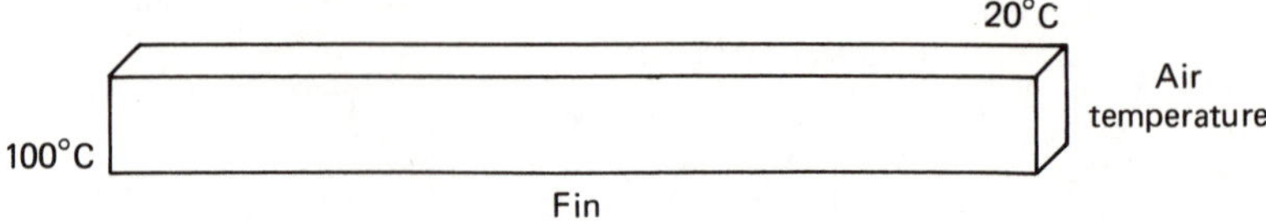

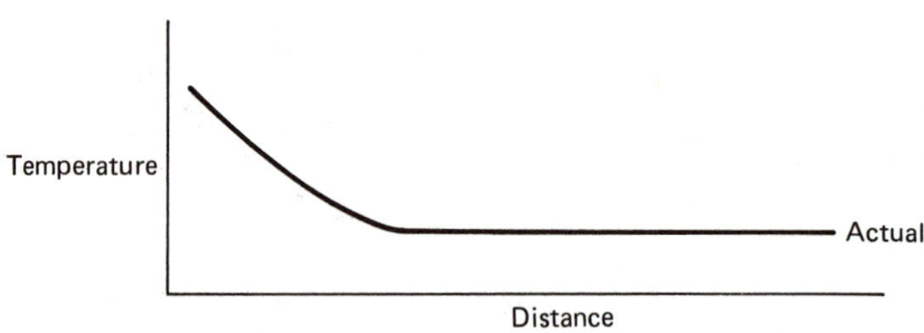

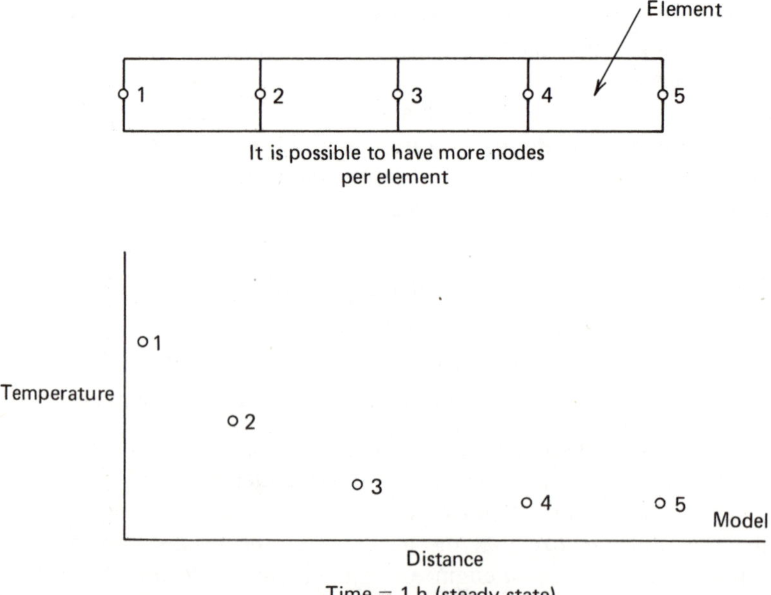

Figure 7.4 A four-element fin model with five nodes.

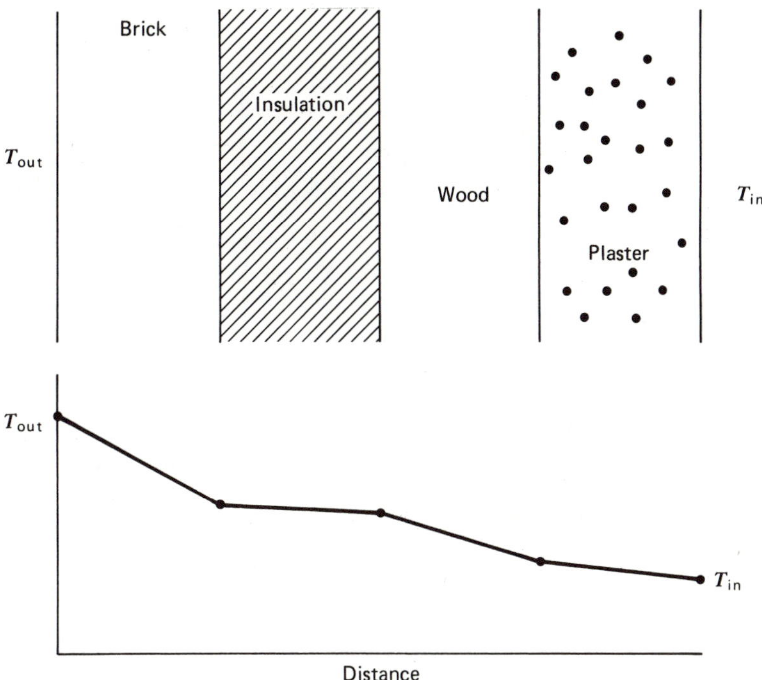

Figure 7.5 Different material properties are modeled using finite elements.

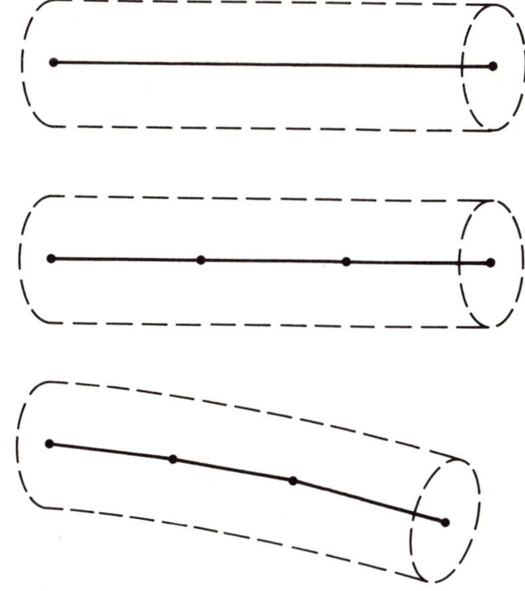

Figure 7.6 Types of one-dimensional elements common in flow and truss problems.

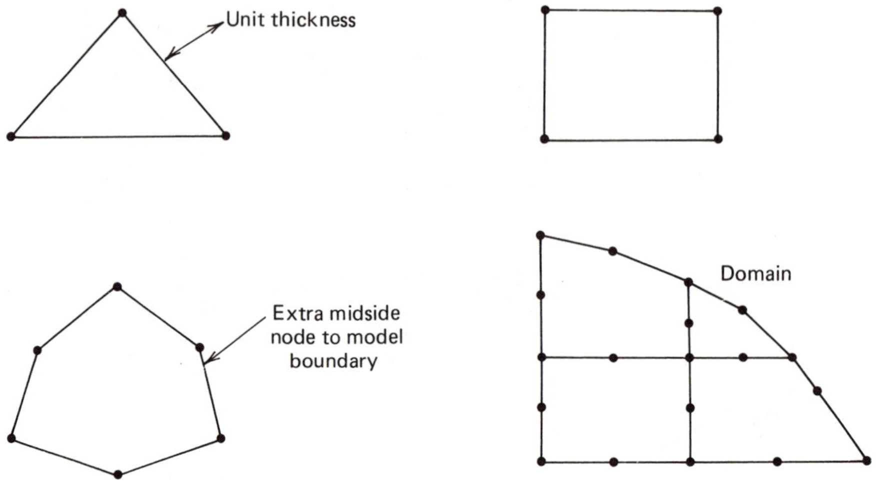

Unit thickness

Extra midside node to model boundary

Domain

(a)

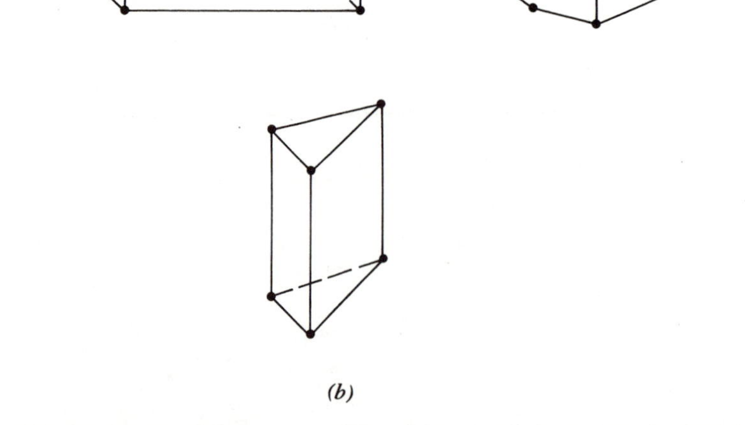

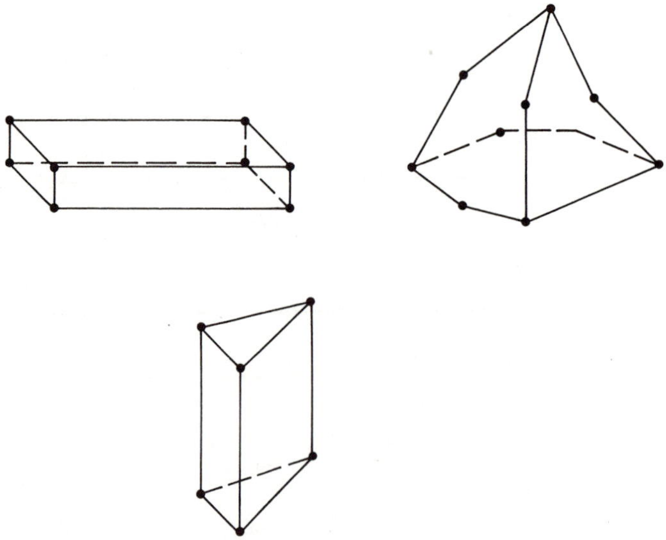

(b)

***Figure** 7.7* Many types of elements. (*a*) Two-dimensional elements and a domain. (*b*) Three-dimensional elements.

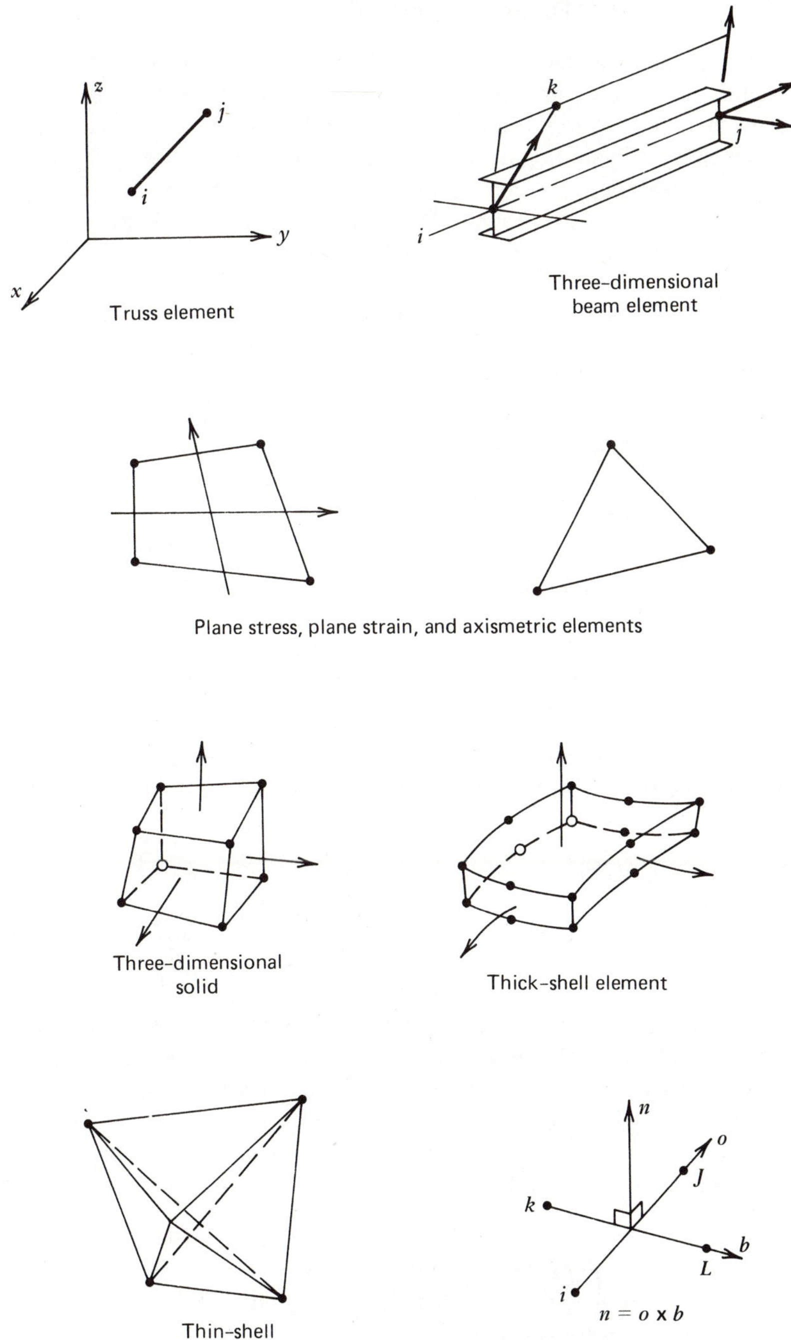

***Figure** 7.7* (Cont.) Many types of elements. (*c*) Types of elements available in structural analysis program version IV.

7.3 TRUSS AND BEAM ELEMENTS

The finite-element technique evolved when structural engineers successfully applied matrices to strength of materials during the early 1960s. Mathematicians then verified solution convergence and the other theoretical aspects, both of which led to variational calculus forms. These variational forms provided a base for developing matrices that model heat transfer and flow problems.

The spring analogy provides the basic concept of the stiffness matrix. The stiffness of the spring in Fig. 7.8 is defined as

$$K = \frac{F}{\delta} \qquad [7.1]$$

where K = stiffness (N/m)
F = force (N)
δ = deflection (m)

Specifying a boundary condition at node 2 of

$$\delta_2 = 0 \qquad [7.2]$$

then the spring stiffness can be found

$$K = F_1/\delta_1$$

Also, from $\sum F = 0$

$$F_1 = -F_2 = K\delta_1 \qquad [7.3]$$

Similarly, setting $\delta_1 = 0$, another set of equilibrium equations can be written.

$$F_2 = -F_1 = K\,\partial_2 \qquad [7.4]$$

Now these equilibrium equations are converted into matrix form.

$$\begin{Bmatrix} F_1 \\ F_2 \end{Bmatrix} = \begin{bmatrix} K & -K \\ -K & K \end{bmatrix} \begin{Bmatrix} \partial_1 \\ \partial_2 \end{Bmatrix} \qquad [7.5]$$

The values in parentheses on the left are called the force vector; those in brackets, the stiffness matrix; and the deflections in parentheses on the right, the displacement vector. A matrix review is included in Section 7.5.

If two springs are connected (see Fig. 7.9), they can be put into matrix form again by writing the equilibrium equations. A simpler method allows direct construction by superposition. The force and the displacement vector will con-

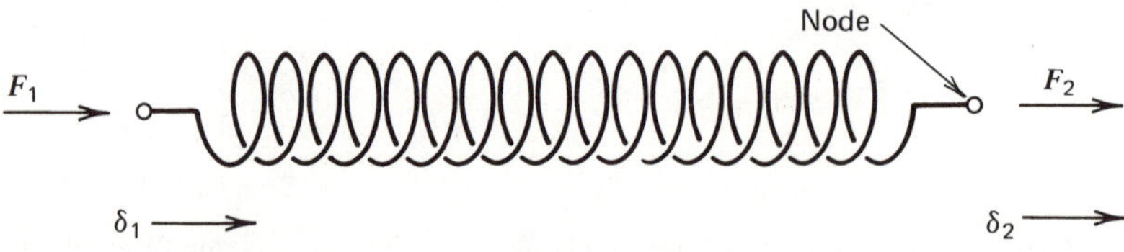

Figure 7.8 A one-dimensional spring element.

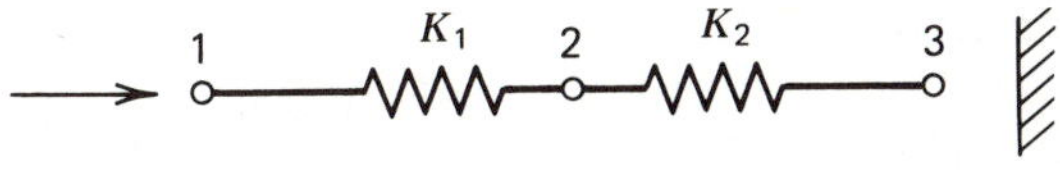

Figure 7.9 A two-spring structural case with three nodes.

tain one variable for each node. The stiffness matrix must be of order equal to the number of nodes, n, and is therefore $n \times n$. Because the spring is a one-dimensional element, its stiffness matrix can easily be written, as in Eq. 7.5, for each element. The matrix for element 2 is

$$\begin{Bmatrix} F_2 \\ F_3 \end{Bmatrix} = \begin{bmatrix} K_2 & -K_2 \\ -K_2 & K_2 \end{bmatrix} \begin{Bmatrix} \partial_2 \\ \partial_3 \end{Bmatrix} \tag{7.6}$$

The general matrix for both elements 1 and 2 is obtained by listing the forces and displacements in their respective vectors. The stiffness matrix constructed by superposition (i.e., location $K_{22} = K_1 + K_2$ and location $K_{13} = 0$)

$$\begin{Bmatrix} F_1 \\ F_2 \\ F_3 \end{Bmatrix} \begin{bmatrix} K_1 & -K_1 & 0 \\ -K_1 & K_1+K_2 & -K_2 \\ 0 & -K_2 & K_2 \end{bmatrix} \begin{Bmatrix} \partial_1 \\ \partial_2 \\ \partial_3 \end{Bmatrix} \tag{7.7}$$

Expanding Eq. 7.7 will give the general equilibrium equations for the two-spring system. Note that the values in the rows or columns add up to zero for this stiffness matrix. The solution of unknown forces and displacements is obtained by inputting the actual spring rates, known forces, and boundary conditions into Eq. 7.7. One of many computer solution techniques can then be used to output the answer. One technique, Gauss elimination, saves on computer time over matrix-inverse methods.

Example 7.1 Setting Up Matrices (see Fig. 7.10)

For the following spring system set up the general matrix equation for the entire system of four nodes and three elements.

For element 1-2

$$\begin{Bmatrix} F_1 \\ F_2 \end{Bmatrix} = \begin{bmatrix} K_{12} & -K_{12} \\ -K_{12} & K_{12} \end{bmatrix} \begin{Bmatrix} \partial_1 \\ \partial_2 \end{Bmatrix} \tag{7.8}$$

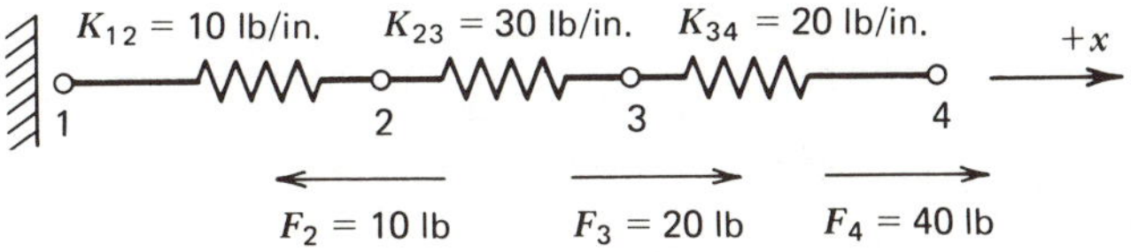

Figure 7.10 Example 7.1.

By direct construction

$$\begin{Bmatrix} F_1 \\ F_2 \\ F_3 \\ F_4 \end{Bmatrix} = \begin{bmatrix} K_{12} & -K_{12} & 0 & 0 \\ -K_{12} & K_{12}+K_{23} & -K_{23} & 0 \\ 0 & -K_{23} & K_{23}+K_{34} & -K_{34} \\ 0 & 0 & -K_{34} & K_{34} \end{bmatrix} \begin{Bmatrix} \partial_1 \\ \partial_2 \\ \partial_3 \\ \partial_4 \end{Bmatrix} \quad [7.9]$$

Substituting numerical values into Eq. 7.9 yields

$$\begin{Bmatrix} F_1 \\ -10 \\ 20 \\ 40 \end{Bmatrix} = \begin{bmatrix} 10 & -10 & 0 & 0 \\ -10 & 40 & -30 & 0 \\ 0 & -30 & 50 & -20 \\ 0 & 0 & -20 & 20 \end{bmatrix} \begin{Bmatrix} 0 \\ \partial_2 \\ \partial_3 \\ \partial_4 \end{Bmatrix} \quad [7.10]$$

Engineers have developed more complex elements than springs to be used in the finite-element modeling process. The simplest, the truss element, allows only one degree of freedom at each node (see Fig. 7.11).

From strength of materials, the deflection of a truss is

$$\partial = \frac{FL}{AE} \quad [7.11]$$

Then, finding the stiffness of a truss element and using Eq. 7.9,

$$K = F/\partial = \frac{AE}{L} \quad [7.12]$$

Since the truss element has uniaxial properties similar to the spring, its stiffness matrix can be written by combining Eqs. 7.5 and 7.12.

$$\frac{AE}{L} \begin{bmatrix} 1 & -1 \\ -1 & 1 \end{bmatrix} \quad [7.13]$$

The stiffness is the force required to produce a unit deflection in the direction specified. The overall set of equilibrium equations for one truss element is

$$\begin{Bmatrix} F_1 \\ F_2 \end{Bmatrix} = \frac{AE}{L} \begin{bmatrix} 1 & -1 \\ -1 & 1 \end{bmatrix} \begin{Bmatrix} \partial_1 \\ \partial_2 \end{Bmatrix} \quad [7.14]$$

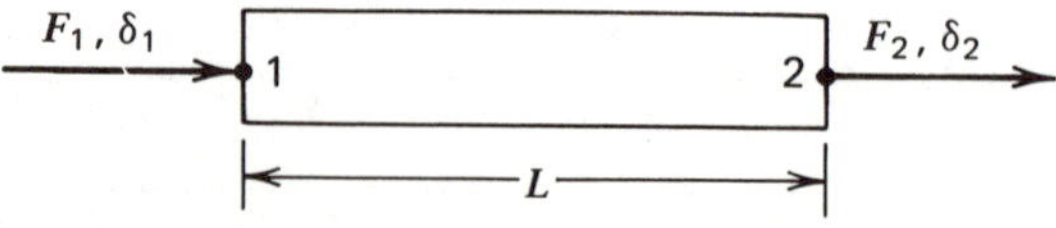

Figure 7.11 Truss element.

A more complex element, called a beam element, has been developed. In two dimensions the beam element has forces and moments at each node, resulting in rotation and deflections. The stiffness matrix for the beam element shown in Fig. 7.12 can be developed from strength of materials.

$$\begin{aligned}\delta_1 &= \frac{F_1L^3}{3EI} + \frac{M_1L^2}{2EI}\\ \theta_1 &= \frac{F_1L^2}{2EI} + \frac{M_1L}{EI}\end{aligned} \qquad [7.15]$$

where θ_1 = rotation at node 1
M_1 = the moment at node 1

Reconstructing the equations in matrix form similar to Eq. 7.5 results in

$$\begin{Bmatrix} F_1 \\ M_1 \end{Bmatrix} = EI \begin{bmatrix} 12/L^3 & -6/L^2 \\ -6/L^2 & 4/L \end{bmatrix} \begin{Bmatrix} \partial_1 \\ \theta_1 \end{Bmatrix} \qquad [7.16]$$

Note that Σ rows or Σ columns do not equal zero because the force vector and deflection vector have mixed terms.
Expanding to include both nodes on the beam, (see Fig. 7.13), a generalized form is written as

$$\begin{Bmatrix} F_1 \\ M_1 \\ F_2 \\ M_2 \end{Bmatrix} = EI \begin{bmatrix} 12/L^3 & & \text{symmetrical} & \\ -6/L^2 & 4/L & & \\ -12/L^3 & 6/L^2 & 12/L^3 & \\ -6/L^2 & 2/L & 6/L^2 & 4/L \end{bmatrix} \begin{Bmatrix} \partial_1 \\ \theta_1 \\ \partial_2 \\ \theta_2 \end{Bmatrix} \qquad [7.17]$$

This beam element has only two degrees of freedom per node, δ_1 (in x direction) and θ_1. Beams with six degrees of freedom are common in most computer programs. They have a ∂_x, ∂_y, ∂_z, θ_x, θ_y, and θ_z in three dimensions, allowing for forces, moments, and torques at each node.

The stiffness matrix can also be constructed using strain energy and Castigliano's theorem. The strain energy is written as

$$U = \frac{1}{2} F' \partial \qquad [7.18]$$

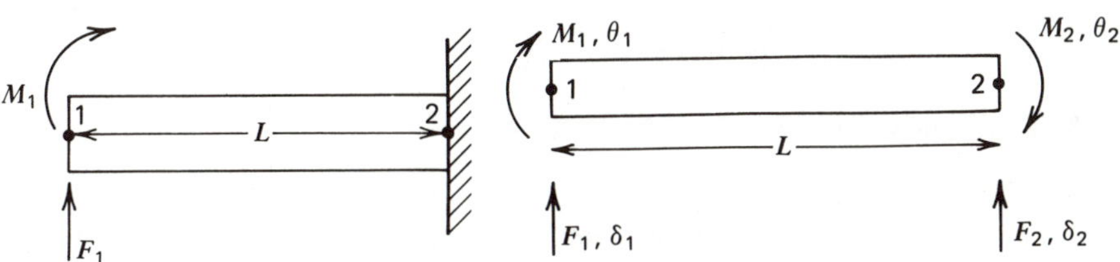

Figure 7.12 Cantilever beam with external moment and force.

Figure 7.13 Generalized beam element.

where F' is the average force. Substituting for F' into Eq. 7.18, the strain energy is now written as

$$U = \frac{1}{2}\ \partial'\ K\ \partial \qquad [7.19]$$

where ∂' = average deflection

and the elements of the stiffness matrix are given by

$$K_{ij} = \frac{\partial^2 U}{\partial_i\ \partial_j} \qquad [7.20]$$

If the strain energy from Castigliano's theorem is expressed as

$$U = \frac{EI}{2} \int_0^L \left[\frac{\partial^2 V}{\partial x^2}\right]^2 dx \qquad [7.21]$$

and the shear V is modeled as a cubic, the stiffness matrix can be derived from Eq. 7.20, as shown in Eq. 7.15. This energy method has become the common practice for developing stiffness matrices for different types of elements.

Example 7.2 Modeling a Structure in Matrix Form

Using the direct construction method, set up the stiffness matrix, force vector, and deflection vector for Fig. 7.14, a three-element structure built of wood blocks. Using a truss element with one-degree of freedom defining area-over-length ratios for each element,

$$\frac{A_{(1)}}{L_{(1)}} = 0.5 \qquad \frac{A_{(2)}}{L_{(2)}} = 0.67 \qquad \frac{A_{(3)}}{L_{(3)}} = 0.75 \qquad [7.22]$$

where (1) = element number

The overall matrix for $E = 2 \times 10^6$ psi for wood can be written as

$$\begin{Bmatrix} F_1 \\ F_2 \\ F_3 \\ F_4 \end{Bmatrix} = E \begin{bmatrix} 0.5 & -0.5 & 0 & 0 \\ -0.5 & 1.17 & -0.67 & 0 \\ 0 & -0.67 & 1.42 & -0.75 \\ 0 & 0 & -0.75 & 0.75 \end{bmatrix} \begin{Bmatrix} \partial_1 \\ \partial_2 \\ \partial_3 \\ \partial_4 \end{Bmatrix} \qquad [7.23]$$

Again, note the symmetry and that Σ row and Σ column values equal zero.

7.4 TRANSFORMATION MATRIX

The stiffness matrix previously developed assumed that all the elements were on one axis. In most engineering finite element problems, elements at different angles to each other are prevalent. Therefore, the local stiffness matrices must be transformed to a general coordinate system before they can be added together to form the general stiffness matrix by the direct-construction method.

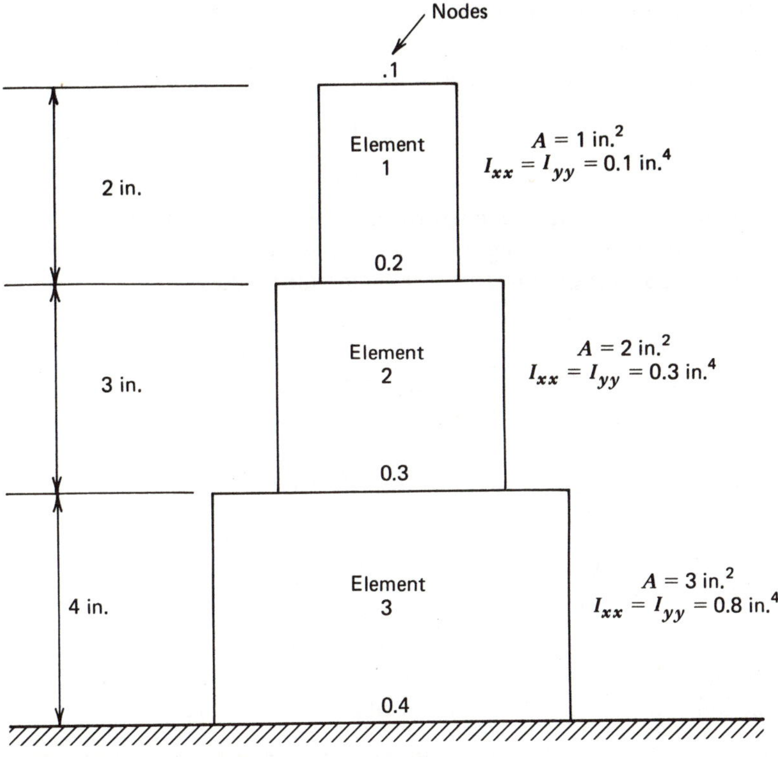

Figure 7.14 Example 7.2.

A truss element is placed at angle θ relative to the general coordinate system (see Fig. 7.15a).

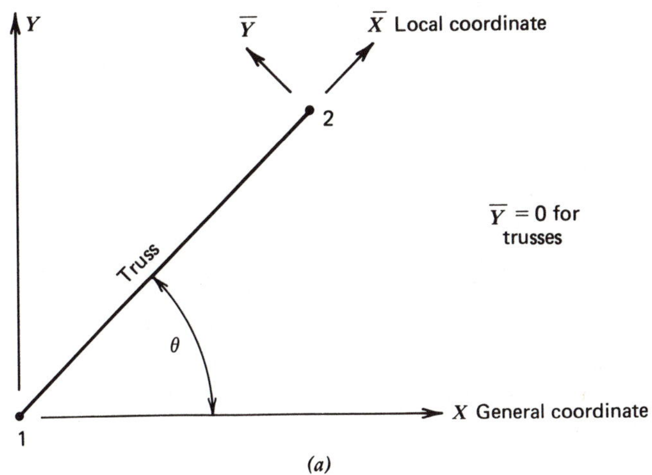

Figure 7.15 Coordinate systems. (*a*) For truss element.

The local coordinate system is defined as $\bar{X}$ and $\bar{Y}$. Writing the equilibrium equations in matrix form, where $\bar{X}$ forces relate to $\bar{U}$ displacements and, similarly, $\bar{Y}$ forces cause $\bar{V}$ displacements, then

$$\begin{matrix} \bar{X}_1 \\ \bar{X}_2 \end{matrix} = \begin{bmatrix} K & -K \\ -K & K \end{bmatrix} \begin{Bmatrix} \bar{U}_1 \\ \bar{U}_2 \end{Bmatrix} \qquad [7.24]$$

Now the equation can be expanded by including $\bar{Y}$, but noting that for trusses $\bar{Y}$ does not exist (only $\bar{X}$ along the axis). A new form is needed similar to Eq. 7.25, but with transformation factors changing the stiffness matrix

$$\begin{Bmatrix} X_1 \\ X_2 \end{Bmatrix} = \begin{bmatrix} K & -K \\ -K & K \end{bmatrix} \begin{Bmatrix} U_1 \\ U_2 \end{Bmatrix} \qquad [7.25]$$

Expanding Eq. 7.24,

$$\begin{Bmatrix} \bar{X}_1 \\ \bar{Y}_1 \\ \bar{X}_2 \\ \bar{Y}_2 \end{Bmatrix} = \begin{bmatrix} K & 0 & -K & 0 \\ 0 & 0 & 0 & 0 \\ & 0 & K & 0 \\ 0 & 0 & 0 & 0 \end{bmatrix} \begin{Bmatrix} \bar{U}_1 \\ \bar{V}_1 \\ \bar{U}_2 \\ \bar{V}_2 \end{Bmatrix} \qquad [7.26]$$

If θ is the angle of rotation between two coordinate systems (local and global), the following relationship can be written (see Fig. 7.15b).

$$\begin{aligned} \bar{Y} &= -X\sin\theta + Y\cos\theta \\ \bar{X} &= X\cos\theta + Y\sin\theta \end{aligned} \qquad [7.27]$$

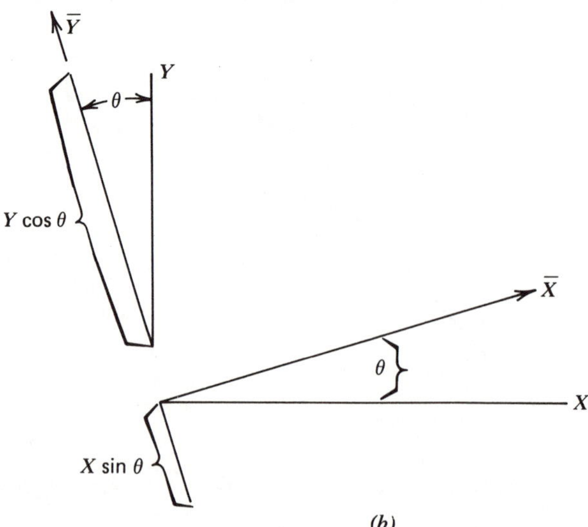

Figure 7.15 (Cont.) (*b*) Generalized case of local and global.

Note: Remember from the laws of trigonometry that sines and cosines change in each quadrant. For the forces in Fig. 7.15b, the following relationship can be written as

$$\underset{\text{local}}{\begin{Bmatrix} \bar{X} \\ \bar{Y} \end{Bmatrix}} = \underset{\text{transformation matrix}}{\begin{bmatrix} \cos\theta & \sin\theta \\ -\sin\theta & \cos\theta \end{bmatrix}} \underset{\text{global}}{\begin{Bmatrix} X \\ Y \end{Bmatrix}} \qquad [7.28]$$

setting

$$\begin{aligned} \lambda &= \cos\theta \\ \alpha &= \sin\theta \end{aligned} \qquad [7.29]$$

and substituting into Eq. 7.28, the transformation matrix $[T]$ is written as

$$\begin{bmatrix} \lambda & \alpha \\ -\alpha & \lambda \end{bmatrix} = [T] \qquad [7.30]$$

Since $[T]^T = [T]^{-1}$,

$$[T]^{-1}\,\bar{X} = [T]^{-1}[T]\,X \qquad [7.31]$$

The general force vector on the truss can be related to the local force vector by

$$\begin{Bmatrix} X_1 \\ Y_1 \\ X_2 \\ Y_2 \end{Bmatrix} = [T]^T \begin{Bmatrix} \bar{X}_1 \\ \bar{Y}_1 \\ \bar{X}_2 \\ \bar{Y}_2 \end{Bmatrix} \qquad [7.32]$$

Substituting Eq. 7.26 for the right side of the local-force vector, Eq. 7.32 becomes

$$\begin{Bmatrix} X_1 \\ Y_1 \\ X_2 \\ Y_2 \end{Bmatrix} = [T]^T \begin{bmatrix} K & 0 & -K & 0 \\ 0 & 0 & 0 & 0 \\ -K & 0 & K & 0 \\ 0 & 0 & 0 & 0 \end{bmatrix} \begin{Bmatrix} \bar{U}_1 \\ \bar{V}_1 \\ \bar{U}_2 \\ \bar{V}_2 \end{Bmatrix} \qquad [7.33]$$

A similar relationship to that shown by Eq. 7.28 can be written for the displacement vectors

$$\underset{\text{local}}{\begin{Bmatrix} \bar{U}_1 \\ \bar{V}_1 \\ \bar{U}_2 \\ \bar{V}_2 \end{Bmatrix}} = [T] \underset{\text{global}}{\begin{Bmatrix} U_1 \\ V_1 \\ U_2 \\ V_2 \end{Bmatrix}} \qquad [7.34]$$

Combining Eqs. 7.33 and 7.34 yields the relationship needed to transform the local stiffness matrix K (if the element is at an angle) into the global coordinates

$$\begin{Bmatrix} X_1 \\ Y_1 \\ X_2 \\ Y_2 \end{Bmatrix} = [T]^T [K][T] \begin{Bmatrix} U_1 \\ V_1 \\ U_2 \\ V_2 \end{Bmatrix} \qquad [7.35]$$

where $[K]$ is in local coordinates.
Equation 7.35 will now allow determination of the global stiffness matrix for angled elements by direct construction.

Example 7.3 Using the Transformation matrix
Write the stiffness matrix for the following structure and solve for the unknown displacements and forces (Fig. 7.16).

$$[K]_{\text{global}} = [T]^T [K]_{\text{local}} [T] \qquad [7.36]$$

or

$$[K]_{\text{global}} = \frac{AE}{L} \begin{bmatrix} \lambda^2 & \lambda\alpha & -\lambda^2 & -\lambda\alpha \\ \lambda\alpha & \alpha^2 & -\lambda\alpha & -\alpha^2 \\ -\lambda^2 & -\lambda\alpha & \lambda^2 & \lambda\alpha \\ -\lambda\alpha & -\alpha^2 & \lambda\alpha & \alpha^2 \end{bmatrix} \qquad [7.37]$$

where in Eq. 7.37, column 1 relates to U_1, column 2 relates to V_1, column 3 relates to U_2, and column 4 relates to V_2. Then the following values are determined for λ and α.

Member	θ	λ	α	λ^2	α^2	$\lambda\alpha$
1-2	0	1	0	1	0	0
1-3	90	0	1	0	1	0
2-3	135	$-1/\sqrt{2}$	$1/\sqrt{2}$	1/2	1/2	−1/2

Element 1-3 has $\alpha^2 = -1$ and $\lambda^2 = 0$. Substituting into Eq. 7.36 to determine the stiffness matrix K_{1-3},

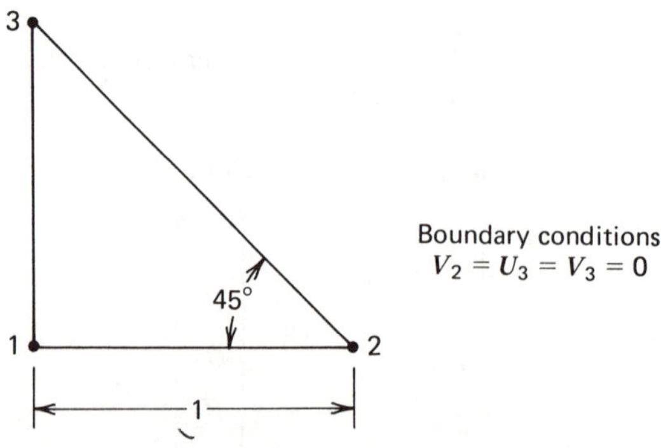

Figure 7.16 Example 7.3: A three-element truss.

$$K_{1-3} = \frac{AE}{L}\begin{bmatrix} 0 & 0 & 0 & 0 \\ 0 & 1 & 0 & -1 \\ 0 & 0 & 0 & 0 \\ 0 & -1 & 0 & 1 \end{bmatrix} \qquad [7.38]$$

also

$$K_{1-2} = \frac{AE}{L}\begin{bmatrix} 1 & 0 & -1 & 0 \\ 0 & 0 & 0 & 0 \\ 1 & 0 & 1 & 0 \\ 0 & 0 & 0 & 0 \end{bmatrix} \qquad [7.39]$$

and

$$K_{2-3} = \frac{AE}{\sqrt{2}L}\begin{bmatrix} 1/2 & -1/2 & -1/2 & 1/2 \\ -1/2 & 1/2 & 1/2 & -1/2 \\ -1/2 & 1/2 & 1/2 & -1/2 \\ 1/2 & -1/2 & -1/2 & 1/2 \end{bmatrix} \qquad [7.40]$$

where the element 2-3 length is $\sqrt{2}L$. Combining all three elements by direct construction and substituting

$$r = \frac{1}{2\sqrt{2}} \qquad [7.41]$$

the general matrix results as

$$\begin{Bmatrix} X_1 \\ Y_1 \\ X_2 \\ Y_2 \\ X_3 \\ Y_3 \end{Bmatrix} = \frac{AE}{L}\begin{bmatrix} 1 & 0 & -1 & 0 & 0 & 0 \\ 0 & 1 & 0 & 0 & 0 & -1 \\ -1 & 0 & 1+r & -r & -r & r \\ 0 & 0 & -r & r & r & -r \\ 0 & 0 & -r & r & r & -r \\ 0 & -1 & r & -r & -r & 1+r \end{bmatrix}\begin{Bmatrix} U_1 \\ V_1 \\ U_2 \\ 0 \\ 0 \\ 0 \end{Bmatrix} \qquad [7.42]$$

Equation 7.41 can be partitioned because some of the boundary conditions in the deflection vector are zero.

$$\begin{Bmatrix} X_1 \\ Y_1 \\ X_2 \end{Bmatrix} = \frac{AE}{L}\begin{bmatrix} 1 & 0 & -1 \\ 0 & 1 & 0 \\ -1 & 0 & 1+\frac{1}{2\sqrt{2}} \end{bmatrix}\begin{Bmatrix} U_1 \\ V_1 \\ U_2 \end{Bmatrix} \qquad [7.43]$$

By using matrix solution techniques the displacements can be written as functions of forces and truss properties.

$$U_1 = \frac{L}{AE}[(2\sqrt{2} + 1)\, X_1 + (2\sqrt{2})X_2] \qquad [7.44]$$

$$U_2 = \frac{L}{AE}\left[2\sqrt{2}\left[X_1 + X_2\right]\right] \qquad [7.45]$$

$$V_1 = \frac{Y_1 L}{AE} \qquad [7.46]$$

7.5 STRESSES AND MATRIX METHODS

Once the displacements are determined, the stress in each element can be calculated by conventional strength of materials. For a truss element and unit displacement

$$\sigma_{12} = E\epsilon_{12} = E\left[\, \partial_1 - \partial_2 \,\right] \tag{7.47}$$

For other types of elements, such as two-dimensional plates, strain (ϵ) is proportional to the deflection and related to Poisson's ratio. Another term $[D]$ enters

$$\epsilon = [D]\,\partial \tag{7.48}$$

where

$$[D] = \frac{1}{1-\nu^2} \begin{bmatrix} 1 & \nu & 0 \\ \nu & 1 & 0 \\ 0 & 0 & \dfrac{1-\nu}{2} \end{bmatrix} \quad \text{and } \nu = \text{Poisson's ratio} \tag{7.49}$$

Plane stress is calculated by

$$\{\sigma\} = [E]\{\epsilon\} = [E][D][\,\partial] \tag{7.50}$$

To solve linear equilibrium equations, matrix operations need to be reviewed. A matrix $[A]$ is defined as an array of elements.

$$[A] = \begin{bmatrix} a_{11} & a_{12} & \cdots & a_{1n} \\ a_{21} & a_{22} & & \cdot \\ a_{m1} & & \cdots & a_{mn} \end{bmatrix} = a_{ij} \tag{7.51}$$

where m = rows
n = columns
if $m \neq n$ (rectangular)

then if

$m = 1; n > 1$ (row matrix [])
$n = 1; m > 1$ (column matrix {})
$m = n$ (square matrix)

Some matrices are symmetric (e.g., $a_{ij} = a_{ji}$). The identity matrix $[I]$ is one symmetric matrix

$$I = \begin{bmatrix} 1 & 0 & 0 \\ 0 & 1 & 0 \\ 0 & 0 & 1 \end{bmatrix} \tag{7.52}$$

The transpose of $[A]$, written as $[A]^T$, is useful in solving simultaneous equations and is constructed by replacing every a_{ij} by a_{ji}.

$$\text{if } [A] = \begin{bmatrix} 1 & 2 & -1 \\ 4 & -3 & 5 \end{bmatrix} \quad \text{then} \quad [A]^T = \begin{bmatrix} 1 & 4 \\ 2 & -3 \\ -1 & 5 \end{bmatrix} \tag{7.53}$$

The order defines the matrix size, stating the number of rows and then the number of columns, such as 2×3; its transpose is 3×2. Matrices can be added only if they are the same size, and they can be multiplied if the number of columns of the first matrix equals the number of rows of the second matrix, as shown in Eq. 7.54.

$$A \times B = C \qquad [7.54]$$

$$(m \times n)\,(n \times p) = (m \times p) \qquad [7.55]$$

where C_{ij}, the resultant of A times B (not B times A), is $\sum_{\beta=1}^{n} a_{i\beta}\, b_{\beta j}$

Example 7.4 Multiplying Matrices
Determine the new C matrix using matrix multiplication.

$$[A][B] = \begin{bmatrix} 1\,2 \\ 3\,4 \end{bmatrix} \begin{bmatrix} 1\,2\,3 \\ 4\,5\,6 \end{bmatrix} = \begin{bmatrix} C_{11} & C_{12} & C_{13} \\ C_{21} & C_{22} & C_{23} \end{bmatrix} \qquad [7.56]$$

The first term is

$$C_{11} = a_{11}b_{11} + a_{12}b_{12} = 1 \times 1 + 2 \times 4 = 9 \qquad [7.57]$$

$$C_{22} = [34]\begin{Bmatrix} 2 \\ 5 \end{Bmatrix} = (3 \times 2) + (4 \times 5) = 26 \qquad [7.58]$$

The final matrix is

$$C = \begin{bmatrix} 9 & 12 & 15 \\ 19 & 26 & 33 \end{bmatrix} \qquad [7.59]$$

Three simultaneous equations can be written in matrix form.

$$x + 2y - z = -5 \qquad [7.60]$$

$$4x - 3y + 4z = 1 \qquad [7.61]$$

$$2x - y + z = -8 \qquad [7.62]$$

$$\begin{bmatrix} 1 & 2 & -1 \\ 4 & -3 & 4 \\ 2 & -1 & 1 \end{bmatrix} \begin{Bmatrix} x \\ y \\ z \end{Bmatrix} = \begin{Bmatrix} -5 \\ 1 \\ -8 \end{Bmatrix} \qquad [7.63]$$

or, generally, as $[A]\{X\} = [B]$
Knowing that $[A]^{-1}[A] = [I]$, then

$$[A]^{-1}[A]\{X\} = [A]^{-1}[B] \qquad [7.64]$$

which is

$$[I][A]\{X\} = [A]^{-1}[B] \qquad [7.65]$$

The unknown $\{X\}$ can then be found if the inverse $[A]^{-1}$ is known.

$$\{X\} = [A]^{-1}[B] \qquad [7.66]$$

Determinant ($|A|$ symbol) is used in determining the $[A]^{-1}$ where

$$[A]^{-1} = \frac{\text{adjoint } A}{|A|} \qquad [7.67]$$

and to find the determinant $|A|$ of a 3×3 matrix for use in Eq. 7.67

$$|A| = \sum_{\substack{i=1 \\ \alpha=1}}^{n} a_{i\alpha}\bar{A}_{i\alpha} = a_{11}\bar{A}_{11} + a_{21}\bar{A}_{21} + a_{31}\bar{A}_{31} \qquad [7.68]$$

when $\alpha = 1$ and $n = 3$, $\bar{A}_{11}$, $\bar{A}_{21}$, and $\bar{A}_{31}$ are the cofactors. The cofactor "$\bar{A}_{ij}$" $= (-1)^{i+j}$ times the first minor $[a_{ij}]$.

Given a matrix A

$$[A] = \begin{bmatrix} 1 & 2 & 1 \\ 4 & -3 & 4 \\ 2 & -1 & 1 \end{bmatrix} \qquad [7.69]$$

The first minor a_{ij} is calculated by crossing out the appropriate row and column and then taking the determinant of the resulting matrix.

For first minor a_{11}, cross out row one and column one, as shown in Eq. 7.70

$$A_{11} \text{ first minor} = \begin{bmatrix} \not{1} & \not{2} & \not{1} \\ \not{4} & -3 & 4 \\ \not{1} & -1 & 1 \end{bmatrix} \qquad [7.70]$$

then take the determinant of the resultant

$$\begin{vmatrix} -3 & 4 \\ -1 & 1 \end{vmatrix} = 1 \qquad [7.71]$$

For Eq. 7.69 the cofactors are

$$\bar{A}_{11} = (-1)^2 \begin{vmatrix} -3 & 4 \\ -1 & 1 \end{vmatrix} = -3 + 4 = 1 \qquad [7.72]$$

$$\bar{A}_{21} = (-1)^3 \begin{vmatrix} 2 & -1 \\ -1 & 1 \end{vmatrix} = (-1)(2-1) = -1 \qquad [7.73]$$

The adjoint A is defined in Eq. 7.74

$$\text{adjoint } A = \begin{bmatrix} \bar{A}_{11} \\ \bar{A}_{21} \\ \bar{A}_{31} \end{bmatrix}^T \qquad [7.74]$$

Example 7.5 Finding the Inverse of a Matrix

For $A = \begin{bmatrix} 1 & 2 \\ 3 & 4 \end{bmatrix}$, find $[A]^{-1}$

First, the adjoints are $\bar{A}_{11} = 4$; $\bar{A}_{12} = -3$; $\bar{A}_{21} = -2$; $\bar{A}_{22} = 1$

$$\text{adj } A = \begin{bmatrix} 4 & -3 \\ -2 & 1 \end{bmatrix}^T = \begin{bmatrix} 4 & -2 \\ -3 & 1 \end{bmatrix}$$

$$A^{-1} = \frac{\text{adj } A}{|A|} \qquad \text{where} \qquad |A| = -2$$

then

$$A^{-1} = \begin{bmatrix} -2 & 1 \\ 3/2 & -1/2 \end{bmatrix}$$

Check A^{-1} by multiplying A by its inverse to get the identity matrix I.
The process of determining inverses is best accomplished by a computer algorithm. For finite-element solutions of large structures, more efficient solution algorithms are used in place of the inverse matrix.

7.6 OTHER FINITE-ELEMENT PROBLEMS AND INTERACTIVE DESIGN

7.6.1 Heat Transfer

Similar stiffness matrices are developed from variational calculus for flow problems. In heat transfer the differential equation is given as

$$K_{xx}\frac{dT}{dx} + q + h(T - T\infty) = 0 \qquad [7.75]$$

or, in matrix form, as

$$[K] \quad \{T\} = \{F\} \qquad [7.76]$$

where q = heat flux
h = convection coefficient
k_{xx} = thermal conductivity
T = unknown temperature
T_∞ = fluid temperature

For the one-dimensional element in Fig. 7.17 the stiffness matrix is

$$\frac{AKxx}{L}\begin{bmatrix} 1 & -1 \\ -1 & 1 \end{bmatrix} + \frac{hPL}{6}\begin{bmatrix} 2 & 1 \\ 1 & 2 \end{bmatrix} \qquad [7.77]$$

where $[K]$ is a function of geometry and material properties.

In vibrational problems the matrix equation is expanded into

$$[C]\{\dot{x}\} + [M]\{\ddot{x}\} + [K]\{x\} = \{F\} \qquad [7.78]$$

where C = damping
M = mass matrix
K = stiffness matrices
F = unit input or sinusoidal forces

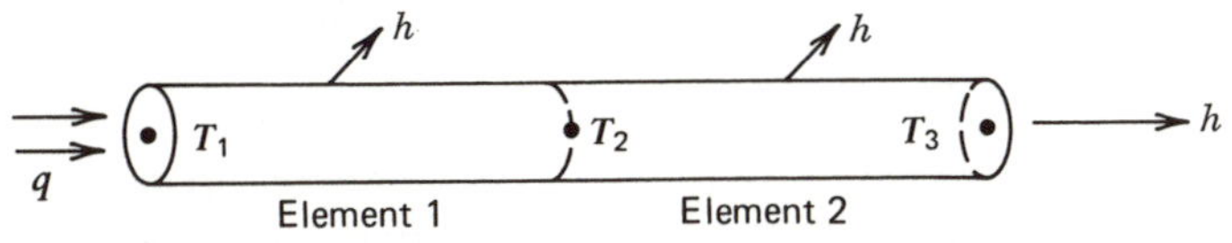

Figure 7.17 A one-dimensional heat transfer finite-element model.

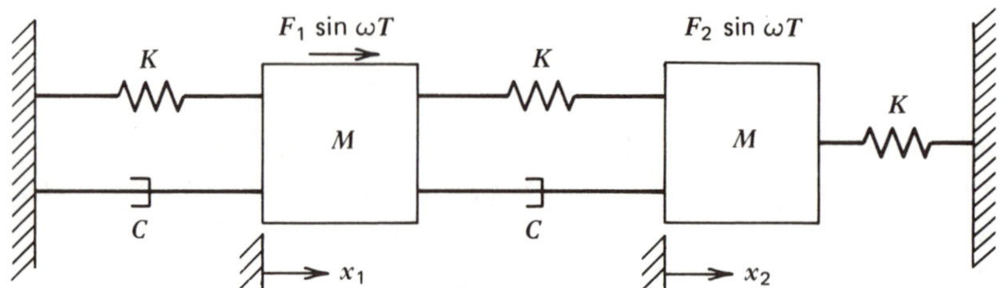

Figure 7.18 Finite-element vibration model.

This form is used in analyzing gears, airplane wings, and machine structures. A typical vibrational component model is seen in Fig. 7.18. The vibrational solution takes the form

$$[A - \lambda I]\{x\} = 0 \qquad [7.79]$$

The main objective of a vibration analysis is to determine the eigenvalue (λ) which is the natural frequency.

$$\lambda = \omega^2 \qquad [7.80]$$

Some sample machine-component, finite-element models are given in Fig. 7.19.

7.6.2 Interactive Graphics

The following descriptions will provide an idea of an interactive graphics capability. Interactive programs have made the finite-element method easier to use. A typical interactive portion takes about six months to develop. The graphics in this example was programmed in Fortran, and a data file was created for the structural analysis program SMIS3. The following is a typical two-page introduction, which prepares the user for modeling and analysis.

ITS is an interactive structural analysis program, designed to obtain solutions to two dimensional truss and frame element structures. ITS is composed of two major sections: an interactive section and the analysis section (SMIS3). The interactive package uses displays to prompt the user to input structure, boundary conditions, and applied forces. Its main task is to generate the datacards required by SMIS3 in a simple and straightforward manner.

Operation proceeds in roughly four parts; create structure, boundary conditions, forces, and solution. After each part, it is possible to branch out to other procedures; it is also easy to backtrack to previous sections, view data, delete elements, move nodes, obtain hard copies of the solution, and so on. ITS is basically self-explanatory; however, documentation is needed for a few important points.

PROBLEMS. To exit ITS, type a CTRL-C key or appropriate break key. To turn off the graphics display and reset the left margin, press function key 1.

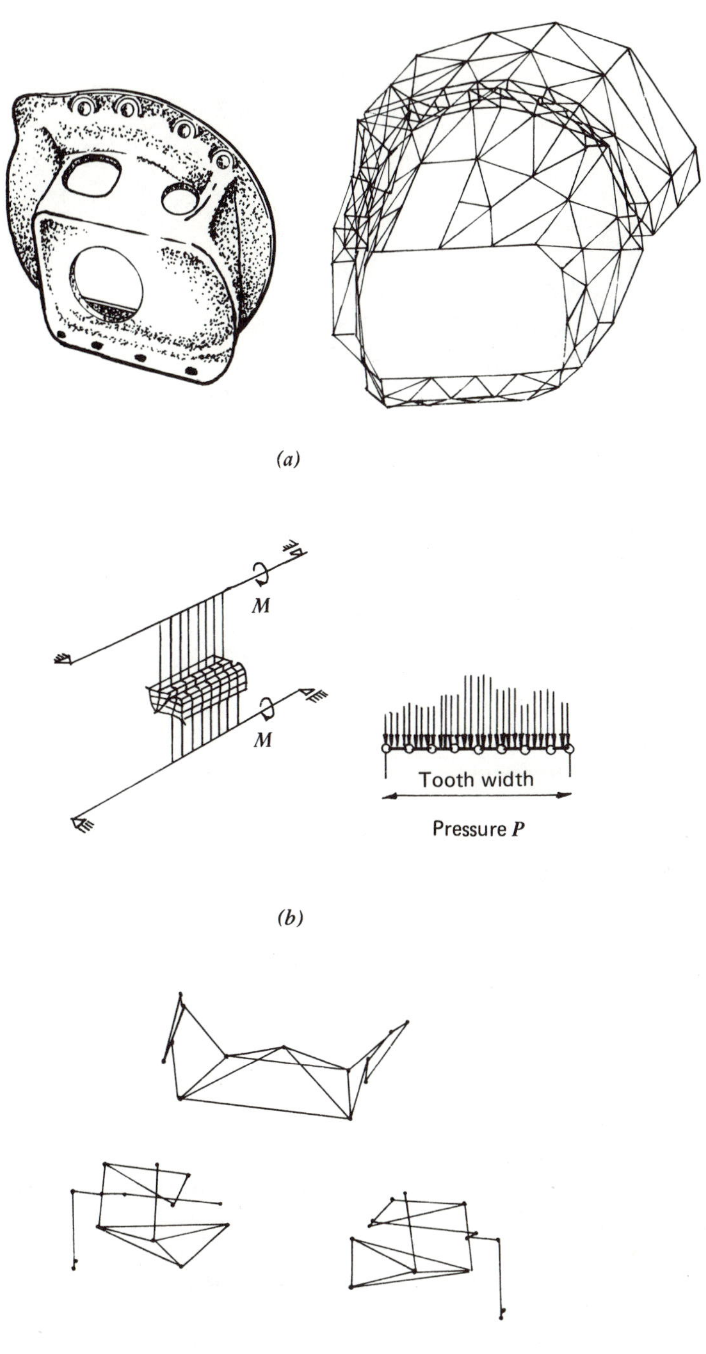

Figure 7.19 Examples of finite-element models. (*a*) Bracket. (*b*) Gear. (*c*) Front axle.

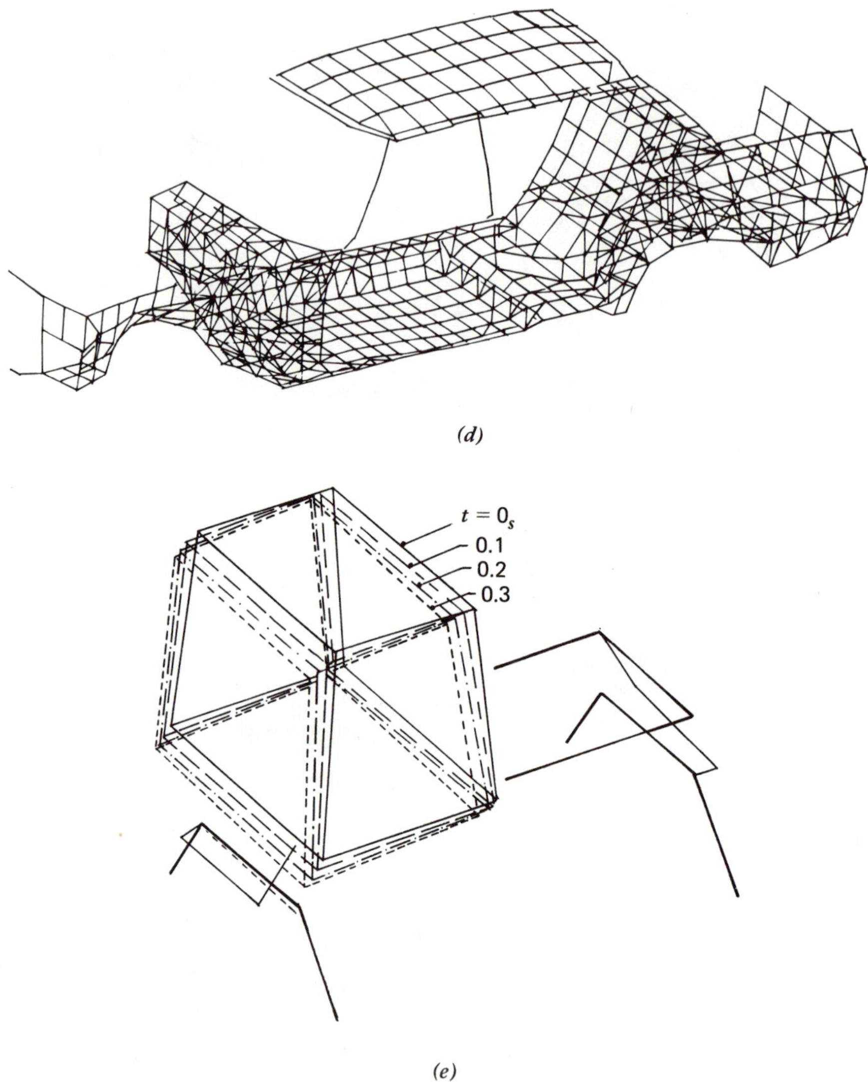

Figure 7.19 (Cont.) (*d*) Car body. (*e*) Rollover bar.

DEFAULTS. Almost all requested inputs are defaulted (to likely or heavily used responses), and only the return key need be pressed. Beware of the habit of quickly entering these convenient defaults before they are requested. Although in many cases the read-ahead feature of the operating system enables you to type ahead, in this case it may stop ITS and ruin your input effort.

If the box in which a cursor is waiting contains a value, that is the default. If you do not want the default, type your response over it. "Do you want to set autoplot manually (y/n)?" If you want to set up your own axes, (see the graphics terminal user's manual) enter a "y"; otherwise press the return key.

CREATE STRUCTURE. A structure can be created using node numbers 1 through 20, inclusive for truss and frame elements. Trusses have four degrees of freedom; frames have six. Only frame elements may be loaded along their lengths. Uniformly distributed loads and point loads may be put on frame elements. To apply a point load, calculate its distance from each nodal endpoint of the frame and use that number when asked.

MOVENODE This moves any node and all its connecting elements. After typing a node number just hit the return key twice (default diagnostics) to move it.

BOUNDARY CONDITIONS. Answer "y" or "n" to fix directions; the default is "n". A zero or a return for the number of conditions will cause you to exit from this procedure. If you make a mistake or need to exit the program, type the -1, which allows you to start over or return.

FORCES. All forces are applied to the structure at nodes. They are assumed zero unless changed. If you make a mistake, reenter the node number and make the change.

SOLUTION. The deflections are given in the units used. The trusses have only one end force, either compression or tension. The first row of end forces for frames refers to the first node listed, the second row to the second node, and so on. If you forget where your nodes are, press return to obtain "Command choices," from which you can "View data" and determine the location of the nodes. The input data and the solutions can be put in a file for later printing.

HOMEWORK PROBLEMS

7.1 Solve for the unknown deflections and forces in Example 7.1.

7.2 If a 50-lb weight is put on top of the block in Example 7.2, what is the maximum deflection?

7.3 Prove Eqs. 7.31 and 7.32.

7.4 What size stiffness matrix ($h \times h$) would the three-dimensional structural element in Fig. 7.20 have?

7.5 Write the stiffness matrix for the structure in Fig. 7.21.

7.6 Put the following equations in matrix form.

$$a + c - g = 1$$
$$2a + b + c = 6$$
$$4b + 6d - f = 8$$
$$e + f + g = 4$$
$$a + g - b = 10$$
$$d - 4f + 8 = 13$$

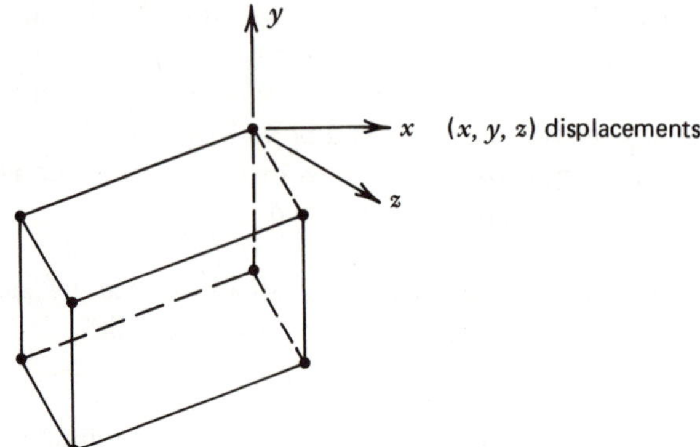

Figure 7.20 Problem 7.4.

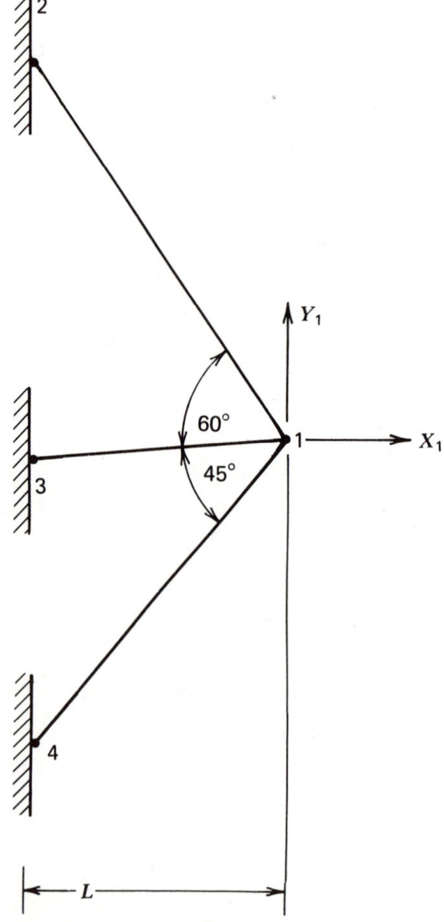

Figure 7.21 Problem 7.5.

7.7 Model a farm tractor's front-end loader with finite-elements.

7.8 Model a rollover bar (three-member space frame) with beam elements.

7.9 Compare an actual finite-element calculated deflection in the elastic range for Problem 7.8 with that of an analytical solution found in a handbook.

7.10 What is the global stiffness matrix K_{12} for the structure given in Fig. 7.22?

REFERENCES

1. Bathe, K.J. and E.L. Wilson. *Numerical Methods in Finite-Element Analysis,* Prentice Hall, Englewood Cliffs, N.J., 1976.
2. Brebbia, C. and J.J. Connor. *Fundamentals of Finite-Element Techniques*, Butterworths, London, 1973.
3. Cook, R.D. *Concepts and Applications of Finite-Element Analysis,* John Wiley and Sons, Inc., New York, N.Y., 1974.
4. Desai, C.S. *Elementary Finite-Element Method,* Prentice-Hall, Englewood Cliffs, N.J., 1979.
5. Desai, C.S. and J.R. Abel. *Introduction to the Finite-Element Method,* Van Norstand Reinhold, New York, N.Y., 1972.
6. Gallagher, R.H. *Finite Element Analysis—Fundamentals,* Prentice-Hall, Englewood Cliffs, N.J., 1975.
7. Huebner, K.H. *The Finite-Element Method for Engineers,* John Wiley and Sons, Inc., New York, N.Y., 1975.
8. Martin, H.D. and G.F. Carey. *Introduction to Finite-Element Analysis—Theory and Applications,* McGraw-Hall, New York, N.Y., 1973.
9. Norrie, D.H. and G. de Vries. *The Finite-Element Method—Fundamentals and Applications,* Academic Press, New York, N.Y., 1973.
10. Oden, J.T. *Finite Elements of Nonlinear Continua,* McGraw-Hill, New York, N.Y., 1972.
11. Pian, T.H.H. and P. Tong. *"Finite-Element Method in Continuum Mechanics,"* Advances in Applied Mechanics, Volume 12, Academic Press, New York, N.Y., 1972.

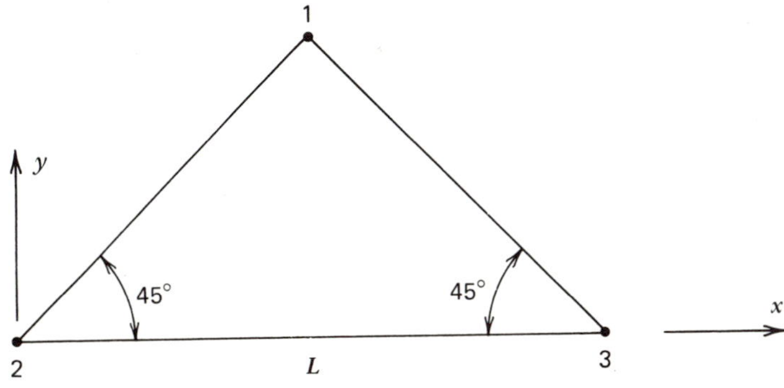

Figure 7.22 Problem 7.10.

12. Segerlind, L. *Applied Finite-Element Analysis,* John Wiley and Sons, Inc., New York, N.Y., 1976.
13. Smith, I.M. *Programming the Finite-Element Method,* John Wiley and Sons, Inc., New York, N.Y., 1982.
14. Strang, Gilbert and George Fix. *An Analysis of the Finite-Element Method,* Prentice-Hall, Englewood Cliffs, N.J., 1973.
15. Zienkiewicz, O.D. *The Finite-Element Method in Engineering Sciences,* McGraw-Hill, London, 1977.

CHAPTER EIGHT

POWER TRANSMISSION

The purpose of this chapter is to cover the design of components that collectively constitute the working parts of power-transmission drive systems on mobile agricultural machinery. Some of the most common power-transmission components are endless V-belt drives and sheaves, roller chains and sprockets, bearings and seals, gears, and planetary drives.

The selection of proper components must take into account customer requirements, cost constraints, manufacturing requirements, field usage, operator safety, and reliability. Design criteria that consider these factors should be identified early in the engineering development program to enhance the functional and operational characteristics of field usage.

8.1 V-BELT DRIVES

A V-belt drive consists of an endless flexible belt that transmits power by contacting and gripping the sheaves, which are keyed to the shafts of the driving and driven mechanisms. The primary reasons for using a V-belt drive are its simplicity and low maintenance costs. In addition, the elastomeric V-belt is able to absorb moderately high shock loads and mitigate the effects of vibratory forces.

The underlying principle for the operation of a V-belt (Fig. 8.1) is to provide a wedging action in the groove of the sheave, which increases the effective coefficient of friction between the sidewalls of the belt and the sheave. The drive transmits the power from the sheave to the belt by the contact between the sidewalls of the belt and the sheave. The bottom of the belt and the sheave groove should never come into contact during operation, like a flat belt drive. The V-belt drive develops a tension ratio (ratio of tight-side to slack-side tensions) that is 5 to 10 times greater than the tension ratio of a flat-belt drive.

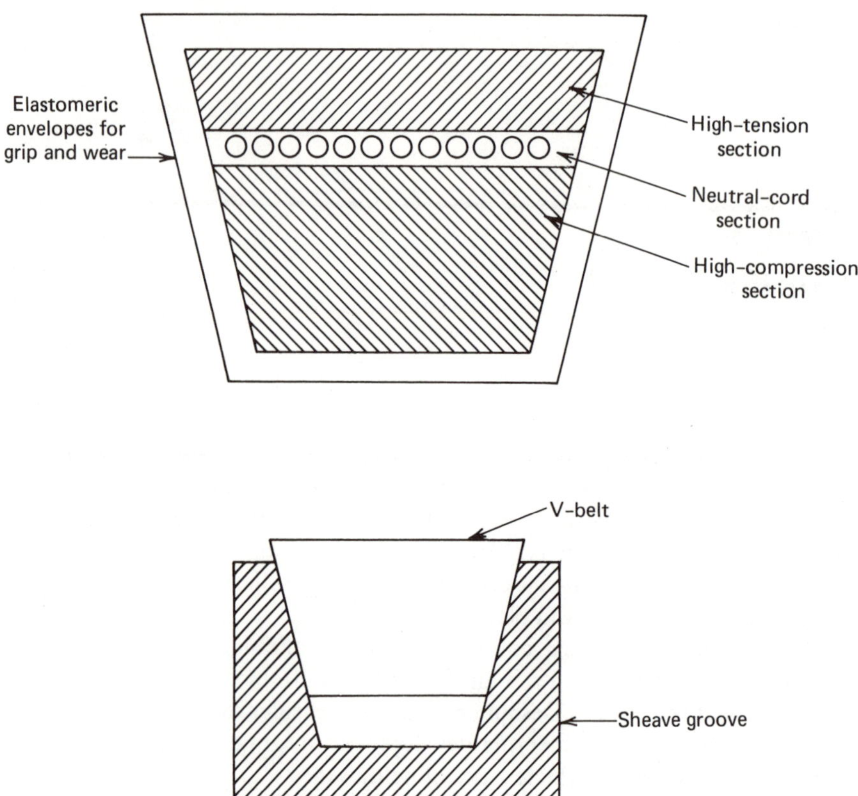

Figure 8.1 V-belt construction.

V-belts have the following characteristics:

1. They can be used for long center distances.
2. They can transmit heavy tensile loads.
3. They are usually used to transmit power between parallel shafts; however, they can be used for nonparallel shaft drives, such as the quarter- or eighth-turn drive and the mule-pulley drive.
4. They may be used singly or in multiple on sheaves.
5. They elongate because of wear and operation; hence a mechanism is needed to maintain the proper tension to take up the belt slack.
6. They rotate with slip and creep conditions; therefore the angular velocity between the two shafts is neither constant nor exactly equal to the ratio of the sheave pitch diameter. The power losses caused by slip and creep range from 3 to 5% for most belt drives.
7. They allow an efficient and easily operated method of varying the angular velocity ratio by employing spring-loaded adjustable-width sheaves. The diameter of the sheave is a function of the belt tension.

ASAE Standard S211.3 establishes the acceptable manufacturing tolerances, the methods of measurement, and the proper application of V-belt drives for agricultural machinery and provides the physical data for specifying the belt and the sheave. This standard is included in the belt design manuals listed in the references.

All V-belts for agricultural equipment are designated by a code that begins with the letter H. An additional letter of the alphabet is included in the desig-

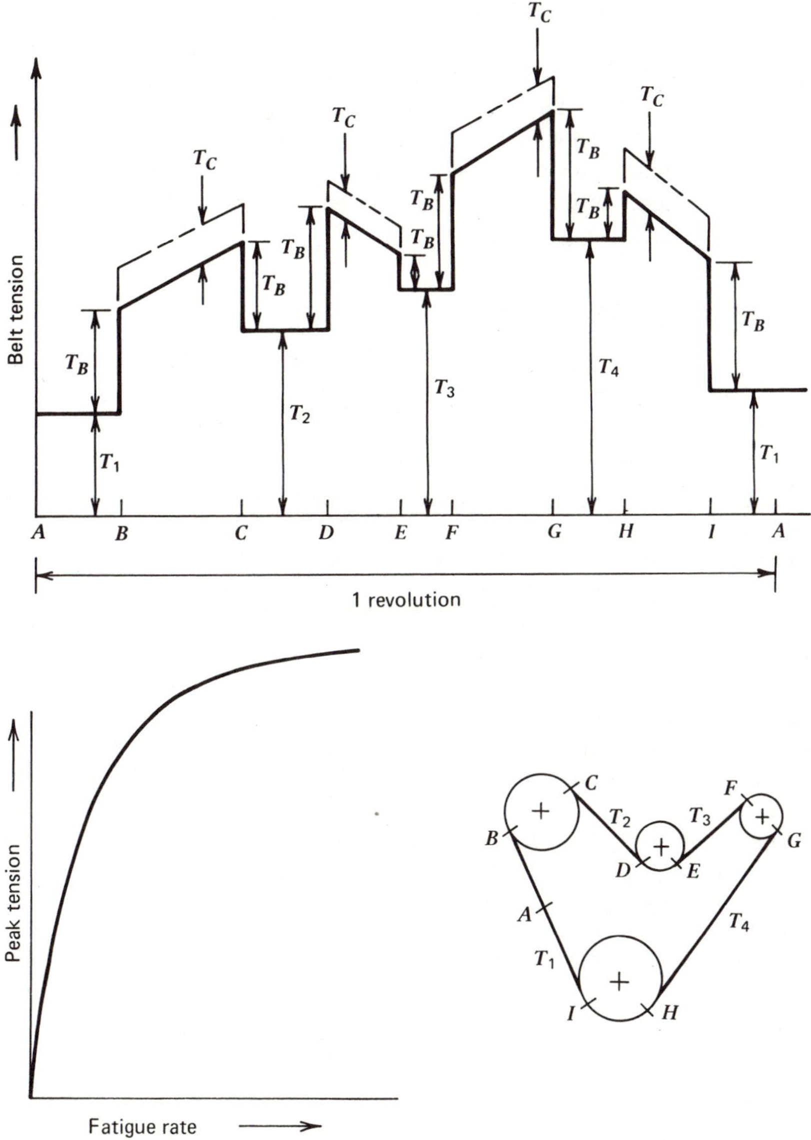

Figure 8.2 Tension levels and fatigue rate in a V-belt drive.

nation code to specify the cross section and application of the V-belt. For example, an HK V-belt is a 38.1 mm (1.5 in.) wide belt used on variable-speed drives.

V-belt drives are designed in terms of service life and stress fatigue rate. Thus, the more accurately the actual load history is known for a particular application, the more accurate the life estimation. From Fig. 8.2 it can be seen that, as the peak tension is increased at any one sheave, the fatigue rate also is increased at that sheave. The fatigue life of a belt represents the rate at which the life potential of a belt will be used when operated on a specific drive under given load and speed conditions.

The initial selection of the proper V-belt cross section is based on the design power requirements and the speed of the fastest shaft. The selection of the proper V-belt type depends on such factors as the level of shock loading, heat buildup, and contact with chemical and abrasive materials. For a complete selection procedure, the reader is referred to one of the V-belt manufacturer's design manuals.

8.2 V-BELT FORCES

When power is transmitted by a V-belt, the tensile load is carried by the elastomeric cord strands, which are located at the belt's neutral axis. As the belt rotates around a typical drive, the upper part of the belt experiences differences in belt tensions adjacent to a particular sheave, as shown in Fig. 8.2. The following tensions are imposed on the belt: 1. the initial tension, T_1; 2. the effective pull or transmitted load, $T_4 - T_1$; 3. the centrifugal tension, T_c; and 4. the bending tension around the sheaves, T_B.

To aid in analysis of a V-belt drive, the free-body diagram of a differential element of a belt is shown in Fig. 8.3. The forces acting on the element are: 1. tensions, T and $(T + dT)$; 2. centrifugal force, $(w\,v^2/g)d\phi$; 3 normal force, $\frac{1}{2}\,dN$; and 4. frictional force, $f\,dN$. The normal forces on the sides of the V-belt are assumed equal on both sides, $\frac{1}{2}\,dN$. The resultant normal force in the $x-y$ plane is $dN\,(\sin \frac{1}{2}\,\theta)$, where θ is the V-belt angle. The sum of the frictional forces is not dependent on the angle θ, but is $2[f(\frac{1}{2}\,dN)]$. The centrifugal force occurs because the belt has weight and travels generally in a circular path and is $(wR\,d\phi/g)\,(v^2/R)$. The element is assumed to be in static equilibrium because the inertial force is included. The summation of the forces in the arbitrary x and y directions then gives

$$\Sigma F_x = 0 \Longrightarrow (T + dT)\cos \tfrac{1}{2}\,d\phi - f\,dN - T\cos \tfrac{1}{2}\,d\phi = 0 \qquad [8.1]$$

$$\Sigma F_y = 0 \Longrightarrow (T + dT)\sin \tfrac{1}{2}\,d\phi + T\sin \tfrac{1}{2}\,d\phi - dN\sin \tfrac{1}{2}\,\theta - (wv^2/g)\,d\phi = 0 \qquad [8.2]$$

Substituting $\cos \frac{1}{2}\,d\phi = 1$ and $\sin \frac{1}{2}\,d\phi = \frac{1}{2}\,d\phi$ into Eqs. 8.1 and 8.2 gives

$$(T + dT) - f\,dN - T = 0 \qquad \text{or} \qquad dN = dT/f \qquad [8.3]$$

$$(T + dT)(\tfrac{1}{2}\, d\phi) + T(\tfrac{1}{2}\, d\phi) - dN \sin \tfrac{1}{2}\, \theta - (wv^2/g)\, d\phi = 0 \qquad [8.4]$$

Substituting Eq. 8.3 into Eq. 8.4 and dropping out the second-order differential terms yields

$$Td\phi - (dT/f) \sin \tfrac{1}{2}\, \theta - (wv^2/g)\, d\phi = 0 \qquad [8.5]$$

Rearranging Eq. 8.5 yields

$$\frac{dT}{T - wv^2/g} = \frac{f}{\sin \tfrac{1}{2}\, \theta}\, d\phi \qquad [8.6]$$

Then, if T changes from T_2 to T_1 as ϕ changes from 0 to α, integration between these limits gives

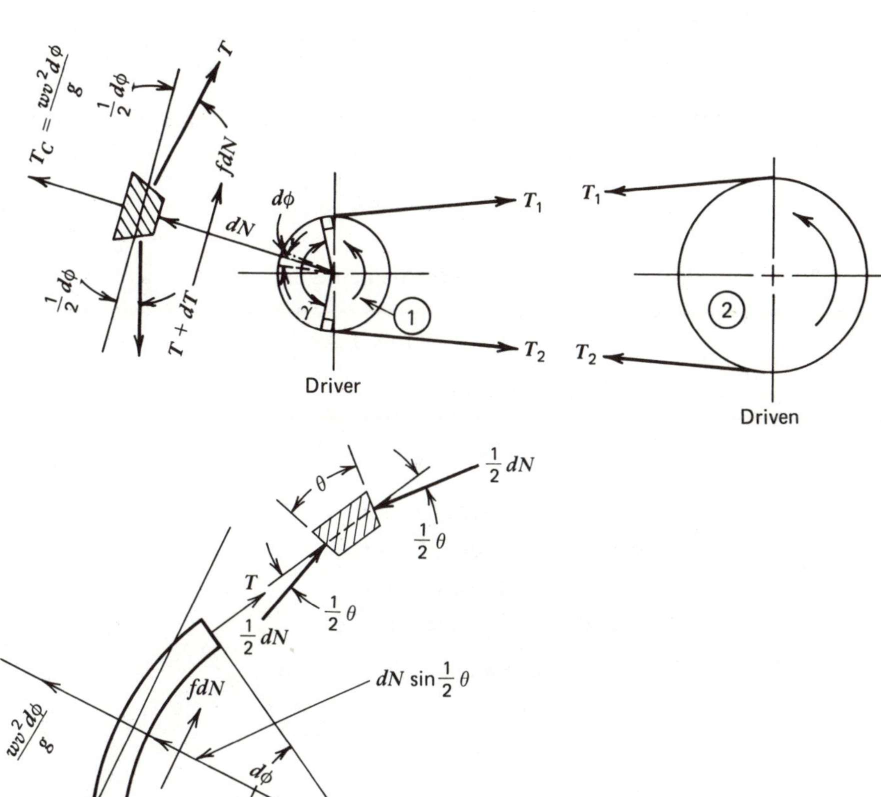

Figure 8.3 V-belt drive force analysis.

$$\int_{T_2}^{T_1} \frac{dT}{T - wv^2/g} = \int_0^{\alpha} \frac{f}{\sin½\,\theta}\, d\phi \qquad [8.7]$$

to yield

$$\ln\left[\frac{T_1 - T_c}{T_2 - T_c}\right] = f\,\alpha/\sin 1/2\,\theta \qquad [8.8]$$

$$or \qquad \frac{T_1 - T_c}{T_2 - T_c} = \exp\left[\frac{f\,\alpha}{\sin 1/2\,\theta}\right]$$

where T_c = centrifugal force, wv^2/g
f = coefficient of friction
w = belt weight per unit length, N/mm
v = belt velocity, m/sec
g = acceleration due to gravity, m/sec^2
α = arc of belt contact, rad

Fig. 8.3 shows the belt forces that act when sheave 1 is the driver in the direction indicated. The torque acting on the driver is

$$T_R = (T_1 - T_2)\,\frac{D_R}{2} \qquad [8.9]$$

and on the driven sheave, or sheaves, the acting torque is

$$T_N = (T_1 - T_2)\,\frac{D_N}{2} \qquad [8.10]$$

where T_1 = tight side tension, N
T_2 = slack side tension, N
$(T_1 - T_2)$ = effective pull, N
T_R = driver sheave torque, N · mm
T_N = driven sheave torque, N · mm
D_R = driver sheave diameter, mm
D_N = driven sheave diameter, mm

The effective pull may be related to the power by

$$P = \frac{(T_1 - T_2)\,V}{1000}$$

or

$$(T_1 - T_2) = \frac{1000 \times P}{V} \qquad [8.11]$$

where P = power, kW
V = belt velocity, m/sec

The allowable tension ratio relates the tight-side tension to the slack side tension of a drive, $R_A = T_1/T_2$. The ratio is based on a design tension ratio of 5 to 1 at 180° arc of contact for V-belts in sheaves and 2.5 to 1 at 180° arc of contact for V-belts on flat pulleys. For any other arc of contact, the allowable tension ratio for V-belts in V-sheaves is

$$R_A = \exp\left[\frac{(0.5123)\,\theta\pi}{180}\right] \qquad [8.12]$$

and for V-belts on flat pulleys it is

$$R_A = \exp\left[(0.2917)\,\frac{\theta\pi}{180}\right] \qquad [8.13]$$

where R_A = allowable tension ratio
θ = arc of contact, deg.
π = 3.14159 rad.

8.3 V-BELT DRIVE KINEMATICS

Belts are generally used to connect parallel shafts so that the sheaves rotate in the same direction (as in the open-belt drive in Fig. 8.4*a*) or in opposite directions (as in the crossed-belt drive in Fig. 8.4*b*).

The speed ratio of a V-belt drive is defined as the ratio of the angular velocity of the driver sheave to the angular velocity of the driven sheave. With V-belts, the pitch diameters must be used in the speed-ratio calculations to provide the desired shaft speeds. The expression for the speed ratio is

$$\frac{PD_R}{PD_N} = \frac{RPM_N}{RPM_R} \qquad [8.14]$$

where PD = pitch diameter, mm
RPM = shaft speed, m/sec or rpm
R = driver sheave
N = driven sheave

The arc of contact is defined as the angle subtended by the arc over which the belt contacts the sheave. For an open-belt drive (Fig. 8.4*a*), the arc of contact is

$$\theta = \pi \pm 2\phi \qquad [8.15]$$

where $\phi = \arcsin\,[(PD_1 - PD_2)/2\,C]$
PD_1 and PD_2 = pitch diameters, with $PD_1 \geqslant PD_2$
C = center-to-center distance of the sheaves

(In Eq. 8.15, the + sign applies to the larger sheave and the − sign applies to the smaller sheave.)

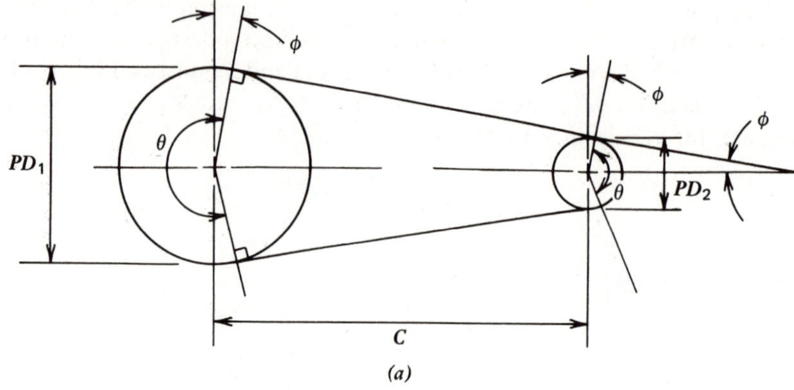

(a)

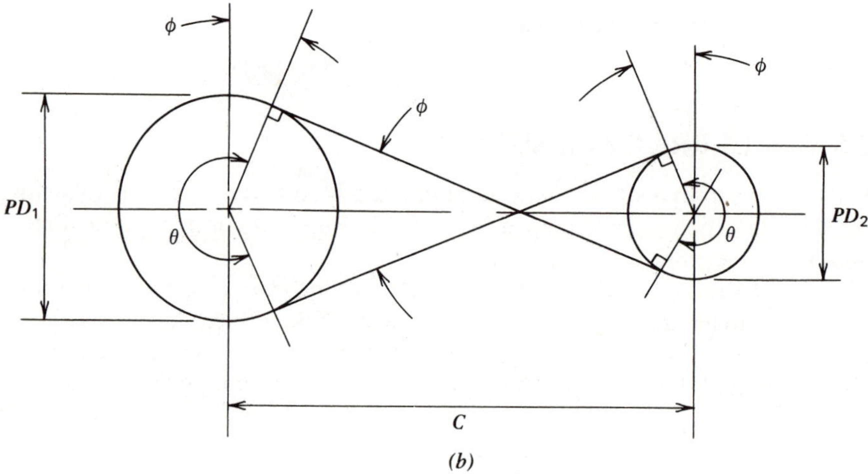

(b)

Figure 8.4 Open and closed belts.

For a crossed-belt drive, the arc of contact for both sheaves is

$$\theta = \pi + 2\phi \tag{8.16}$$

where

$$\phi = \text{arsin}\left[\frac{PD_1 + PD_2}{2C}\right] \tag{8.17}$$

The length of an open-belt drive is

$$L = 2\left[C^2 - \left[\frac{PD_1 - PD_2}{2}\right]^2\right]^{1/2} + \pi\left[\frac{PD_1 + PD_2}{2}\right] + (PD_1 - PD_2)\,\phi, \text{ mm} \tag{8.18}$$

An approximation to Eq. 8.18, which avoids the calculations of the square root and arcsin functions, is

$$L = 2C + \pi \left[\frac{PD_1 + PD_2}{2} \right] + \frac{(PD_1 - PD_2)^2}{4C}, \text{ mm} \qquad [8.19]$$

The length for a closed (crossed) belt drive is

$$L = 2 \left[C^2 - \left[\frac{PD_1 + PD_2}{2} \right]^2 \right]^{1/2} + \left[\frac{\pi}{2} + \phi \right] (PD_1 + PD_2), \text{ mm} \qquad [8.20]$$

An approximation to Eq. 8.20 is

$$L = 2C + \frac{\pi}{2} (PD_1 + PD_2) + \frac{(PD_1 + PD_2)^2}{4C}, \text{ mm} \qquad [8.21]$$

A method to calculate the belt length and the arcs of contact for drives with more than two sheaves, such as that shown in Fig. 8.5, is presented in Eq. 8.22. First, the belt length for each straight section is calculated with the following equation.

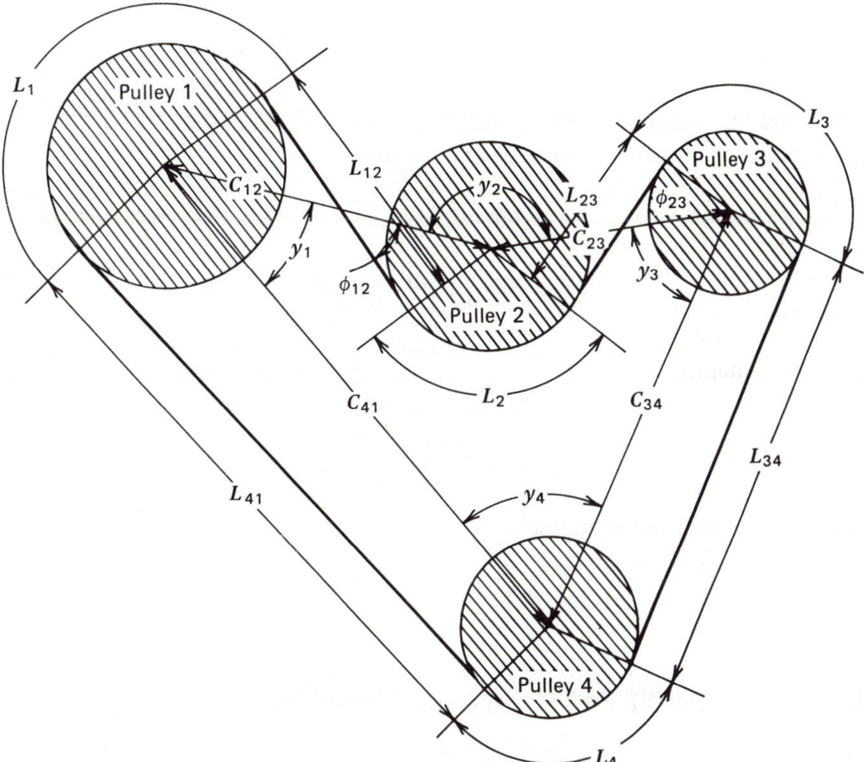

Figure 8.5 Belt-length calculation for multisheave system.

$$L_{ij} = \left[C_{ij}^2 \pm \left[\frac{PD_i \pm PD_j}{2} \right]^2 \right]^{1/2}, \text{mm} \qquad [8.22]$$

where i and j are the two sheaves. The $-$ sign is used for open sections and the $+$ sign is used for closed sections. The section is called a closed section if the straight section between two sheaves crosses the center-to-section line between those sheaves. Otherwise, the straight section is called an open section.

Next, the belt length of the curved sections and the arcs of contact is calculated. The angle that each straight section between sheaves makes with the center-to-center line between these same sheaves is calculated with the following equation.

$$\phi_{ij} = \arcsin \left[\frac{PD_j \pm PD_i}{2C_{ij}} \right] \qquad [8.23]$$

Again, the $-$ sign is used for open sections and the $+$ sign is used for closed sections. The arc of contact at each sheave is calculated with the following equation.

$$\theta_j = \pi - \gamma_j \pm \phi_{ij} \pm \phi_{jk} \qquad [8.24]$$

where i is the sheave before sheave j and k is the sheave after sheave j. The $-$ sign is used for ϕ_{ij} only if L_{ij} is an open section and if sheave j is smaller than sheave i. Likewise, the $-$ sign for ϕ_{jk} is used if PD_j is less than PD_k. The angle between center-to-section lines about each sheave is γ_j. The belt length of each curved section is calculated next.

$$L_j = \frac{\theta_j \, PD_j}{2} \qquad [8.25]$$

Finally, the total belt length is the sum of the total of the straight sections plus the total of the curved sections.

The belt speed of agricultural V-belt drives ranges from ½ to more than 30 m/sec. Generally, the drive is designed so that the belt speed is in the range of 5 to 20 m/sec. The expression for belt speed is

$$V = \frac{RPM \times PD}{19{,}100} \qquad [8.26]$$

where V = belt speed, m/sec
RPM = shaft speed, rpm
PD = pitch diameter of sheave, mm

8.4 V-BELT DRIVE FACTORS AND TENSIONING

Generally, the arrangement of the sheaves and the sheave diameters are limited by structural and functional constraints for the machine. However, optimum placement of the drive shafts with respect to one another and the sheave diameters increases belt life and effective pull.

The following conditions reduce belt life and capacity.

1. Bent or broken sheaves.
2. Abrasive environments, such as dust and dirt.
3. Sheave wobble.
4. Sheave groove worn or not uniform.
5. Misaligned sheaves.
6. Improper belt tension.
7. Sheave diameter too small.

The use of larger sheave diameters results in:

1. Lower shaft bearing loads.
2. Less face width because of reduced belt cross section and number of strands.
3. Better belt service life.
4. Better overall drive economy and efficiency.

The main factors to consider when positioning shafts with respect to one another are:

1. The path of the belt should enter the driven sheaves in order of increasing power loads. This arrangement allows the tight-side tension to increase gradually as the belt moves on to the more highly loaded sheaves.
2. Small-diameter sheaves should be located in spans of the drive with less tension. This arrangement prevents the combination of higher bending tension with a higher tight-side tension.
3. The idler should be located in the span with the least tension.
4. The arc of contact of a given sheave should be increased to reduce adjacent span tensions and to increase belt life.

Ideally, the best belt tension for a drive is the lowest tension at which the belt will not slip when the highest load condition is imposed. Two basic systems are available to tension a V-belt drive: 1. the locked-center method, and 2. the automatic tensioning (constant tension) method. In the locked-center method, the sheave is moved so as to tighten the drive and then locked in place. In the automatic-tensioning method, spring-loaded idlers or sheaves maintain a near-constant tension in the span. This latter method offers the following advantages: 1. it eliminates the need for manual take-up tensioning; 2. it provides correct and constant tension at all times; and 3. it allows the drive to operate at minimum tension. Also, this method automatically compensates for sheave sidewall wear, belt sidewall wear, and belt elongation and prolongs the life of the drive components.

Agricultural-machinery drives generally use spring-loaded idlers to maintain constant tension. If at all possible, these idlers should be located at the midpoint of slack span in the drive to minimize the required spring force. These idlers should never be located in a drive that reverses.

8.5 V-BELT DRIVE DESIGN PROCEDURES

The reader is referred to any belt manufacturer's design manual for the following information: (1) a detailed procedure to design an agricultural V-belt drive; (2) design application and service factors; (3) description of belt types and cross sections; (4) data for checking belt tension and speed and determining the fatigue rate and expected life of a belt [1-3].

Briefly, the design procedure includes the following steps:

1. Establishing the desired speed ratios and the acceptable belt life.
2. Selecting the different speed and load conditions.
3. Determining the driven load for each condition.
4. Selecting the belt type and cross section, and the sheave pitch diameters.
5. Calculating the approximate belt length, the belt speed, effective pull, and total tight-side tension.
6. Finding the fatigue rate and calculating the belt life.
7. Evaluating the calculated belt design to the acceptable design life.

Example 8.1

A V-belt drive must fit within the area, as shown in Fig. 8.6. The driver sheave rotates at 3820 RPM in a clockwise direction, while the driven sheave rotates at 1273 RPM also in a clockwise direction. The drive transmits an average of 4.5 kW but must transmit 9.0 kW peak for 10 percent of the time.

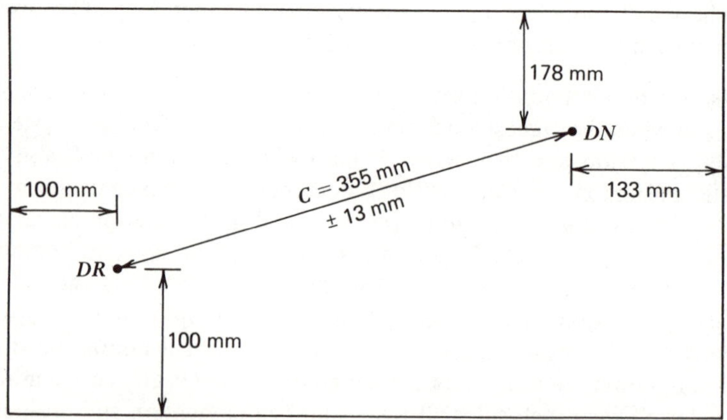

Figure 8.6 Belt example 8.1.

Determine

a. The speed ratio and the maximum sheave diameters:
From Eq. 8.14,

$$\text{Speed Ratio} = \frac{RPM_N}{RPM_R} = \frac{1273}{3820} = \frac{1}{3}$$

$$\text{Maximum driven diameter} = 2\left[133 - \frac{13}{2}\right] = 253 \text{ mm}$$

Therefore, maximum driver diameter $= 253/3 = 84$ mm

b. The approximate belt length:
From Eq. 8.18 for an open-belt drive the length is

$$L = 2(355) + 1.57(252 + 84) + \frac{(253 - 84)^2}{4(355)} = 1257 \text{ mm}$$

c. The belt speed is

$$V = \frac{RPM \times PD}{19.1} = \frac{(1272)(0.253)}{19.1} = 16.8 \text{ m/s}$$

d. The effective pull from Eq. 8.11 is for the 4.5 kW load

$$T_1 - T_2 = 1000\frac{P}{V} = \frac{(1000)(4.5)}{16.8} = 268\text{N}$$

For the 9.0 kW load

$$T_1 - T_2 = \frac{1000(9.0)}{16.8} = 536\text{N}$$

Example 8.2

An open V-belt drive, as shown in Fig. 8.4*a*, transmits 12 kW from a 255-mm pitch diameter sheave operating at 1800 RPM to a 635-mm diameter flat pulley. The center distance between the input and output pulleys is 1000 mm. The V-belt sheave groove angle is 40°. The coefficient of friction for the belt and sheave is 0.2 while the coefficient of friction between the belt and the flat pulley is 0.2. The cross-section of the belt is $b_2 = 38$ mm at the top and $b_1 = 19$ mm at the bottom by $d = 25$ mm deep. Each belt weighs 1086 N/m^3 and has an allowable tension of 900 N. How many belts are required for this drive? Determine:

a. The pitch diameter of the V-belt on the flat pulley by assuming the pitch diameter is measured to the centroid of the belt section:

Distance from belt base to the CG,

$$\bar{x} = \frac{d(b_1 + 2b_2)}{3(b_1 + b_2)} = \frac{25(19 + 2 \times 38)}{3(19 + 38)} = 13.9 \text{ mm}$$

b. Pitch diameter of larger pulley:

$$PD = 635 + 2(13.9) = 662.8 \text{ mm}$$

c. Arc of contact for sheave:

$$\theta = 180 - 2 \arcsin \frac{(662.8 - 255)}{2 \times 1000} = 156.5^\circ$$

d. Arc of contact for pulley:

$$\theta = 180 + 2 \arcsin \frac{(662.8 - 255)}{2 \times 1000} = 203.5^\circ$$

e. Allowable tension ratio for V-belt on sheave:

$$R_A = \exp \left[\frac{0.2(156.5\pi/180)}{\sin 20^\circ} \right] = 4.94$$

f. Allowable tension ratio for V-belt on pulley:

$$R_A = \exp \left[\frac{0.2(203.5\pi)}{180} \sin 90 \right] = 2.03$$

g. Area of the V-belt:

$$A = ½(b_1 + b_2)d = \frac{1}{2}(19 + 38)(25) = 712.5 \text{ mm}^2$$

h. Weight of the V-belt:

$$w = (712.5)(108)(10^{-6}) = 0.78 \text{ kg/mm}$$

i. Velocity of the V-belt:

$$v = \pi \left(\frac{255}{1000} \right) \left(\frac{1800}{60} \right) = 24 \text{ m/sec}$$

j. Centrifugal force in the V-belt:

$$T_c = \frac{wv^2}{g} = \frac{(0.78)(9.8)(24)^2}{9.8} = 449\text{N}$$

k. Belt tension on the slack side:

$$\frac{900 - 449}{T_2 - 449} = 2.03$$

therefore $T_2 = 671$ N

l. Power transmitted by belt:

$$P = \frac{(900 - 671)(124)}{1000}$$

$$P = 2.8 \text{ kW/belt}$$

m. Number of belts required:

$$n = 12 \text{ kW}/2.8 = 4.3$$

Use four belts.

8.6 CHAIN DRIVES

The chain drive consists of an endless chain whose links engage the teeth of sprockets keyed to the shafts of the driving and driven mechanisms. The basic functions of a chain drive are to transmit power and to synchronize motion or maintain a fixed-speed ratio between rotating shafts. The advantages of chain drives are:

1. Shaft distances are unrestricted; that is, the drive is suited for long- and short-center distances.
2. Chain drives do not creep or slip; therefore, they maintain a positive speed ratio between the driver and the driven shafts.
3. Chain drives are physically more compact than a belt drive, but they do require more accurate alignment of the shafts and the sprockets.
4. The chain rollers engage the sprocket with a rolling action.
5. The arc of contact is smaller for chains than for belts.
6. Chain drives are more practical for low speeds.

The types of chains that are available for power transmission and materials conveyance are:

1. Detachable chain.
2. Pintle chain.
3. Engineering steel chain.
4. Inverted-tooth, or silent, chain.
5. Roller chain.

The roller chain is the most important chain type for agricultural use and is shown in Fig. 8.7.

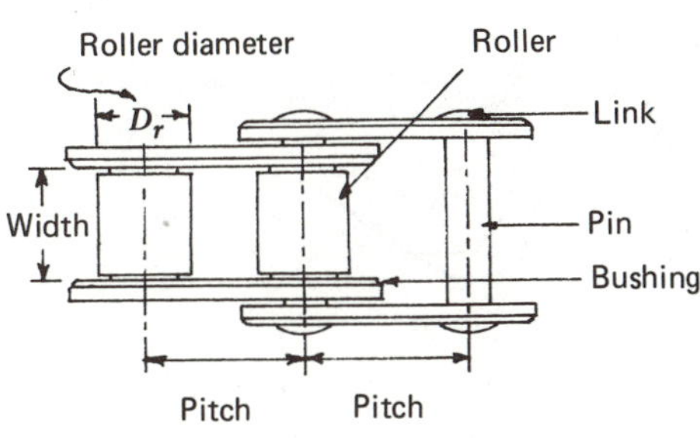

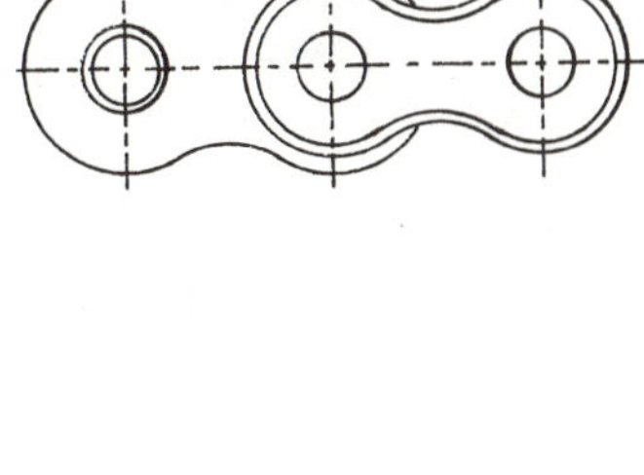

Figure 8.7 Roller chain components.

8.7 ROLLER CHAIN

Roller chain is available in two arrangements: the single-strand or the multiple-strand type. Standard roller chain is identified and specified by a two-digit designation code. The first number denotes the number of eighth-inches (3.175 mm) in the chain pitch, which is the distance between the centers of the adjacent joint members. The second number denotes the type of construction–zero for regular construction roller chain, 1 for light-weight chain, and 5 for rollerless bushing chain. The hyphenated number suffixed to the chain number denotes the number of chain strands. The letter H following the chain number denotes a chain of heavy construction. For example, the chain number 60-2H indicates a double strand, 6/8-in. pitch chain of heavy proportions.

Roller chain is composed of roller, links, and pin links, alternately spaced throughout the length of the chain. The roller link consists of two sets of rollers and bushings, and two link plates, and the pin links consist of two pins and two link plates.

The dimensions for roller chain and toothed sprockets are standardized by the American National Standards Institute (ANSI). Roller chain is identified by three principal dimensions: pitch, chain width, and roller diameter, which are identified in Fig. 8.7. The technical data for the design of roller chain and sprockets is found in the Design Manual for Roller and Silent Chain Drives.

The components of roller chain are subjected to wear, fatigue, impact, and galling during actual operation. Wear is generally the limiting factor for roller chain in agricultural use. Wear is caused by articulation of the pins in the bushings and is accelerated by dirt and inadequate lubrication. Wear results in elongation of the chain, as illustrated in Fig. 8.8. The rate of chain elongation depends on:

1. Chain tension.
2. Projected bearing area.
3. Smoothness and hardness of the contacting surfaces.
4. Frequency and type of lubrication.

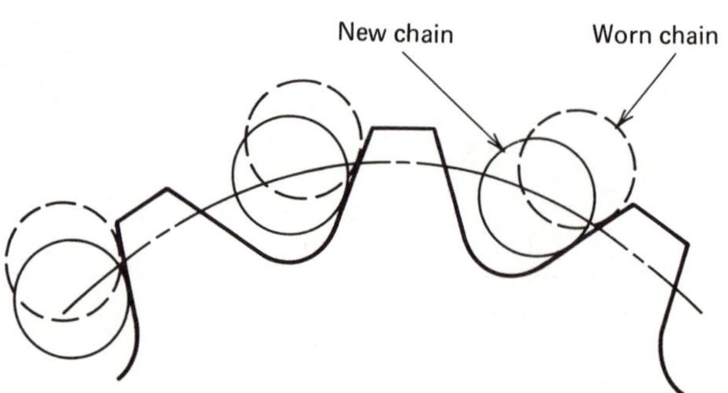

Figure 8.8 Diagram showing chain elongation caused by wear.

5. Frequency and degree of articulation in the chain joints.
6. Dust, dirt, and other abrasive or corrosive materials.

Chain tension in the drive is caused by:

1. The pull required for power transmission.
2. The centrifugal force generated as the chain travels around the sprocket and caused by its travel over a curved path between sprockets.
3. The catenary effect caused by the weight of the portion of the chain between sprockets.
4. The chordal action.

The pull required to transmit power is equal to the tension in the tight side. There is no tension in the slack side of the drive as well as no initial tension.

8.8 ROLLER CHAIN DRIVE KINEMATICS

As the chain passes around the sprocket as a series of chordal links, the centerline of the chain is not at a uniform radius, as is the centerline of a belt passing around a sheave. The chain pitch approaches and leaves the sprocket at a continually varying radius. The rise and fall of each chain pitch as it engages a sprocket is termed chordal action. As illustrated by Fig. 8.9, the *chordal action* and speed variation decrease as the number of teeth on the smaller sprocket is increased. Chordal action becomes negligible when sprockets with 25 or more teeth are used.

If the driver sprocket is assumed to rotate at a uniform speed, the driven sprocket rotates at a varying speed that fluctuates above and below some average value. Because of chordal action, the average speed of a sprocket cannot be calculated in terms of the pitch diameters but is equal to the length of the chain passing around the sprocket in a unit time. Thus,

$$V = \frac{p \times N \times RPM}{376} \tag{8.27}$$

where V = average chain velocity, m/sec
p = chain pitch, in.
N = number of teeth on sprocket
RPM = sprocket speed, rpm

Similarly, the speed ratio of a chain cannot be calculated in terms of the pitch diameters, but must be calculated from

$$\frac{RPM_1}{RPM_2} = \frac{N_2}{N_1} \tag{8.28}$$

For simplicity, the chain length is calculated in terms of pitches, and the result is multiplied by the chain pitch to obtain the actual chain length. The

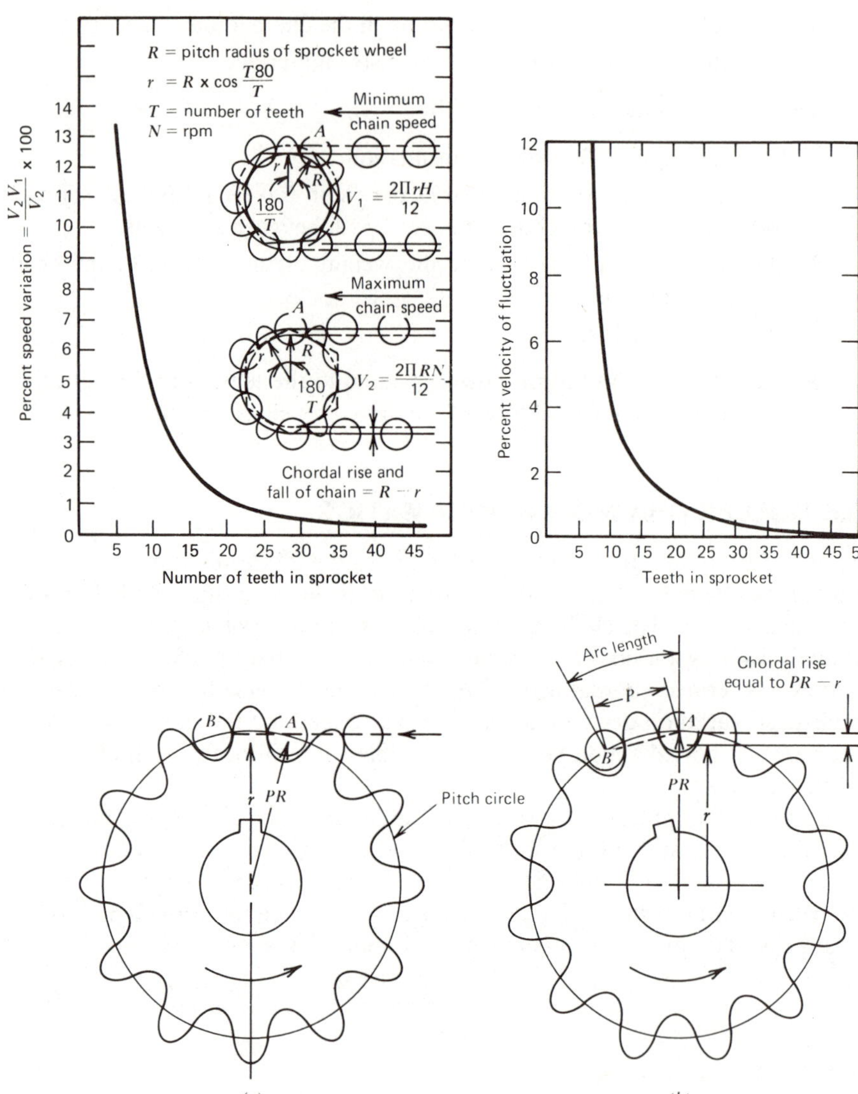

Figure 8.9 Chordal action.

approximate length of a chain drive in pitches is

$$L = 2C + \left[\frac{N_2 + N_1}{2}\right] + \frac{(N_2 - N_1)}{4\pi^2 C} \qquad [8.29]$$

where C = center distance in chain pitches

N_2 = number of teeth on larger sprocket

N_1 = number of teeth on smaller sprocket

The approximated length, if fractional, must be modified to a whole, and, preferably, to an even number of pitches. The center distance may be corrected for the revised chain length by solving Eq. 8.29 for C. The exact center distance of a chain drive in pitches is

$$C = \frac{L - \left[\frac{N_2 + N_1}{2}\right] + \left\{\left[L - \left[\frac{N_2 + N_1}{2}\right]\right]^2 - \frac{8(N_2 - N_1)^2}{4\pi^2}\right\}^{1/2}}{4} \qquad [8.30]$$

The pitch diameter of a sprocket is

$$PD = \frac{p}{\sin\left[\frac{180}{N}\right]} \qquad [8.31]$$

where p = chain pitch, in.
N = number of teeth on the sprocket

8.9 CHAIN FORCES

The power transmitted by a chain drive can be calculated by Eq. 8.11, except that the slack-side tension T_2 is assumed to be zero. Thus, the chain pull or working load can be calculated as follows

$$CP = 1000\,\frac{P}{V} \qquad [8.32]$$

where P = power, KW
V = chain velocity, m/sec

The force rating for a single strand chain can be determined by the following empirical equation.

$$F = 0.273\,p^2\left\{5100 - 115\,V^{0.41}\left[1 + 25\left[1 - \frac{\cos 180}{N}\right]\right]\right\} \qquad [8.33]$$

where F = force, lb
p = chain pitch, in.
V = chain velocity, ft/min
N = number of teeth on the sprocket

At lower speeds, the horsepower capacity is determined by the fatigue life of the link plates, and is calculated as follows:

$$HP = 0.004\,N_1^{1.08}\,n_1^{0.9}\,p^{3.0} - 0.007p \qquad [8.34]$$

where N_1 = number of teeth in the smaller sprocket
n_1 = speed of the smaller sprocket, rpm
p = chain pitch, in

At higher speeds, the horsepower is determined by the roller bushing fatigue life and is calculated as follows:

$$HP = \frac{17000\, N_1^{1.5}\, p^{0.8}}{n_1^{1.5}} \qquad [8.35]$$

8.10 ROLLER CHAIN DRIVE SELECTION

The selection of a chain and sprockets for a drive is based on: 1. the power requirements and type of drive, 2. the speeds and sizes of the shafting, and 3. the surrounding conditions.

Roller-chain drives for agricultural use are often overstressed and under designed according to the recommendations of the American Chain Association (ACA). To reduce costs, these chain drives are underdesigned for the recommendations specified in Fig. 8.10. This practice, called *underchaining* is used when the design life of a machine is less than 2000 hours, instead of the 15,000 hours used by the ACA. The degree of underchaining is established by engineering experience for each particular drive application.

The limiting factor for the drive is the speed of the smaller sprocket, normally the driver. The following factors are also considered:

1. *Single or multiple-strand chains.* The rule of thumb is to use the smallest chain pitch that will carry the power and the load requirements to maximize the number of teeth on the sprocket. Smaller chain pitches are used on moderate to high-speed drives and when quietness is essential. Large chain pitches are used on slow-speed drives.
2. *Sprocket size.* To improve life expectancy sprockets with 19 or 21 teeth should be considered. To obtain smooth operation, the normal minimum number of teeth is 17 and the normal maximum number of teeth is 120.
3. *Center distances.* The center distance must be more than the sum of one-half the smaller sprocket diameter plus one-half the larger sprocket diameter and is preferably 30 to 50 times the pitch of chain used. Eighty times the pitch is considered maximum.

8.11 RECOMMENDED CHAIN DRIVE ARRANGEMENTS

The arrangement or relative position of the driver and driven sprockets and the tight- and slack-side strands have great influence on the life of a drive. Means should be provided for adjusting the slack in a chain, preferably by having the one shaft adjustable so that the center distance may be varied. The same effect may be accomplished by an idler sprocket on a movable shaft or a spring-loaded sprocket for engaging the slack side of a drive. Chain-drive arrangements should include the following features:

1. *Horizontal centers.* The slack strand should be on the lower side, although it may be on either side for normal center distances.
2. *Long center distances.* The slack strand should always be on the lower side because wear could cause a slack upper strand to rub against the lower strand.
3. *Short center distances.* The slack strand should preferably be on the lower side because of a tendency for the loose upper strand to be pushed out of proper arrangement with the teeth on the smaller sprocket.
4. *Vertical center distances.* The chain should be run fairly taut to prevent its sagging and to prevent disengagement from the teeth on the lower sprocket. The drive should be placed slightly off the vertical plane.

8.12 CHAIN DRIVE DESIGN

The following items are required for specifying a chain drive: 1. power requirements; 2. shaft speeds; 3. load and power source; 4. shaft diameters; 5. center distance; 6. space limitations; and 7. surrounding conditions. The step-by-step procedure for the selection of the chain and sprockets is outlined in the following example.

Example 8.3

Consider the selection of a roller-chain drive for a shelling mechanism. The sheller must operate at 1000 rpm and is driven by a shaft that rotates at 600 rpm. The design center distance for the mechanical drive is 0.61 m. The power transmitted by the drive is 15 kW.

Step 1: Service Factor

Service factors are used to determine the additional chain capacity to account for fluctuating load and atmospheric conditions. The service factors for chain loading, SF_1, are found in Table 8.1. For this example, SF_1 is 1.4 for

Table 8.1 **Service Factors for Chain Loading,** SF_1

		Drive	
Loading	*Agricultural Machine*	*Mechanical*	*Hydraulic*
Smooth, nonreversing loads	Fan	1.2	1.0
Moderate-shock, nonreversing loads	Combines, forage harvesters	1.4	1.2
Heavy shock loads, Severe speed variations, reversing loads	Hammermills	1.7	1.4

a moderate-shock, nonreversing load. The service factor for atmospheric conditions, SF_2, is 1.4, because the chain drive for agricultural use is exposed to weather and dirty conditions, use Table 8.2.

Step 2: ***Computation of Design Power***
The design power is obtained by multiplying the power by the service factors.

$$P = P \times SF_1 \times SF_2 \qquad [8.36]$$

$$P = 15 \times 1.4 \times 1.4 = 29.4 \text{ kW}$$

Step 3: ***Tentative Chain Selection***
The tentative required chain pitch is selected from Fig. 8.10. The horizontal line representing the design power is followed until it intersects the vertical line representing the specified rpm for the smaller sprocket.

For this example, the suggested chain pitch is 60.

Step 4: ***Final Selection of Chain and Small Sprocket***
For specific applications, the power rating tables should be checked to find the required number of teeth on the smaller sprocket.

For this example, the table shows that a 35-tooth sprocket is required. However, because of physical constraints for the drive, a 25-tooth sprocket is selected.

Step 5: ***Selection of the Large Sprocket***
The number of teeth for the large sprocket is computed from Eq. 8.28, substituting values for N_1, RPM_1, and RPM_2.

$$N_2 = \frac{N_1 \times RPM_1}{RPM_2} = \frac{25 \times 1000}{540} = 46 \text{ teeth}$$

Step 6: ***Chain Length***
The approximate chain length is calculated from Eq. 8.29, substituting values for C, N_1, and N_2:

$$L = 2\left[\frac{610/25.4}{0.75}\right] + \left[\frac{46 + 25}{2}\right] + \frac{(46 - 25)}{4\pi^2\left[\dfrac{610/25.4}{0.75}\right]} = 99.59 \text{ pitches}$$

$$\text{or} \qquad 1.9 \text{ m}(74.7 \text{ in})$$

This chain length exceeds the recommended maximum center distance by 20 pitches.

Step 7: ***Correction of the Center Distance***
The center distance is recalculated by Eq. 8.30 to obtain an even number of pitches.

Table 8.2 **Service Factors for Environment, SF_2**

	Relatively clean and moderate temperature	1.0
Atmospheric Conditions	Moderately dirty and moderate temperature	1.2
	Exposed to weather, very dirty, abrasive, mildly corrosive and reasonably high temperatures	1.4

$$C = \frac{100 - \left[\dfrac{46 + 25}{2}\right] + \left\{\left[100 - \dfrac{(46 + 25)}{2}\right]^2 - \dfrac{8(46 - 25)^2}{4\pi^2}\right\}^{1/2}}{4}$$

$$= 34.09 \text{ pitches} \quad \text{or} \quad 0.65 \text{ m}(25.6 \text{ in.})$$

8.13 GEARS

Gears are toothed wheels or multilobed cams used to transmit power and motion at a constant ratio from one rotating shaft to another by means of the positive contact of successively engaging teeth. Compared with V-belt and chain drives, gear drives are more compact, can operate at higher speeds, and provide precise timing of motion. Gear drives also require better lubrication and more cleanliness and are affected by shaft misalignment.

8.14 GEAR TOOTH PROFILES

When considering two curved surfaces in direct contact, the selection of the proper curves for gear-tooth profiles is essential for smooth transfer of force and motion. This defines the basic criterion for gearing; that is, to transmit motion in such a way that the angular velocity ratio is a constant. For all mechanisms with constant angular velocity ratio, the line of action always intersects the line of centers at a fixed point. It is possible to assume the profile of one tooth, apply the basic law of gearing, and determine the outline of the mating tooth. These teeth would be considered conjugate teeth. Industry has standardized on the involute profile for two reasons: 1. ease of manufacture, and 2. the center distance between two involute gears may vary without changing the velocity ratio.

An involute is the locus of a point on a straight line as that line rolls without slipping around a circle called the base circle. The construction of an involute is shown in Fig. 8.11. A straight line drawn tangent to the base circle is normal to any surface in contact with the involute and it must be tangent to the base circle. When two involutes are in contact, as shown in Fig. 8.12, the common

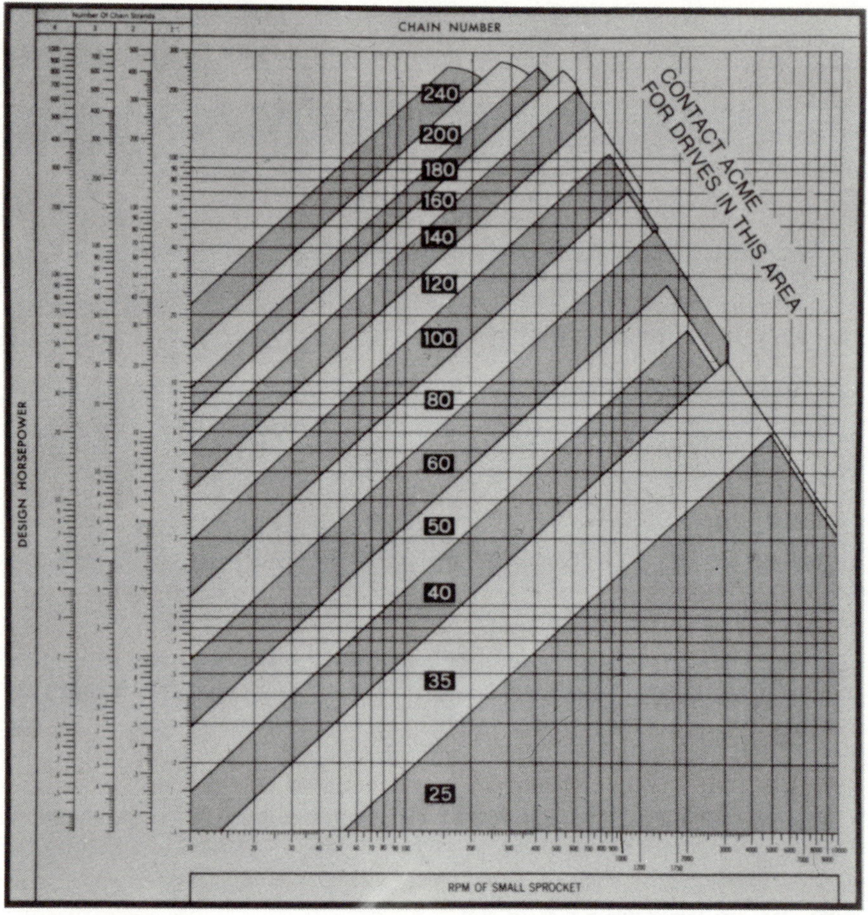

Figure 8.10 Roller chain pitch selection chart. Courtesy American Chain Association.

normal is the straight line tangent to both base circles, and contact between two involutes always occurs on this line, called the line of action. As shown in Fig. 8.13, variation in the center distance causes the common normal to the surface in contact to always intersect the line of centers at the same point, which results in a constant angular velocity ratio. Any variation in angular velocity occurs as a result of error in tooth profile and loaded tooth deflection.

Because the force transmitted along the common normal is that acting perpendicular to the line of centers, the angle ϕ between the line of centers and the perpendicular to the line of centers is important to the calculation of gear forces. The pressure angle ϕ is the angle between the line of centers and the radial line from the center to the point at which the line of action is tangent to the base circle, as shown in Fig. 8.14. The transmitted force F_t results in torque transmission, and the separating force F_s contributes no work and increases bearing loads. These forces are related as follows.

$$F_s = F_t \tan \phi \tag{8.37}$$

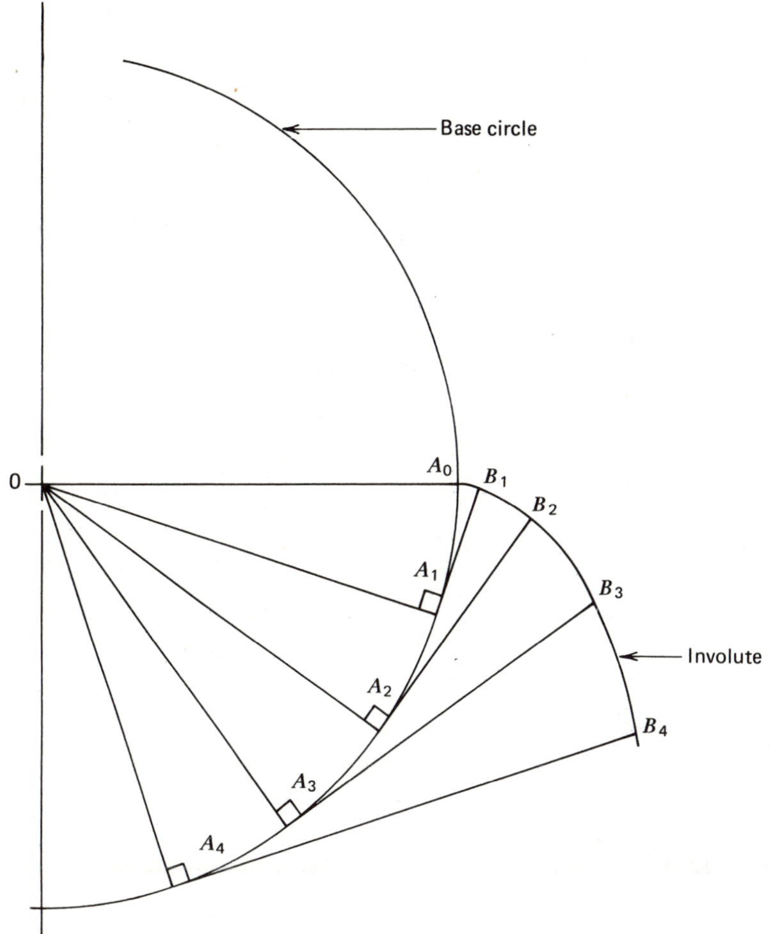

Figure 8.11 Construction of involute curve.

This formula is important in calculating the strength and wear of the gear teeth and in determining what forces act on the gear shaft and bearings.

8.15 GEAR NOMENCLATURE

To study involute gearing, it is necessary to define the terminology applied to the basic elements of a gear, which are shown in Fig. 8.15. The smaller of two gears in mesh is called the pinion and is generally the driver.

1. Pitch surfaces are imaginary cylinders or cones that roll together without slippage.
2. A pitch circle is the imaginary circle that rolls without slippage with the pitch circle of a mating gear.

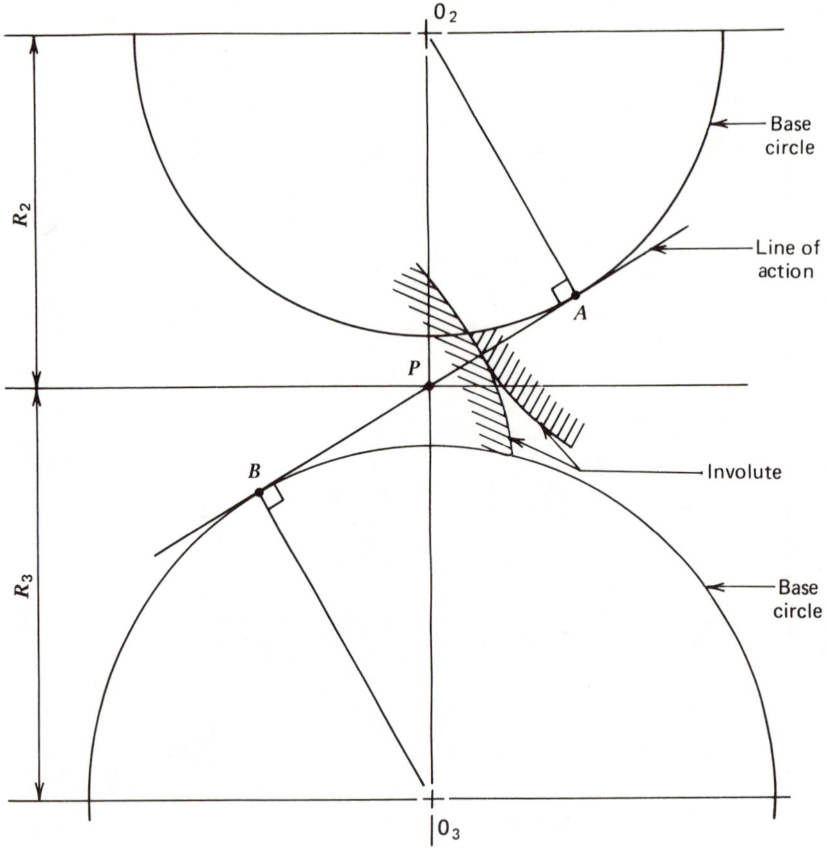

Figure 8.12 Involute action.

3. Pitch diameter, D, is the diameter of the pitch circle.
4. Circular pitch, P_c, is the distance between corresponding profiles of adjacent teeth and is calculated by dividing the circumference of the reference cylinder by the number of teeth.

$$P_c = \frac{\pi D}{N} \qquad [8.38]$$

5. Diametral pitch, P, is the number of teeth per length of reference diameter and is the number of teeth divided by the reference diameter.

$$P = \frac{N}{D} \qquad [8.39]$$

From Eqs. 8.38 and 8.39, the relation between diametral pitch and circular pitch is

$$P_c\, P = \pi \qquad [8.40]$$

6. Addendum, a, is the radial distance between the pitch circle and the outside diameter.

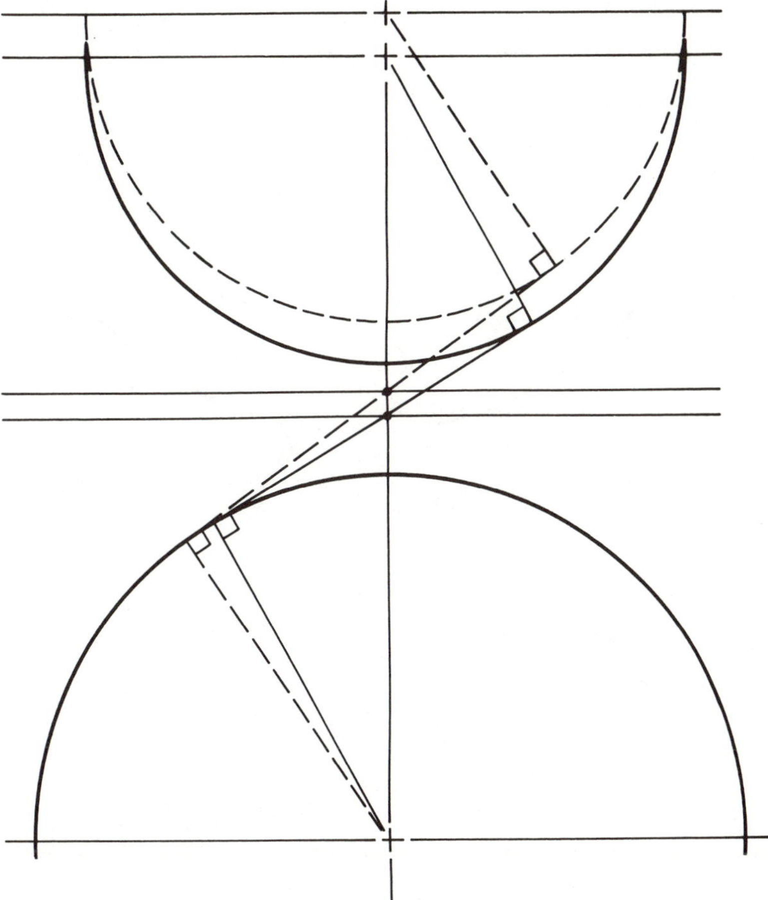

Figure 8.13 Zone of meshing gear.

7. Dedendum, d, is the radial distance between the pitch circle and the root diameter.
8. Clearance is the amount by which the dedendum in a gear exceeds the addendum of its mating gear.
9. Working depth is the depth of engagement of two mating gears, or the sum of their addendums.
10. Whole depth is the total depth of a tooth space, equal to addendum plus dedendum, or the working depth plus the clearance.
11. Backlash is the amount by which the width of a tooth space exceeds the thickness of the engaging tooth on the pitch circles.
12. Face width is the length of the teeth in an axial plane.

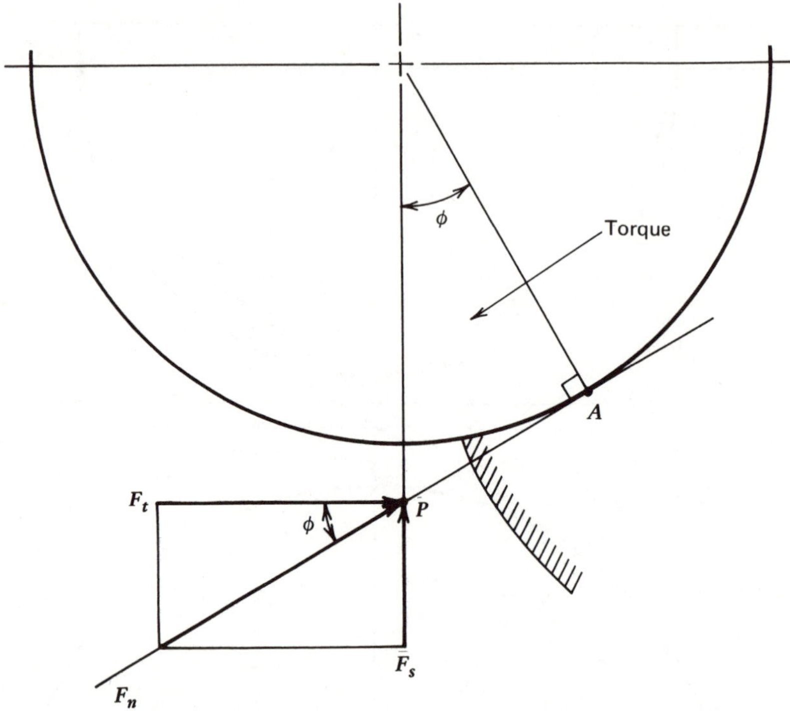

Figure 8.14 Spur gear tooth forces.

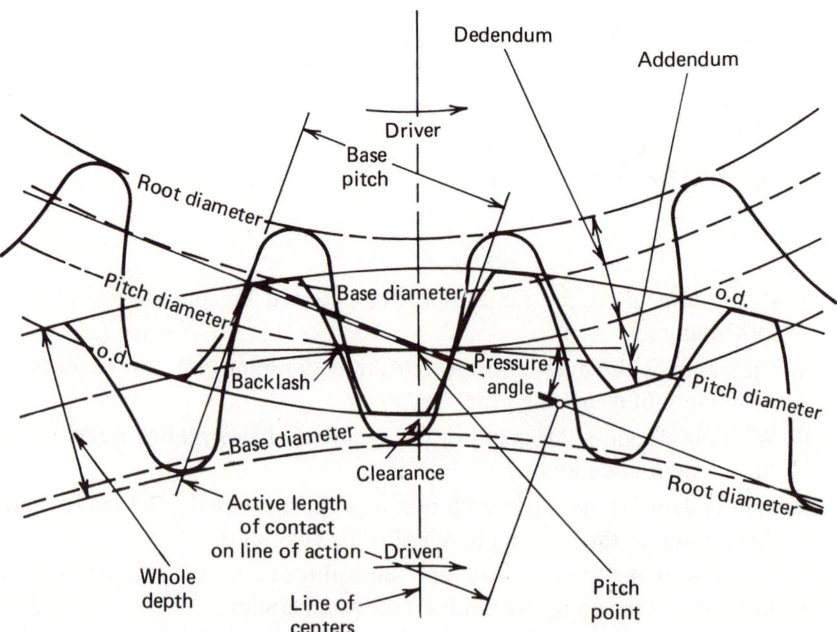

Figure 8.15 Gear nomenclature.

8.16 INVOLUTE ACTION

In the prior section on the generation of the involute, it was shown that the common normal to the two involute surfaces is tangent to the base circles. This common normal is referred to as the line of action. The beginning of contact occurs when the line of action intersects the outside diameter of the gear, and the end of contact occurs when the line of action intersects the outside diameter of the pinion. The ratio of the utilized length of the path of contact to the circular base pitch is referred to as the involute contact ratio; it is also the mean number of pairs of teeth carrying the load. Spur gears normally operate at an involute contact ratio of 1 : 3 or more. As pitch line velocities approach a limit, an involute contact ratio of 2 is desirable for quiet operation.

An equation for the utilized length of the path of contact, Za, is derived from Fig. 8.16

where P_1 = beginning of contact
P_2 = end of contact
P_3 and P_4 = tangency points of line of action and base circle
R_o = outside radius
R_b = base radius
C = center distance
ϕ = pressure angle

Therefore,

$$Za = P_1 P_2 = P_3 P_4 + P_3 P_1 + P_2 P_4 \tag{8.41}$$

then

$$Za = (R_{o_1}^2 - R_{b_1}^2)^{1/2} + (R_{o_2}^2 - R_{b_2}^2)^{1/2} - C \sin\phi$$

The base pitch

$$P_b = \frac{2\pi R_b}{N} \tag{8.42}$$

where R_b = base radius
N = number of teeth

The contact ratio, r_i, is then

$$r_i = \frac{Za}{P_b} \tag{8.43}$$

Since an involute starts at the base circle and is generated outward, it is impossible to have an involute inside the base circle. The line of action is tangent to the two base circles of a pair of gears in mesh, and these points represent the extreme limits of the length of action, as shown in Fig. 8.16. These two points are known as interference points. If the beginning of contact occurs before the interference point occurs, then the involute portion of the

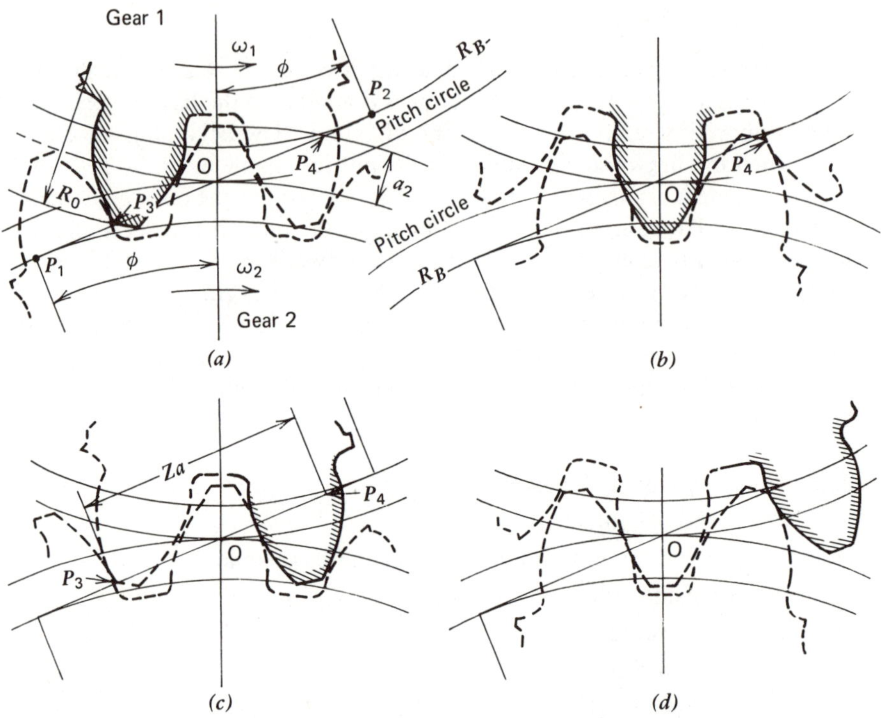

Figure 8.16 Contact zone for involute gears.

driven gear will contact a noninvolute portion of the driving gear, resulting in involute interference. To eliminate interference, the outside diameter must be decreased so that it passes through the start of the line of action, resulting in a new start of contact. This interference can be avoided if the maximum addendum radius for each gear is equal to or less than $[(\text{base circle radius})^2 + (\text{center distance})^2 (\sin\phi)^2]^{1/2}$. Involute interference is of major importance in regards to the manufacture and use of gears with small numbers of teeth.

Interchangeable involute gears must have the same pitch, the same pressure angle, the same addendum, and a circular thickness equal to one-half the circular pitch.

8.17 GEAR TOOTH LOADS

Figure 8.17 shows an involute gear with its center at O_2, rotating at a speed of RPM_1 and driving a second gear at a speed of RPM_2 with its center at O_3. The resultant forces occur along the pressure line $P_1 - P_2$. In Fig. 8.17, the two spur gears are separated from each other and their action replaced by the resultant load F, which is directed along the pressure line. Because the two gears are supported by the shafts, the action of the shafts may be replaced by another load F, oppositely directed and at the centers of the shafts.

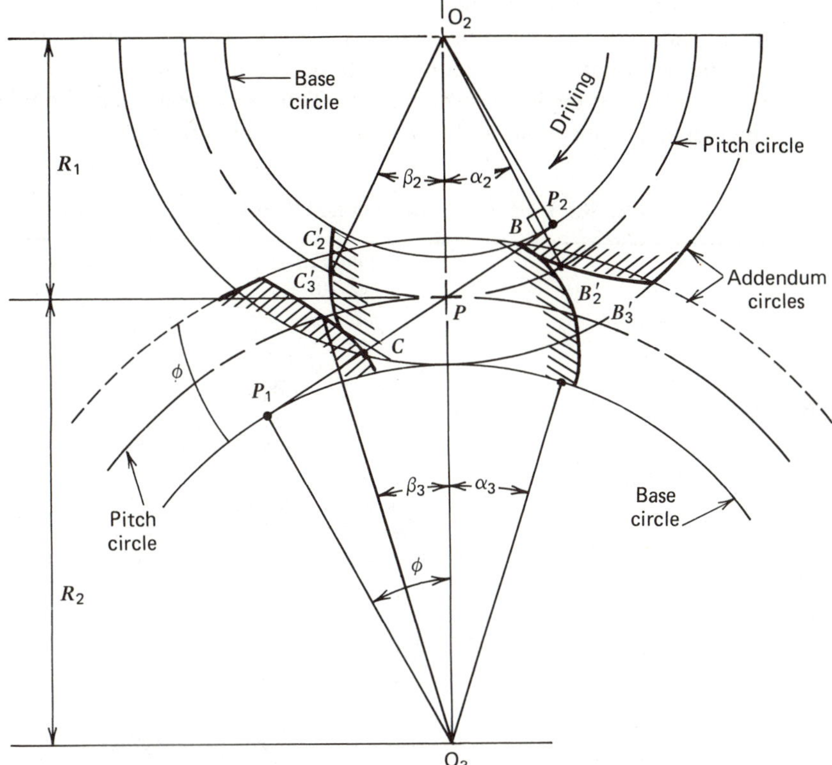

Figure 8.17 Gear tooth action.

The resultant load may be resolved into tangential and radial load components. The radial component acts to separate the gears while the tangential component transmits the power. The useful load on a gear tooth, called the transmitted load, F_t, is calculated from

$$F_t = \frac{T}{D/2} = \frac{19.1 \times 10^6\, KW}{D \times RPM} \qquad [8.44]$$

where T = torque, N · m or J

D = pitch diameter, mm

KW = power, kW

RPM = speed of shaft, rpm

8.18 INVOLUTE SPUR GEARS

Spur gears have the teeth parallel to the axis of the gears. They transmit power between parallel shafts at a constant angular velocity. The basic pressure angle is 20°, but a pressure angle of 14½° may be used. The proportions for standard spur gears are found in Table 8.3.

In the operation of spur gears, the tooth on one gear instantaneously makes full-length contact with a tooth on the mating gear. This sudden contact means that a sudden deformation of the one tooth occurs to compensate for the deflection under load of the teeth just engaged. Thus, spur gears should not be used for high-speed applications and when noise levels are critical.

Table 8.3 **Proportions for Standard Spur Gears**

Quantity	*Coarse Pitch** ($< 20P$) *full depth*		*Fine Pitch* ($\geqslant 20P$) *full depth*
Pressure angle	20°	25°	20°
Addendum	$\frac{1.000}{P}$	$\frac{1.000}{P}$	$\frac{1.000}{P}$
Dedendum	$\frac{1.250}{P}$	$\frac{1.250}{P}$	$\frac{1.200}{P} + 0.002\ in.$
Working depth	$\frac{2.000}{P}$	$\frac{2.000}{P}$	$\frac{2.000}{P}$
Whole depth (min)	$\frac{2.25}{P}$	$\frac{2.25}{P}$	$\frac{2.200}{P} + 0.002\ in.$
Circular tooth thickness	$\frac{\pi}{2P}$	$\frac{\pi}{2P}$	$\frac{1.5708}{P}$
Fillet radius of basic rack	$\frac{0.300}{P}$	$\frac{0.300}{P}$	Not standardized
Basic clearance (min)	$\frac{0.250}{P}$	$\frac{0.250}{P}$	$\frac{0.200}{P} + 0.002\ in.$
Clearance c (shaved or ground teeth)	$\frac{0.350}{P}$	$\frac{0.350}{P}$	$\frac{0.3500}{P} + 0.002\ in.$
Minimum number of pinion teeth	18	12	18
Minimum number of teeth per pair	36	24	
Minimum width of top load	$\frac{0.25}{P}$	$\frac{0.25}{P}$	Not standardized

* But not including $20P$; in inches.

8.19 INVOLUTE HELICAL GEARS

In helical gears, the line of contact is diagonal across the face of the tooth. This gradual engagement between the teeth of mating gears allows the load to be transferred from one tooth to another at high speeds and makes it possible to use helical gears in transmissions to reduce noise levels.

As the name implies, the teeth of helical gears form true involute helices, which make a constant angle with the axis of rotation, helix angle ψ. The helix angle is the same for mating gears, but one gear has a right-hand helix while the other has a left-hand helix.

With helical gears, the loads are transmitted along the normal between mating teeth. From Fig. 8.18, the relationship between the angles for the normal

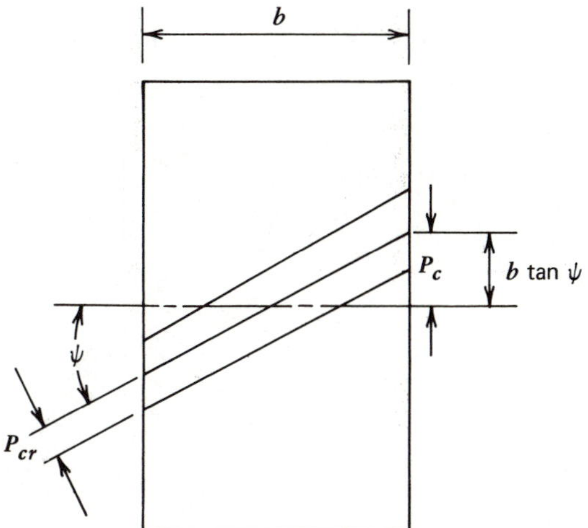

Figure 8.18 Overlap of helical gear teeth.

plane and the plane of rotation is

$$P_n = P_c \cos\psi \tag{8.45}$$

where P_n = normal circular pitch.

Since $P_c\ P = \pi$, the normal diametral pitch is

$$P_n = \frac{\pi\cos\theta}{P} \tag{8.46}$$

From Fig. 8.19, the normal pressure angle ϕ_n is related to the pressure angle ϕ and the helix angle ψ by the equation

$$\tan\phi_n = \tan\phi\cos\psi \tag{8.47}$$

Thus the load transmitted by the gear is

$$F_t = F_n \cos\phi_n \cos\psi \tag{8.48}$$

where F_n = normal load, in N.

The separating force between the gears is

$$F_s = F_t \tan\phi \tag{8.49}$$

The axial-thrust load component is

$$F_a = F_t \tan\psi \tag{8.50}$$

A pair of teeth should be in contact because of the twist of the helix. From Fig. 8.18 this relationship is

$$b \tan\psi = P_c \tag{8.51}$$

where b = the helical gear width.

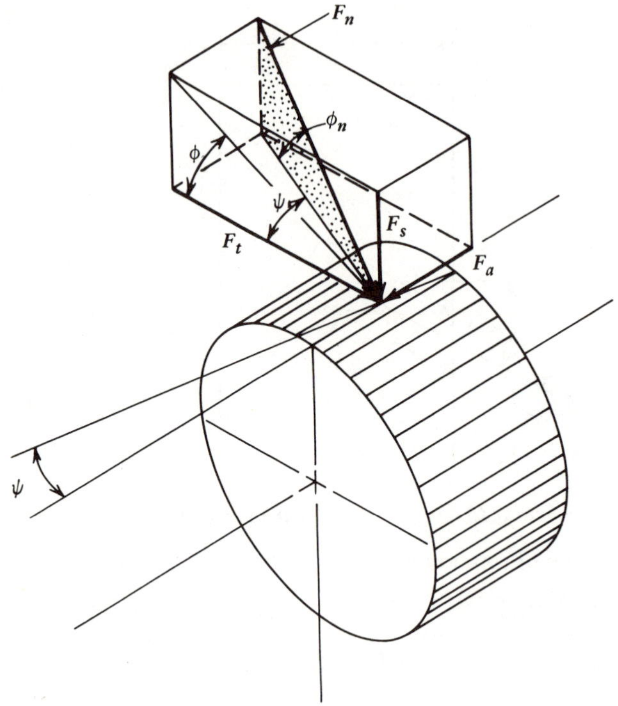

Figure 8.19 Helical gear tooth forces.

It is recommended that the gears overlap approximately 15% of the circular pitch.

The helix angle introduces a thrust load that results in an additional bending moment on the shaft. This condition requires at least one bearing to carry the thrust load.

8.20 BEVEL GEARS

Bevel gears are the most common means of transmitting power and motion between intersecting shafts. These gears are based on pitch surfaces that are frustums of cones. The basic pressure angle is 20°, but angles from 14 ½ to 25° may be used.

The three basic types of bevel gears are 1. straight, 2. spiral, and 3. Zerol. On straight bevel gears, the teeth are straight and tapered and, if extended inward, would intersect the gear axis. Spiral bevel gears have oblique teeth, which contact each other gradually and smoothly from one end to the other. Zerol bevel gears have curved teeth, (similar to the spiral bevels) with zero-degree spiral angle at the middle of the face width, resulting in little end thrust.

Figure 8.20 shows a cross-sectional view of two bevel gears in mesh to illustrate the nomenclature and dimensions. The elements of the pitch and root

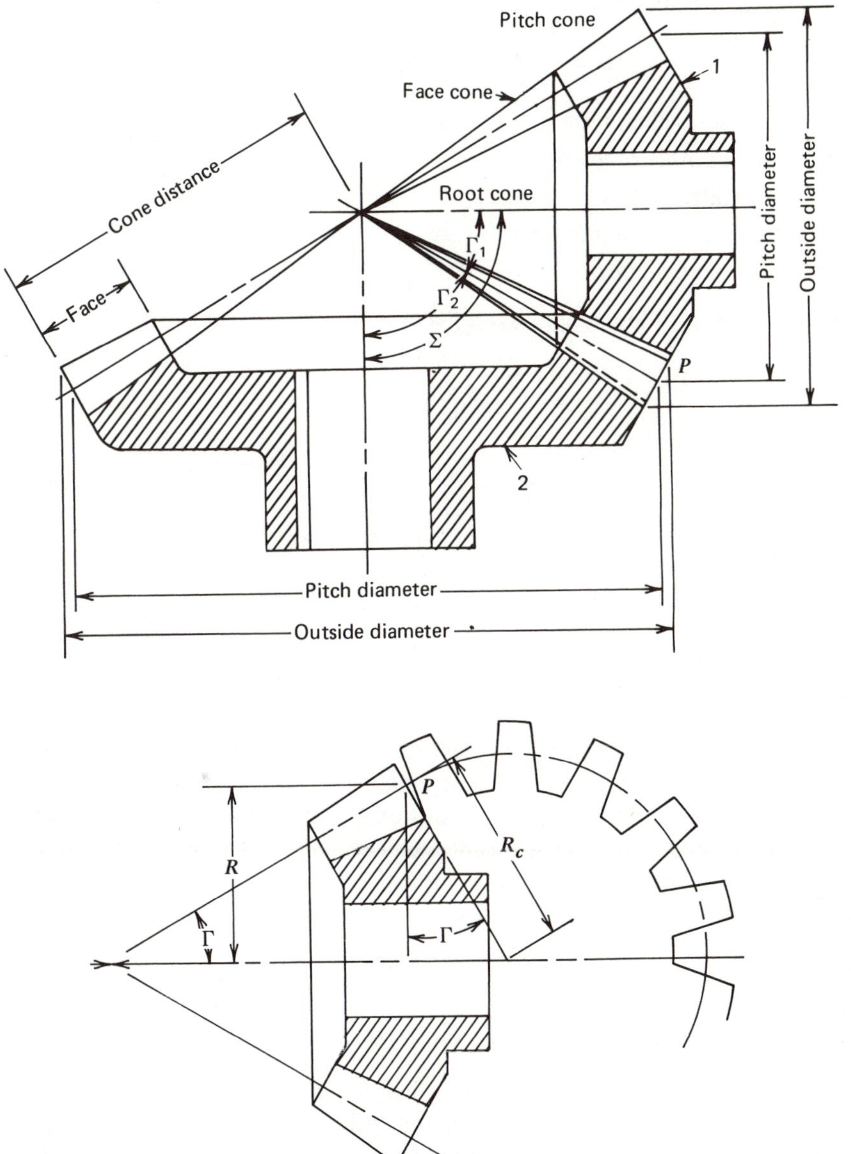

Figure 8.20 Bevel gear nomenclature.

cones intersect at the pitch apex or shaft intersection. The face cone does not interact at this location; rather, the face cone of one gear is parallel to the root cone of the other.

The action of bevel gears is the same as that of equivalent spur gears. A pair of bevel gears has a greater contact ratio and runs more smoothly than a pair of spur gears with the same number of teeth because the equivalent number of teeth is greater than the actual number of teeth. This relationship is

$$Ne = 2PR_e = \frac{2PR}{\cos\Gamma} = \frac{N}{\cos\Gamma} \quad [8.52]$$

where Ne = equivalent number of teeth
N = number of teeth
R = pitch radius
R_e = equivalent pitch radius
Γ = pitch angle

The pitch angle for the mating bevel gears is found from

$$\tan\Gamma_1 = \frac{\sin\Sigma}{N_2/N_1 + \cos\Sigma} \quad [8.53]$$

and

$$\Gamma_2 = \Sigma - \Gamma_1 \quad [8.54]$$

where Σ = angle of intersection. For the usual case of $\Sigma = 90^\circ$, Eq. 8.53 becomes

$$\tan\Gamma_1 = \frac{N_1}{N_2} \quad [8.55]$$

It is recommended that the face width not exceed 0.3 of the cone distance, or $10/P$, whichever is the smaller.

The straight bevel gear system is the most common for use in agricultural machinery. It may be used with gears containing at least 13 teeth. For the purpose of calculating shaft bending moments and bearing loads, the transmitted force F_t is assumed to act at the mean pitch diameter. As shown in Fig. 8.21, the radial and axial separating forces are calculated from the pressure, pitch angles, and the transmitted force as

$$F_R = F_t \tan\phi \cos\Gamma \quad [8.56]$$

and

$$F_s = F_t \tan\phi \cos\Gamma \quad [8.57]$$

Straight bevel gears are not recommended for use with peripheral speeds greater than 5 m/sec, or 1000 rpm, along the entire face width.

With spiral bevel gears the teeth are circular arcs, and the spiral angle θ is measured at the midpoint of the face, as shown in Fig. 8.21. The gear-tooth-force analysis must consider the additional radial and axial separating force components. The equations for the total radial and separating forces are

$$F_R = \frac{F_t}{\cos\theta}(\tan\phi\cos\Gamma \pm \sin\theta\sin\Gamma) \quad [8.58]$$

and

$$F_s = \frac{F_t}{\cos\theta}(\tan\phi\sin\Gamma \pm \sin\theta\cos\Gamma) \quad [8.59]$$

The hand of the gears, the direction of rotation, and the driving gear must be considered in deciding whether the plus or minus is used.

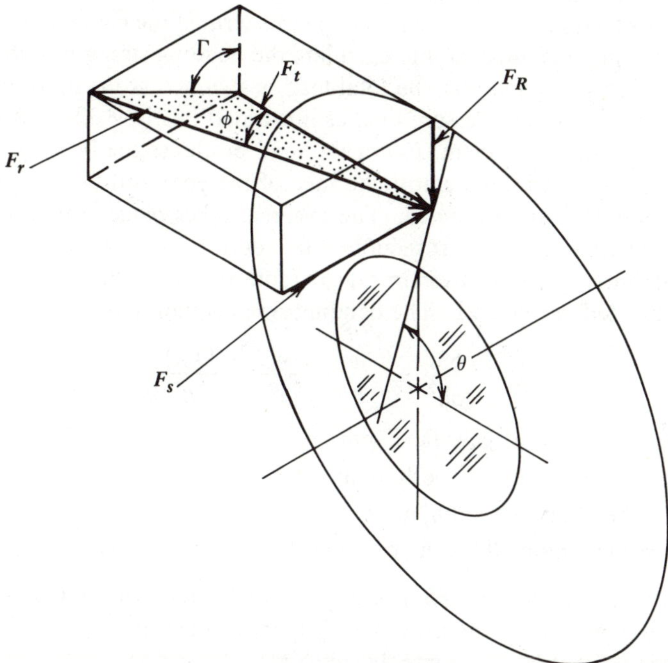

Figure 8.21 Straight bevel gear tooth forces.

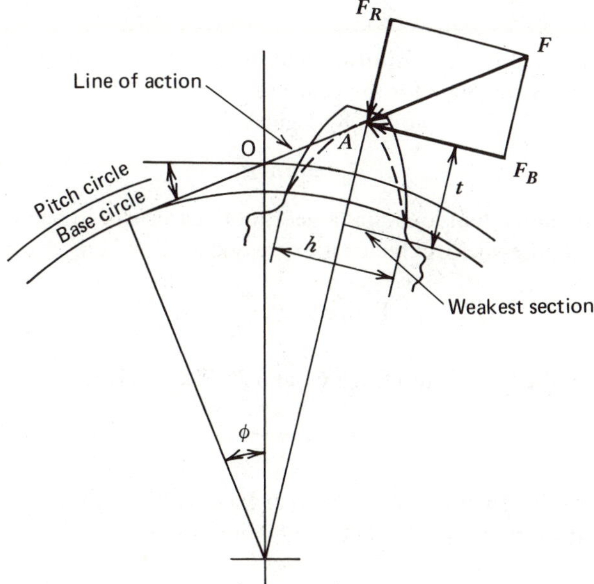

Figure 8.22 Gear tooth loads.

8.21 GEAR TOOTH STRENGTH

The number of pairs of teeth on mating gears varies from one to two or more as the teeth rotate through the load-contact zone. The number of teeth on one gear in contact with the teeth on another gear is termed the contact ratio.

In gear design, it is important to calculate the bending strength of the loaded gear tooth. For stress analysis, the total load is assumed at the tip of one tooth and is applied along the pressure line, as illustrated in Fig. 8.22. It is assumed that the tangential component, F_B, of the load produces the bending moment and is uniformly distributed across the face of the gear and that the effects of stress concentration are neglected. The moment takes place at the tooth base, or at the weakest and narrowest section of the tooth.

The maximum bending stress in a gear tooth occurs at the weakest section line, as illustrated in Fig. 8.22. The bending stress equation is

$$\sigma = \frac{6M}{bh^2} = \frac{F_B}{b} \times \frac{6t}{h^2} = \frac{6F_B t}{bh^2} \qquad [8.60]$$

where b = width of the gear face, mm
h = thickness of the tooth, mm
t = height of the tooth, mm
σ = maximum allowable bending stress, μPa

The geometric factor ($6t/bh^2$) is dependent on the size and shape of the gear, the number of teeth, and whether the gear is internal or external and may be written as a function of the circular pitch as

$$P_c\, y = \frac{h_2}{6t} \quad or \quad y = \frac{h^2}{6t\ P_c} \qquad [8.61]$$

The term y is called the Lewis form factor and is dependent on the number of teeth and the pressure angle for the gear used.

Substitution of Eq. 8.60 into Eq. 8.61 gives

$$F_B = \sigma b y P_c \qquad [8.62]$$

which gives the tangential load that a gear tooth can carry in beam action.

Sometimes the Lewis form factor is expressed as Y, where

$$Y = \pi y = \frac{\pi h^2}{6tP_c} \qquad [8.63]$$

Substitution of Eq. 8.62 into Eq. 8.61 and $P_c P = \pi$ gives

$$F_B = \frac{\sigma b y}{P} \qquad [8.64]$$

For ordinary design conditions, the gear face width is limited to a maximum of four times the circular pitch. The transmitted force is

$$F_B = \Gamma P_c^2 k y = \Gamma \pi^2 \frac{ky}{P^2} \qquad [8.65]$$

where $b = kP_c$
$k \leqslant 4$

$$F_B = 2M_t/D = (\text{gear torque}/\text{gear pitch radius}) \tag{8.66}$$

If the pitch diameter is known, the Lewis equation is written as

$$\frac{P^2}{y} = \frac{\Gamma k \pi^2}{F_B} \tag{8.67}$$

where $k = 4$. If the pitch diameter is unknown, the Lewis equation is written as

$$\Gamma = \frac{2M_t P^3}{k\pi^2 yN} \tag{8.68}$$

where Γ = stress $\leqslant$ allowable stress
M_t = torque or weaker gear
k = 4, upper limit
N = number of teeth on weaker gear, which is equal to or greater than 15

Example 8.4

A spur gear train is shown in Fig. 8.23. Spur gear A receives 3 kW of power through its shaft and rotates clockwise at 600 rpm; gear B is an idler, and gear C is driven. The gears are 20° full depth and have a diametral pitch of 0.2 teeth per millimeter. Determine 1. the torque that each shaft must transmit, 2. the tooth load to be designed for each gear; and 3. the force applied to the idler shaft as a result of the gear-tooth loads.

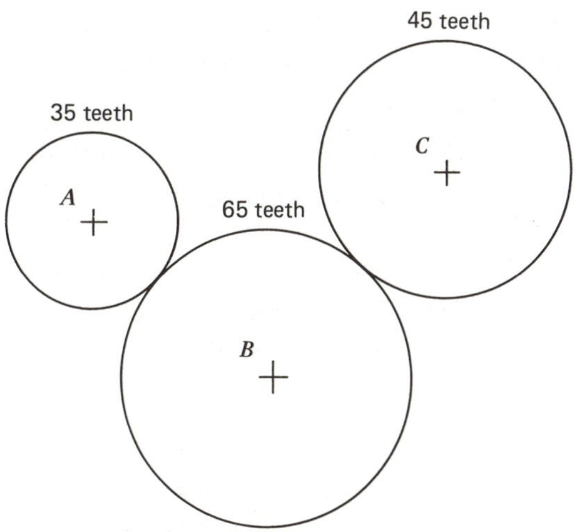

Figure 8.23 Spur gear train.

a. Gear diameters.

$$D_A = \frac{35}{0.2} = 175 \text{ mm}$$
$$D_B = \frac{65}{0.2} = 325 \text{ mm}$$
$$D_C = \frac{45}{0.2} = 225 \text{ mm}$$

b. Torque on shafts of gears.

$$T = 9.549 \times 10^6 \frac{KW}{N} \quad [8.69]$$

where T = shaft torque, N · mm
KW = power, kW
N = shaft speed, rpm

$$T_A = \frac{(3)(9.549 \times 10^6)}{600} = 47{,}745 \text{ N} \cdot \text{mm}$$
$$T_B = 0$$
$$T_C = \frac{3(9.549 \times 10^6)}{600(35/45)} = 61{,}390 \text{ N} \cdot \text{mm}$$

c. Tangential force on gear A.

$$F_t = \frac{2M_t}{D_A} = \frac{2(47{,}745)}{175} = 545.7 \text{ N}$$

d. Separating force on gear A.

$$F_S = 545.7 \tan 20 = 198.6 \text{ N}$$

e. The same tangential and separating forces occur between gears A and B and between gears B and C.

f. The tooth load for each gear must be designed for 545.7N.

g. The force applied to the idler gear shaft is the resultant of the forces applied to gear B by gears A and C.

$$F_B = \sqrt{(545.7 + 198.6)^2 + (545.7 + 198.6)^2}$$
$$F_B = 1052.6 \text{ N}$$

Example 8.5

A pair of mating spur gears have 14½° full depth involute teeth of 0.1 (teeth/mm) diameteral pitch. The pitch diameter of the smaller gear is 160 mm. If the transmission ratio is 3 to 2, determine

a. Number of teeth for each gear.

$$D_1 = 160 \text{ mm}$$
$$D_2 = 160\,(3/2) = 240 \text{ mm}$$
$$N_1 = D_1 P = 160(0.1) = 16 \text{ teeth}$$
$$N_2 = 240(0.1) = 24 \text{ teeth}$$

b. Addendum.

$$\frac{1}{P} = \frac{1}{0.1} = 10 \text{ mm}$$

c. Whole depth.

$$\frac{2.157}{P} = \frac{2.157}{0.1} = 21.57 \text{ mm}$$

d. Clearance.

$$\frac{0.157}{P} = \frac{0.157}{0.1} = 1.57 \text{ mm}$$

e. Outside diameters.

$$(\text{pitch} + 2 \times \text{ addendum})$$

$$OD_1 = 160 + 2(10) = 180\text{mm}$$
$$OD_2 = 240 + 2(10) = 260\text{mm}$$

f. Root diameters.

$$(\text{outside diameter} - 2 \times \text{whole depth})$$

$$D_{R1} = 180 - 2(21.57) = 136.86 \text{ mm}$$
$$D_{R2} = 240 - 2(21.57) = 198.86 \text{ mm}$$

g. Dedendum.

$$\frac{1.157}{P_D} = \frac{1.157}{0.1} = 11.57 \text{ mm}$$

h. Radius of base circle.

$$(\text{pitch radius} \times \cos\phi)$$

$$R_1 = \frac{160}{2} \cos 14\tfrac{1}{2} = 77.45 \text{ mm}$$

$$R_2 = \frac{240}{2} \cos 14\tfrac{1}{2} = 116.18 \text{ mm}$$

i. Interference is avoided if the addendum radius is

$$\leqslant \sqrt{(116.18)^2 + [\tfrac{1}{2}(160 + 180)]^2(\sin 14\tfrac{1}{2}^\circ)} = 123.73 \text{ mm}$$

Since the addendum radius $= (260 + 240)/4 = 125$ mm, interference will exist. Note: For $14\frac{1}{2}^\circ$ pressure angle, gear proportions are given in this example.

8.22 GEAR TOOTH FAILURES

Gears fail for many reasons, often random. The majority of failures result from fatigue failures caused by repeated loading of the individual teeth. The most important types of gear failure are:

1. Pitting, identified by small pits in the tooth surface below the pitch line.

2. Scoring and galling, identified by seized areas on the tooth surface. This failure is caused by metal-to-metal contact, welding, or the failure of the lubricant.
3. Spalling, identified by flaked out areas, does not have the longitudinal cracks across the face of the tooth.
4. Abrasive wear, identified by satiny, lapped appearance and irregular tooth profile.
5. Plastic yielding, identified by uneven lip ledges on the ends or tips of the gear teeth. Yielding occurs because the gears are subjected to heavy continuous or intermittent heavy loading conditions.
6. Fractures, identified by tooth breakage.

8.23 PARALLEL-AXIS GEAR TRAINS

The speed of a gear train can be calculated if the number of teeth on each gear and the shaft speeds are known. The gear trains illustrated in Fig. 8.24 are known as simple gear trains because only two gears are involved. The velocity ratio of a gear train is the ratio of angular velocity of the driving gear to that of the driven gear. Thus, the speed ratios for the gear trains are given by the following equation.

$$\frac{RPM_1}{RPM_2} = \frac{N_2}{N_1} \qquad [8.70]$$

where RPM = speed or angular velocity of shaft
N = number of teeth on the gear

The gear train illustrated in Fig. 8.25*a* is a compound gear train because more than two gears are involved. The intermediate gears between the first and last gears are called idler gears, and the train is known as an idler gear train. Idlers are used to fill the space between the first and last shafts or to change the direction of rotation of the last shaft relative to the first. If an odd

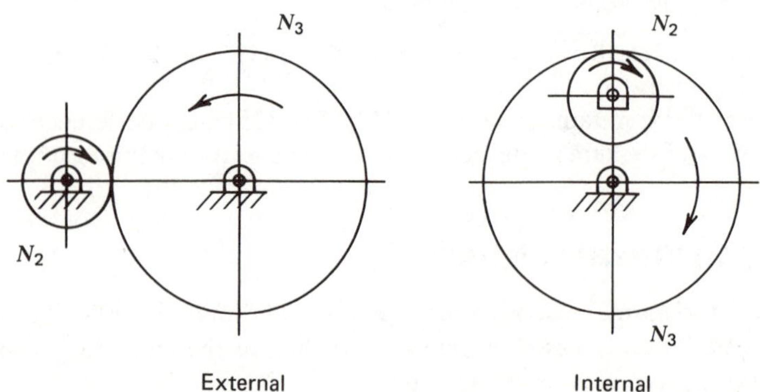

Figure 8.24 Simple gear trains.

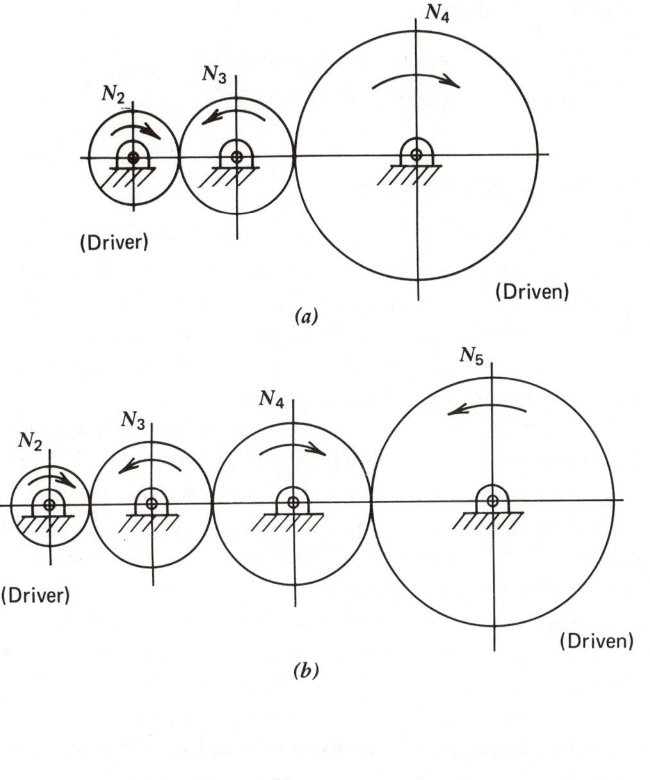

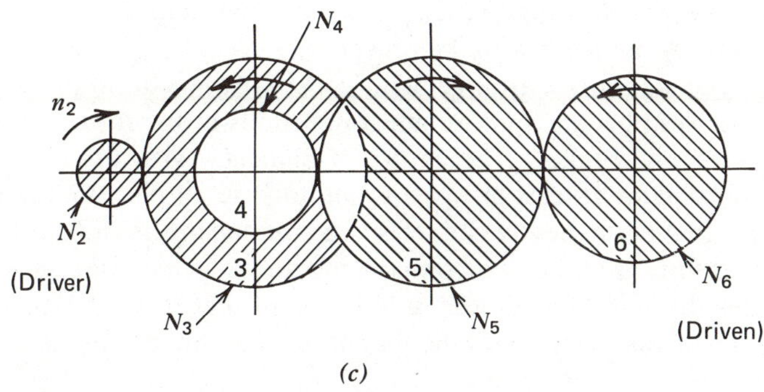

Figure 8.25 Compound gear trains.

number of gears are used, as illustrated in Fig. 8.25*a*, the driver and driven gears rotate in the same direction; if an even number of gears are used, as illustrated in Fig. 8.25*b*, the driver and driven gears rotate in the opposite directions. If one shaft has two gears mounted on it and an odd number of gears are used as illustrated in Fig. 8.25*c*, the driver and driven gears rotate in the opposite directions.

The velocity ratio for a compound gear train is the ratio of the products of the driven tooth numbers to the product of the driver tooth numbers. For example, the speed ratio of the train in Fig. 8.25*c* is

$$\frac{RPM_2}{RPM_6} = \frac{\text{Product of driven tooth numbers}}{\text{Product of driving tooth numbers}} = \frac{N_3 N_5 N_6}{N_2 N_4 N_5} \qquad [8.71]$$

8.24 PLANETARY GEAR TRAINS

Planetary gear drives offer several advantages over conventional stationary-drive gear mechanisms for transmitting power and motion. These advantages are 1. higher speed ratios, 2. compactness, 3. in-line arrangement of input and output shafts, 4. lower overall weight, 5. lower pitch-line velocities, and 6. lower loadings on the teeth. Examples of planetary gear drives are found in the transmissions and final drives of agricultural tractors.

As illustrated in Fig. 8.26*a*, a simple planetary drive arrangement consists of the following basic components: 1. planetary ring, 2. sun gear, 3. three planet pinions, and 4. carrier. In this arrangement of components, power can enter into the system through the sun gear, the ring gear, or the carrier. In addition, these components may rotate in either the clockwise or counterclockwise direction. This simple planetary drive constitutes a row because the gears mesh in one plane. In Fig. 8.26*a*, the sun gear is the center gear and the ring gear is the outer gear. The planet pinion gear meshes with the sun gear and the ring gear. The carrier is the device to which the planet pinions are attached.

A reversing planetary drive arrangement consists of a simple planetary drive with a set of reversing planet pinions, as illustrated in Fig. 8.26*b*. The reversing pinion always meshes with the planet pinion and never meshes with the sun gear. It may or may not mesh with the ring gear.

Compound planetary drive arrangements consist of rows of simple or reversing planetary drives plus the appropriate connections of the rows with the use of links, clutches, and brakes. The nature of planetary gear drives mandates that every row contains at least one carrier and one set of planet pinions. The link is a device that permanently connects two components in different rows at all times and ensures that the two components rotate at the same speed. The clutch provides the same function as a link except that it can be engaged or disengaged. The brake selectively holds components in the planetary drive arrangement.

Most planetary drive systems are complex kinematic arrangements used to provide numerous combinations of input and output speeds. It is thus difficult to calculate the velocity, force, and torque relationships between the various components.

Many methods have been used to analyze planetary drive arrangements. The best method is the building block approach, described by Hanson. This method is a simple and systematic approach and is adaptable to computer programming.

The building-block approach establishes the following eight planetary building blocks: 1. carrier, 2. pinion gear, 3. reversing pinion gear, 4. sun gear, 5.

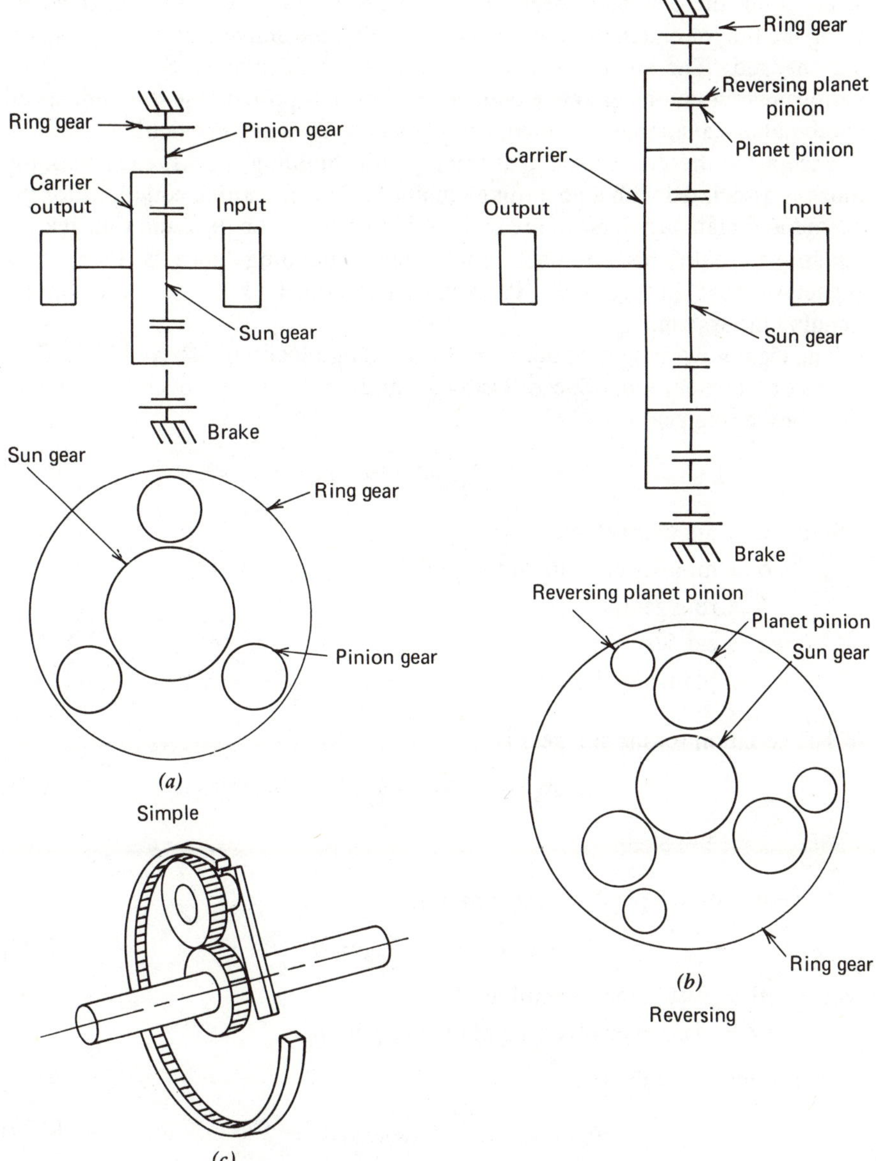

Figure 8.26 Planetary gear arrangements.

ring gear, 6. link, 7. clutch, and 8. brake. The first five building blocks provide the capability to define a simple planetary drive and a reversing planetary drive; the last three building blocks provide the capability to define compound planetary drive arrangements.

The process involves first defining the single planetary row with the first five building blocks. The process of defining the planetary rows is repeated successively until all rows have been described. Next, the rows are connected by using the link or clutch building blocks. Finally, the active clutches and brakes are engaged. The angular velocities can now be calculated. Selection of the various combinations of active clutches and brakes provide the different speed combinations available with the given planetary drive arrangement.

Except for the carrier and planetary pinion building blocks, each building block is associated with a governing equation. The governing equations for the carrier and planetary pinions are included in both the sun gear and ring gear. equations. With this approach, each time a building block is added to a planetary drive arrangement, its equation is added to the set of equations defining the system.

The eight governing equations for the building blocks are derived in "Calculation of Planetary Gear Speeds Using a Building Block Approach." The equation for the ring gear is

$$\omega_R N_R - \omega_c N_R - \omega_c N_p = 0 \qquad [8.72]$$

where ω = angular velocity
N = number of teeth on the gear
R = ring gear
c = carrier
p = planetary pinion

The equation for the sun gear is

$$\omega_S N_S - \omega_c N_S + \omega_p N_p = 0 \qquad [8.73]$$

where S = sun gear.

The equation for the reversing pinion is

$$\omega_{p1} N_{p1} + \omega_{p2} N_{p2} = 0 \qquad [8.74]$$

where $p1$ = first planetary pinion
$p2$ = second or reversing planetary pinion

The equation for the link is

$$\omega_{(\text{component } 1)} - \omega_{(\text{component } 2)} = 0 \qquad [8.75]$$

The equation for the clutch is

$$\omega_{(\text{component } 1)} - \omega_{(\text{component } 2)} = 0 \qquad [8.76]$$

The equation for the brake is

$$\omega_{(\text{component})} = 0 \qquad [8.77]$$

The following example illustrates how the building-block approach is applied.

Example 8.6

The simple planetary drive shown in Fig. 8.26*a*, has 75 teeth on the ring gear, 40 teeth on the sun gear, and 15 teeth on the planetary pinion gear. The sun gear is driven by the input, the carrrier drives the output, and the ring gear is braked. Determine the angular velocities of the output and the components.

For this system, six equations are required as follows.

$$\omega_R N_R - \omega_c N_R - \omega_p N_p = 0 \qquad [8.78]$$

$$\omega_S N_S - \omega_c N_S + \omega_p N_p = 0 \qquad [8.79]$$

$$\omega_S - \omega_{\text{in}} = 0 \qquad [8.80]$$

$$\omega_c - \omega_{\text{out}} = 0 \qquad [8.81]$$

$$\omega_R = 0 \qquad [8.82]$$

$$\omega_{\text{in}} = 1 \qquad [8.83]$$

Substituting the number of teeth into Eqs. 8.78 to 8.81 and assuming an arbitrary angular velocity of 1 for the input, the soution is

$$\begin{aligned} \omega_{\text{in}} &= 1.0 \\ \omega_{\text{out}} &= 0.3478 \\ \omega_R &= 0.0 \\ \omega_S &= +1.0 \\ \omega_c &= 0.3478 \\ \omega_p &= 1.7391 \end{aligned}$$

8.25 BEARINGS

Bearings for agricultural machinery permit relative motion of two parts in one or two directions with a minimum of friction while resisting motion in the direction of the applied loads. In general, bearings may be separated into two major classes: 1. plain or journal bearings, which pass the load through the surfaces of the parts by sliding contact or separated by a film of lubricant, and 2. antifriction bearings, which pass the load through the surfaces of the parts by rolling contact.

8.26 PLAIN BEARINGS

Plain bearings are designed either to carry a radial load or to carry an axial or thrust load. Plain bearings have two major components, as illustrated in Fig. 8.27: 1. a journal, the inside cylindrical part, which is a rotating or oscillating shaft, and 2. a bearing, which is the stationary or mobile surrounding shell.

Journal bearings develop hydrodynamic pressure to carry the loads and separate the bearing elements in order to minimize friction. The load-carrying

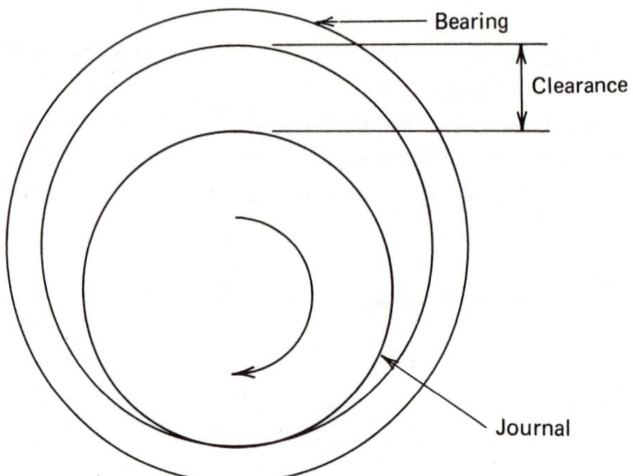

Figure 8.27 Journal bearing.

capacity arises because the lubricant resists being pushed around in the radial clearance zone. The bearing load per unit of projected area is directly proportional to the viscosity of the lubricant and the relative speed of rotation and inversely proportional to the diametral clearance. This clearance is between the diameter of the bearing and the journal. A complete discussion of hydrodynamic lubrication is found in Design of Machine Elements, Mechanical Engineering Design.

Many of the plain or journal bearings in agricultural machinery do not have thick-film hydrodynamic lubrication but have only thin-film or boundary lubrications.

Advantages of plain bearings are:

1. Little or no service requirements.
2. Lower cost.
3. Greater tolerances between the journal and the bearing.
4. No elaborate enclosure requirements.

An example of the plain bearing is the babbit-metal bushing. This bushing has the ability to conform to misalignment, to resist corrosion and grit, and to operate against steel without scoring.

8.27 ROLLING CONTACT BEARINGS

Rolling-contact bearings carry the load from the rotating shaft on balls or rollers. Rolling-contact bearings must fulfill the following requirements.

1. The rolling elements must be properly aligned in their motion.
2. The rolling elements must not be overloaded.
3. Unavoidable sliding should be minimized.

4. All rolling elements must be exactly the same size.
5. All rolling elements and the races must be extremely hard and polished.
6. The load should be approximately normal to the surface of contact.

Some advantages of rolling-contact bearings are:

1. Relatively accurate shaft alignment can be maintained over long periods of time.
2. Friction is low except at high operating speeds.
3. Heavy momentary loads can be carried without failure.
4. Lubrication is simple and requires little attention.
5. They are suitable for low speeds.
6. Replacement in case of failure is easy.
7. Starting friction and required torque are low.

Some disadvantages of rolling-contact bearings are:

1. The expense is generally greater because of the cost of the bearing and the necessary provision for mounting.
2. Failure can occur without warning.
3. Resistance to shock loads is lower.
4. Sensitivity to dirt and foreign matter is high.

Rolling-contact bearings may be divided into two main classes: 1. ball bearings and 2. roller bearings with cylindrical, conical, spherical, or concave rollers. Each class may be subdivided into the following load-carrying types.

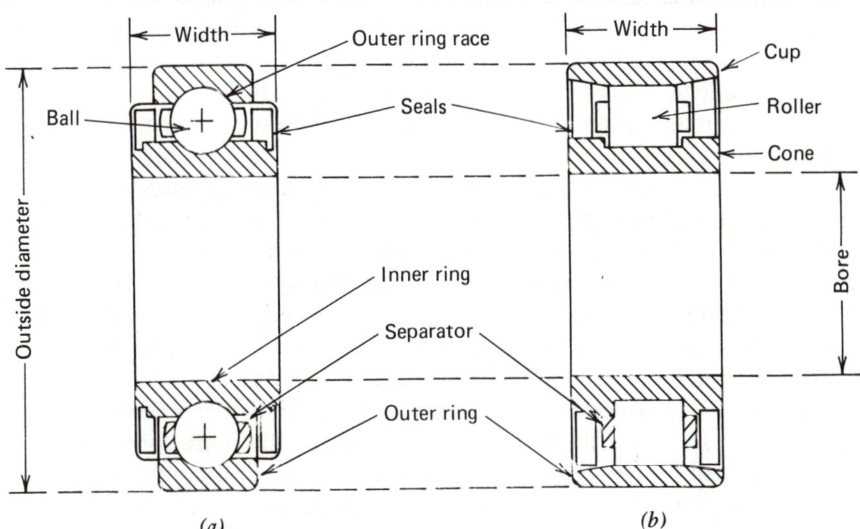

Figure 8.28 Rolling-contact bearings. (*a*) Ball bearing. (*b*) Roller bearing.

1. Radial.
2. Thrust or axial.
3. Combination of radial and thrust.

The two main classes of bearings and their nomenclature are illustrated in Fig. 8.28. The four components of ball bearings are the inner race, grooved on its outer surface; the outer race, grooved on its inner surface; the balls; and the separator cage, which spaces the balls and reduces wear and noise. The four components of roller bearings are the cone, the cup, the rollers, and the cage. Seals may be added to roller or ball bearings to prevent the entrance of dirt and the loss of the lubricant.

Ball and roller bearings are available in a wide selection of sizes and types to handle any load condition or service application. Some of the different types of ball and roller bearings are illustrated in Fig. 8.29. Bearing manufacturers' catalogs provide descriptions of the bearing types and technical data on the various sizes of each type.

8.28 BEARING LIFE

The basic parameters for selecting the bearing size are 1. the radial load, 2. the thrust load, 3. speed, 4. the required life, 5. race rotation, and 6. shock or vibration conditions. Other factors that have an indeterminate effect on service life are 1. misalignment, 2. abnormal temperature, and 3. contamination and poor lubrication.

When bearings are loaded, elastic deformations occur on the surfaces at the regions of contact between the bearing elements. The contact areas depend on the elastic properties of the materials, the size of the balls or rollers, and the shape of the races. The Hertz contact stresses are calculated from the geometry of the bearing.

The theoretical basis for bearing design is found in Rolling Bearing Analysis, but the factors for speed, life, and shock are based on performance data. For satisfactory bearing service, the elements must be accurately made from high-quality, heat-treated steels and must have finely finished surfaces. Bearings generally fail in fatigue from the stress applications to the elements.

The *rated load* of a bearing is the load that a bearing will carry for a specific time period when operating at a given speed. The *life* of a bearing is defined as the total number of revolutions or hours that a bearing can withstand at a given speed before any evidence of fatigue develops on the rolling elements or the races. The rated life of a group of bearings is defined as the number of hours or revolutions at which 90% of the bearing population will exceed a given speed and load before the first evidence of fatigue develops. This *rated life* is called the *B-10 life* and *minimum life.* The *average life* of a group of bearings is the number of revolutions or hours that 50% of the population will complete at a given speed and before the first evidence of fatigue develops.

The standard bearing rating is established without consideration for speed.

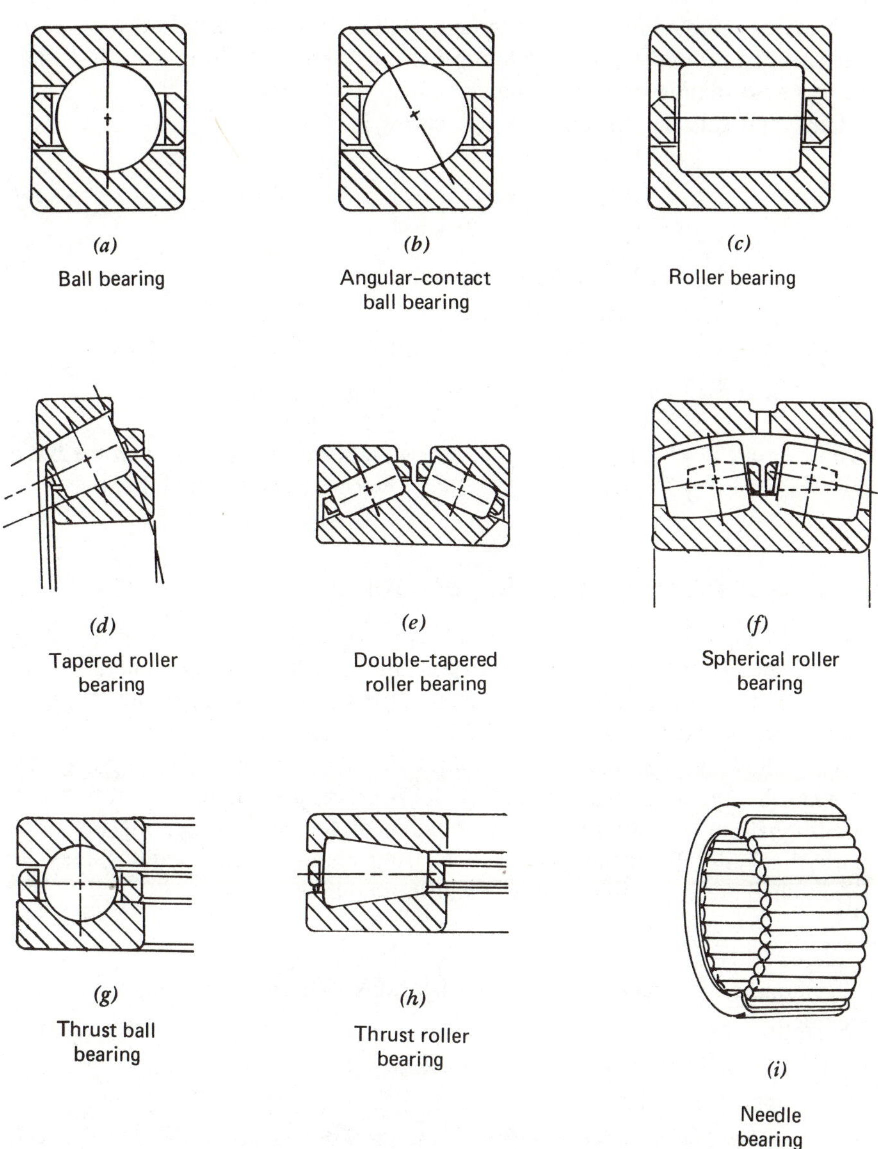

Figure 8.29 Types of ball and roller bearings. (*a*) Ball bearing. (*b*) Angular contact ball bearing. (*c*) Roller bearing. (*d*) Tapered roller bearing. (*e*) Double-tapered roller bearing. (*f*) Spherical roller bearing. (*g*) Thrust ball bearing. (*h*) Thrust roller bearing. (*i*) Needle bearing.

The *basic dynamic capacity* is the constant radial load that any bearing, in 90% of the bearing population, will carry for 500 hours at 33⅓ rpm (1 million revolutions) without evidence of fatigue. The *basic dynamic thrust capacity* is the constant central thrust load that 90% of the bearings will carry for 1 million revolutions.

The *basic static radial* or *thrust capacity* is the maximum radial or thrust load that can be imposed on a nonrotating bearing without causing excessive permanent deformation of the bearing elements.

The empirical relationship between bearing load and life is

$$L = \left[\frac{C}{F}\right]^n \qquad [8.84]$$

where L = bearing life, millions of revolutions
C = basic dynamic capacity, N
F = actual radial load, N
n = 3.33 for roller bearings, and 3 for ball bearings

For example, if a bearing life of 8 million revolutions is desired, then the basic dynamic capacity is $C = (8)^{1/3} F = 2F$, or twice the actual radial load.

8.29 BALL BEARING SIZE SELECTION

Optimum bearing performance is achieved by selection of the appropriate bearing and shaft to suit the service application. Bearing size is generally controlled by the shaft-bending and torsional stresses. In other applications, however, bearing load-carrying capacity is the determining design factor.

The following procedure is used to determine the required bearing capacity to meet the application requirements. This procedure converts all forces and conditions into a resultant equivalent radial forces on each bearing that supports the shaft. With these loads, a bearing manufacturer's catalog may be used to find the appropriate bearing load-rating table.

Step 1

Determine all forces acting on the bearings. The following load types are included:

1. *Static loads.* Weight of the shaft assembly and other machine elements supported by the bearings.
2. *Applied loads.* Forces applied on the shaft during normal operation, such as thrust loads on the shaft.
3. *Drive loads.* Forces transmitted from the power source to the shaft through V-belt or roller-chain drives and gears.

To determine the radial load acting on a shaft, the power transmitted by the drive is

$$F = \frac{KW \times 19.1 \times 10^6 \times K}{PD \times RPM} \qquad [8.85]$$

where F = radial force on the shaft, N
KW = power transmitted, kW
PD = pitch diameter of sheave, sprocket, and so on, mm
RPM = speed of shaft, rpm
K = drive tension factor with $K = 1$ for chain drives and gears and $K = 1.5$ for V-belt drives

For gear drives, the gear tooth separating force, F_s, must be calculated from Eqs. 8.37, 8.49, and 8.57. In eccentric loads an unbalanced shaft or eccentric mass generates a centrifugal force which acts radially on the shaft. The equation for eccentric loads is

$$CF = 0.00316 \times R \times W \times (RPM)^2 \quad [8.86]$$

where CF = centrifugal force, N
R = radius of gyration of eccentric loads, mm
W = weight of eccentric loads, N
RPM = speed of shaft, rpm

Step 2
Resolve the radial forces acting on the shaft into radial loads acting on each bearing. With engineering mechanics, the bearing's support forces may be calculated for the shaft with two bearings systems by means of simple free-body diagrams and the method of superposition.

Step 3
Combine the radial forces vectorially to obtain the resultant radial force acting at each bearing.

Step 4
Combine all thrust forces to determine the total thrust load acting on the shaft.

Step 5
Calculate the equivalent radial force using the total resultant radial and thrust forces. The method of converting thrust loads to equivalent radial forces depends on the bearing type and construction and varies among manufacturers. For shafts less than 45 mm (1.75 in.) in diameter, the maximum thrust factor is 2.2. If the thrust force is less than 46% of the radial force, the equivalent radial force is

$$P_e = P_R + 1.47\, P_t \quad [8.87]$$

where P_e = equivalent radial force, N
P_R = resultant radial force, N
P_t = resultant thrust force, N

Step 6

Apply the service modification factors and obtain the required radial load capacity to reflect the actual application conditions. For bearings with only radial loads, the required radial load capacity is

$$P = P_R \times M \tag{8.88}$$

while for bearings with combined radial and thrust loads, the required load capacity is

$$P = P_e \times M \tag{8.89}$$

where M is the operational and/or shock factors from Tables 8.2 and 8.3.

Step 7

Select the bearing size and type from the appropriate load-rating table in a bearing manufacturer's catalog.

Example 8.7

Determine the load rating for the two bearings supporting the shaft in Fig. 8.30. The drive transmits 15 kW and rotates at 200 rpm. The sprocket pitch diameter is 250 mm, and the sheave pitch diameter is 380 mm. The distances between the bearings and from the bearings to the sheave and sprocket are shown in Fig. 8.30.

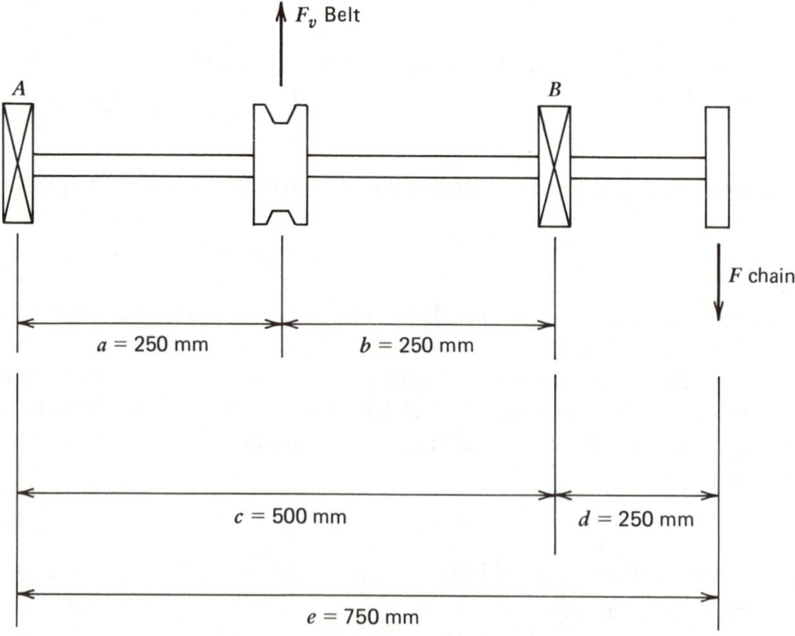

Figure 8.30 Bearing design example 8.7.

The radial force on the shaft from the chain drive is calculated by using Eq. 8.85.

$$F_{\text{chain}} = \frac{(15\,kW)(19.1 \times 10^6)(1)}{(250)(200)} = 5730 \text{ N}$$

The radial force on the shaft from the V-belt drive is calculated by using Eq. 8.85.

$$F_{\text{v-belt}} = \frac{(15\,kW)(19.1 \times 10^6)(1.5)}{(380)(200)} = 5655 \text{ N}$$

The pull of the chain and the pull of the V-belt are downward.

The radial loads on the bearings are computed from both the chain pull and the V-belt pull, using basic force and moment equations for static equilibrium from engineering mechanics.

Bearing A

Load from the chain pull

$$F_A = \frac{F_{\text{chain}} \times d}{c} = \frac{(5730)(250)}{(500)} = 2856 \text{ N}$$

where A denotes the downward direction.

Load from the V-belt pull

$$F_A = \frac{F_{\text{V-belt}} \times b}{c} = \frac{(5655)(250)}{500} = 2828 \text{ N}$$

Combining the two loads, the total load at bearing A is

$$F_A = 2856 + 2828 = 5693 \text{ N}$$

Bearing B

Load from the chain pull

$$F_B = \frac{F_{\text{chain}} \times e}{c} = \frac{(5730)(750)}{500} = 8595 \text{ N}$$

Load from the V-belt pull

$$F_B = \frac{F_{\text{V-belt}} \times a}{c} = \frac{(5655)(250)}{500} = 2828 \text{ N}$$

Combining the two loads, the total load at bearing B is

$$F_B = 8595 - 2828 = 5767 \text{ N}$$

where B denotes the upward direction.

The load at bearing B would be used to select the appropriate-sized bearing because it is greater than the load at bearing A.

In this example, all loads were in the same plane, which permitted the loads to be added or subtracted. In most applications, the radial forces are not all in the same direction or plane. Thus the radial forces must be resolved to vertical and horizontal components as

$$F_V = F_R \sin\theta \qquad [8.90]$$

$$F_H = F_R \cos\theta \qquad [8.91]$$

where F_V = vertical component, N
F_H = horizontal component, N
θ = angle between the force F and the base line

These vertical and horizontal force components are combined to obtain a resultant vertical and horizontal force component. The resultant radial force on a bearing is

$$P_R = \left[P_{RV}^2 + P_{RH}^2\right]^1/2 \qquad [8.92]$$

where P_R = resultant radial force
P_{RV} = total vertical radial force
P_{RH} = total horizontal radial force

8.30 MOUNTING AND PRELOADING AND ROLLING-CONTACT BEARINGS

The proper mounting of rolling-contact bearings to shafts is essential to obtain maximum bearing service reliability and to provide for all the required restraint without introducing any additional restraint. Various bearing manufacturers' manuals give many recommended mounting details in almost every design application.

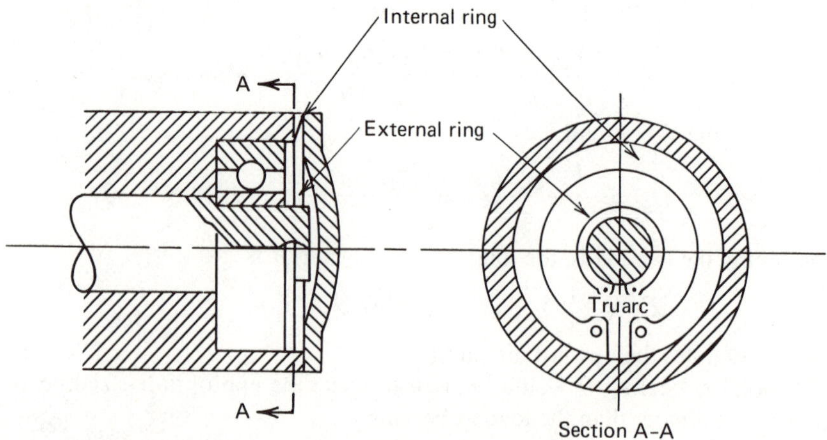

Figure 8.31 Self-locking retaining rings for bearing system.

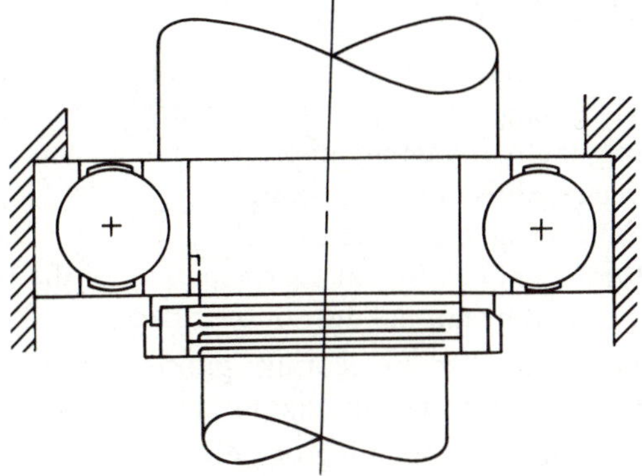

Figure 8.32 Mounting ball bearing.

Figures 8.31 and 8.32 illustrate typical mounting arrangements for bearings on a rotating shaft. The fit between the inner race and the shaft is an interference fit to prevent relative rotation and wear between the race and the shaft. The outer race is also an interference fit in the housing. The inner and outer races are positioned axially by a positive means.

Excessively tight interference fits cause injurious preloading of the bearing, which results in the fatigue failure of the bearing and lubricant failure because of film breakdown or overheating. Excessively loose sliding fits result in slipping of the bearing races in the mounting, which causes overheating, vibration, and detrimental wear.

Bearings are usually mounted with an interference fit in which the bearing ring rotates with respect to the direction of the load that is acting on it. Where the bearing ring does not rotate in respect to the direction of load, the bearing ring is slide-fitted in the mounting.

Preloading is a means to minimize radial deflection of the bearing, which can contribute to the system's eccentricity. Preloading is an internal load applied to the bearing assembly while the bearing is in an unloaded, stationary condition. The two types of preloading are radial and axial.

Preloading removes the internal clearances that normally exist between the balls or rollers and one of the raceways. Preloading results in an interference fit between the rolling elements and the races. The advantages of preloading are:

1. Provides more accurate axial and radial shaft positioning.
2. Reduces the shaft deflection under load and improves the bearing assembly stiffness characteristics.
3. Increases the bearing fatigue life.
4. Decreases the bearing noise.
5. Permits the bearing to take higher shock loads.
6. Provides for identical deflections in the axial and radial directions (isoelasticity).

8.31 BEARING FAILURES

Bearing failures have many causes. Some of these failures are attributed to material fatigue. The majority of premature bearing failures are caused by improper lubrication, faulty mounting, improper maintenance or handling, and the intrusion of foreign debris into the bearing.

The basic types of bearing failure are:

1. Fatigue failure, identified by flaking or spalling of the raceway or noisy running and caused by normal duty or overload.
2. Contamination, identified by scoring, pitting, scratching, or rust and caused by dirty or damp surroundings or abrasive waste materials.
3. Brinelling, identified by mounting or radial indentations and caused by force incorrectly exerted or radial shock load.
4. False brinelling, identified by axial or circumferential indentations and caused by vibration or static bearing.
5. Thrust failure, identified by counterbored bearing or maximum bearing thrust capacity and caused by improper mounting or misapplication.
6. Misalignment, identified by ball path and retainer and caused by shaft and housing misalignment or shaft bowing.
7. Electric arcing, identified by electric arc erosion, granular race surfaces, and pitting or cratering and caused by static electricity, electric leakage, or short circuit.
8. Lubrication failure, identified by grease appearance, abnormal temperature use, noise, bearing discoloration, and retainer failure and caused by dirty or wrong kind of lubrication or too much or inadequate lubrication.
9. Cam failure, identified by broken cam, ball path, or wobble of bearing and caused by undersized shafting or outer ball unable to align.

8.32 BEARING LUBRICATION

For bearings, the lubricant accomplishes the following:

1. Provides lubrication to the contacting surfaces in the bearing.
2. Helps prevent metal-to-metal contact, which would lead to scoring and seizure.
3. Acts as a coolant and a heat exchange medium.
4. Provides a means to flush dirt and other abrasives from the bearing.
5. Reduces friction of the rolling and sliding elements within the bearing.
6. Protects bearing surfaces from corrosion.

For successful bearing operation, thin lubricant films must exist between the contacting surfaces in the bearing. The contacting surfaces in anti-friction bearings have relative motion that is both sliding and rolling. Elastohydrodynamic lubrication is the name given when the lubricant is entrained between the rolling contacts of the bearing surfaces. The bearing components deform elastically under a load, permitting the hydrodynamic lubrication action to separate the bearing components. When the lubricant is trapped between the contacting surfaces, the pressure within the lubricant film increases, thereby increasing the viscosity of the lubricant.

Oil and grease may be used as lubricants. The following stated rules may help in deciding between them:

Use grease when

1. The temperature is not over 93°C.
2. Speeds are low.
3. Unusual protection is required from the entrance of debris.
4. Simple bearing seals are desired.
5. Operation for long periods without attention is desired.

Use oil when

1. The temperature is over 93°C.
2. Speeds are high.
3. Oiltight seals are readily employed.
4. Bearing type is not suitable for grease.
5. The bearing is lubricated from a central supply that is also used for other machine parts.

8.33 SPRINGS

Springs are machine elements designed and fabricated to provide large elastic deflections under load. The major functions for springs on agricultural machinery are 1. to apply force and to control motion; 2. to alter the vibratory characteristics of a structural system; 3. to reduce the magnitude or transmitted force caused by impact or shock loading; 4. to store energy; and 5. to measure force.

Springs are generally classified as wire springs or flat springs. The wire-spring type includes the helical tension, compression, and torsional springs of round or square wire. The flat-spring type includes the cored disk or Belleville spring, the leaf spring, the motor spring, the spiral torsion spring, and the retaining ring.

8.34 HELICAL COMPRESSION SPRINGS

A helical compression spring fabricated from round wire is shown in Fig. 8.33*a*. The spring is deflected by an axial load P, as shown in Fig. 8.33*b*. If the axial load is sufficiently great, the spring is compressed to its solid height, as

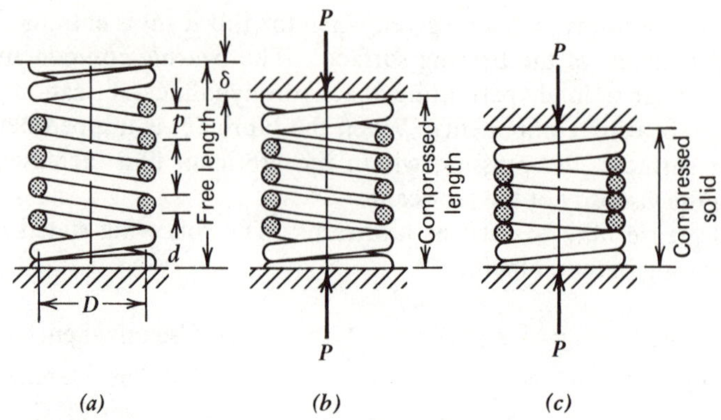

Figure 8.33 Terminology for helical compression springs.

shown in Fig. 8.33*c*. The design of the spring must satisfy two requirements:

1. The spring must carry the load within the elastic limits of the material.
2. The spring rate, (the force-deflection characteristics) must be satisfactory for the given application.

The following notation applies to a helical spring:

Let P = axial load, N
D = mean diameter of the spring coils, mm
d = diameter of the wire, mm
p = pitch of the coils, mm
S = deflection of the spring, mm
n = number of active coils that contribute to the deflection of the spring
C = spring index = D/d
G = torsional modulus of elasticity, Pa
S_s = shearing stress, Pa
T = torsional mount, N · mm
J = polar moment of inertia, mm^4

A free-body diagram of a part of the spring supporting the compression load P is shown in Fig. 8.34. The part of the spring is in equilibrium under the action of two forces P and the resisting torque T. Since the maximum shearing stress in torsion S_s is $TD/2J$, and J for a circular section is $\pi d_i^4/32$, the shearing stress caused by the torque T is

$$S_s = \frac{8PD}{\pi d^3} \qquad [8.93]$$

By adding the direct shearing stress, $4P/\pi d^2$, to the torsional shearing stress in Eq. 8.93, the maximum shearing stress at the inner side of the curved wire is

$$S_s = \frac{8PD}{\pi d^3} + \frac{4P}{\pi d^2} = \frac{8PD}{\pi d^3}\left[1 + \frac{1}{2C}\right] \qquad [8.94]$$

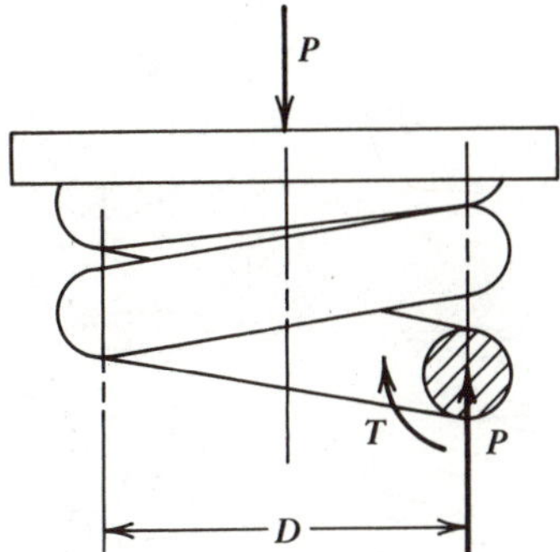

Figure 8.34 Loading on helical compression springs.

Wahl established stress factor K to include the effects of curvature as a result of wire coiling and direct shear. The stress factor is used with Eq. 8.93 to determine the maximum shearing stress as follows:

$$S_s = K\frac{8PD}{\pi d^3} = K\frac{8PC}{\pi d^2} \qquad [8.95]$$

where $K = (4C-1)/(4C-4) + 0.615/C$ for round wire.

The optimum spring design should have an index from 5 to 12. Index values less than 5 result in higher stresses being applied to the spring, and index values greater than 12 result in flimsiness and in difficulty holding the manufacturing tolerances.

The spring deflection is calculated from

$$S = \frac{8PD^3n}{Gd^4} = \frac{8PC^3n}{Gd} \qquad [8.96]$$

(The equation assumes that the helix angle is small.)

A rearrangement of the above equation shows that the spring rate, or constant K_S is

$$K_S = \frac{P}{\delta} = \frac{Gd^4}{8D^3n} = \frac{Gd}{8C^3n}, \frac{\text{N}}{\text{mm}} \qquad [8.97]$$

The following suggestions should help to reduce the fabrication costs.

1. Allow sufficient space so that a low-stressed spring may be used.
2. Design within the optimum range of the spring index.
3. Do not specify a maximum solid height unless dictated by the installation requirements.
4. Avoid the extra cost of presetting the free length of the spring.
5. Use commercial tolerances if at all possible.

Spring ends for helical compression springs may be either plain, plain ground, squared, or squared and ground, as shown in Fig. 8.35. This choice of ends results in a decrease of the active coils and affects the free length and solid height of the spring, as shown in Table 8.4.

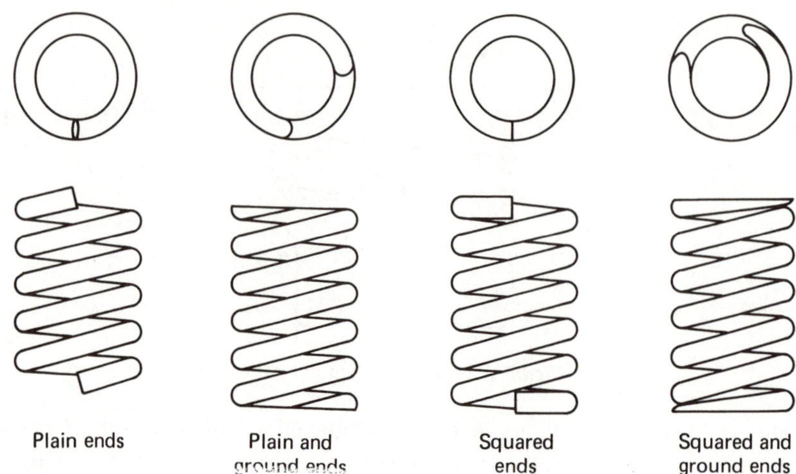

Figure 8.35 Types of ends for helical compression springs.

Table 8.4. **Guidelines for Dimensional Characteristics of Compression Springs.**

Type of Ends	*Total Coils*	*Solid Length*	*Free Length*
Plain	n	$(n+1)d$	$np+d$
Plain ground	n	nd	np
Squared	$n+2$	$(n+3)d$	$np+3d$
Squared and ground	$n+2$	$(n+2)d$	$np+2d$

8.35 HELICAL EXTENSION SPRINGS

A helical extension spring fabricated from round wire is shown in Fig. 8.36. Such a spring is wound with all coils in contact to obtain a built-in load or initial tension. The tension is a function of the spring index—the force increases as the index decreases.

The procedure followed in designing an extension spring is essentially the same as that used in designing a compression spring. In extension-spring applications, several inherent limitations must be considered. First, the residual stresses induced by deflecting (compression springs) cannot be obtained except by coiling to a high value of initial tension and stretching the spring until set occurs. Second, a stress-relieving operation is necessary to obtain the highest torsional yield point of the material.

Most extension-spring failures occur in the hook. Thus the stresses in the hook must be within acceptable limits.

Both torsional and bending stresses occur in the hook. The maximum stress in bending and torsion will not occur at the same place in the hook. The nominal bending stress caused by the moment at point A on the hook in Fig. 8.36 is

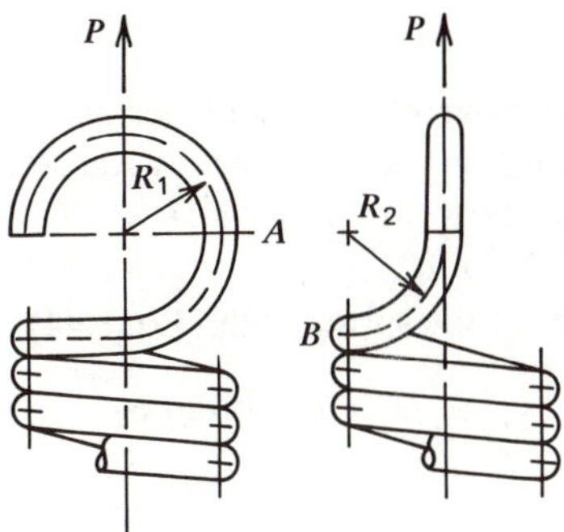

Figure 8.36 Location of maximum bending and torsion stresses in twist loops.

equal to

$$S_B = \frac{32PR_1}{\pi d^3} \qquad [8.98]$$

where R_1 = mean coil radius, mm. This bending stress is multiplied by a stress-concentration factor K, which is equal to

$$K = \frac{4C^2 - C - 1}{4C(C - 1)} \qquad [8.99]$$

where $C = 2R_1/d$. By adding the direct tension stress, which is equal to $4P/\pi d^2$, to the nominal bending stress in Eq. 8.98, the maximum bending stress S_B at point A is equal to

$$S_B = K\frac{32PR_1}{\pi d^3} + \frac{4P}{\pi d^2} \qquad [8.100]$$

At point B, the stress is in tension. Thus, the stress-concentration factor K is equal to

$$K = \frac{4C - 1}{4C - 4} \qquad [8.101]$$

The maximum torsional stress S_T caused by the torsional moment PR_2 is equal to

$$S_T = \frac{16PR_2}{\pi d^3}\left[\frac{4C - 1}{4C - 4}\right] \qquad [8.102]$$

A combined torsional and bending stress will occur between points A and B. Because the peak stresses will be at different locations, this combined stress is not considered.

A standard form may be used to submit the spring design data to the manufacturer.

Example 8.8

1. Determine the number of coils and permissible deflection in a helical spring made of 1.6 mm diameter steel wire. Assume a spring index of 6 and an allowable shear stress of 344.7×10^6 Pa (N/m^2) in shear. The spring rate is 1.75 N/mm.
   a. Mean diameter of the spring coils (Eq. 8.44):

$$D = cd = 6(1.6)$$

$$D = 9.6\text{mm}$$

   b. Shear or Wahl factor:

$$K = \frac{4(6) - 1}{4(6) - 4} + \frac{0.615}{6}$$

$$K = 1.25$$

   c. Number of spring turns:

$$n = \frac{Gd}{8C^3(P/\delta)} = \frac{(11.5 \times 10^6 \times 6894.76 \times 10^{-6})(1.6)}{8(6)^3(1.75)}$$

$$n = 41.95 \text{ turns}$$

   d. Force carried by spring:

$$P = \frac{S_S \pi d^3}{8KD} = \frac{(344.7 \times 10^6 \times 10^{-6})(1.6)^3 \pi}{8(1.25)(9.6)}$$

$$P = 46.2 \text{ N}$$

   e. Permissible deflection:

$$\delta = \frac{P}{K_S} = \frac{46.2}{1.75}$$

$$\delta = 26.4 \text{ mm}$$

Example 8.9

2. When a coil spring with a spring rate of 17.5 N/mm is compressed 32 mm, the coils are closed or compressed to the spring's solid height. The spring index is 8; the allowable shear stress is 344.7×10^6 Pa; the torsional modulus of elasticity is 792.90×10^6 Pa; and the spring ends are squared and ground. Calculate the required wire diameter, the required coil diameter, and the closed length of the spring.
   a. Stress or Wahl factor:

$$K = \frac{4(8) - 1}{4(8) - 3} + \frac{0.615}{8}$$

$$K = 1.15$$

b. Wire diameter:

$$d^2 = K\frac{8PC}{\pi S_S} = \frac{(1.15)(8)(17.5)(32)(8)}{\pi(344.7 \times 10^6 \times 10^{-6})}$$

$$d = 6.2 \text{ mm}$$

c. Coil diameter:

$$D = dC = (6.2)(8)$$

$$D = 49.6 \text{ mm}$$

d. Number of spring turns:

$$n = \frac{dG}{8C^3K_S} = \frac{(6.2)(79{,}290)}{8(8)^3(17.5)}$$

$$n = 6.86$$

e. Closed length:

$$L_C = (n + 2)d = (6.86 + 2)(6.2)$$

$$L_C = 54.9\text{mm}$$

8.36 COUPLINGS

Couplings are used to connect sections of shafts or to connect a driver shaft end to end to a driven shaft on a machine. The coupling should be capable of transmitting the rated torque capacity of the shaft while accommodating any misalignment between the shafts.

Rigid couplings are suitable for low speeds and accurately aligned shafts. Several rigid types are the flange coupling, compression coupling, and tapered-sleeve coupling.

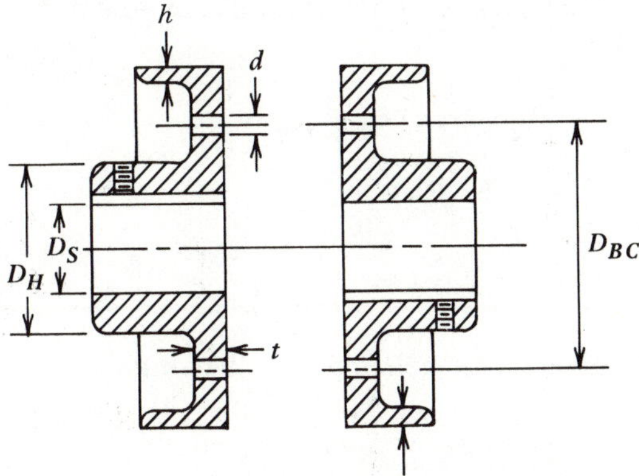

Figure 8.37 Design terminology for a rigid-flange coupling.

Flexible couplings are used to 1. take care of a small amount of unintentional misalignment; 2. provide for axial movement of a shaft (end float); and 3. alleviate shock by providing transfer of power through springs or absorb some of the vibration in the coupling. Several flexible types are the Falk flexible coupling, the Oldham coupling, and the roller-chain coupling.

Rigid couplings can transmit bending in a shaft and can therefore cause stresses, which may result in fatigue failure. It is desirable to provide for good alignment and location of the coupling where the bending moment is practically zero. Thus, rigid couplings are analyzed for torsion only.

It is essential to have the necessary proportions for the various parts of a rigid-flange coupling, as shown in Fig. 8.37. The following notation applies to the coupling:

D_{BC} = diameter of the bolt circle, mm
D_H = diameter of the hub, mm
D_s = diameter of the shaft, mm
M_t = torque capacity, N · m
S_S = allowable shear stress, Pa (N/m^2)
S_B = allowable bearing pressure for the bolt or web, Pa
t = thickness of the web, mm
d = bolt diameter, mm
n = number of effective bolts, taken as all the bolts if finished bolts are used in reamed holes and taken as half the number of bolts if bolts are set in clearance holes

The diameter of the hub should be approximately 1.75 to 2 times the diameter of the shaft, or

$$D_H = 1.75\,D_S \text{ to } 2.0\,D_S \tag{8.103}$$

The minimum thickness of the web is based on two considerations: 1. shear strength of the web, and 2. bearing of the web and bolt, assuming that the bolts are finger tight and are pressed against the web. The shear of the web, the torque capacity, which is based on the shear of the minimum area at the junction of the hub and web, is

$$M_t = S_S\,(\pi D_H\,t)\,\frac{D_H}{2} \tag{8.104}$$

or

$$t = \frac{2M_t}{\pi S_S\,D_H^2}$$

For bearing load caused by the bolt and web, the torque capacity is

$$M_t = S_B\,(dt)\,\frac{D_{BC}}{2}n \tag{8.105}$$

or

$$t = \frac{2M_t}{S_B d D_{BC} n}$$

For rigid-flange couplings, the analysis of the bolts can be made with one of several different assumptions.

1. The bolts are just finger tight, and the load is transferred from one half of the coupling to the other half by a uniform shear stress in the shank of the bolt.
2. The bolts are just finger tight, and the load is transferred from one half of the coupling to the other half with a maximum shear stress in the shank of the bolt equal to 1.33 times the average shear stress.
3. The bolts are tightened sufficiently so that the power is transmitted from one half of the coupling to the other half by means of friction.
4. The bolts are tightened; that part of the power is transmitted by means of friction, and the rest of the power is transmitted by shear in the bolts.

Example 8.10

1. A rigid-flange coupling has a bore of 50 mm. Six bolts are positioned on a 125-mm diameter bolt circle. The shaft and bolts are made from SAE 1030 steel with an ultimate stress of 551.6 MPa and a yield point in tension of 344.7 MPa. Determine the size of the bolts needed to have the same capacity as the shaft in torsion. Assume a shock and fatigue factor of 1. The stress concentration for a keyway is 0.75.

   a. If the allowable shear stress is the smaller of 18% of the ultimate stress or 30% of the yield point, use the lower value.

   Thus $S_S = 0.18(551.6) = 99.3 \times 10^6$ Pa
   $S_S = 0.3(344.7) = 103.4 \times 10^6$ Pa

   b. The shaft capacity:

$$M_t = \frac{\pi S_S D_S^3}{16} = \frac{\pi(99.3 \times 10^6)(50)^3(10^{-6})}{16} = 2437.2 \text{N} \cdot \text{m}$$

   c. Using assumption 1, the bolt diameter is

$$M_t = S_S(1/4\pi d^2)(1/2 D_{BC})n$$

$$2437.2 = (99.3)(\frac{\pi}{4})(d^2)(\frac{1}{2})(125)(6)(10^6 \times 10^{-6})$$

$$d = 8.0\text{mm}$$

d. Using assumption 2, the bolt diameter is

$$2437.2 = \frac{3}{4}(99.3)(\frac{\pi}{4})d^2(\frac{1}{2})(125)(6)$$

$$d = 10.5\text{mm}$$

Example 8.11

2. Assume that the rigid-flange coupling design is based on friction (assumption 3). The coupling has six 12.7-mm diameter bolts. The bolts have a preload of 22,240 N. The inner diameter of contact is 178 mm, and the outer diameter of contact is 203 mm. The coefficient of friction between the faces of the coupling is 0.15. The coupling rotates at 300 rpm. The 50-mm diameter shaft has an ultimate tensile strength of 586×10^6 Pa and a yield point in tension of 310.3×10^6 Pa. The bolts are set in large clearance holes in the coupling.

   Determine the maximum power capacity based on the friction such that slip occurs between the faces of contact. Compare the shaft power capacity with the friction power capacity. Assume that the shaft is in torsion only and operates under steady load conditions.

   a. Torque capacity based on friction:

$$F = \text{axial force by bolt loading} = 6(22{,}240) = 13{,}3440 \text{ N}$$

$$R_f = \text{friction radius} = \frac{2}{3}\left(\frac{R_o^3 - R_i^3}{R_o^2 - R_i^2}\right)$$

$$= \frac{2}{3}\left[\frac{(101.5)^3 - (89)^3}{(101.5)^2 - (89)^2}\right] = 95.25 \text{ mm}$$

$$M_t = FR_f f = (133{,}440)(95.25)(0.15) = 1{,}906{,}524 \text{ N} \cdot \text{mm}$$

$$P = \text{friction power} = \frac{M_t N \pi}{30{,}000{,}000} = \frac{(1906.524)(10)\pi}{30{,}000{,}000} = 59.9 \text{ kW}$$

   b. Shaft torque capacity:

$$M_t = \frac{S_S \pi D_S^3 K}{16} = \frac{(93.1)\pi(50)^3(0.75)}{16} = 1{,}796{,}460 \text{ N} \cdot \text{mm}$$

   Where S_S is the smaller of $0.18(586 \times 10^6) = 105.5 \times 10^6$ Pa and $0.3(310.3 \times 10^6) = 93.1 \times 10^6$ Pa, and K is the stress concentration of 0.75.

$$\text{shaft power} = \frac{(1{,}796{,}460)(300)}{30{,}000{,}000} = 56.4 \text{ kW}$$

   Therefore, the coupling has a greater power capacity based on friction than on shaft capacity.

8.37 OVERLOAD PROTECTION FOR DRIVELINES

The need for overload protection devices has arisen because the growth of agricultural tractor capacity has exceeded the power normally required to drive its components.

When the implement is plugged, the PTO driveline becomes overloaded. As a result: 1. the PTO shafting fails if it does not have sufficient structural strength; 2. one of the drive components (such as a gear set, belt, or chain) fails because it becomes the weak link; or 3. the tractor stalls or breaks down.

The approach to the overloading problems generally lies with a redesign of the PTO shafting or the implement drive components. A better approach is to design all driveline components to transmit normal operating power and to protect the tractor and the implement from high torque loads or plugging with overload mechanisms.

Torque overloads in PTO driveline systems are best prevented by incorporating mechanisms to disconnect or limit the power transfer from the tractor to the implement. The most common mechanisms used to protect the driveline are (1) the overload clutch, (2) the overrunning clutch, and (3) the elastic clutch.

The selection of overload or overrunning clutch for PTO drivelines is determined by (1) the type of implement to be overprotected, (2) the anticipated operating conditions, (3) the torque characteristics that act on the driveline during normal operation and overload, (4) the load limits of all driveline components, (5) PTO drive-shaft speed, and (6) the joint operating angles. The torque characteristics are usually obtained by operating the implement under actual field conditions.

Universal joint manufacturers' catalogs give the various types and details for the clutches in numerous design applications.

Example 8.12

A plate-clutch overload mechanism with a single friction surface has a 250-mm outside diameter and a 100-mm inner diameter. The coefficient of friction is 0.2.

a. Determine the required axial force for a maximum clutch pressure of 700,000 Pa and the torque for the clutch. Assume that the uniform wear theory is valid.

$$F_n = 2\pi p_{\max} R_i (R_o - R_i) \qquad [8.106]$$

where F_n = normal axial force to the surface, N

$p_{\max}$ = maximum clutch-plate pressure, Pa

R_i = inner clutch radius, mm

R_o = outer clutch radius, mm

$$F_n = 2\pi(700{,}000)(50)(125 - 50)(10^{-6}) = 16{,}770 \text{ N}$$

$$T = \pi \mu p_{\max} R_i \left[R_o^2 - R_i^2\right] \qquad [8.107]$$

where T = clutch torque, N·mm

μ = coefficient of friction

$$T = \pi(0.2)(7 \times 10^5)(50)[(125)^2 - (50)^2](10^{-6}) = 288{,}635 \text{ N}\cdot\text{mm}$$

b. Determine the required axial force and torque for a clutch pressure of 7×10^5 Pa. Assume that the uniform plate pressure theory is valid.

$$F = \pi p \left[R_o^2 - R_i^2\right] \qquad [8.108]$$

$$F = \pi(7 \times 10^5)[(125)^2 - (50)^2](10^{-6})$$

$$F = 28{,}865 \text{ N}$$

$$T = \frac{2}{3}\pi\mu p(R_o^3 - R_i^3) = \frac{2}{3}\pi(0.2)(7 \times 10^5)[(125)^3 - (50)^3](10^{-6})$$

$$= 536{,}035 \text{ N}\cdot\text{mm}$$

c. Determine the torque the clutch will carry for an axial force of 22,200 N and the value of maximum plate pressure, if the uniform wear theory is valid.

$$P_{\max} = \frac{22{,}200}{2\pi(50)(125 - 50)} = 9.4 \times 10^5 \text{ Pa}$$

$$T = \frac{1}{2}\mu(R_o + R_i)\, F_n$$

$$T = \frac{(0.2)(125 + 50)(22{,}200)}{2} = 388{,}500 \text{ N}\cdot\text{mm}$$

d. Determine the torque the clutch will carry for an axial force of 22,200 N and the maximum plate pressure, if the uniform plate pressure theory is valid.

$$p = \frac{22{,}200}{\pi[(125)^2 - (50)^2]} = 5.38 \times 10^5 \text{ Pa}$$

$$T = \frac{2\mu(R_o^3 - R_i^3)}{3(R_o^2 - R_i^2)} F_n = \frac{2(0.2)[(125)^3 - (50)^3]}{3[(125)^2 - (50)^2]}(5.38 \times 10^5)(10^{-6})$$

$$= 9.99 \text{ N}\cdot\text{mm}$$

8.38 UNIVERSAL JOINTS

A Cardan universal joint is a nonconstant velocity mechanism consisting of two yokes connected by a cross through four bearings, as illustrated in Fig. 8.38. The primary use of the universal joint is to connect intersecting shafts to transmit power and motion. The principal advantages of the joint are (1) low manufacturing cost, (2) simple and rugged construction, and (3) long life and ease of service.

The kinematics of motion of a joint is unusual. When the joint operates at an angle, the motion of the output yoke does not follow the motion of the input

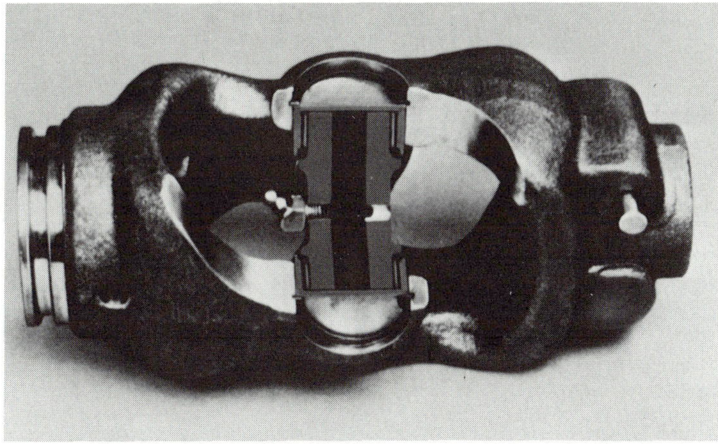

Figure 8.38 Cardan universal joint parts. Courtesy Walterscheid, Inc.

yoke in angular displacement, velocity, and acceleration. The variation between the input and output motions depends on the operating angle between the two shafts. The motion has the following characteristics:

1. The average angular displacement and velocity is uniform.
2. The angular displacement of the driven or output yoke during one revolution lags and leads the driver or input yoke twice.
3. The maximum instantaneous angular acceleration or deacceleration of the driven yoke occurs when the angular velocity of the two yokes is the same.
4. The maximum angular acceleration or deacceleration coincides with the maximum lag and lead angles, respectively.
5. The incremental angular displacement, velocity, and acceleration increase as the joint angle increases, but at an increasing rate.

Figure 8.39 shows the basic angular relationship of a universal joint operating at an arbitrary angle. The output displacement β_o is governed by the operating angle θ and is

$$\tan \beta_o = \cos \theta \tan \beta_i \quad [8.109]$$

By differentiating Eq. 8.109 with respect to time, the output angular velocity is

$$\omega_o = \frac{\omega_i \cos \theta}{1 - \sin^2 \theta \sin^2 \beta_i} \quad [8.110]$$

The operating angle is assumed constant with time. The maximum angular velocity ratio is

$$\omega_m = (\frac{\omega_o}{\omega_i})_m = \frac{\cos \theta}{1 - \sin^2 \theta} = \frac{1}{\cos \theta} \quad [8.111]$$

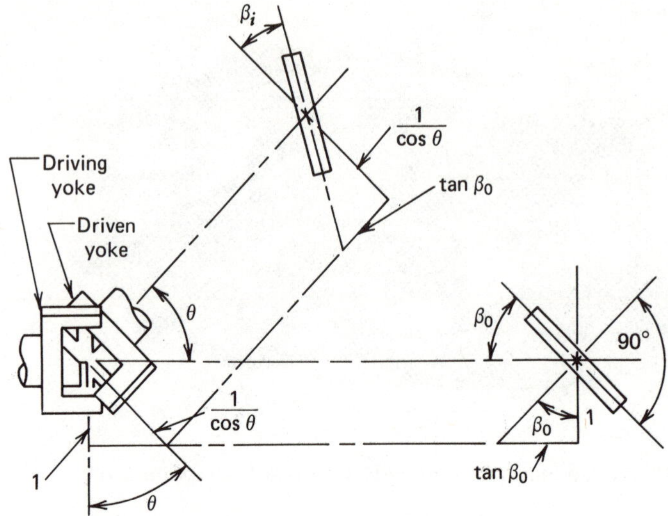

Figure 8.39 Joint operating angle relationship.

By differentiating Eq. 8.110 with respect to time, the output angular acceleration is

$$\alpha_o = 2\frac{\omega_i^2 \cos\theta \sin^2\theta \sin 2\beta_i}{(1 - \sin^2\theta \sin^2\beta_i)^2} \qquad [8.112]$$

The maximum angular acceleration is approximately

$$(\alpha_o)_m = \omega_i^2 \sin^2\beta_i \qquad [8.113]$$

The maximum lag or lead between the input and output displacements is

$$(\Delta)_m = 2\left|45 - Y_i\right| \qquad [8.114]$$

where

$$Y_i = \frac{1}{(cos\ \beta_0)^{1/2}} \qquad [8.115]$$

A simplified derivation of the universal joint's nonconstant velocity is illustrated in Fig. 8.40. When the joint is not operated at an angle, the output angular velocity is the same as the input angular velocity. However, when the joint is operated at an angle, the output angular velocity is not the same as the input angular velocity. To analyze the speed variation, it is assumed that a point on the driving yoke rotates around the shaft centerline in a circular path of radius R at a constant linear velocity V. The input angular velocity of the driving yoke is

$$\omega_i = \frac{V}{R} \qquad [8.116]$$

When viewed along the rotational axis of the driven shaft or yoke, the circular path of the point appears as an ellipse with a major axis of R and a minor axis of $R \cos\theta$. The distance of the point from the driven yoke rotational axis is V. Thus the angular velocity of the driven yoke in terms of the input angular velocity is

$$\omega_o = \frac{V}{R\cos\theta} = \frac{R\,\omega_i}{R\cos\theta} = \frac{\omega_i}{\cos\theta} \qquad [8.117]$$

If the driving yoke is 90° from the position in Fig. 8.40, the distance from the point to the rotational axis of the driven yoke is R, and the linear velocity around the driven rotational shaft axis is $V \cos\theta$. At this instant of drive orientation, the angular velocity of the driven yoke, in terms of the input angular velocity, is

$$\omega_o = \frac{V\cos\theta}{R} = \frac{\omega_i\,R\cos\theta}{R} = \omega_i\cos\theta \qquad [8.118]$$

During the 90° of rotation of the driving yoke, the instantaneous driven yoke output angular velocity varied from $\omega_i/\cos\theta$ to $\omega_i \cos\theta$. The output angular velocity varies continually between these maximum and minimum rotational speeds of the driven yoke. As observed from Fig. 8.40, the driven yoke undergoes two complete cycles of speed variation for every revolution of the driving yoke.

The lag or lead angle between the driving yoke and the driven yoke is

$$\tan\sigma = \frac{\tan\beta_2 - \tan\beta_1}{1 + \tan\beta_1 \tan\beta_2} \qquad [8.119]$$

where β_1, β_2 = joint angles.

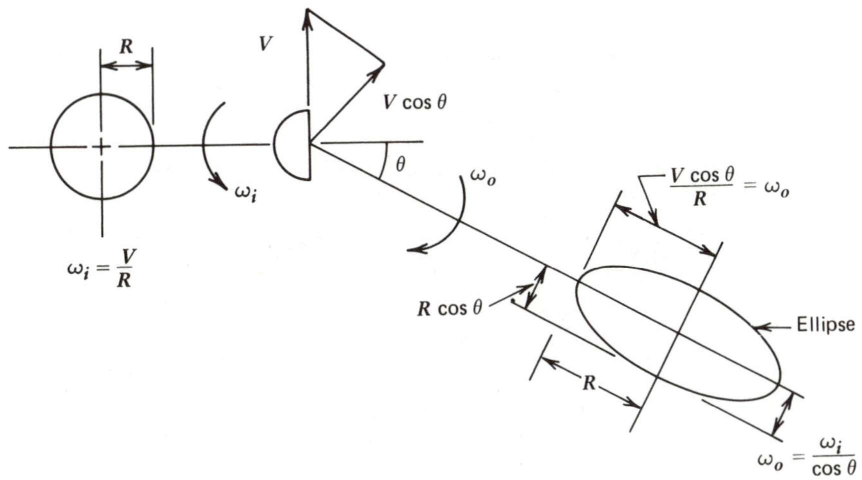

Figure 8.40 Joint kinematic motion characteristics.

8.39 PHASING JOINTS AND DRIVELINE OPERATING CHARACTERISTICS

The relative rotational positioning of the yokes on the connecting shafts so that the output motion is nearly uniform with the input motion is termed phasing.

For two-joint systems on agricultural machinery, phasing is accomplished by placing the yokes on the connecting shafts so that they are in their true joint planes simultaneously. With two joints in a drive-shaft system, the velocity variation between the input and output shafts can be canceled with the proper yoke-phase relationship.

Two basic arrangements that cancel the velocity variation are shown in Fig. 8.41. In Fig. 8.41*a*, the driving yoke D_1 of joint 1 lies in the plane of its joint angle at the same instant the driving yoke D_2 of joint 2 is perpendicular to the

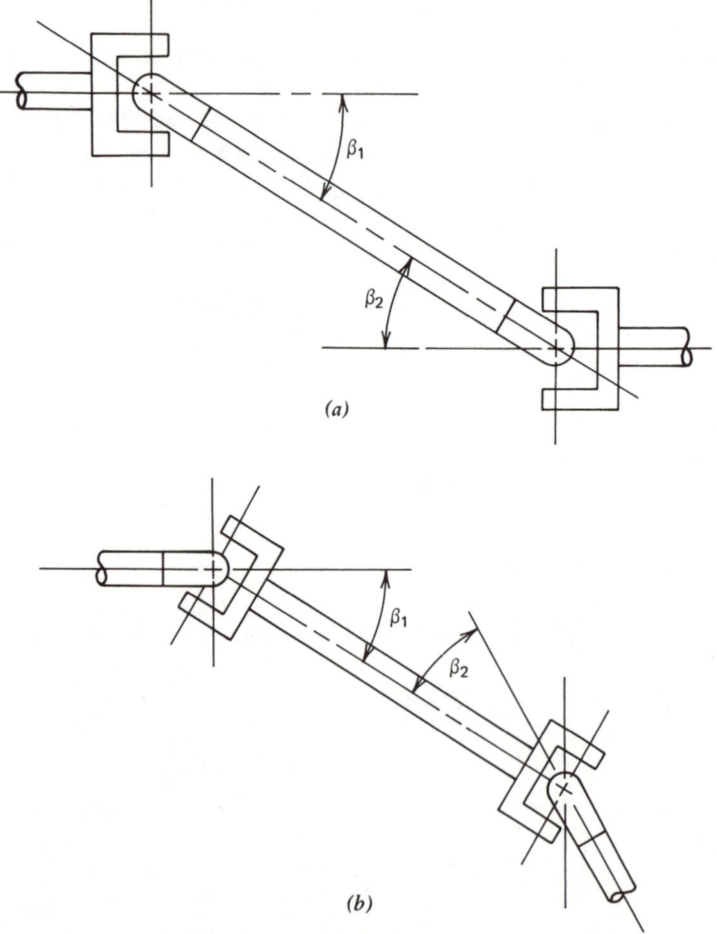

Figure 8.41 Driveline arrangements with all shafts in common plane.

plane of its joint angle. The parallel arrangement of the inboard yokes D_1 and D_2 is expressed as a zero degree phase angle. The residual joint angle of the drive system is

$$\beta_{res} = (\beta_1^2 - \beta_2^2)^{1/2} \qquad [8.120]$$

where β_1 = joint angle of the first joint
β_2 = joint angle of the second joint
β_{res} = equivalent angle of a single-joint drive

When β_{res} is zero, constant velocity is transmitted between the input and output shafts because the joint angles are equal.

If the driving yoke D_1 of joint 1 is crossed or phased 90° to the driving yoke D_2 of joint 2, the equivalent joint angle of the drive system is

$$\beta_{res} = \left[\beta_1^2 + \beta_2^2\right]^{1/2} \qquad [8.121]$$

When the two joints are operating at an angle, the residual joint angle is never zero. Hence, the velocity transmitted between the input and output angles is never constant.

In Fig. 8.41*b*, the intersecting shafts arrangement produces the same kinematic characteristics produced by the parallel shafts arrangement in Fig. 8.41*a*. The cancellation effect of one joint by another is identical at the same joint phase condition.

With agricultural machinery, the location of the PTO shaft relative to the hitch point, either drawbar or three-point hitch, is specified in ASAE Standard S203.9 and SAE Standard J1170 and permits any implement to be used with any tractor having the required 540 or 1000 rpm PTO shaft speed. Single-purpose implements have input-shaft-to-hitch-point relationships that vary considerably and are not always in line with the tractor PTO shaft because of the design configuration.

The joint angles of the driveline vary continuously and over a wide range depending on the turning radius of the tractor-implement configuration and the contour of the terrain. Thus, when a pull-type implement that has a nonstandard input-shaft-to-hitch-point distance is used with a tractor that has a standard PTO-shaft-to-hitch-point distance, unequal joint angles, the same in both left and right turns, are produced. If the input shaft is offset relative to the tractor PTO shaft, however, the unequal joint angle relationship in a right turn differs from that produced in a left turn.

For the ideal driveline geometry, the horizontal distance from the implement input shaft to the hitch point is the same as the horizontal distance from the tractor PTO shaft to the hitch point. This geometric relationship is referred to as the hitch-point spacing, or split. Also, the implement input shaft is in line with the tractor PTO shaft. This ideal geometric arrangement produces equal joint angles for any tractor-implement position during a turn. Under actual conditions, the ideal geometry of equal joint angles is seldom attained because of the implement configuration or insufficient drive-shaft telescoping length.

Figures 8.42, 8.43, and 8.44 illustrate schematically three driveline arrangements that are commonly found in pull-type implement applications. The first two arrangements are for 1000-rpm applications, and the last arrangement is for a 540-rpm application.

As previously stated, certain maximum levels of torsional and inertial vibrations can be tolerated throughout the entire tractor-implement turning-angle position. The maximum angular acceleration is used as the design parameter to express the limits of these disturbances. The maximum angular acceleration of the driven joint $(\alpha_o)_{max}$ is

$$(\alpha_o)_{max} = \beta^2 \omega_i^2 \qquad [8.122]$$

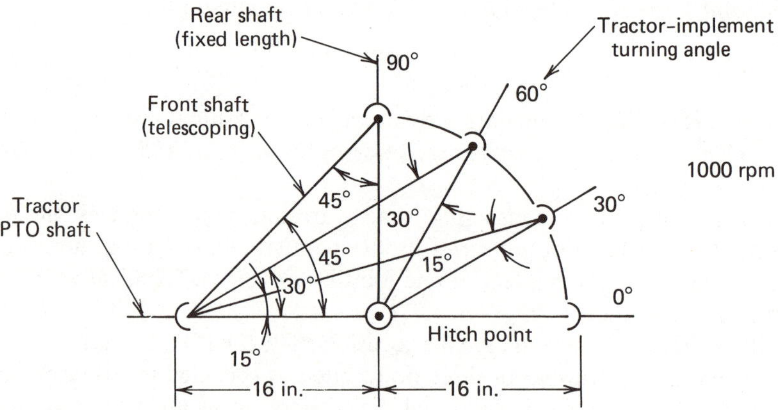

Figure 8.42 Driveline geometry for three-joint driveline (1000 rpm). (*a*) Parallel shafts arrangement. (*b*) Intersecting shafts arrangement.

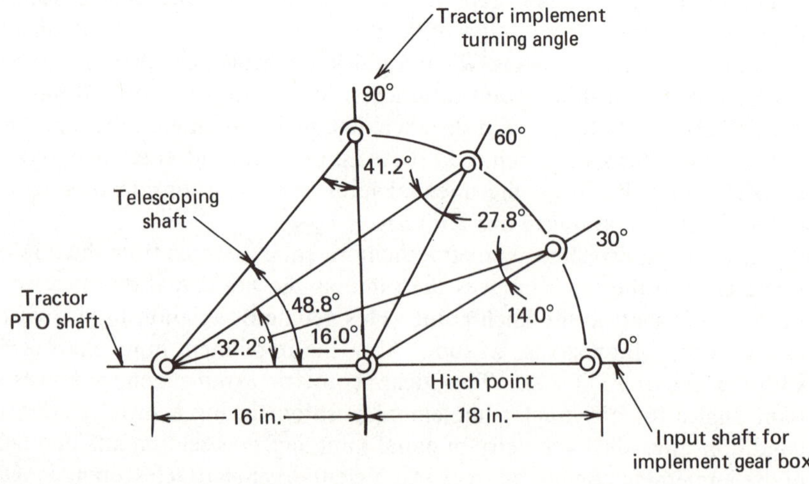

Figure 8.43 Driveline geometry for two-joint driveline.

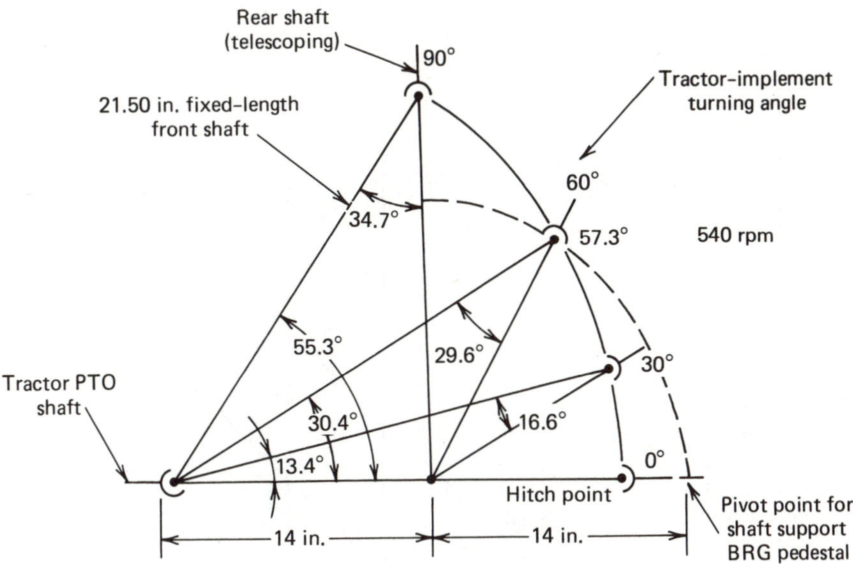

Figure 8.44 Driveline geometry for three-joint driveline (540 rpm).

where β = joint angle, rad

ω_i = angular velocity of driving yoke, rad/sec

The design chart in Fig. 8.45 is developed from Eq. 8.122 to analyze driveline operating geometries of tractor-implement applications in terms of joint angle and maximum angular acceleration. The practical limits of maximum angular acceleration are 1400 rad/sec^2 at the implement input shaft for the straight-ahead operating position of the implement, and 3000 rad/sec^2 in the connecting shaft of the driveline. When cornering with an implement, the angular acceleration in the drive-shaft connecting member may be up to approximately 7000 rad/sec^2. This momentary condition and resulting vibration is unavoidable but can be tolerated.

Example 8.13

Determine the acceptability of the driveline geometry in Fig. 8.44 with regard to torsional and inertial vibrations for the 60° turn position.

The torsional equivalent angle is

$$\beta_{\text{eq}} = [(32)^2 - (28)^2]^{1/2} = 15.5°$$

From Fig. 8.45, the angular acceleration is approximately 800 rad/sec^2. Therefore, the torsional and inertial vibration are satisfactory.

8.40 SECONDARY COUPLES IN JOINTS

In a cornering position, the torque transmitted through the driveline produces forces and moments above and beyond the external loads acting on the joint.

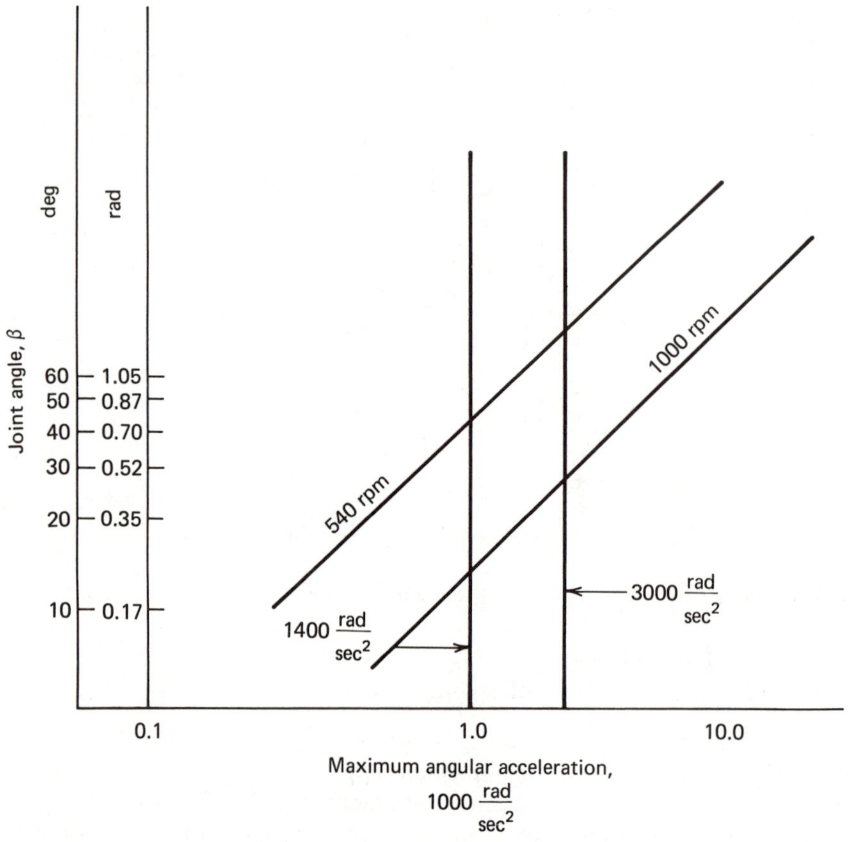

Figure 8.45 Limits for maximum angular acceleration.

The bending moment is always in the plane of the ears of the yokes. For the driving shaft, the approximate bending moment is

$$(M)_i = T \cos\theta \tan\beta_i \qquad [8.123]$$

where $(M)_i$ = bending N · m
T = transmitted torque, N · m
θ = operating angle, °
β_i = input joint angle, °

For the driven shaft, the approximate bending moment is

$$(M)_o = T \cos\theta \sin\beta_0 \qquad [8.124]$$

8.41 CONSTANT-VELOCITY UNIVERSAL JOINTS

Constant-velocity universal joints eliminate the velocity fluctuations of conventional universal joints. The constant-velocity joint consists of two Cardan conventional universal joints. The crosses of the two joints are connected by a coupling yoke with an internal supporting and centering means, and has intersecting shaft axes, as illustrated in Fig. 8.46. With these joints, the instantaneous angular velocity ratio is unity at the design joint angle and zero and is near unity at all other joint angles.

The kinematic advantages of the constant-velocity universal joints are:

1. Torsional vibration is canceled within the joint.
2. The drive-shaft rotates at an effective uniform velocity.
3. The secondary dynamic couples are eliminated.

Other advantages of constant-velocity universal joints for PTO drivelines are:

1. They can be arranged so that the operating angles are equal in opposite directions or in the same direction.
2. They are typically used where the joint operating angles are too large for a single Cardan joint.
3. The centering mechanism essentially compensates for the angular-velocity fluctuations of the two Cardan joints, thereby providing a constant output velocity.
4. They can operate at constant joint operating angles of 30 to 35° and intermittently at angles to 70°.
5. They have relatively long life.
6. They use the common bearing and cross components of the Cardan joint.

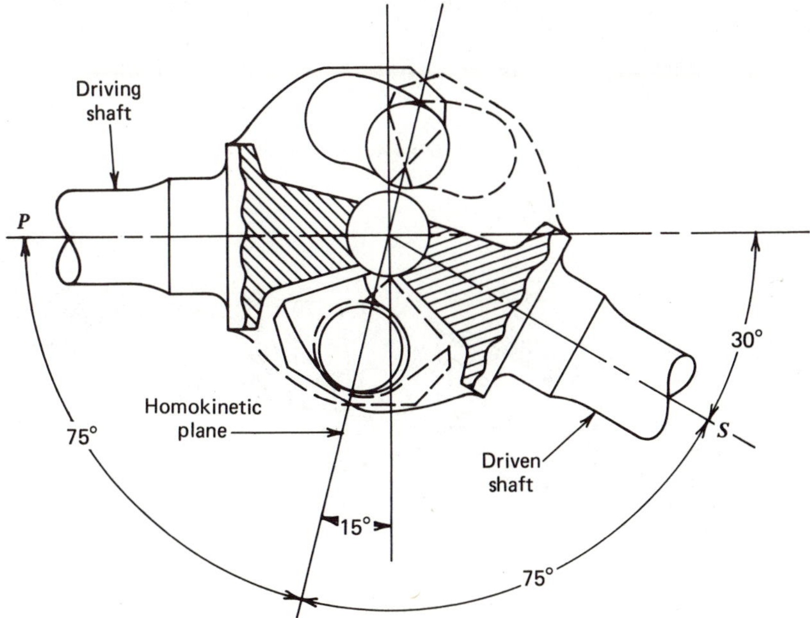

Figure 8.46 Constant-velocity universal joint.

7. They permit greater flexibility in the PTO driveline layout within the tractor-implement system because the location of the hitch point relative to the PTO input shaft and the implement input shaft is not as critical.
8. They allow the implement to be operated with greater efficiency than do Cardan joints.

Example 8.14

A torque of 40,000 N · mm is applied to shaft S_1 of a universal joint in which S_1 and the output shaft S_2 are in the same horizontal plane.

1. Determine the torque of shaft S_2 for the position shown in Fig. 8.47.

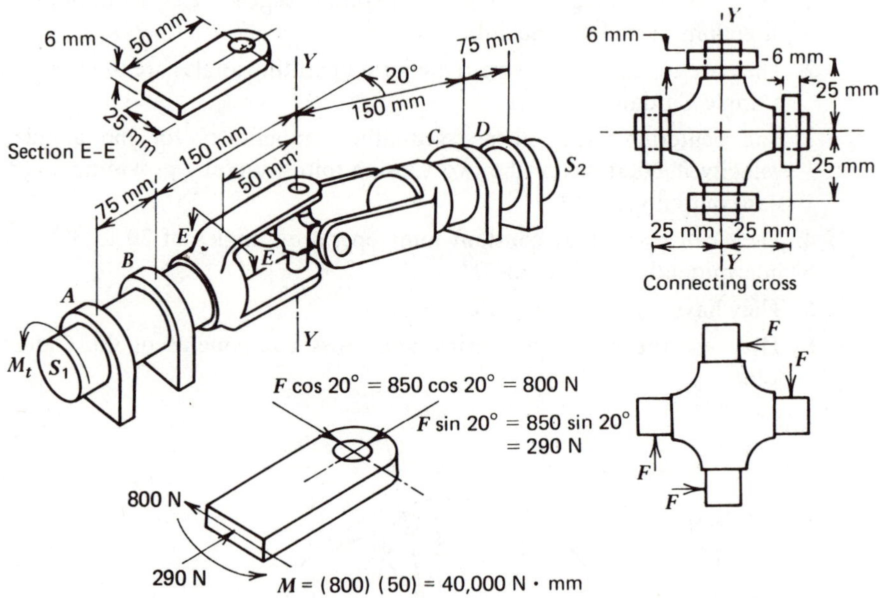

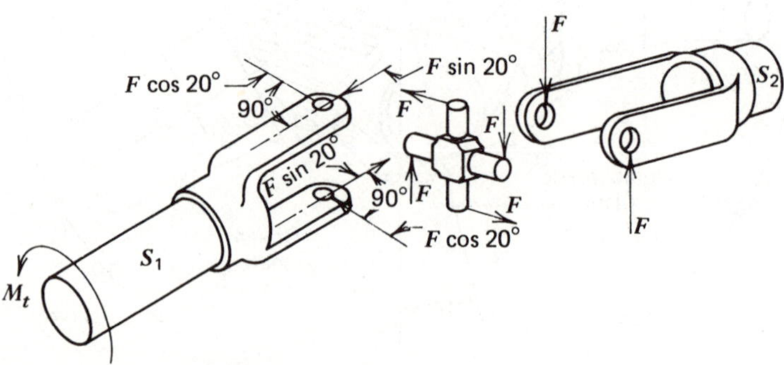

Figure 8.47 Design analysis of a universal joint.

The relation of angular speeds may be used to obtain the torque. Assuming no frictional loss, the power in must equal the power out.

$$T_{S1}\,\omega_{S1} = T_{S2}\,\omega_{S2}$$

or

$$T_{S1}\,\omega_{S1} = T_{S2}\,\omega_{S1}\cos\theta$$

For $\theta = 20°$ and $\beta = 90°$ for the position shown,

$$T_{S2} = \frac{40{,}000}{\cos 2\,0}$$

$$T_{S2} = 42{,}500\ \text{N}\cdot\text{mm}$$

Another method of finding the torque is to examine the forces applied to the cross by the application of the equations of equilibrium. The components of the force F acting on the shaft S are $F\cos 20°$ and $F\sin 20°$. The torque acting on the shaft S caused by the action of the cross, is

$$M_t = 40{,}000 = (F\cos 20)(50);\ \text{therefore, } F = 850\ \text{N}$$

The torque on the shaft S_2 is 50(F), or 42,500 N · mm.

For the position, shaft S_2 is in torsion only while shaft S_1 is in bending as well as torsion. If shaft S_1 is rotated 90°, shaft S_1 will be in torsion only while shaft S_2 is subjected to bending and torsion.

2. Determine the size of the pins of the connecting cross for an allowable bearing stress of 13.79×10^6 Pa (per projected area), an allowable tensile stress of 137.895×10^6 Pa, and an allowable shear stress of 68.947×10^6 Pa.

   a. Diameter of pin based on bearing:

$$S_b = \frac{F}{A} = \frac{F}{1/4d}$$

$$13.789 \times 10^6 = \frac{850}{6d}$$

$$d = 10.3\ \text{mm}$$

   b. Diameter of pin based on bending:

$$S = \frac{MC}{I}$$

$$137.895 \times 10^6 = \frac{(850)(6)(12d)}{(\pi d^4/64)}$$

$$d = 20.8\ \text{mm}$$

c. Diameter of pin based on transverse shear:

$$S_S = \frac{4}{3}\frac{F}{A}$$

$$68.947 \times 10^6 = \frac{(4)(850)}{(3)(\frac{1}{4}\pi d^2)}$$

$$d = 4.6 \text{ mm}$$

3. Determine maximum shear stress in section E-E, which is 50 mm from axis $Y-Y$.
   Maximum compressive stress:

$$S = \frac{MC}{I} + \frac{P}{A} = \frac{(850 \cos 20)(50)(12)}{\frac{6(25)^3}{12}} + \frac{850 \sin 20}{(6)(225)}$$

$$S = 7.05 \times 10^6 \text{ Pa}$$

$$\text{Maximum shear stress} = \frac{1}{2}(7.05 \times 10^6)$$

$$= 3.53 \times 10^6 \text{ Pa}$$

HOMEWORK PROBLEMS

8.1 For a given open, two-sheave, V-belt drive system with a center distance of 600 mm, the driving shaft rotating at 300 rpm, and the driven shaft rotating at 450 rpm, determine
   a. The speed ratio.
   b. The pitch diameter of the drive sheave, if the pitch diameter of the driver sheave is 200 mm.
   c. The arc of contact.
   d. The length of the belt.
   e. The allowable tension ratios.

8.2 For the V-belt drive system in Problem 8.1, the drive transmits 15 kW of power. Determine
   a. The belt speed.
   b. The effective pull.
   c. The tight-side tension.

8.3 The V-belt drive system, as illustrated in Fig. 8.48, consists of a 100-mm PD driver sheave that transmits 11 kW, two 100-mm PD driven sheaves that require 5.5 kW at each shaft, and a 100-mm PD outside-idler pulley. The driver shaft rotates at 3500 rpm. Determine
   a. The arcs of contact for all sheaves and pulley.
   b. The belt length.

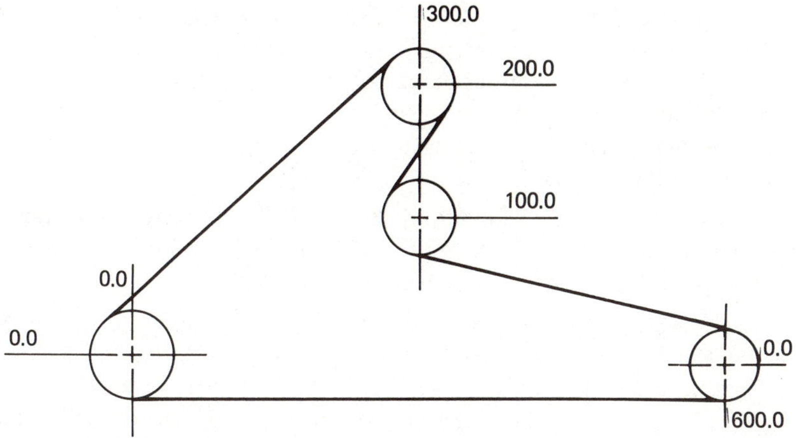

Figure 8.48 Problem 8.3.

c. The allowable tension ratio.
d. The belt speed.
e. The effective pull.

8.4 A No. 50 roller is used to transmit 5 kW of power between a 15-tooth driving sprocket rotating at 250 rpm and a 50-tooth drive sprocket. Determine

a. The approximate center distance if the chain length is 80 pitches.
b. Pitch diameters of the sprockets.
c. The tension in the chain.
d. The shaft torque.
e. The chain velocity.

8.5 The roller chain drive is No. 60, illustrated in Fig. 8.49. The center distances are expressed in terms of the pitch. The 11-tooth sprocket rotates

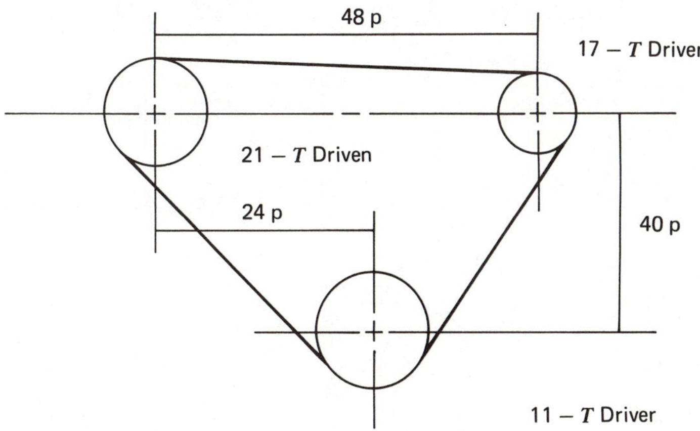

Figure 8.49 Problem 8.5.

at 150 rpm and transmits 10 kW to the 21-tooth sprocket and 10 kW to the 17-tooth sprocket. Calculate

a. The tension in the various portions of the chain.
b. The chain velocity.
c. The chain length.
d. All shaft reactions.

8.6 In Problem 8.5, the 17-tooth and the 21-tooth sprockets are to transfer equal torques to their shafts. For counterclockwise rotation of the driver what is the tension in the various portions of the chain?

8.7 Determine the load ratings for two bearings supporting the shaft, illustrated in Fig. 8.50. The drive transmits 10 kW and rotates at 350 rpm. The center distances are shown in Fig. 8.50. The sheave pitch diameter is 300 mm, and the sprocket pitch diameter is 250 mm. The dead load F_2 is 500 N.

8.8 Determine the power transmitted by two spur gears. The driver turns at 750 rpm. $N_1 = 30$; $N_2 = 75$ b $= 60$ mm, h $= 15$ mm, and $\sigma_{ALL} = 8.3$ kPa.

8.9 Determine the transmitted load and the separating force for the gear set in Problem 8.8.

8.10 Determine the load rating for two bearings supporting the shaft, illustrated in Fig. 8.51. The spur-gear drive transmits 15 kW and rotates at 300 rpm. The pitch diameter is 150 mm, and the tooth-pressure angle is 20°.

8.11 Two helical gears with 90° shafts have a center distance of 300 mm. The gears have 21 and 77 teeth. The normal diametral pitch is 0.3. Find the

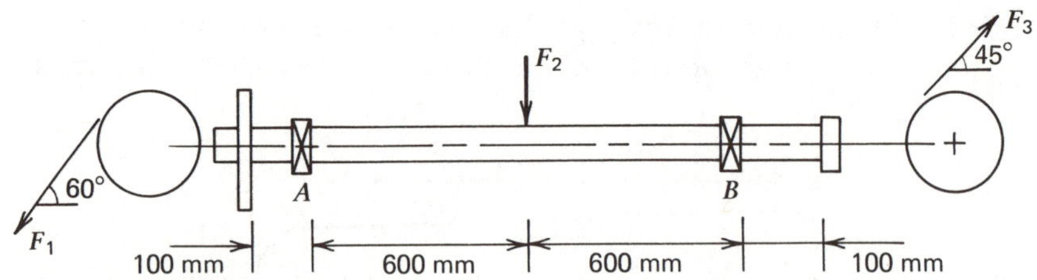

Figure 8.50 Problem 8.7.

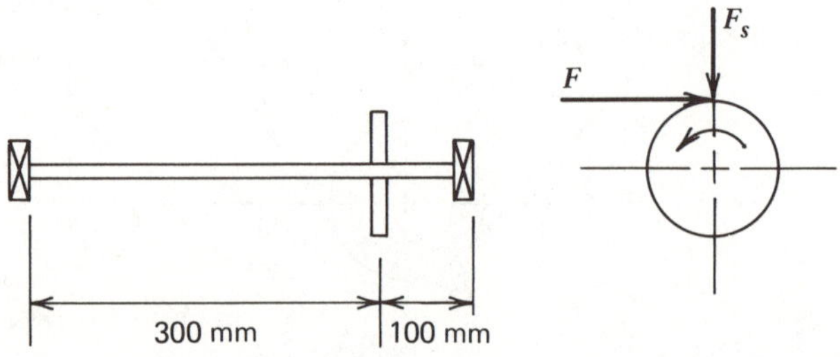

Figure 8.51 Problem 8.10.

required values for the helix angles and the corresponding pitch diameters of the two gears.

8.12 Two helical gears have parallel shafts. Normal pressure angle is 20°, and the face width is 60 mm. The diametral pitch is 0.2. Numbers of teeth are 32 and 44. The center distance is 250 mm. Determine the helix angle and the power the gears are carrying. The driver rotates at 540 rpm.

8.13 Find the power carried by a pair of bevel gears with a pressure angle of 20° and $N_1 = 40$ and $N_2 = 60$ teeth. Diametral pitch is equal to 100mm. The face width is 40mm. The driver rotates at 600 rpm.

8.14 Determine the gear tooth forces and the bearing reactions for the gear pair in Problem 8.13. The pitch cone of the larger gear is located at the midpoint of a 600-mm simply supported shaft.

8.15 Determine the speed ratio, the tooth forces of all gears, and the bearing reactions for the gear train illustrated in Fig. 8.52.

8.16 Determine the angular velocities of the reversing planetary gear system in Fig. 8.26*b* shown in this chapter. The ring gear has 82 teeth, the sun gear has 25 teeth, the pinion gear has 25 teeth, and the reversing pinion gear has 17 teeth.

8.17 Determine the angular velocities for the compound planetary gear arrangement illustrated in Fig. 8.53. On the first row the sun gear has 38 teeth and the pinion gear has 16 teeth. On the second row the ring gear has 70 teeth and the pinion gear has 25 teeth.

8.18 Determine the radial rating required at given speed for a radial ball bearing with a radial load of 100 N, a thrust load of 20 N, and an operating speed of 1500 rpm for an expected average life of 5000 hours.
Use the equation

$$R_R = \frac{RF_c\ F_L}{F_S}$$

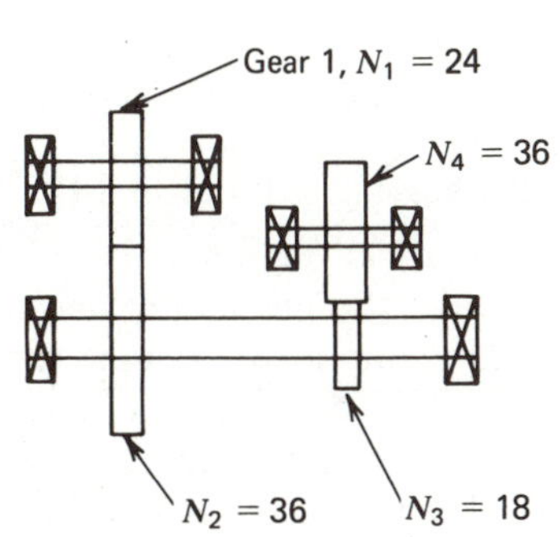

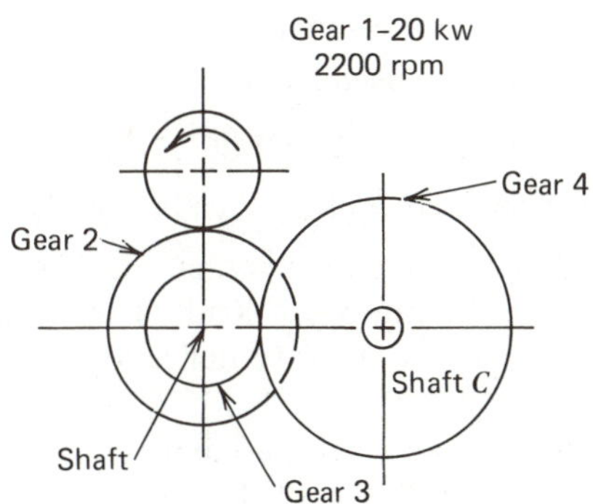

Figure 8.52 Problem 8.15.

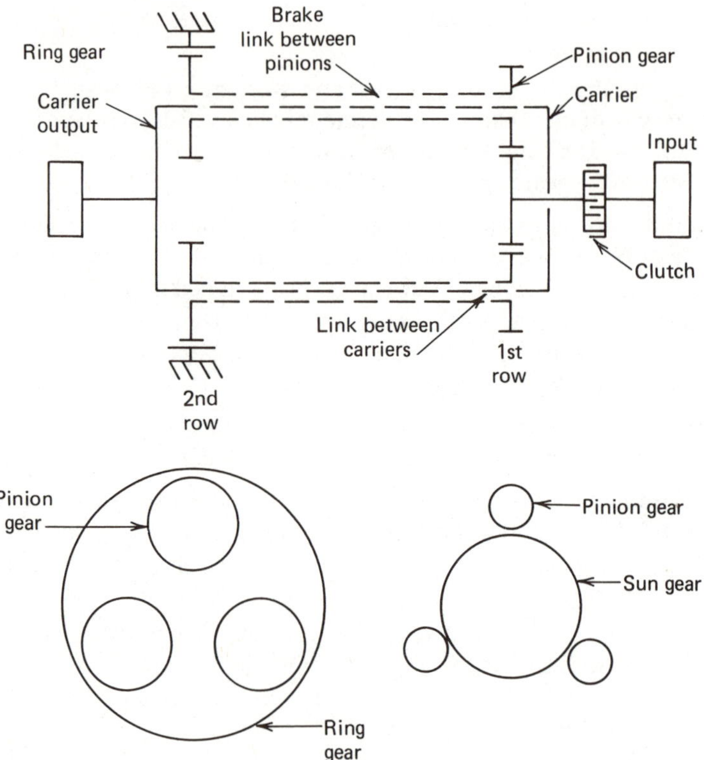

Figure 8.53 Problem 8.17.

where R_R = radial rating required at given speed

R = radial load equivalent

F_c = combined load factor for the conversion of both radial and thrust into the equivalent radial load

F_L = life-modifying factor

F_S = speed factor

Assume that $F_c = 1.06$, $F_L = 1.1$, and $F_S = 1$.

8.19 Determine the basic dynamic capacity of the bearing in Problem 8.18.

8.20 A helical compression spring has a maximum load that is 20 N greater than the minimum load. The deflection under the maximum load is 6 mm greater than the deflection under the minimum load. Assume tentatively that the number of active coils is 10 and the design factor of safety is 1.5. The mean coil diameter is 12 mm. The torsional modulus of elasticity is $79{,}290 \times 10^6$ Pa, and the allowable shear stress is 344.7×10^6 Pa. Determine the spring wire size, the exact number of active coils, and the initial deflection of the spring.

8.21 A helical extension spring is wound from 2-mm diameter wire with the mean coil diameter of 12 mm. Find the approximate value of the load P that the spring can sustain before deflection is noticeable.

8.22 An engine valve spring must exert a force of 240 N when the valve is closed and 450 N when the valve is open. The lift is 8 mm. The spring index is 6, and the design factor of safety is 1.5. Assume that the same values for allowable shear stress and torsional modulus of elasticity as in Problem 8.20.

8.23 An automotive-type clutch utilizes friction in transmitting a torque from one member to another. The normal force on the friction surfaces is supplied by helical compression springs, so that the clutch is always engaged unless an external force is applied to compress further the spring and release the clutch. The specifications for a heavy-duty tractor clutch are that the normal force must be less than 9000 N when the clutch facing has worn 3 mm, and the normal force should not be greater than 11,000 N when the clutch is new. A total of 16 springs will be used to supply the normal force. Size the springs.

8.24 Design a helical extension spring to maintain an approximately uniform force as the part moves 25 mm. The force is to be at least 75 N and not to exceed 90 N. Service conditions may be considered average. Space limitations are not critical, but it is desirable that the spring be wound with the maximum possible initial tension.

8.25 A flange coupling connects two 50-mm diameter lengths of commercial shafting. The coupling webs are bolted together with four bolts of the same material as the shaft. The bolts are set in clearance holes. The diameter of the bolt is 250 mm, and the web thickness is 20 mm. Determine the minimum bolt diameter required to transmit the same torque that the shaft can transmit. What power may be transmitted at 200 rpm under steady load conditions?

8.26 Determine the output angular velocity and acceleration for a 100-rpm PTO drive-shaft as the input displacement varies between zero and 360°. Assume that the joint operating angle is 30°.

8.27 Determine the output angular velocity and acceleration for input displacements of zero, 90, 180 and 270° as the joint operating angle varies between zero and 90°.

8.28 Determine the lag angle between the driving yoke and the driven yoke if $\beta_1 = 49°$ and $\beta_2 = 41°$.

8.29 Determine the acceptibility of the driveline geometry in Fig. 8.44 with regard to torsional and inertial vibrations for the 30, 60, and 90° turn positions.

REFERENCES

1. *Agricultural V-Belt Drive Design Manual*, The Gates Rubber Co., Denver, Col., 1976.
2. *Agricultural V-Belt Engineering and Design Manual*, The Goodyear Tire and Rubber Co., Industrial Products Division, Lincoln, Nebr., 1978.
3. *Analyzing Gear Failures*, Caterpillar Tractor, Peoria, Ill., 1974.
4. *Bearing Failure Prevention Guide*, Fafnir Bearing Division, New Britain, Conn., 1980.
5. Borillon, G. and G. V. Toridion. "On Polygonal Action in Roller Chain Drives," *ASME Transactions of the Journal of Engineering for Industry* 87(2):965, 1965.
6. Buckingham, E. *Analytical Mechanics of Gears*, McGraw-Hill Book Co., New York, N.Y., 1949.
7. Burr, A. H. *Mechanical Analysis and Design*, Elsevier North-Holland, New York, 1981.
8. Cameron, R. "Tractor Bearing System Analysis," ASAE Paper 74-1516, American Society of Agricultural Engineers, St. Joseph, Mich., 1974.
9. *Design Handbook-Engineering Guide to Spring Design*, Associated Spring--Barnes Group, Bristol, Conn., 1981.
10. *Design Manual Roller and Silent Chain Drives*, American Chain Association, St. Petersburg, Fla., 1974.
11. Ferguson, J. H., Jr. and F. Woodruff. "The Fourteen Forces in Couplings," *Machine Design* 45(21):146-150, 1973.
12. Garrett, P. L. "Equations for Computing Creep in Belt Drives," *Product Engineering* 35(5):86-89, 1963.
13. Gerbert, B. G. "Adjustable Speed V-Belt Drives: Mechanical Properties and Design," SAE Paper 740747, Society of Automotive Engineers, Inc., Warrendale, Pa., 1974.
14. Gerbert, B. G. "Pressure Distribution and Belt Deformation in V-Belt Drives," *ASME Transactions of the Journal of Engineering for Industry* 97(3):976-982, 1975.
15. Gerbert, B. G. "Tensile Stress Distribution in the Cords of V-Belts," *ASME Transactions of the Journal of Engineering for Industry*, 97(1):14-22, 1975.
16. Hanson, R. A. "Calculation of Planetary Gear Speeds Using a Building Block Approach," SAE Paper 810992, Society of Automotive Engineers, Inc., Warrendale, Pa., 1981.
17. Harris, T. A. "How to Compute the Effects of Preloaded Bearing," *Product Engineering* 36(15):84-93, 1965.
18. Harris, T. A. *Rolling Bearing Analysis*, John Wiley and Sons, Inc., New York, 1966.
19. Hunt, D. H. and L. W. Garver. *Farm Machinery Mechanisms*, Iowa State University Press, Ames, Iowa, 1973.
20. Keller, D. L. and R. E. Wilson. "Design and Development of a High Horsepower Torque-Sensing Variable-Speed Drive," SAE Paper 720709, Society of Automotive Engineers, Inc., Warrendale, Pa., 1972.
21. Kravitz, S. "Belt Length for Several Pulleys," *Design News* 27(15):101-102, 1971.
22. Marco, S. M., W. L. Starkey, and K.G. Hornung. "A Quantitative Investigation of the Factors which Influence the Fatigue Life of a V-Belt," *ASME Transactions of the Journal for Engineering in Industry* 82(1):47-59, 1980.
23. Martin, J. H. "Constant Velocity PTO Drives," *Machine Design* 53(3):101-105, 1981.

24. Martin, J. H. "Overload Protection for PTO Drives," *Machine Design* 53(17):119-126, 1981.
25. Mihal, D. "Planetary Speeds Made Easy—A Practical Method," SAE Paper 780785, Society of Automotive Engineers, Inc., Warrendale, Pa., 1978.
26. Moyar, C. A. "The Use of Elastohydrodynamic Lubrication in Understanding Bearing Performance," SAE Paper 710733, Society of Automotive Engineers, Inc., Warrendale, Pa., 1971.
27. Oliver, L. R., C. O. Johnson, and W. F. Breig. *Agricultural V-Belt Drive Design*, Dayco Co., Dayton, Ohio, 1977.
28. Oliver, L. R. and D. D. Henderson. "Torque-Sensing Variable-Speed V-Belt Drive," SAE Paper 720708, Society of Automotive Enginers, Inc., Warrendale, Pa., 1972.
29. Oliver, L. R., K. G. Hornung, and H. N. Shapiro. "An Automatic V-Belt Transmission with an Asymmetrical Belt," *ASME Transactions of the Journal of Engineering for Industry* 95(3):771-779, 1973.
30. Oliver, L. R., K. G. Hornung, J. L. Swenson, and H. N. Shapiro. "Design Equations for a Speed and Torque-Controlled Variable Ratio V-Belt Transmission," SAE Paper 730003, Society of Automotive Engineers, Inc., Warrendale, Pa., 1973.
31. Oliver, L. R., C. O. Johnson, and W. F. Breig. "V-Belt Life Prediction and Power Rating," *ASME Transactions for the Journal for Engineering in Industry* 98(1):340-347, 1976.
32. Radzimovsky, E. I. "Eliminating Pulsations in Chain Drives," *Product Engineering* 27(7):153-157, 1955.
33. Radzimovsky, E. I. "Planetary Gear Drives," *Machine Design* 31(12):144-153, 1959.
34. Robinson, C. "The Influence of Roller Geometry and Shaft Misalignment on Contact and Subsurface Stress," ASAE Paper 81-1596, American Society of Agricultural Engineers, St. Joseph, Mich., 1981.
35. Shigley, J. E. *Mechanical Engineering Design*, 3rd edition, McGraw-Hill Book Co., New York, 1977.
36. Spotts, M. F. *Design of Machine Elements*, 4th edition, Prentice-Hall, Englewood Cliffs, N. J., 1971.
37. Spotts, M. F. *Mechanical Engineering Analysis*, Prentice-Hall, Englewood Cliffs, N. J., 1964.
38. Sutherland, G. H. "Finding Bearing Loads Caused by Constant-Velocity U-Joints," *Machine Design* 50(9):55-59, 1978.
39. Turnbull, S. R. and J. N. Fawcett. "Dynamic Behavior of Roller Chain Drives," *Mechanisms*, Institute of Mechanical Engineers, London, 1972.
40. Universal Joint and Drive-Shaft Design Manual, Advances in Engineering Series, no. 7, Society of Automotive Engineers, Warrendale, Pa., 1979.
41. Universal Joint Layout and Selection Data Book, Rockwell International Automotive Division, Troy, Mich., 1981.
42. Wahl, A. M. *Mechanical Springs*, 2nd edition, McGraw-Hill Book Co., New York, 1963.
43. Worley, W. S. "Designing Adjustable-Speed V-Belt Drives for Farm Implements," *SAE Transactions* 63:321-333, 1955.
44. Yeaple, F. "Planetary Gears Take on New Jobs," *Product Engineering*, 50(5):37-41, 1979.

CHAPTER NINE

LINKAGES IN FARM MACHINERY

9.1 VELOCITY AND ACCELERATION DETERMINATION

The design of mechanical motion into farm machinery can be accomplished by using linkages. Linkages are generally composed of two or more inflexible bodies, pinned together at a joint in order to change:

1. Rotation into constant or varying velocity rotation.
2. Rotation into partial rotation or straight-line motion.
3. Partial rotation into straight-line or a different partial rotating motion.

With properly dimensioned links, these conversions may be reversed.

A general knowledge of vectors is a prerequisite to an analysis of linkages. Vectors have both magnitude and direction. Examples of vector quantities are displacement, velocity, acceleration, and force.

9.1.1 Vector Manipulation

Vectors may be added graphically or mathematically. Graphical methods require careful attention to scale and direction, and mathematical methods require consistency in sign conventions to obtain correct results.

Vectors may be added graphically as follows.

1. By constructing a parallelogram and drawing the diagonal from the origin to obtain the resultant R (see Fig. 9.1*a*).
2. By placing either vector with its tail at the origin and placing the tail of the other vector on the tip of the first; the resultant R is the vector obtained by drawing a line from the origin to the tip of the last vector (see Fig. 9.1*b*).

   If more than two vectors are to be added, method 2 provides a suitable procedure for expansion. A polygon is formed by placing vectors tail to tip, starting at the origin, until all vectors have been used. Given magnitudes and directions must be maintained, but the order of addition does

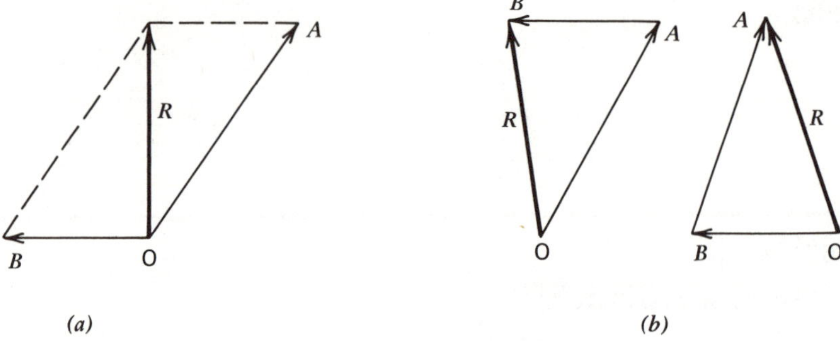

Figure 9.1 Adding vectors graphically.

not affect the resultant. The resultant is always directed outward from the origin and is the closing side of the polygon.

Mathematically, vectors are usually broken into *components* before manipulation is possible. A Cartesian-coordinate system with three axes is normally used. For plane motion, only components in the x and y directions are required to fully describe vectors. Since any vector can be described by the addition of two more vectors of proper magnitude and direction, the mathematical method works as follows:

To add vectors A and B in Fig. 9.2, the vectors are broken down into their x and y components. The x components are added and the y components are added, giving attention to signs, and the resultant is obtained by combining the two.

The angles θ are taken from the positive x-axis in a counterclockwise direction to determine a suitable sign convention. A and B are magnitudes.

$$\theta_A = 45^\circ \ \theta_B = 180^\circ; \ A = 20 \ B = 10$$

For the x components:

$$Ax = 20 \cos 45^\circ = 14.14$$

$$Bx = 10 \cos 180^\circ = -10.00$$

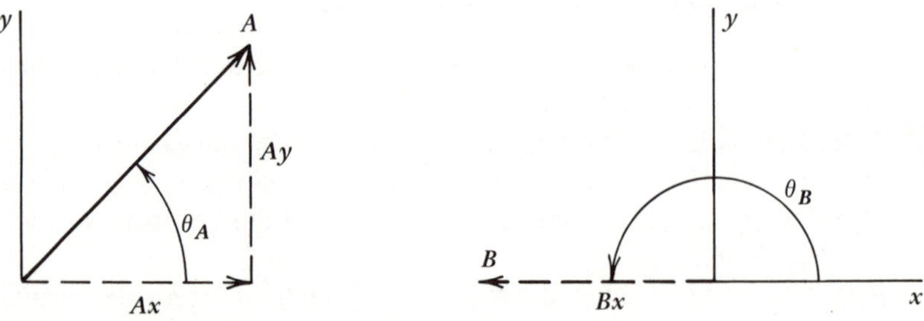

Figure 9.2 Adding vectors mathematically.

For the y components:

$$Ay = 20 \sin 45^\circ = 14.14$$

$$BY = 10 \sin 180^\circ = 0$$

Adding x components yields

$$14.142 - 10.00 = 4.142$$

Adding y components yields

$$14.142 + 0 = 14.142$$

The new vector (resultant) is then

$$4.142\,x + 14.142\,y$$

To obtain the magnitude of the resultant, use the Pythagorean theorem:

$$R = [(4.142)^2 + (14.142)^2]^{1/2}$$

$$R = 14.736$$

The direction of the resultant is obtained by finding the inverse tangent of the x and y components:

$$\tan^{-1}\frac{y}{x} = \tan^{-1}\frac{(14.14)}{(4.14)} = 73.68^\circ \qquad [9.1]$$

The resultant vector is now completely described by 14.74 θ 73.68°

Any number of vectors can be added or subtracted with this method.

9.1.2 Motion and Spatial Relationships

Two kinds of motion are found in linkages: absolute motion and relative motion. Absolute motion is the motion of a point in relation to some other point at rest. In most cases, the earth is regarded as fixed. With regard to linkages, it is sometimes expedient to consider the frame or mounting to which the links are fixed as the reference; the reference must be specified during linkage analysis. Points exhibit relative motion only if there is a difference in their absolute motions.

The displacement of a body C relative to another body D is the absolute displacement of C minus the absolute displacement of D. Velocity and acceleration vectors are treated similarly. The vector notation used to denote velocities, for example, is as follows:

1. V_C = absolute velocity of point C.
2. V_D = absolute velocity of point D.
3. $V_{D/C}$ = the relative velocity vector denoting the velocity of D with respect to C.

Using vector addition (symbolized by $+\!\!\rightarrow$), the following equation defining relative velocity is obtained.

$$V_{D/C} = V_D - V_C \qquad \text{or} \qquad V_D = V_C +\!\!\rightarrow V_{D/C} \qquad [9.2]$$

This matter of relative velocities may be clarified by a simple example, consisting of a single link in motion (see Fig. 9.3).

In a given time, link *CD* has moved from position 1 to position 2 at a constant velocity. Although the bar was pivoting and translating simultaneously, it can be assumed, for reasons of clarification, that the bar first moved to C_1D_0 and then pivoted on point *C*, until D_0 reached D_1.

During the translation phase of this movement from *CD* to C_1D_0, no relative motion is present between the two points. The relative motion of D with respect to *C* ($V_{D/C}$) occurs in the rotation phase from C_1D_0 to the final position of C_1D_1. This rotation through an angle of θ has taken place during the time elapsed between the initial position (*CD*) and the final position C_1D_1. The rotational velocity component can be equated to $r\omega$, where r is the distance between *C* and *D* (always constant) and ω is the angular velocity $d\theta/dt$. Since points *C* and *D* are spatially linked, the rotational velocity component of *D* with respect to *C* is always perpendicular to the line passing through the two points.

Acceleration vectors may be treated in the same way. Both absolute linear and angular accelerations and relative linear and angular accelerations must be found.

Instantaneous linear acceleration is

$$A = \frac{dV}{dt} = \frac{d^2x}{dt^2} \qquad [9.3]$$

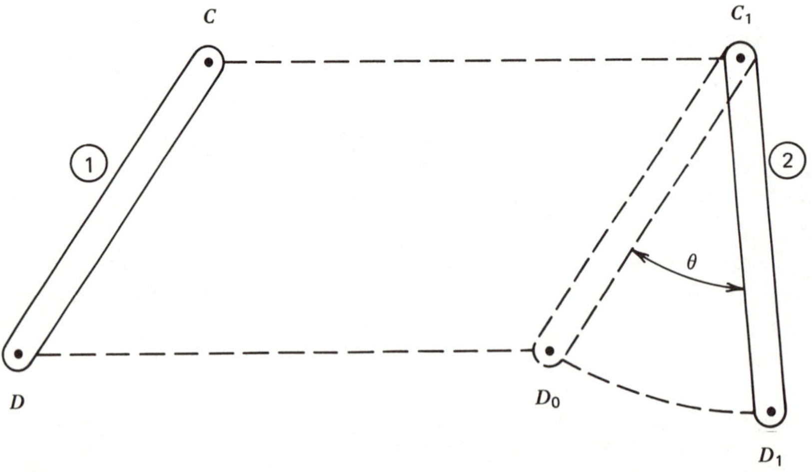

Figure 9.3 Relative velocities.

and angular acceleration is

$$\alpha = \frac{d\omega}{dt} = \frac{d^2\theta}{dt^2} \qquad [9.4]$$

A point can have acceleration normal to its path of motion, tangential to its path of motion, or both. Normal acceleration results in a change in the direction of a point's linear velocity. Tangential acceleration results from a change in the magnitude of a point's linear velocity. A point with rectilinear motion has no normal acceleration because its linear velocity does not change direction. It will have a tangential acceleration if the linear velocity changes in magnitude.

As an illustration, consider the following vectors in Fig. 9.4. A particle at point B moves to point C in time t, along path MN. The original velocity vector V and the original radius of curvature R are shown. After time t the velocity vector is $V \mathrel{+\!\!\!\rightarrow} \Delta V$ and the radius of curvature is (R'). The particle has moved through an angle of $\Delta\,\theta$.

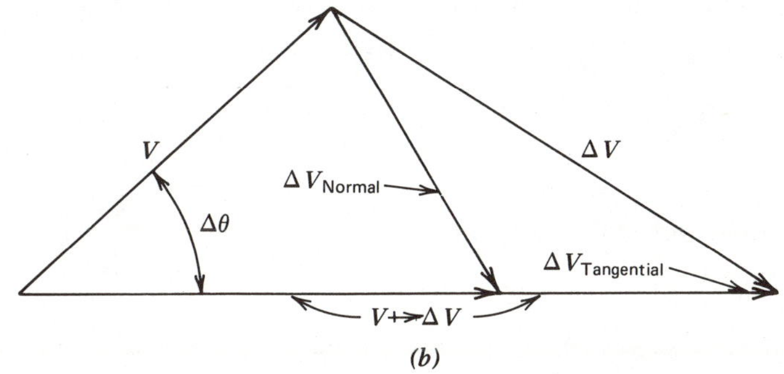

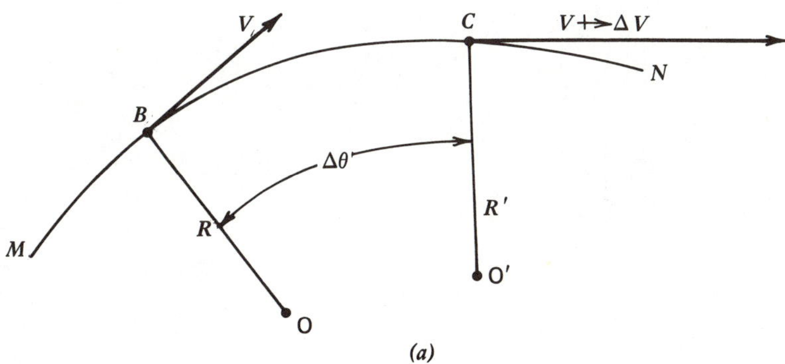

Figure 9.4 Normal and tangential velocity components.

A velocity polygon is constructed where ΔV, the change in velocity, is the sum of ΔV_{normal} and $\Delta V_{tangential}$. Component ΔV_{normal} results from the change in direction of vector V, and $\Delta V_{tangential}$ is the change in magnitude of V.

At point B

$$A_{tangential} = \lim_{\Delta t \to 0} \frac{\Delta V_{tangential}}{\Delta t} = R\frac{d\omega}{dt} = R\alpha \qquad [9.5]$$

$$A_{normal} = \lim_{\Delta t \to 0} \frac{\Delta V_{normal}}{\Delta t} = V\frac{d\theta}{dt} = R\omega^2 = \frac{V^2}{R} \qquad [9.6]$$

Figure 9.4 shows how a graphical method can be utilized to good advantage when analyzing linkages in plane motion. Velocity and acceleration polygons can be carefully drawn to scale to obtain directly measurable values for velocities and accelerations of various points on the mechanism.

9.1.3 Methods in Vector Analysis

The relative-velocity method of graphical analysis, using a rigid linkage with one point limited to rectilinear motion, is illustrated in Fig. 9.5.

Given: V_x is known and point Y is constrained to move along line AA.

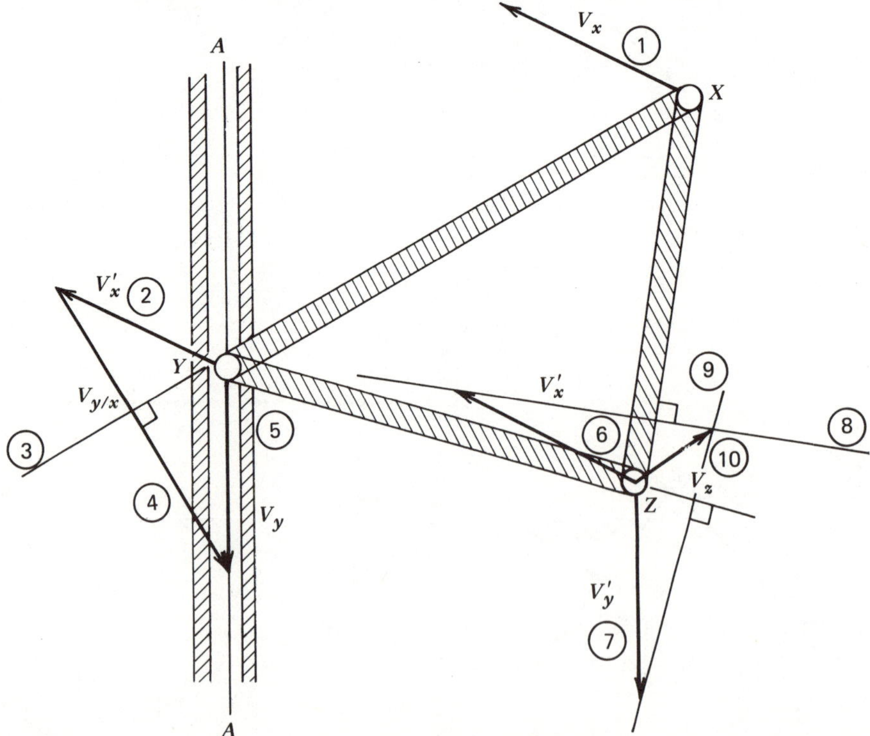

Figure 9.5 Relative velocity method by graphical means.

Draw the body to scale and construct $V_x(1)$ paying careful attention to direction and establishing a scale (such as 1 in. = 1 ft/sec) for the magnitude. At point Y construct vector $V'_X(2)$ with the same magnitude and direction as V_x. The velocity of Y with respect to X must be perpendicular to member XY, so extend XY (3) and draw a perpendicular line to XY which passes through the tip of vector V'_x and intersects with line AA; this is vector $V_{y/x}$. (4) The point where this line intersects AA is the tip of the velocity vector of point Y. (5) Using vector addition, note that V_y is indeed the sum of $V'_x \leftrightarrow V_{y/x}$.

Now that both V_x and V_y are known, V_z may be constructed using the following equations:

$$V_z = V_x \;\leftrightarrow\; V_{z/x} \qquad [9.7]$$

$$V_z = V_y \;\leftrightarrow\; V_{z/y} \qquad [9.8]$$

V'_x (6) and V'_y (7) are both drawn, starting at point Z. The line of action of the velocity of Z with respect to X ($V_{z/x}$) (8) is drawn perpendicular to XZ and through the tip of V'_x. Likewise, the line of action of $V_{z/y}$ (9) is drawn perpendicular to ZY, passing through the tip of V_y. In order for both of the above equations to be true, the vector V_z must extend from point Z to the intersection of $V_{z/x}$ and $V_{z/y}$. Therefore, V_z can be drawn (10) with the direction as indicated and magnitude proportional to the scale used on V_x.

9.1.4 Acceleration Vectors—A Graphical Approach

In many mechanisms, the greatest loads and stresses on moving parts are caused by acceleration forces. The graphical procedure for solving accelerations in a set of linked bodies is the same as that previously described for velocity vectors. Solving for a four-bar linkage with the angular velocity and acceleration of one member known, requires a working knowledge of these earlier equations (see Fig. 9.6).

Let X and Y be two points, where A^n is the normal acceleration and A^t is the tangential acceleration. Then

$$A_x = A_x^n + A_x^t \qquad [9.9]$$

$$A_x^n = r\omega^2 \qquad [9.10]$$

$$A_x^t = r\alpha \qquad [9.11]$$

9.1.5 Relative Acceleration

$$A_x = A_y \;\leftrightarrow\; A_{x/y} \qquad [9.12]$$

Therefore,

$$A_{x/y} = A_{x/y}^n \;\leftrightarrow\; A_{x/y}^t \qquad [9.13]$$

From Eqs. 9.12 and 9.13 it can be concluded that

$$A_x = A_y \;+\!\rightarrow\; A_{x/y}^n \;\leftrightarrow\; A_{x/y}^t \qquad [9.14]$$

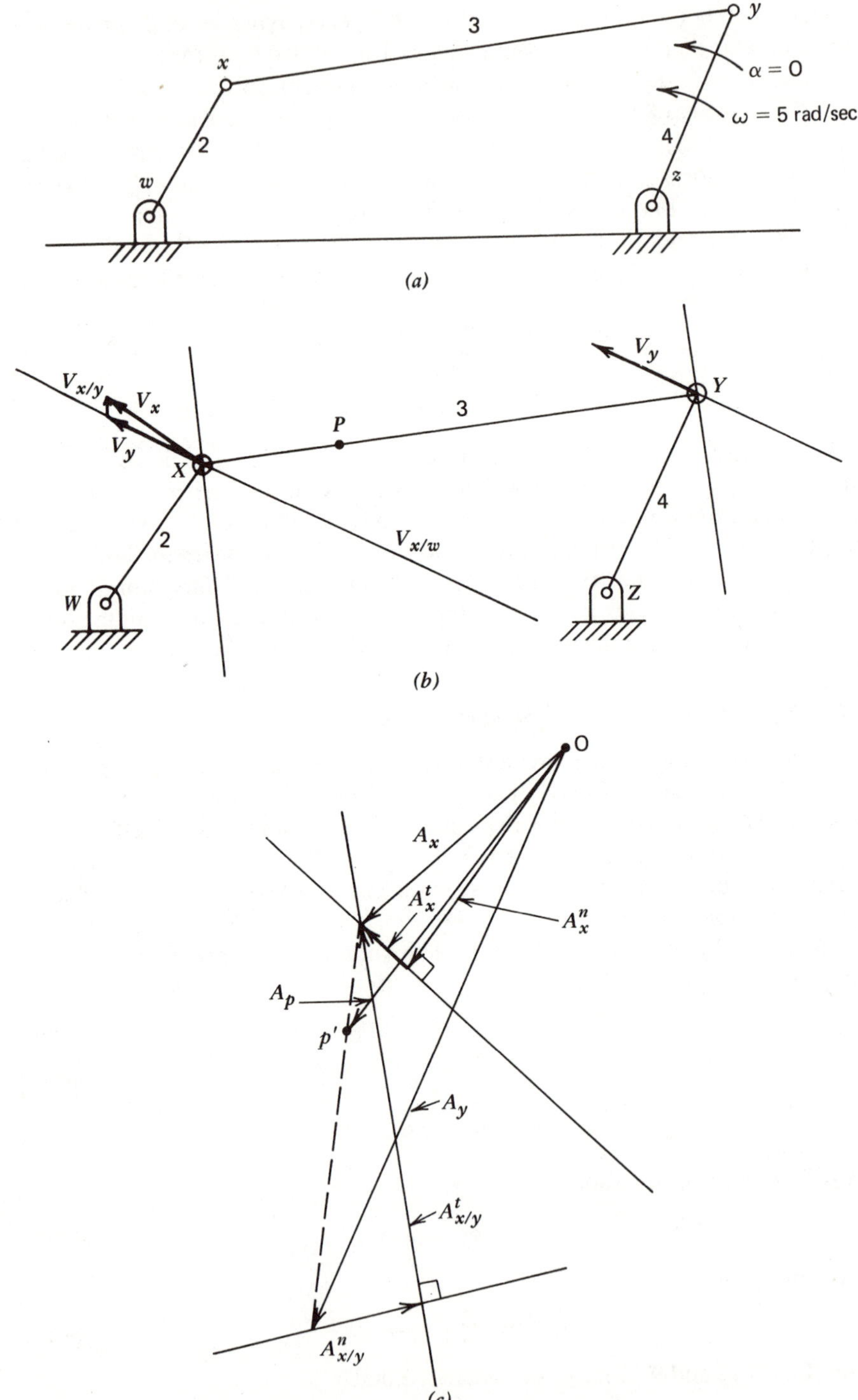

Figure 9.6 Graphically determining acceleration and velocity.

If point Y in Fig. 9.6 is in constant velocity rotation about point Z, then $A_y^t = 0$. Assume that YZ = 0.91 m (3 ft); XY = 2.134 m (7 ft); WX = 0.76 m (2.5 ft). The links are numbered to clarify notations.

Point Y has an instantaneous velocity of $r\,\omega_4$ or $YZ\,\omega_4$. Then, from Eq. 9.9,

$$A_y = YZ\,\omega_4^2 \leftrightarrow YZ\,\alpha_4$$

$$A_y = 0.914\text{ m}(5^2) \leftrightarrow 0 = 22.85\text{ m/sec}^2 = 75\text{ ft/sec}^2$$

A_y is purely normal acceleration toward point z. Because both magnitude and direction of A_y are known, the graphical polygon may be drawn. Using the velocity polygon obtained before, velocity values can be defined for each link. For example, ω_3 is simply the velocity of X with respect to Y ($V_{x/y}$) divided by the length of link 3 (XY); in other words, it is the linear velocity of point X as it rotates about Y, divided by the radius of rotation (the distance from X to Y).

$$V_y = r\,\omega = (0.914\text{m})\,(5\text{ rad/sec})$$

$$V_y = 4.572\text{m/sec}\,(15\text{ ft/sec});\; V_x = 5.182\text{m/sec}\,(17\text{ ft/sec})$$

By scaling the graphical resultant

$$V_{x/y} = 0.914\text{m/sec}\,(3\text{ ft/sec})$$

Therefore,

$$\omega_2 = \frac{5.18\text{m/sec}}{0.76\text{m}} = 6.80\text{ rad/sec}$$

$$\omega_3 = \frac{0.914\text{m/sec}}{2.134\text{m}} = 0.43\text{ rad/sec}$$

Because the velocities of all points, and hence the ω's of all links, can be determined, the following plotting method can be used to complete the acceleration polygon. Using Eq. 9.10

$$A_{x/y}^n = XY\,\omega_3^2$$

Simultaneous equations can then be solved to determine the acceleration of X.

$$A_x = A_x^t \leftrightarrow A_x^n$$

$$A_x = A_y \leftrightarrow A_{x/y} = A_y \leftrightarrow A_{x/y}^t \leftrightarrow A_{x/y}^n$$

where $A_x^n = WX\,\omega_2^2$ and can be plotted from point 0.

The direction of A_x^t is perpendicular to A_x^n and must pass through the tip of A_x^n. The point where vectors A_x^t and $A_{x/y}^t$ intersect defines the tip of vector A_x.

$$A_x = 37.2\text{m/sec}^2(122\text{ ft/sec}^2)$$

$$A_{x/y}^t \text{ is perpendicular to } A_{x/y}^n$$

Since $A_x = A_y \leftrightarrow A_{x/y}$, $A_{x/y}^t$ and $A_{x/y}^n$ must be perpendicular.

The acceleration vector for any point on XY can be determined quite easily now. For instance, point P is one-third of the way along XY from point X. Sketch a line from the tip of A_x to the tip of A_y and then determine a point one-third of the way along this line from the tip of A_x. The vector drawn from this point (P') to the origin (0) is the acceleration vector of point P.

A practical graphical vector analysis of a specialized four-bar linkage known as a slider-crank mechanism follows. This mechanism is widely used because it can convert reciprocating motion into rotary motion or vice-versa. Common examples of slidercranks are piston–connecting rod–crankshaft assemblies in gasoline and diesel engines and in many types of air compressors.

Example 9.1

Determine graphically the velocity and acceleration vectors of points A and B when the mechanism is in the positions shown in Fig. 9.7.

The crankshaft of an automobile engine has crankpins with a throw of 44.45 mm (1.75 in.). The connecting rods are 152.4 mm (6 in.) from the center of the crankpin to the center of the piston pin. The crankshaft is rotating at 3000 rpm (314 rad / sec) in a counterclockwise direction.

Because the piston must move in a straight line, link 4 in the four-bar linkage is assumed to have an infinite length.

$$\omega_{OA} = 3000 \frac{\text{rev}}{\text{min}} \times \frac{2\pi \text{ radians}}{\text{rev}} \times \frac{\text{min}}{60 \text{ sec}} = 100\pi \frac{\text{radians}}{\text{sec}}$$

$$\alpha = 0$$

A slider-crank mechanism has two dead-center positions during one cycle, one at either extreme of the piston travel. To carry the mechanism past these dead points, rotational momentum, usually provided by a flywheel, is required. The solution is shown in Fig. 9.8.

Example 9.2 Checking Graphical Results by Analytical Methods

Using the linkage of Example 9.1 (see Fig. 9.9) the solution is achieved by complex numbers. This method provides a complete kinematic analysis of the slider crank in any position.

Two successive differentiations of the position equation yield the following expressions for velocity and acceleration.

$$V_B = \dot{r}_1 = r_2\omega_2 \left[ie^{i\theta_2}\right] + r_3\omega_3 \left[ie^{i\theta_3}\right] \qquad [9.15]$$

$$A_B = \ddot{r}_1 = r_2 (i\alpha_2 - \omega_2^2) \left[e^{i\theta_2}\right] + r_3 \left[i\alpha_3 - \omega_3^2\right] \left[e^{i\theta_3}\right] \qquad [9.16]$$

By separately equating real and imaginary parts of the position equation, the following results:

$$r_1 = r_2 (\cos\theta_2 + i \sin\theta_2) + r_3 (\cos\theta_3 + i \sin\theta_3)$$

$$r_1 = r_2 \cos\theta_2 + r_3 \cos\theta_3$$

$$0 = r_2 \sin\theta_2 + r_3 \sin\theta_3$$

$$\theta_3 = \sin^{-1} \left[\frac{-r_2}{r_3} \sin\theta_2\right]$$

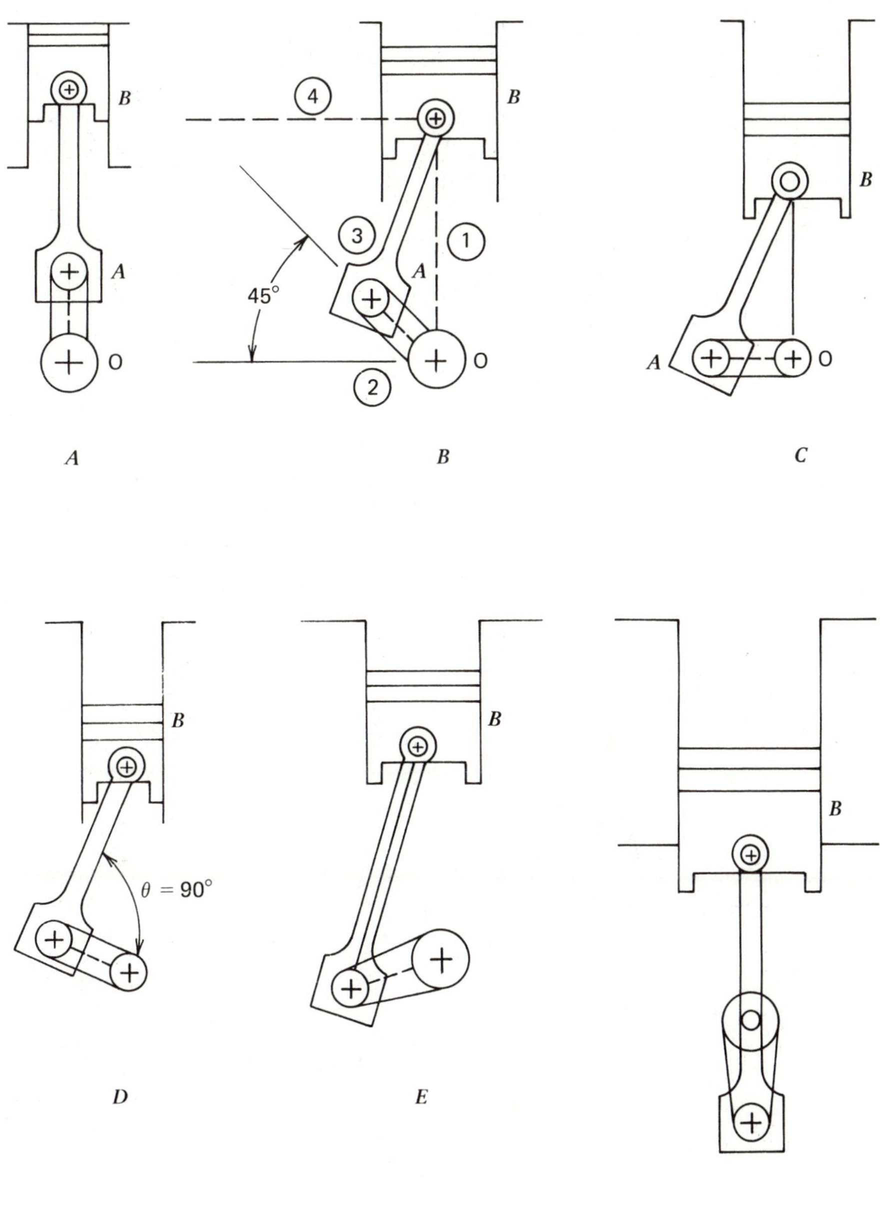

Scale: Velocity diagrams .75 in. = 20 ft/sec = 6.096 m/sec
Acceleration diagrams .75 in. = 10,000 ft/sec^2 = 3048 m/sec^2

Figure *9.7* Engine piston slider crank mechanism.

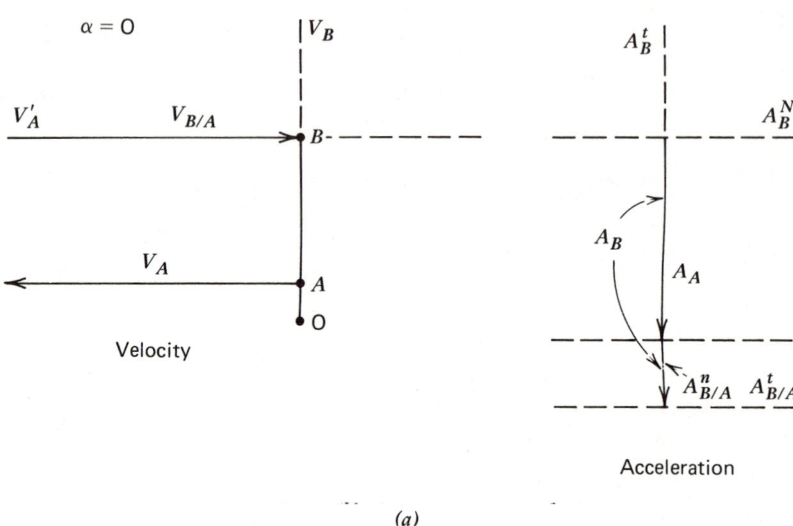

(a)

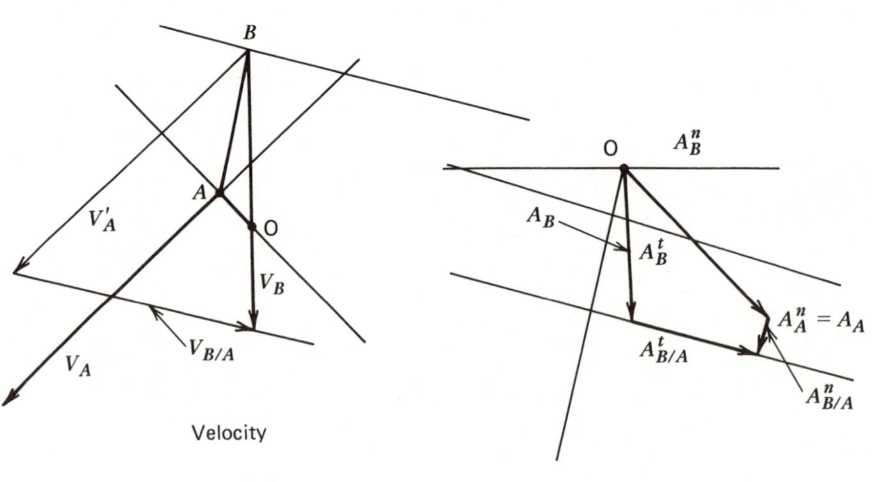

(b)

Figure 9.8 Velocity and acceleration vectors for Example 9.1.

Position C

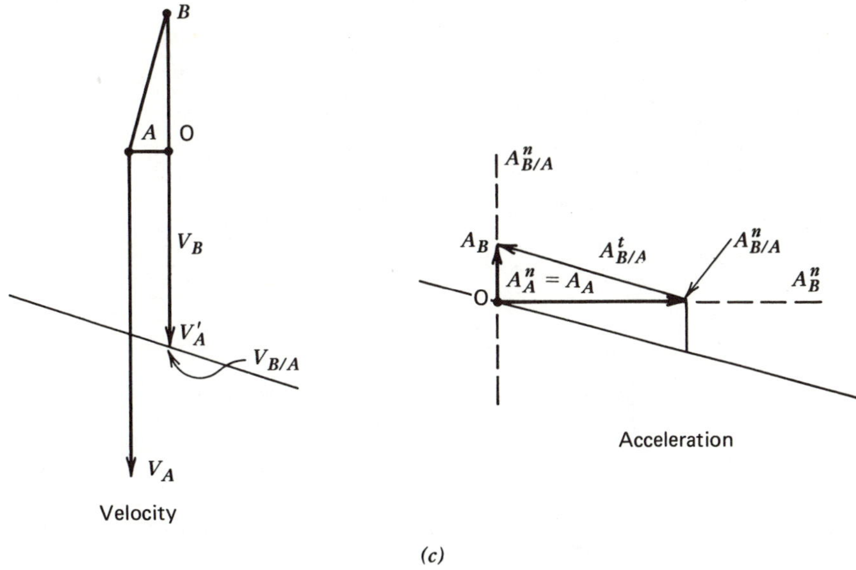

(c)

Position D

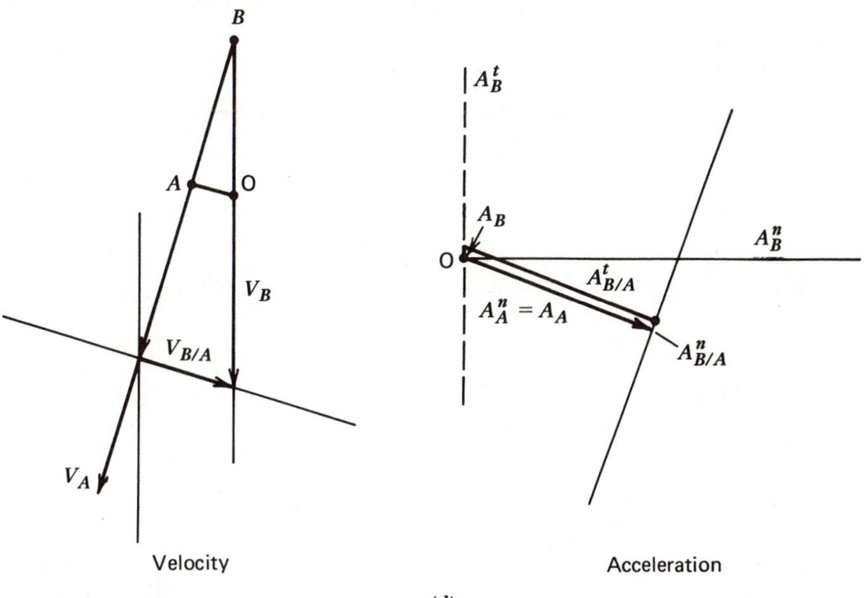

(d)

Figure 9.8 (Cont.).

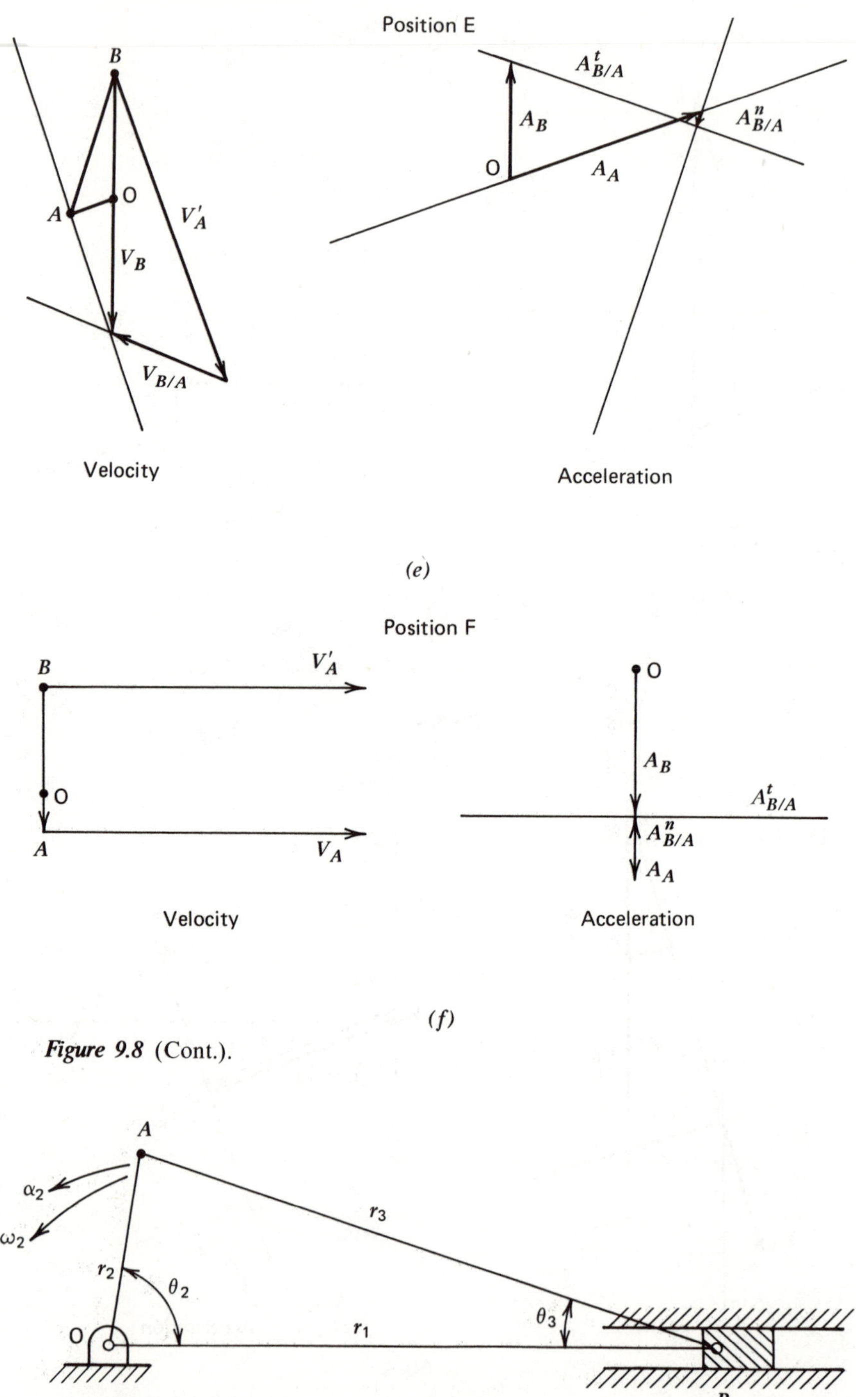

Figure 9.8 (Cont.).

Figure 9.9 Slider crank analytical nomenclature.

For velocity:

$$\dot{r}_1 = r_2\omega_2 (i \cos\theta_2 - \sin\theta_2) + r_3\omega_3 (i \cos\theta_3 - \sin\theta_3)$$

$$\dot{r}_1 = - r_2\omega_2 \sin\theta_2 - r_3\omega_3 \sin\theta_3$$

$$0 = r_2\omega_2 \cos\theta_2 + r_3\omega_3 \cos\theta_3$$

$$\omega_3 = - \omega_2 \frac{r_2 \cos\theta_2}{r_3 \cos\theta_3} \qquad [9.17]$$

For acceleration:

$$\ddot{r}_1 = - r_2 (\omega_2^2 \cos\theta_2 + \alpha_2 \sin\theta_2) - r_3 (\omega_3^2 \cos\theta_3 + \alpha_3 \sin\theta_3)$$

$$0 = r_2 (\alpha_2 \cos\theta_2 - \omega_2^2 \sin\theta_2) + r_3 (\alpha_3 \cos\theta_3 - \omega_3^2 \sin\theta_3)$$

$$\alpha_3 = r_2 \frac{(\omega_2^2 \sin\theta_2 - \alpha_2 \cos\theta_2)}{r_3 \cos\theta_3} + \frac{\omega_3^2 \sin\theta_3}{\cos\theta_3} \qquad [9.18]$$

Velocity and accelerations for the crank in position D are now calculated.

$$r_2 = 44.45 \text{ mm } (1.75 \text{ in.})$$

$$r_3 = 152.4 \text{ mm } (6 \text{ in.});\ \theta_2 = 73.74°;\ \omega_2 = 314 \text{ rad/sec}$$

$$\theta_3 = \sin^{-1} (44.45 \frac{\text{mm}}{152.4\text{m}} \sin 73.74) = -16.26$$

$$r_1 = 158.8 \text{ mm } (6.25 \text{ in.})$$

$$\dot{r} = 44.45\text{mm } (314) \sin 73.74 - (6)\ (\omega_3) \sin -16.26$$

$$\omega_3 = (-314)\ \frac{(44.45) \cos 73.74}{(152.4) \cos -16.26} = -26.71 \text{ rad/sec}$$

$$V_B = \dot{r}_1 = -14.55 \text{ m/sec}$$

$$\ddot{r}_1 = (44.45\text{mm})\ [(314^2) \cos 73.74 + (0)] -$$

$$152.4\text{mm } [(-26.71)^2 \cos -16.26 + \alpha_3 \sin -16.26]$$

$$\alpha_3 = \frac{44.45\text{mm}[(314^2) \sin 73.74 - 0]}{152.4\text{mm} \cos - 16.26} + \frac{(-26.71)^2 \sin -16.26}{\cos -16.26}$$

$$\alpha_3 = 28{,}578 \text{ rad/sec}^2$$

$$A_B = \ddot{r}_1 = -113 \text{ m/sec}_2\ (-4459 \text{ in./sec}^2)$$

The graphical values in Example 9.1 compare favorably with the exact mathematical analysis.

It is interesting to note that the maximum velocity of point B occurs when links OA and AB are perpendicular, at which point the acceleration of point B is zero.

9.2 COMMON LINKAGE TYPES

9.2.1 Four Bar Linkages

The simplest complete kinematic chain has four links. All four-bar kinematic chains can be divided into classes, distinguished from each other by differences in the possible relative motions of the links.

A large variety of motions can be obtained by varying proportions of the links and by combining several simple linkages into compound mechanisms. Several linkage mechanisms are of particular interest to designers because they are applicable to many mechanical problems; the examples that follow should be considered when designing agricultural machines, and existing mechanisms should be used whenever possible.

9.2.2 Eccentric Mechanism

The motion of this mechanism (see Fig. 9.10) is identical to that of a slider-crank with equivalent crank length *OA* and connecting rod length *AB*. One serious disadvantage of this mechanism is that it is difficult to properly lubricate between the eccentric and the rod; therefore the amount of power that can be transmitted is limited.

9.2.3 Scotch Yoke and Quick-Return Mechanisms

The Scotch yoke (Fig. 9.11) will give simple harmonic motion to the slider portion and it has been used on engines and pumps where compactness is important. The Scotch yoke is also used in testing machines to simulate vibrations. One major problem is that rapid wear is generally experienced between the slide and the drive pin.

Quick-return mechanisms are used on machine tools to give a slow cutting stroke and a quick return stroke with a constant input velocity of the driving crank.

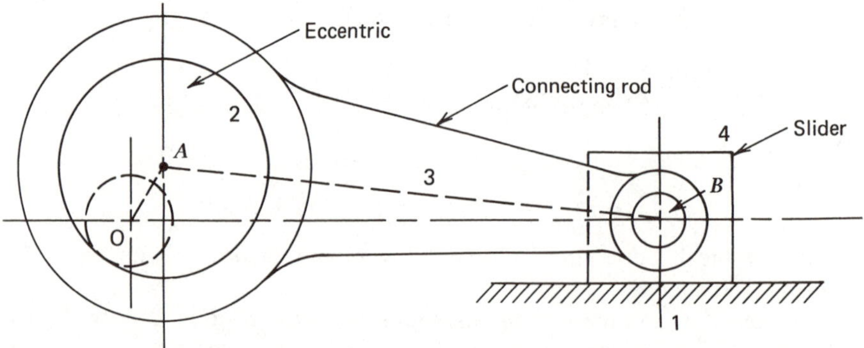

Figure 9.10 Eccentric mechanism.

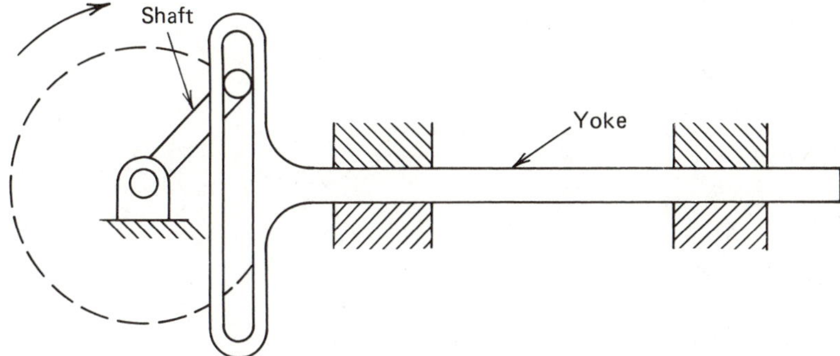

Figure 9.11 Scotch yoke.

Crankshaper (Fig. 9.12). If the driver link 2 rotates counterclockwise at constant velocity, slider 6 will have a slow stroke to the left and a fast return stroke to the right. The time ratio of the forward-to-backward strokes equals θ_1/θ_2.

Whitworth (Fig. 9.13). The links require a ratio of $0_2 0_4 < 0_2 B$ for the mechanism to work.

When the driver crank 2 rotates counterclockwise with constant angular velocity, the slider 6 will move from D' to D'' with a slow motion while 2

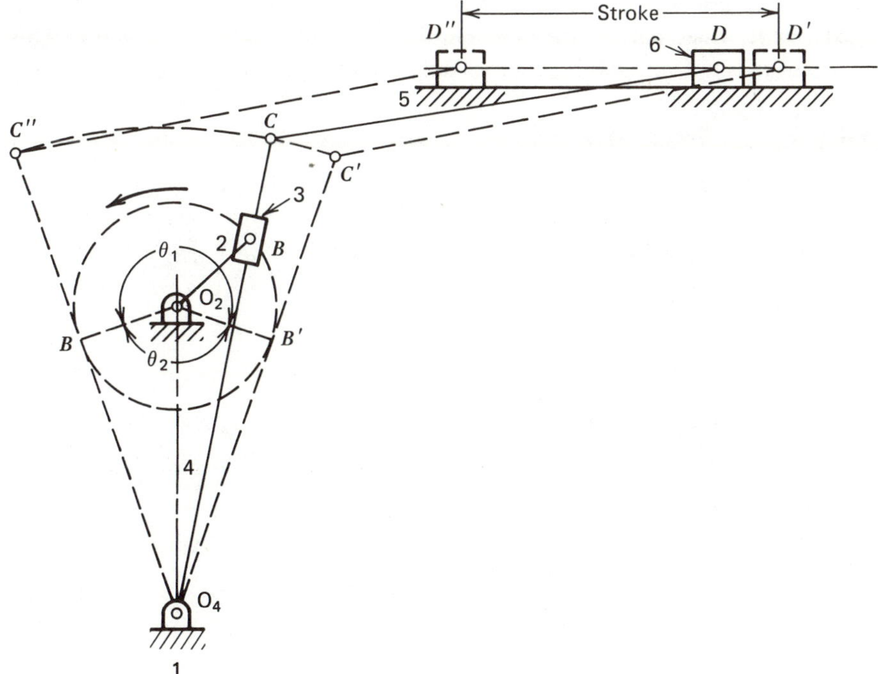

Figure 9.12 Crank shaper.

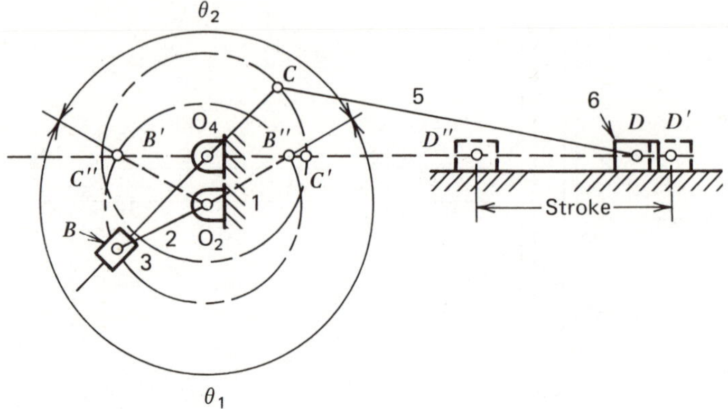

Figure 9.13 Whitworth.

rotates through θ_1. Then as 2 rotates through θ_2, slider 6 will have a quick return from D'' to D'. The forward-to-backward time ratio is θ_1/θ_2.

Drag link (Fig. 9.14). If link 2 (the driver) rotates counterclockwise with constant angular velocity, then slider 6 makes a nearly constant-velocity slow stroke to the left and returns with a quick stroke to the right. The time ratio is θ_1/θ_2.

Offset slider crank (Fig. 9.15). A slidercrank has an offset such that the path of the slider does not intersect the crank axis. The time ratio is θ_1/θ_2. This mechanism is not very effective and is only used when space is limited.

9.2.4 Toggle Mechanism

This mechanism (Fig. 9.16) can overcome large resistances (over short distances) with a small driving force. It is used in applications such as rock crushers, riveting machines, punch presses, clutches, and vice-grip pliers.

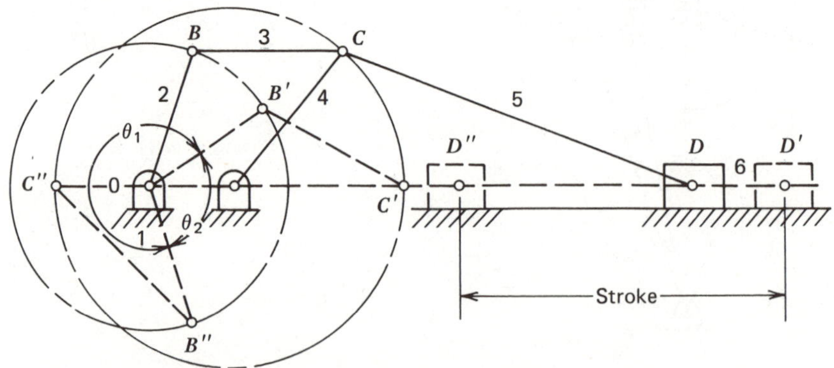

Figure 9.14 Drag link.

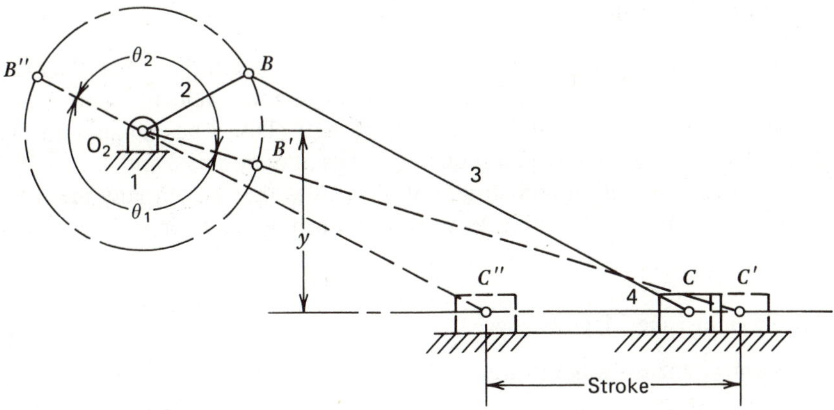

Figure 9.15 Offset slider crank.

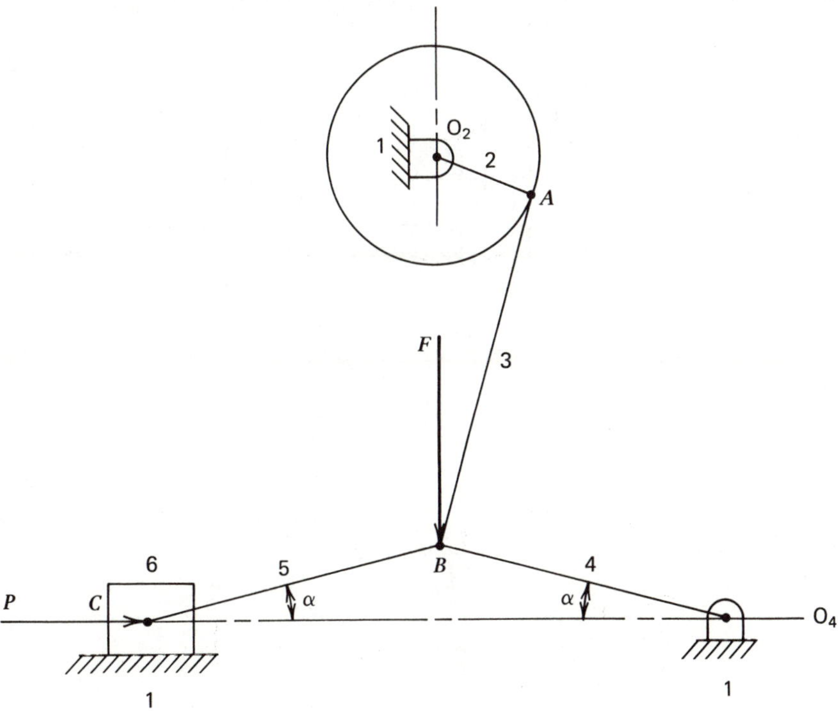

Figure 9.16 Toggle mechanism.

The force F required to overcome a resistance P is

$$\frac{F}{P} = 2 \tan \alpha$$

Theoretically, infinite mechanical advantage is possible with this mechanism.

9.2.5 Oldham Coupling

This mechanism (Fig. 9.17) provides a means for connecting two parallel shafts that are slightly misaligned so that a constant angular velocity ratio can be transmitted from the drive shaft to the driven shaft. Disc 3 has tongues on each side at 90° to one another, which slide in grooves in members 2 and 4.

Quite often the friction and structural difficulties that accompany the use of guides make it highly desirable to substitute turning pairs for sliding pairs in design.

9.2.6 Straight-Line Mechanisms

In these mechanisms a point moves along a straight line without being guided by a plane surface.

Watt's Mechanism. This mechanism (Fig. 9.18) produces approximate straight-line motion. Point P traces a figure-eight-shaped path, a considerable

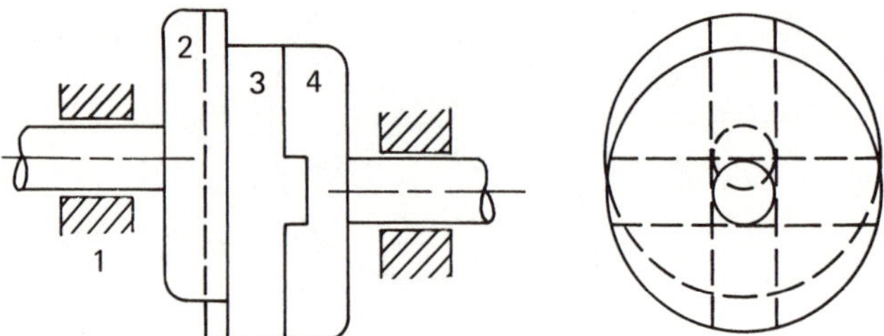

Figure 9.17 Oldham coupling.

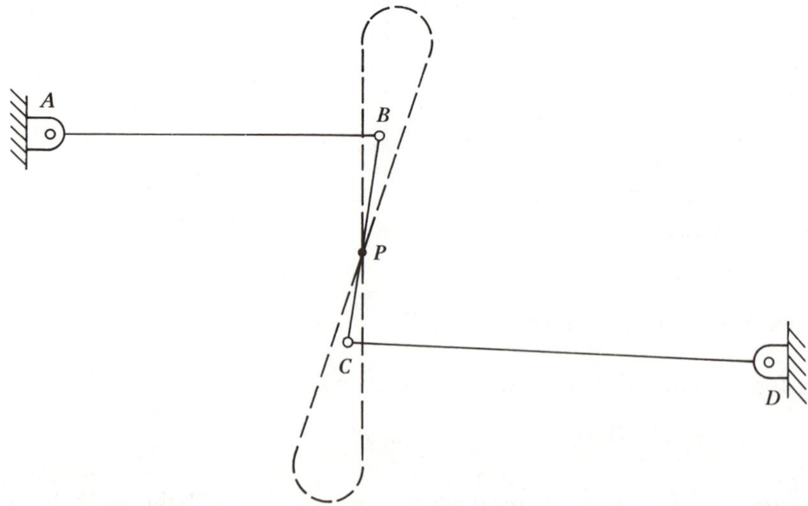

Figure 9.18 Watt's mechanism.

portion of which is approximately a straight line. The lengths must be proportional so that

$$\frac{BP}{PC} = \frac{DC}{AB}$$

Robert's Mechanism. This mechanism (Fig. 9.19) produces approximate straight-line motion. Point P moves very nearly along line AB, and $AC = CP = PD = DB$ and $CD = AP = PB$. The accuracy of the motion can be increased by increasing the ratio of the height of the mechanism to its width.

Scott-Russell Mechanism. This mechanism (Fig. 9.20) gives exact straight-line motion of point P. Length $AC = BC = CP$.

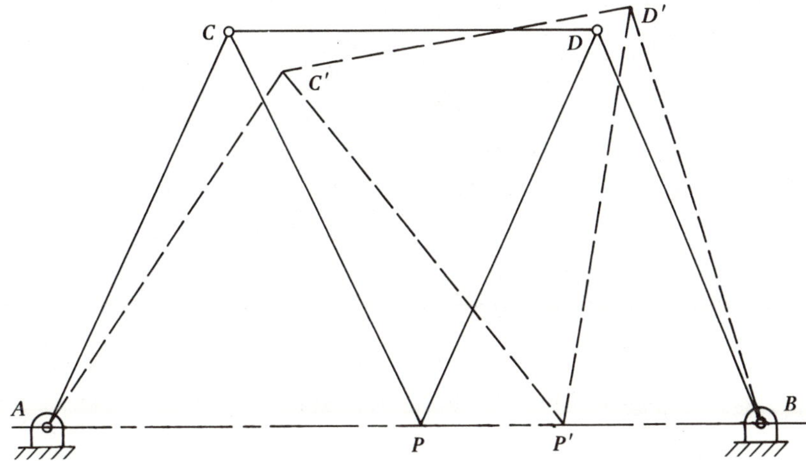

Figure 9.19 Robert's mechanism.

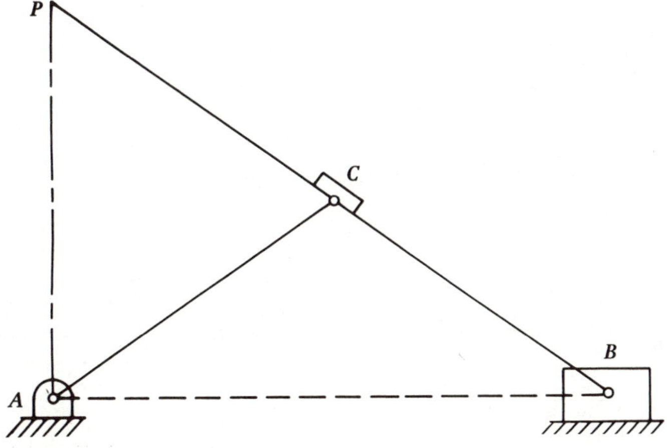

Figure 9.20 Scott-Russell mechanism.

Tchebysheff's Mechanism. This mechanism (Fig. 9.21) gives approximate straight-line motion. Point *P*, the midpoint of *CB*, moves very nearly along line CB. Length AB = *CD* = 1.25 *AD*, and *AD* = 2*CB*.

Peaucillier's Mechanism. This mechanism (Fig. 9.22) produces exact straight-line motion for point *P*. *AB* = *AE*; *BC* = *BD*; and *PC* = *PD* = *CE* = *DE*.

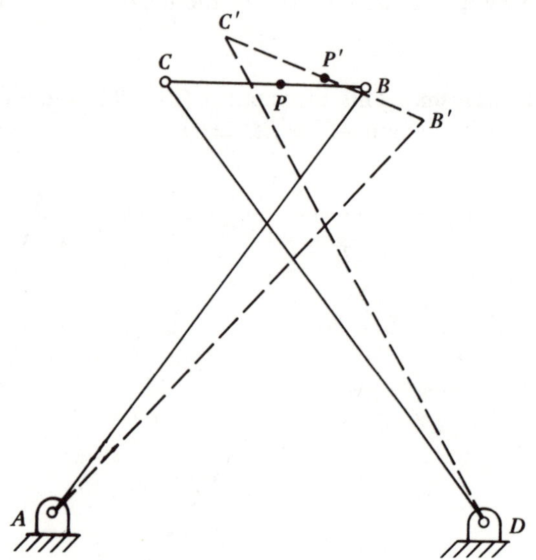

Figure 9.21 Tchebysheff's mechanism.

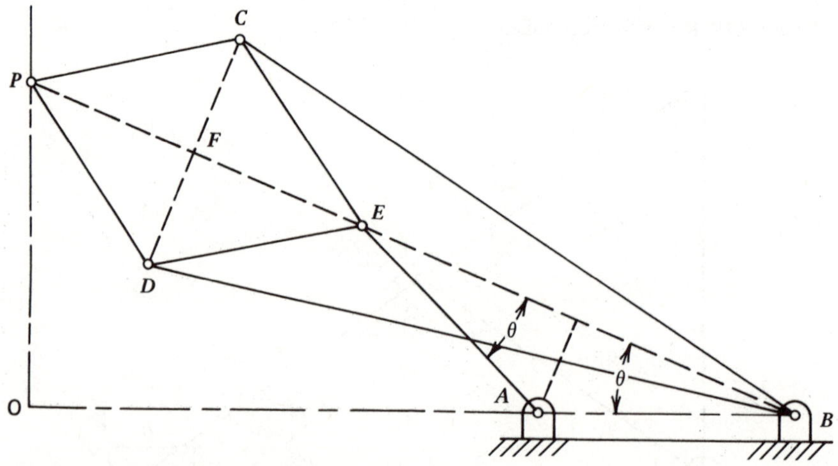

Figure 9.22 Peaucillier's mechanism.

9.2.7 Pantograph

This mechanism (Fig. 9.23) is used to duplicate some motion or surface exactly, but to a reduced or enlarged scale. Links 2, 3, 4, and 5 form a parallelogram, and point P is on an extension of link 4. Point Q is on link 5 at the intersection of a line drawn from O to P. Pantographs are used for reducing or enlarging drawings and maps; to guide cutting tools and torches; in the duplication of complicated shapes, such as turbine blades and cams; and to permit more convenient recording of small or large motions. The following relationship exists for a pantograph:

$$\frac{\text{size of figure at } P}{\text{size of figure at } Q} = \frac{OP}{OQ}$$

9.2.8 Parallel Linkages

A class of four-bar linkages with somewhat limited use consists of those linkages which appear in the form of a parallelogram. Drafting machines are a good example of this type of mechanism. The major disadvantage of this design is that in extreme positions, where the parallelograms become narrow, any lost motion in the pin connections becomes noticeably detrimental to the accuracy of the drawing.

Other types of linkages used extensively in machinery design are:

1. The Hooke-type universal joint, which is commonly found in automotive applications. This type has the disadvantage of producing varying angular velocity ratios between two nonparallel shafts.
2. Constant-velocity universal joints, such as the Bendix-Weiss, Rzeppa, or Tracta joints, which transmit a constant angular velocity ratio between nonparallel shafts.
3. Intermittent-motion mechanisms, such as the Geneva wheel and ratchet mechanisms, which convert continuous motion into intermittent motion.

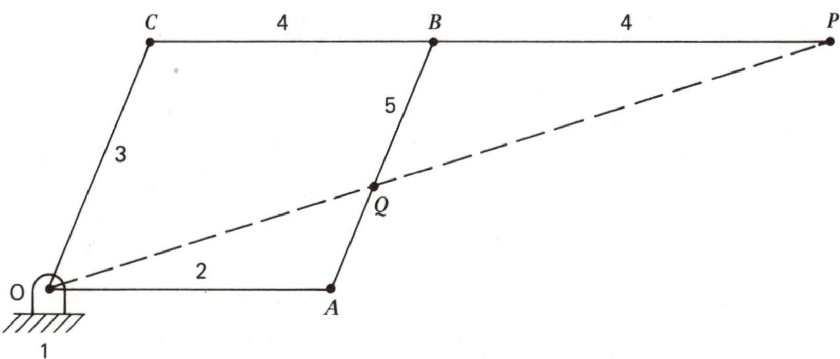

Figure 9.23 Pantograph.

One example is the indexing of a work table on a machine tool so as to bring a new work piece into position for machining by the cutters.

9.2.9 Machinery Mechanisms

Other machinery mechanisms include reversing, dwell, adjustable-output, special-motion, adjustable-stroke, function, and rotary-to-reciprocating mechanisms as well as force-and-stroke and angle-multiplier mechanisms.

One special-motion mechanism, the walking-link drive, is used on large excavator shovels and experimental agricultural tractors and planters. This mechanism is a four-bar linkage with a triangular coupler link that supports a shoe or pad that can pivot. The crank and supporting link are connected to a block or a vehicle chassis, which is moved forward by the linkage as shown in Fig. 9.24.

9.2.10 Spatial or Three-Dimensional Mechanisms

Spatial mechanisms are linkages that move in three-dimensional space, transmit motion between nonparallel shafts, and generate other types of helpful motion. Three of the most useful spatial mechanisms are the spatial crank oscillator, the spatial four bar, and the spatial slider crank.

The spatial crank oscillator consists of a universal joint that is driven by a bent shaft to produce an oscillating motion. This mechanism is used success-

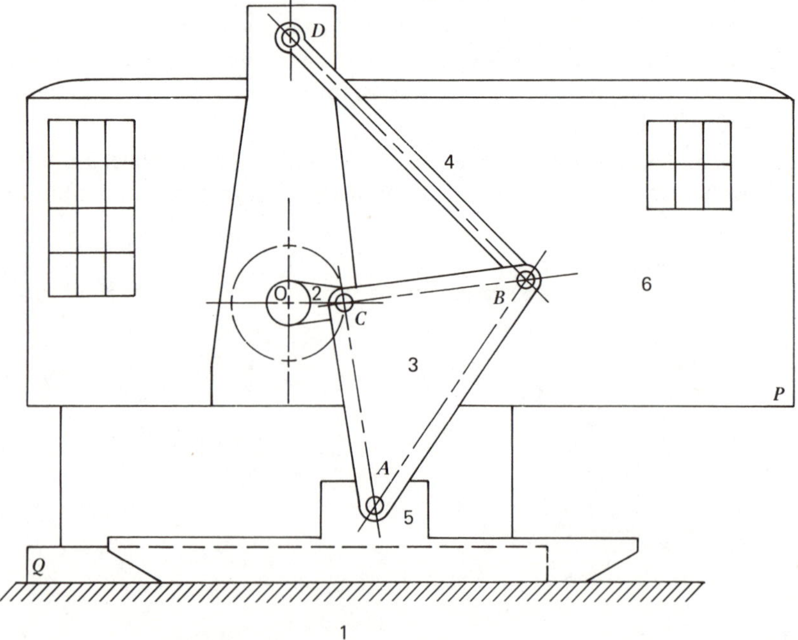

Figure 9.24 Walking link drive for a power shovel.

fully on agricultural machinery, such as the wobble drive for reciprocating cutter bars on combine grain heads and mower-conditioners.

Another application of spatial mechanisms is robot manipulators, which may eventually be used on agricultural machinery.

9.3 LINKAGES APPLICABLE TO AGRICULTURAL AND OTHER OFF-ROAD EQUIPMENT

Linkages are used extensively on agricultural equipment to generate motion and force. Hain provides the best information on linkage design with applications to agricultural equipment. Working models of the linkages described in Hain's references are located at the Institute of Basic Agricultural Engineering Research in West Germany.

The simple four-bar linkages, four bars with two points that move through space, are common on farm equipment. When the opposing links are the same length , a parallel linkage formed, that is, the bars are always parallel to each other. Such a mechanism is useful on agricultural machines. Parallel linkages are most often used to keep implements orientated properly to the soil surface as the implement moves up and down with the soil contour. Parallel linkages are used to keep planters properly orientated (see Fig. 9.25) and on cultivator units so that the shovels remain correctly tilted. John Deere uses a linkage that is almost parallel to control the orientation of the flexible-combine cutter bar. A parallel linkage can keep lights orientated properly (see Fig. 9.26) and can be used on a hay tedder (see Fig. 9.27).

Nonparallel, four-bar linkages are used to raise the wings of tillage implements, such as discs or field cultivators (see Fig. 9.28). The linkages have uneven link lengths to magnify the rotation of one link into a larger rotation of wings. A truck hoist moved by a hydraulic cylinder may function in a similar manner. Many other lifting mechanisms also use four-bar linkages. Bucket control is often accomplished through linkages on loaders and the like (see Fig. 9.29). The three-point hitch is a four-bar linkage whose dimensions are an ASAE standard.

Many control linkages use four-bar linkages. Steering systems use linkages to keep the left and right wheels orientated properly (Fig. 9.30), and linkages are often used to control a hydraulic system (Fig. 9.31).

9.3.1 Rotary to Linear Linkages

This form of four-bar linkage is ubiquitous. In the slidercrank from the four-bar linkage it forms the basis of most internal-combustion engines currently in use. It is also the basis of the pitman mower drive (see Fig. 9.32). The mechanism is also used on tree shakers and baling plungers. A shuttle stroke cleaner uses a pitman on a chain for barn cleaners. A beam-and-crank mechanism changes rotation into oscillation, and this rotating-to-oscillating movement (or almost linear movement) is seen in straw walkers and feeder shakers.

Figure 9.25 Planter four-bar linkage. Courtesy Allis-Chalmers Corporation.

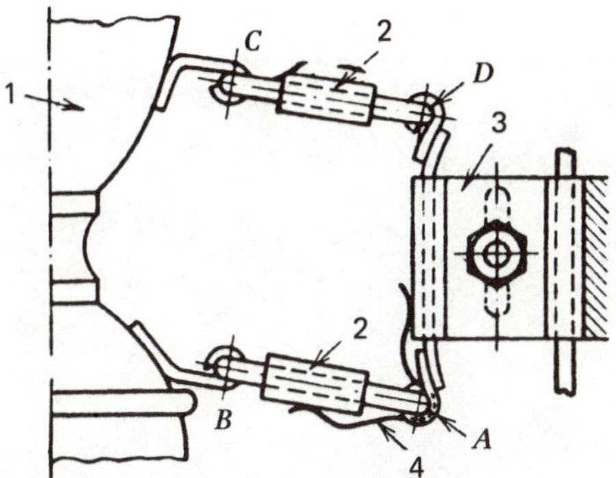

Figure 9.26 Four-bar mechanism of a light bracket.

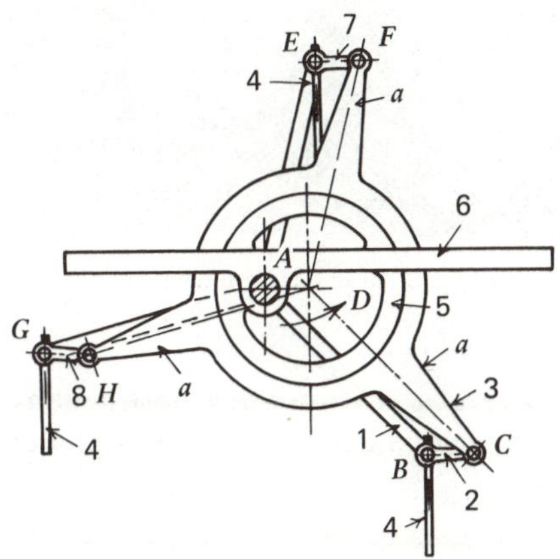

Figure 9.27 Hay tedder and drafting machine four-bar linkage.

The eccentric performs a similar function. It is used on some pieces of equipment where rotational speeds are not as high and it is popular in front of a rachet mechanism on manure spreaders and forage wagons.

The offset crank is not often used, although it could be put to good use on spreaders and forage-wagon rachet mechanisms, where its unequal forward and reverse speed might provide a more even material flow.

9.3.2 Straight-Line Mechanisms

Straight-line mechanisms are used in self-registering instruments and to feed film into projectors. There are no apparent uses of straight-line mechanisms on agricultural equipment.

9.3.3 Toggle Mechanisms

Toggle mechanisms can give mechanical advantage by providing a high clamping force to hold things together, much like a vice grip. The toggle-trip

Figure 9.28 Nonparallel linkages on a field cultivator. Courtesy Allis-Chalmers Corporation.

plow uses a toggle in reverse to trip when obstructions are encountered. A stone crusher is an example of a toggle mechanism that enables a large force to be applied.

9.3.4 Pantograph

This mechanism can accurately enlarge or decrease a motion. It is not used much on agricultural equipment.

9.3.5 Other Common Linkages

The Oldham linkage is not widely used. The Hooke joint however, is often used for PTO drives and to transmit power to feed rolls, axles, and other power users. Constant velocity prevents the inherent torsional speed variation in Hooke joints.

The various intermittent mechanisms, such as Geneva mechanisms, are not widely used in agriculture.

9.3.6 Complex Mechanisms

The various mechanisms can be combined in many ways to form elements that can perform almost any motion. Links can be combined to form five-bar or six-bar linkages. The resulting mechanisms have complex motions and are common on some side-delivery rakes and pickup units for hay and grain straw.

Figure 9.29 Bucket loader linkage. Courtesy Allis-Chalmers Corporation.

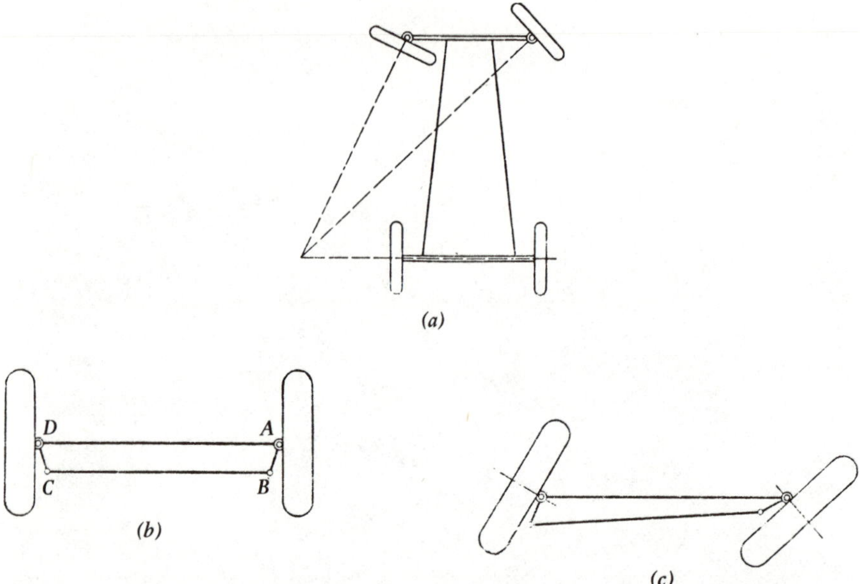

Figure 9.30 Steering four-bar linkage. Reprinted W.M. Carleton, P.K. Turnquist, and D.W. Smith, with permission © 1979, J.B. Liljedahl, *Tractors and Their Power Units*, John Wiley and Sons, Inc.

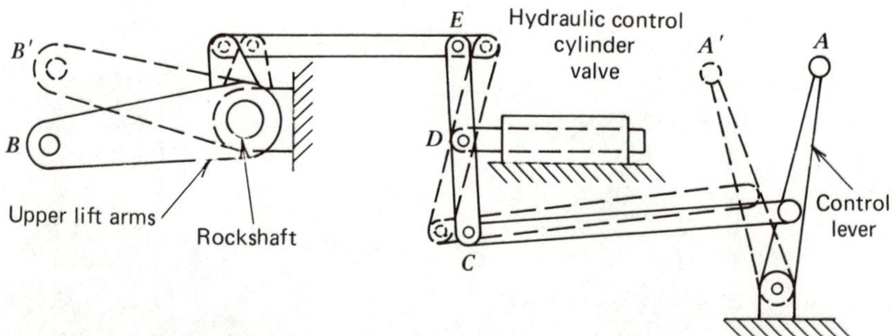

Figure 9.31 Hydraulic control linkage. Reprinted W.M. Carleton, P.K. Turnquist, and D.W. Smith, with permission © 1979, J.B. Liljedahl, *Tractors and Their Power Units*, John Wiley and Sons, Inc.

9.4 LINKAGE ANALYSIS BY COMPUTER SIMULATION

The design of complex linkages and other mechanisms for agricultural machinery is a demanding task in terms of time and complex mathematical analysis. Computers can now be used to synthesize and analyze complex mechanisms that cannot be solved by manual methods.

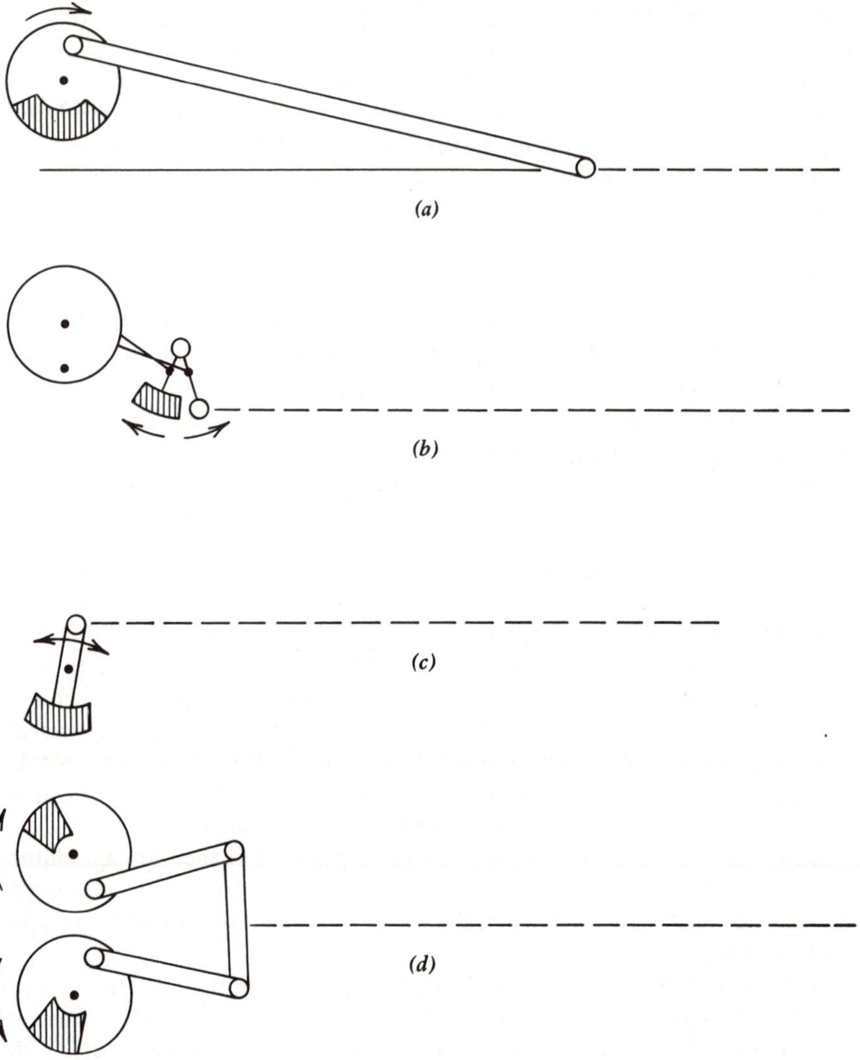

Figure 9.32 Slider cranks for mowers. Reprinted with permission © 1973, D. Hunt, *Farm Machinery Mechanisms*, Iowa State University Press.

Computer programs are available to analyze the motion and force reactions of geometrically constrained, generalized mechanical systems or mechanisms. These programs have the following capabilities: they can automatically formulate the system equations of motion or the mathematical expressions of the constraint joints and input functions; and they can simulate the response of any two- or three-dimensional discrete system of constrained rigid bodies.

One of these analytical simulation programs is the Integrated Mechanisms Program (IMP), which can analyze two- and three-dimensional, motion-constrained, closed-loop kinematic chain, rigid-body mechanisms. IMP uses a problem-oriented language, consisting of definition, data, request, control, delete, and graphics statements, to formulate and analyze the system.

With IMP, the user defines the topological connectivity of the system's links with the necessary numerical data to describe the relative orientation and location of the links. IMP then performs a topological analysis to recognize the number of links, the number and order of the kinematic loops, and other characteristics that are solely determined by the connectivity of the mechanism.

IMP has the following simulation modes:

1. Kinematic mode determines the mechanism geometry required to produce the intended kinematics. The user specifies the input motion, such as velocities or accelerations, and the step size for the inputs.
2. Static mode seeks a minimum potential energy state for the mechanism at its stable static equilibrium configuration.
3. Dynamic mode drives the mechanism with externally applied forces or input motions as a function of time.

Any or all of the following entities may be determined by IMP: the positions, velocities, and accelerations of the joints and points of interest on the links; the static and dynamic forces acting at the constraint joints as well as cyclic variations in the driving source characteristics; the forces acting in the springs and dampers; the system's natural and damped frequencies and the damping ratios; the graphical display of the system at specified time intervals; the small oscillation-system transfer functions resulting from sinusoidally-varying motion, force, and torque inputs; the vibrational mode shapes of the system; and the frequency response magnitude and phase angles.

IMP has been used to analyze several types of linkages for agricultural equipment–such as a folding tool-bar linkage and loader linkages–and vehicle systems–such as a tractor-moldboard plow system and a tractor-transport trailer system.

A John Deere six-bar folding tool-bar linkage, shown in Fig. 9.33, is modeled, as shown in Fig. 9.34, to illustrate the features of IMP. The IMP computer printout that contains the input statements and data to define this linkage and the numerical results of the kinematic analysis is shown in the appendix. The results include the forces that act on the linkage pins MHNG, HNG1, HNG2, HNG3, HNG4, and HNG5 and in the hydraulic cylinder as well as the displacements, velocities, and accelerations of two points of interest CGOR and FLWR on the linkage. The hydraulic cylinder is folding the linkage with an initial velocity of 0.045 m/sec (3 in./sec). IMP drives the cylinder to retract the tool-bar wing from the operating field position to the transport position.

Another design analysis program, similar to IMP, is the Design Analysis and Design Sensitivity (DADS) program for analyzing linkage mechanisms and vehicle systems.

Figure 9.33 Folding tool-bar six-linkage. Courtesy John Deere and Company.

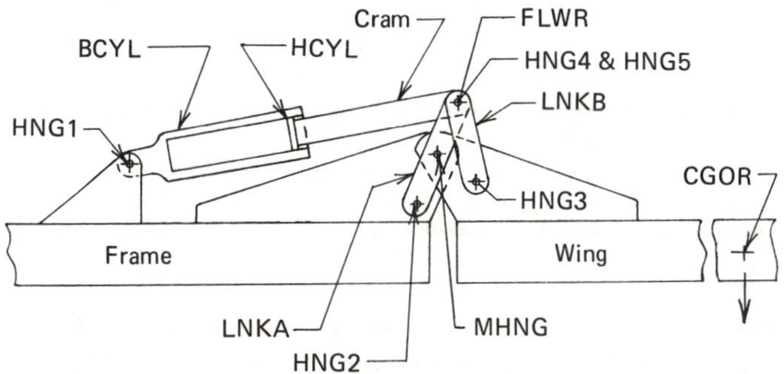

Figure 9.34 IMP model of the folding tool-bar.

Another type of computer simulation program interactively synthesizes planar four-bar and six-bar linkages. Two of the programs included in this category are the Linkage Interactive Computer Analysis and Graphically Enhanced Synthesis (LINCAGES) package and the Mechanisms Synthesis (MECHSYN) program. The skid loader linkage shown in Fig. 9.35 is synthesized with the LINCAGES package.

9.5 FORCES ON PLOWS AND DISCS

The design of tillage machinery requires knowledge of the soil forces that act as the soil is engaged by tools and creates an external pull on the tractor.

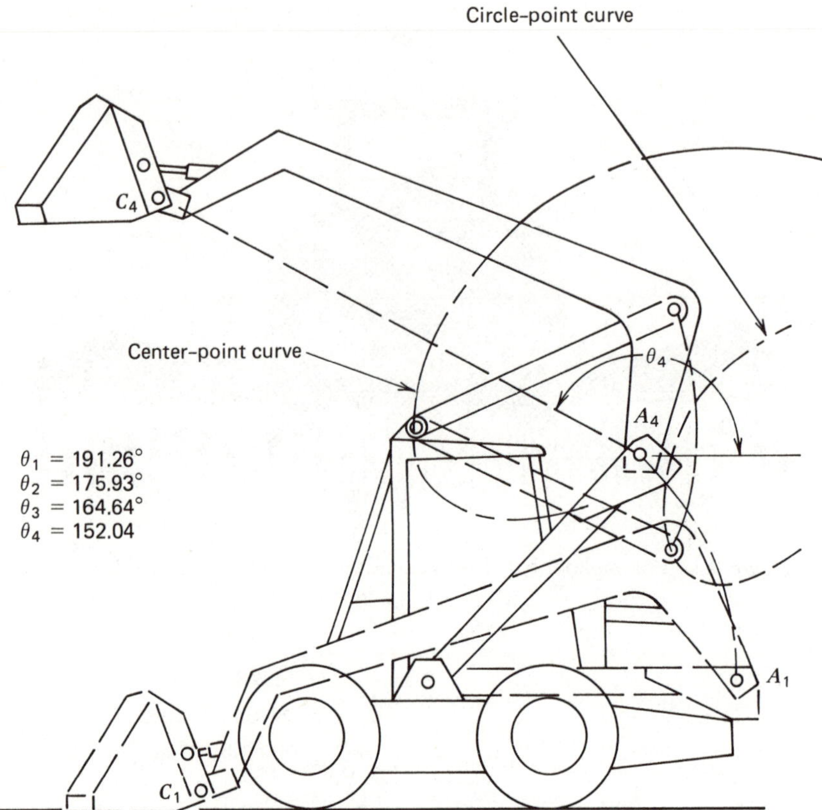

Figure 9.35 Skid loader linkage synthesis.

Plows usually experience the highest forces, and disc harrows the next highest. The greatest force is the soil force on the plow bottom or disc blade as shown in Fig. 9.36. Other large forces that require that members be designed for combined stresses include the pull forces on the drawbar or lower links. Once the free-body diagram of a machine is determined, the size of the structural members can be designed.

The plow's interaction with the tractor's center of pull can increase or decrease landside forces or tail-wheel forces. Trends vary between most U.S. designs with smaller landside forces (i.e., less wear) but larger tail-wheel structures, and European designs, which in the past have been designed for larger landside wear (Fig. 9.37).

The design process of balancing forces can reduce structure and cost, as is common with a tandem disc harrow, where side forces are equal and opposite on the right and left gangs resulting in only straight-ahead drawbar pull for staight-ahead travel (Fig. 9.38). Turning the tractor with the implement in the ground can result in high lateral forces, and in the past has resulted in lower link failures on the tractor or in yielding of the implement's structure.

Forces on a Fully Mounted Three-Point Hitch Plow

1. Weight of plow.
2. Top link force (either direction).
3. Soil force on bottoms.
4. Lower link draft forces.
5. Lower link lifting forces (either direction).
6. Landside horizontal forces (small or large).
7. Soil forces on coulters.
8. Tail-wheel vertical force.
9. Tail-wheel horizontal force (small or large).

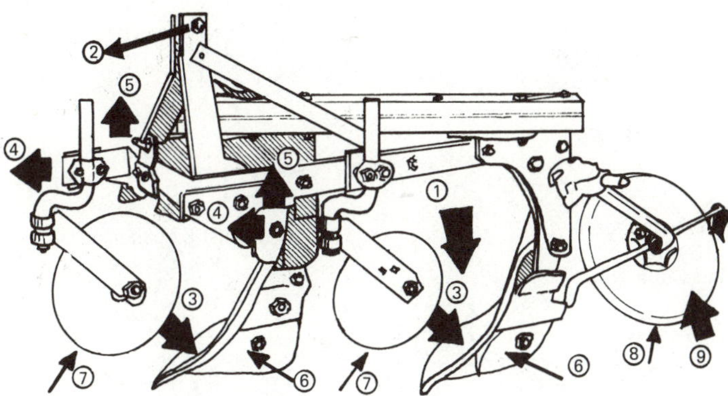

Figure 9.36 Plow forces.

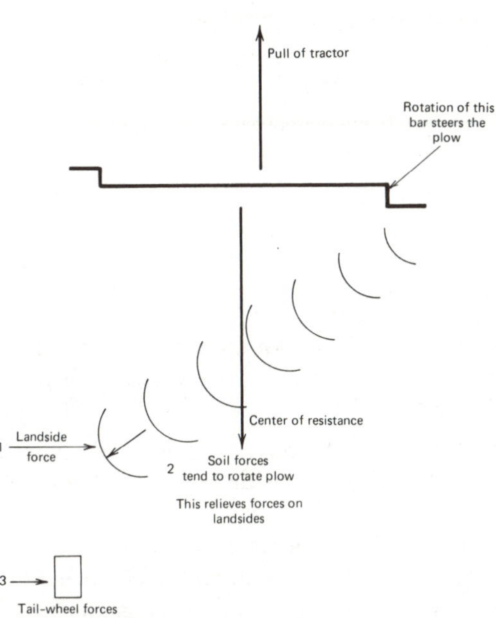

Figure 9.37 Offset of plow to tractor attachment results in side forces.

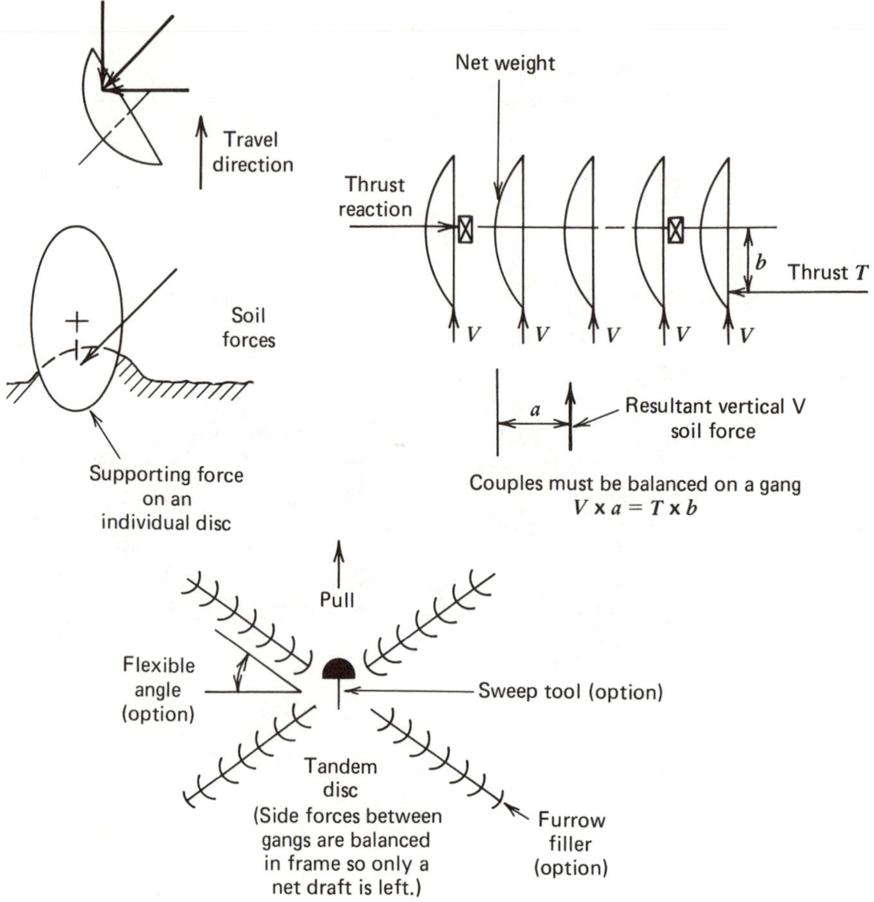

Figure 9.38 Disc and harrow forces.

HOMEWORK PROBLEMS

9.1 Member *AB* in Fig. 9.39 has an angular velocity of 6 rad/sec counterclockwise in the position shown. Determine the angular velocities of members *BC* and *CD*.

9.2 The mechanism in Fig. 9.40 moves in a vertical plane, and member *CD*, when in the position shown, has an angular velocity of 8 rad/sec clockwise. Determine:

a. The angular velocities of members *AB* and *AC*.

b. The velocity of point *M*, the midpoint of member *BC*.

9.3 Member *AB* in Fig. 9.41 has an angular velocity of 4 rad/sec counterclockwise in the position shown. Determine:

a. The velocity of point *C*.

b. The velocity of point *E*.

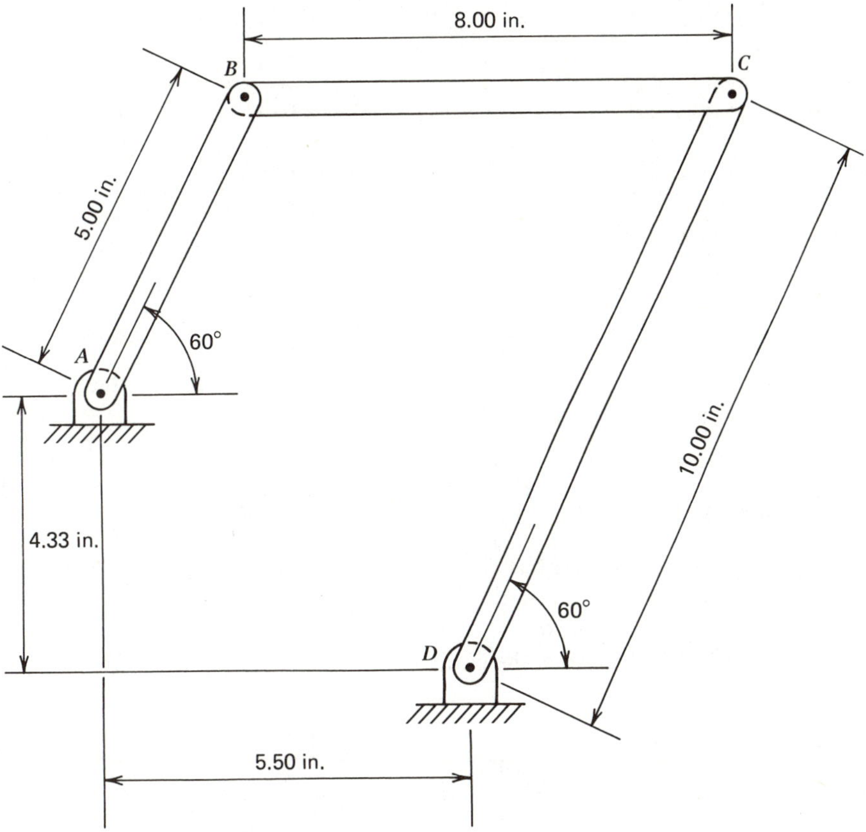

Figure 9.39 Problem 9.1

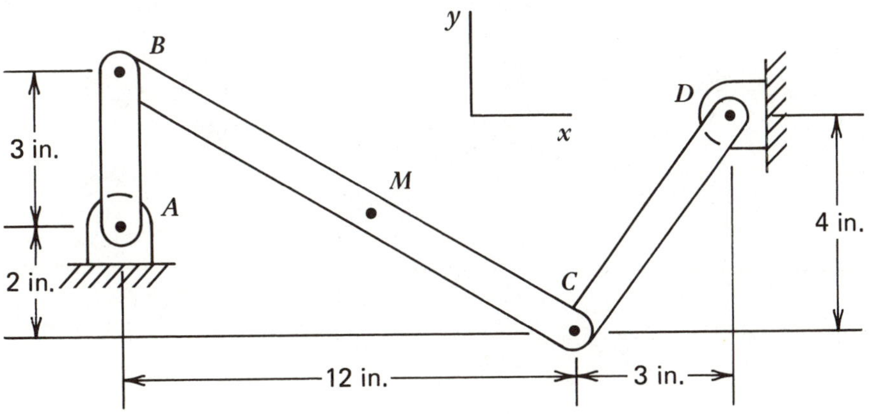

Figure 9.40 Problem 9.2

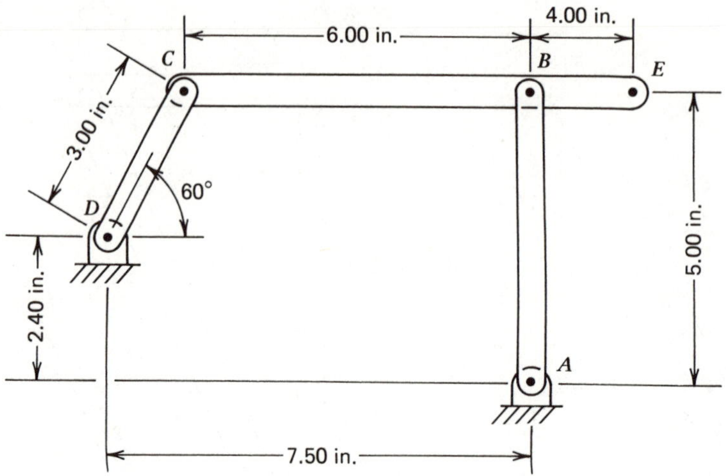

Figure 9.41 Problem 9.3

9.4 Member *AB* in Fig. 9.42 has an angular velocity of 3 rad/sec clockwise in the position shown. Determine:

a. The velocity of point *B*.
b. The angular velocity of the crank *OA*.

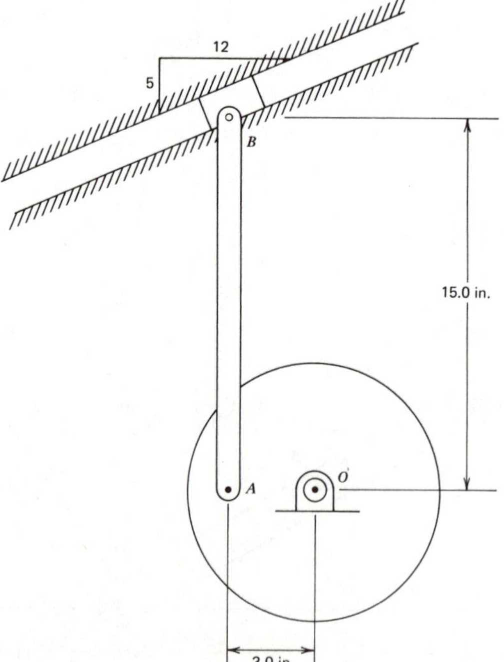

Figure 9.42 Problem 9.4

9.5 For the half-scale model of the complex linkage in Fig. 9.43, construct a velocity polygon using graphical methods and determine magnitude and direction of points A through E. Piston F has an instantaneous velocity of 10 ft/sec.

9.6 Crank AB of the engine system shown in Fig. 9.44 has a constant angular velocity of 2000 rpm. For the crank position shown, determine the angular acceleration of the connecting rod BD and the acceleration of point D.

9.7 Plate $ABCD$ in Fig. 9.45 is pinned to OA at A, and the roller at C slides in the horizontal slot. In the position shown, the velocity of C is 35 in/sec clockwise. Determine the velocities of points A, B and D.

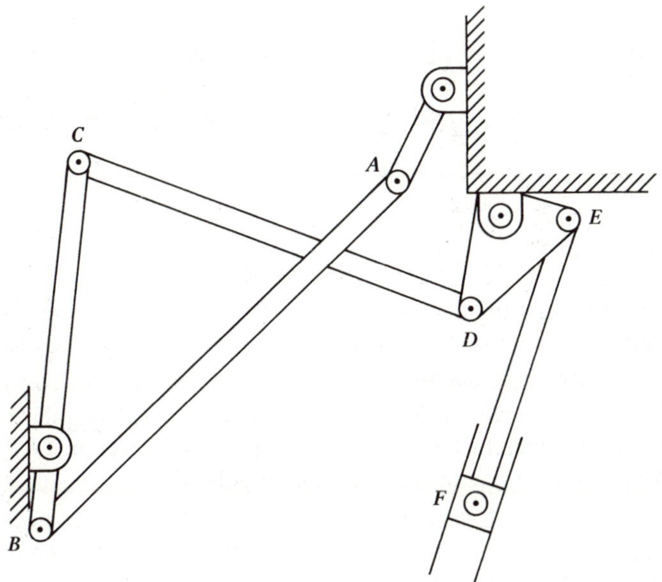

Figure 9.43 Problem 9.5.

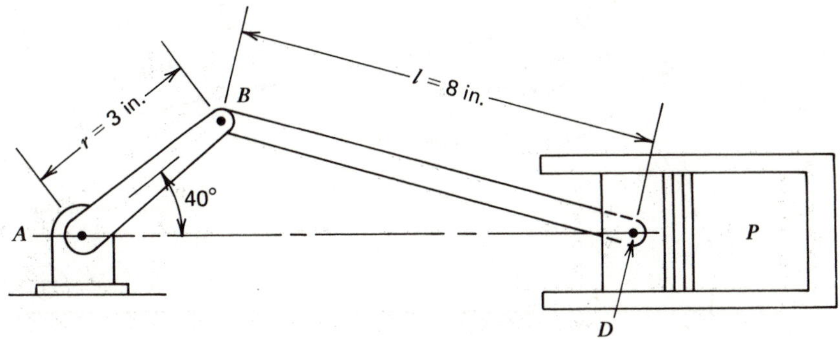

Figure 9.44 Problem 9.6.

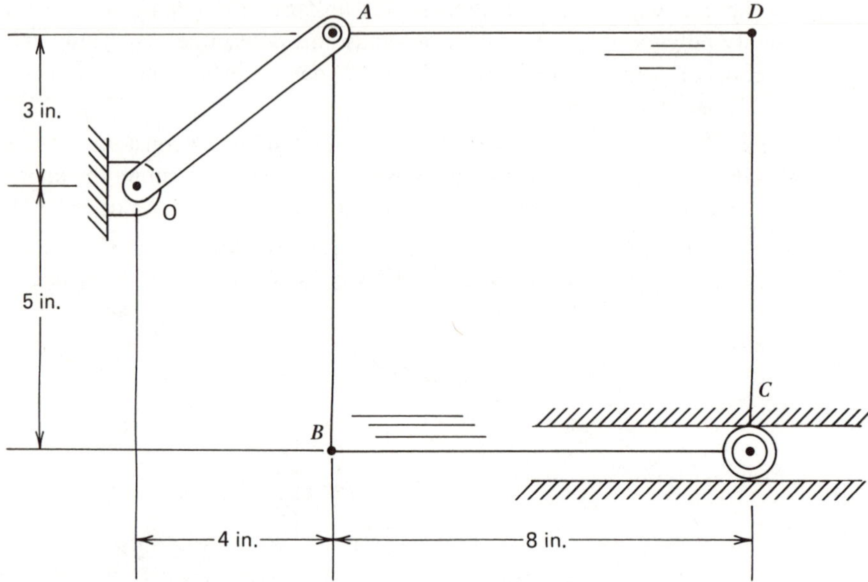

Figure 9.45 Problem 9.7.

9.8 In Fig. 9.46 the angular velocity of *AB* is 10 rad/sec and the angular acceleration is zero. Graphically determine velocity and acceleration vectors for points *B*, *C*, and *E*.

9.9 In the position shown in Fig. 9.47, member *CD* has an angular velocity of 2 rad/sec counterclockwise and an angular acceleration of 2 rad/sec^2 clockwise. Determine the acceleration of point *B*.

9.10 Arm *AB* in Fig. 9.48 has an angular velocity of 2 rad/sec clockwise and an angular acceleration of 10 rad/sec^2 counterclockwise when in the position shown. Determine the velocity and acceleration of block *C*.

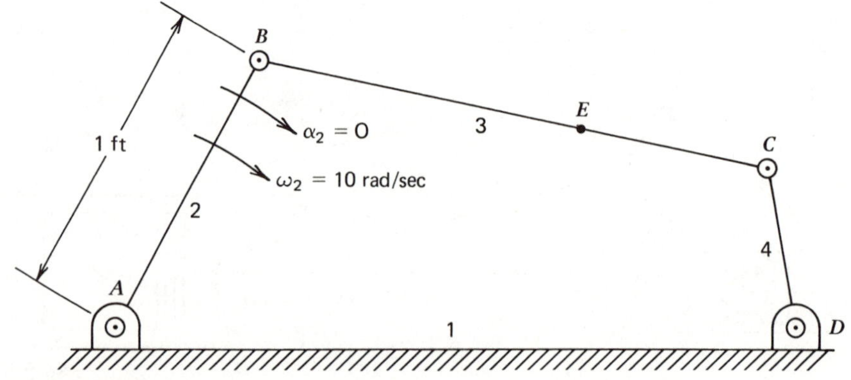

Figure 9.46 Problem 9.8.

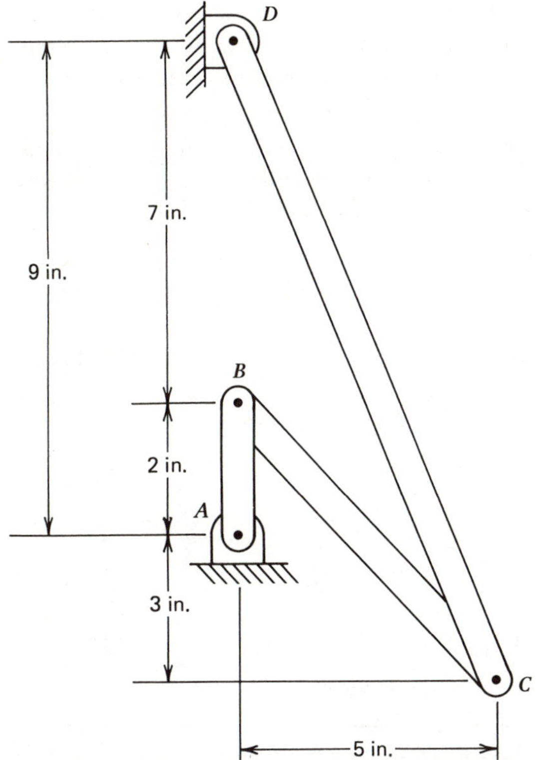

Figure 9.47 Problem 9.9.

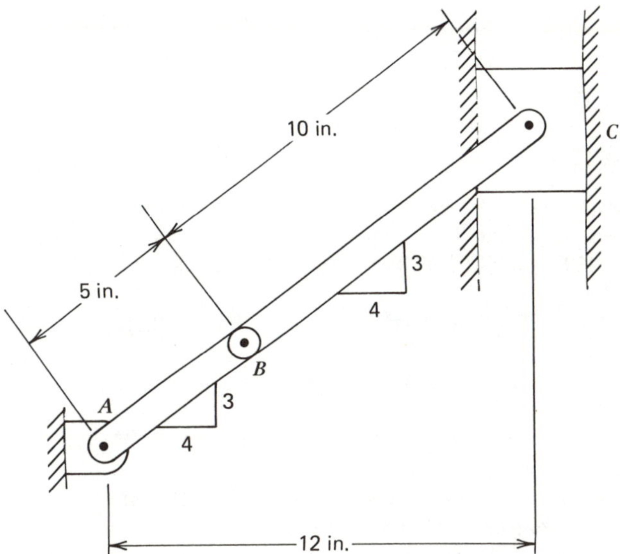

Figure 9.48 Problem 9.10.

9.11 Roller *C* of the mechanism in Fig. 9.45 has a velocity of 42 in./sec to the right and is slowing down at the rate of 100 in./sec^2 in the position shown. Determine the required angular velocity and acceleration of member *OA*.

9.12 A simplified version of the mechanism for operating a hydraulic dump truck is shown in Fig. 9.49. When the dump bed is in the position shown, the angular velocity of link *A* is 0.035 rad/sec clockwise. Determine the angular velocity of the dump bed for this position.

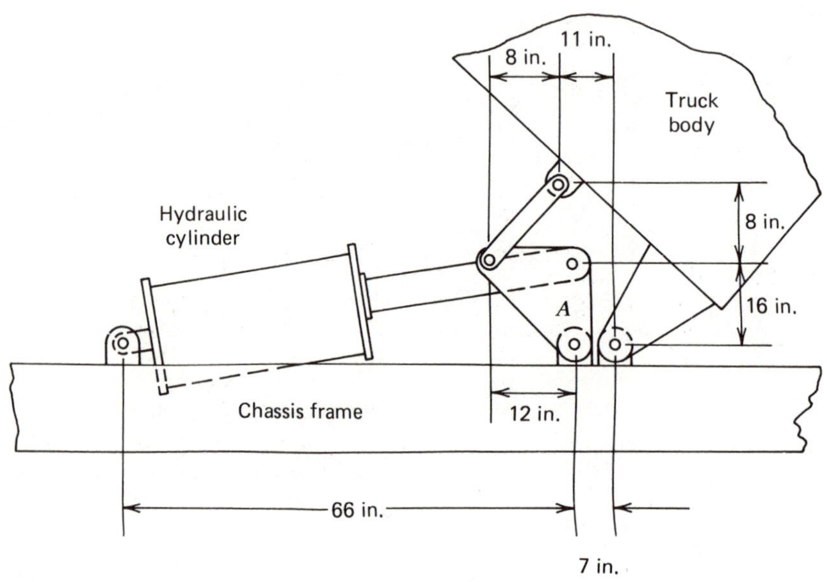

Figure 9.49 Problem 9.12.

REFERENCES

1. Artobolevsky, I.I. Lever Mechanisms, Mechanisms in Modern Engineering Design, vol. 1, Mir Publishers, Moscow, 1975.
2. Barnacle, H.E. and G.E. Walker. *Mechanics of Machines*, vol. 1, Pergammon Press, Oxford, 1965.
3. Barton, L.O. "Painless Analysis of Four-Bar Linkages," *Machine Design* 51 (17):124-127, 1979.
4. Barton, L.O. "Simplifying Velocity Analysis for Mechanisms," *Machine Design*, 53(13):123-127, 1981.
5. Barton, L.O. "A New Way to Analyze Slider Cranks," *Machine Design* 54(17):49-53, 1982.
6. Baumeister, T., E.A. Avallone, and T. Baumeister III. *Marks' Standard Handbook for Mechanical Engineers*, Mc-Graw Hill Book Co., New York, N.Y., 1978.
7. Beer, F.P. and E.R. Johnson, Jr. *Vector Mechanics for Engineers: Dynamics*, McGraw-Hill Book Co., New York, N.Y., 1972.

8. Chen, P. "Application of Spatial Mechanisms to Agricultural Machinery," *Transactions of the ASAE* 16(2):214-217, 1973.
9. Chironis, N.P., ed. *Mechanisms, Linkages, and Mechanical Controls*, McGraw-Hill Book Co., New York, N.Y., 1965.
10. Claar, P.W., II, W.F. Buchele, S.J. Marley, and P.N. Sheth. "Agricultural Tractor Chasis Suspension System for Improved Ride Comfort," SAE Paper 801020, Society of Automotive Engineers, Inc., Warrendale, Pa., 1980.
11. Davidson, J.K. "Programmable Calculators Take the Drudgery out of Linkage Analysis," *Machine Design* 49(13):93-97, 1977.
12. Doughtie, V.L. and W.H. James. *Elements of Mechanism*, John Wiley and Sons, Inc., New York, N.Y., 1954.
13. Erdman, A.G. and D.R. Riley. "Computer-Aided Linkage Design Using the LINCAGES Package," ASME Paper 81-DET-121, The American Society of Mechanical Engineers, New York, N.Y., 1981.
14. Hain K. *Applied Kinematics*, 2d edition, McGraw-Hill Book Co., New York, N.Y., 1967.
15. Hain, K. *Getriebebeispiele Atlas*, VDI Verlag, Dusseldorf, 1973.
16. Hain, K., H. Schaedler, and G. Marx. "Mechanisms A 9-Step Refresher Course," *Product Engineering* 32(1):25-28; 32(2):62-64; 32(3):46-47, 32(4):48-49; 32(5):36-37; 32(6):44-46; 32(7):52-53; 32(8):80-81; 32(9):54-56, 1961.
17. Higdon, A., W.B. Stiles, A.W. Davis, and C.R. Evces. *Dynamics*, Engineering Mechanics, vol. 2, Prentice-Hall, Englewood Cliffs, N.J., 1976.
18. Hunt, D.R. and L.W. Garver. *Farm Machinery Mechanisms*, Iowa State University Press, Ames, Iowa, 1970.
19. Jensen, P.W. "From German Sources: A Compendium of Machinery Mechanisms," *Product Engineering* 35:66-74; 108-115, 1964.
20. Kaufman, R.E. "Mechanism Design by Computer," *Machine Design* 50(24)94-100, 1978.
21. Kepner, R.A., R. Bainer, and E.L. Barger. *Principles of Farm Machinery*, AVI Publishing Company, Westport, Conn., 1978.
22. Liljedahl, J.B., W.M. Carleton, P.K. Turnquist, and D.W. Smith. *Tractors and Their Power Units*, 3d ed., John Wiley and Sons, Inc., New York, N.Y., 1979.
23. Meyer zur Capellen, W. "Seven Popular Types of Three-Dimensional Drives," *Product Engineering* 31(25):76–80. 1960.
24. Nikravesh, P.E. and I.S. Chung. "Application of Euler Parameters to the Dynamic Analysis of Three-Dimensional Constrained Mechanical Systems," ASME Paper 81-DET-123, The American Society of Mechanical Enginers, New York, N.Y., 1981.
25. Paul, B. *Kinematics and Dynamics of Planar Machinery*, Prentice-Hall, Englewood Cliffs, N.J., 1979.
26. Paul, R.P. *Robot Manipulators: Mathematics, Programming, and Control*, The MIT Press, Cambridge, Mass., 1981.
27. Phelan, R.M. *Dynamics of Machinery*, McGraw-Hill Book Co., New York, N.Y., 1962.
28. Rice, S.L. "3-D Linkages Simplify Complex Mechanisms," *Machine Design* 48(14):100-104, 1976.
29. Sheth, P.N. and J.J. Uicker, Jr. "IMP (Integrated Mechanisms Program), A Computer-Aided Design Analysis System for Mechanisms and Linkages," *ASME Transactions for the Journal for Engineering in Industry* 94(2):454-464, 1972.

30. Siversten, O. and A. Myklebust. "MECSYN: An Interactive Computer Graphics System for Mechanism Synthesis by Algebraic Means," ASME Paper 80-DET-86, The American Society of Mechanical Engineers, New York, N.Y., 1980.
31. Smith, D.W., R.A. Light, B.E. Romig, T.A. Berenji, N.V. Orlandea, J.C. Wiley, N. Portillo, S.E. O'Brien, and D.J. Hertema. "Automated Simulation and Display of Mechanism and Vehicle Dynamics," ASAE Paper 82-5019, 1982.
32. Soni, A.H. *Mechanism Synthesis and Analysis*, McGraw-Hill, Book Co., New York, N.Y., 1974.
33. Wehage, R.A. and E.J. Haug. "Generalized Coordinate Partitioning for Dimension Reduction in Analysis of Constrained Dynamic Systems," ASME Paper 80-DET-106, The American Society of Mechanical Engineers, New York, N.Y., 1980.
34. Wolford, J.C. and L. Kersten. "The Applicaton of Chace's Vector Equations in the Kinematic Analysis of Mechanisms." *Proceedings of the Fifth OSU Applied Mechanisms Conference*, 1977.

CHAPTER TEN

HYDRAULIC POWER SYSTEMS

Fluid power is defined as energy transmitted and controlled through the use of a pressurized fluid. A fluid power system is one that transmits and controls power through the use of a pressurized fluid within an enclosed circuit. A pump is a device that converts mechanical force and motion into hydraulic fluid power in the form of the flow of a pressurized fluid. A hydraulic power system is composed of a pump, to create fluid flow; an actuator, which converts flow of pressurized fluid to mechanical work; a means to control the fluid flow; and the necessary conduits to connect the pump and actuator.

Fluid power standards have been developed to improve communications regarding hydraulic-power transmission systems. The definitions of terms used to describe the functional use of various hydraulic components and graphical symbols to schematically show their interrelation within a circuit are found in standards published by the following organizations:

NFPA: (National Fluid Power Association). A nonprofit trade association of U.S. companies that design, manufacture, and market fluid power components.

ANSI-B93: (American National Standards Institute Committee). This institute is a federation of trade associations, professional and scientific societies, and individual companies. The B93 committee for fluid power systems and components reviews and coordinates proposed standards for fluid power submitted for acceptance as American National Standards.

USA TAG: (United States Technical Advisory Group). This is an ANSI fluid power committee that establishes U.S. policy on proposed international standards.

ISO: (International Organization for Standardization). An agency composed of the national standard bodies of 81 nations. ISO works toward agreement on international standards with a view toward expansion of trade, the improvement of quality, and the increase of productivity.

An international system of units (Système International d'Unités), referred to as SI units, is used to promote international trade. SI units avoid the problem of various interpretations of words and terms from one language to another. With the use of SI units and graphical symbols to represent component parts, a universal understanding of fluid power systems can be achieved.

Standards for representing hydraulic power system components have been developed to show symbolically the functions and method of operation of each component. Symbols that represent hydraulic components are simple to draw and show the connections, flow paths, and functions of the hardware they represent. The hydraulic power system and the relationship of the operation of various components can readily be shown in a circuit diagram by depicting each component by its symbol and connecting the flow paths from one symbol to another. The construction of components is not shown, and relative dimensional sizes are not represented.

The circuit diagram, composed of individual-component symbols, is readily understood and provides a simple, clear method of communication for educational purposes and understanding industrial applications. Some of the most basic symbols for hydraulic fluid circuits are shown in Table 10.1. Standard graphic symbols published by ANSI and the Joint Industry Conference (JIC) are far more comprehensive and include composite symbols to represent complex components.

Following are descriptions of the individual components incorporated into a design of a hydraulic power system. The types, function, use, and characteristics of operation of the components are presented. System design requires selecting appropriate components and matching all components for the flow and pressure requirements of the hydraulic circuit.

10.1 HYDROSTATIC DRIVES

The enclosed circuit that transmits fluid power is defined as an open or closed loop. The pump in an open loop circuit draws hydraulic fluid from a reservoir. The pump output flow is conducted to a hydraulic motor or linear actuator, and the return fluid is directed back to the reservoir. The pump inlet volume does not depend on the return flow. A reservoir included in an otherwise closed circuit leads to the terminology of open loop. Open loop circuits, with flow and pressure regulations, are shown in Fig. 10.1. Provisions to filter the oil as it leaves the reservoir and to cool the fluid as it returns from the work function are included. The motor rotation is controlled by a directional flow valve. The speed of the motor depends on the pump flow rate.

A closed loop hydrostatic drive, as shown in Fig. 10.2, circulates the fluid directly from the pump to the motor and returns the motor outlet flow back to the pump inlet. The various combinations of pump and motor fixed or variable displacements that can be incorporated in this drive are shown in Fig. 10.3. Closed loop circuits develop only the hydraulic power required by the

Table 10.1 **Basic Fluid Power Symbols (Partial List)**

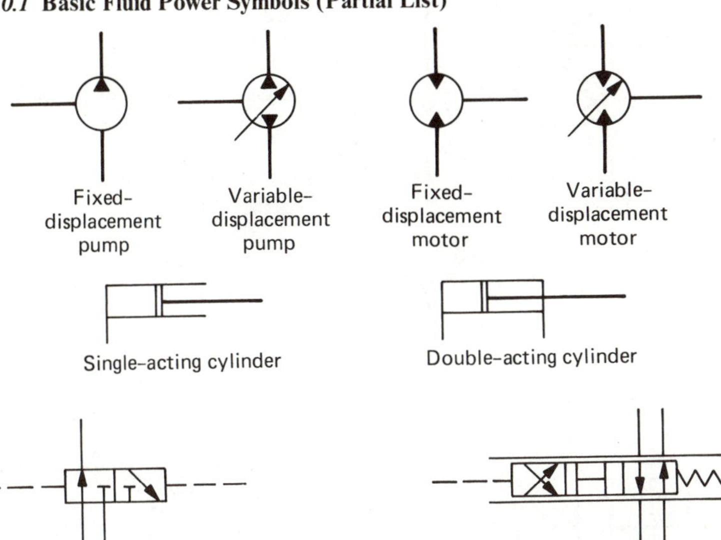

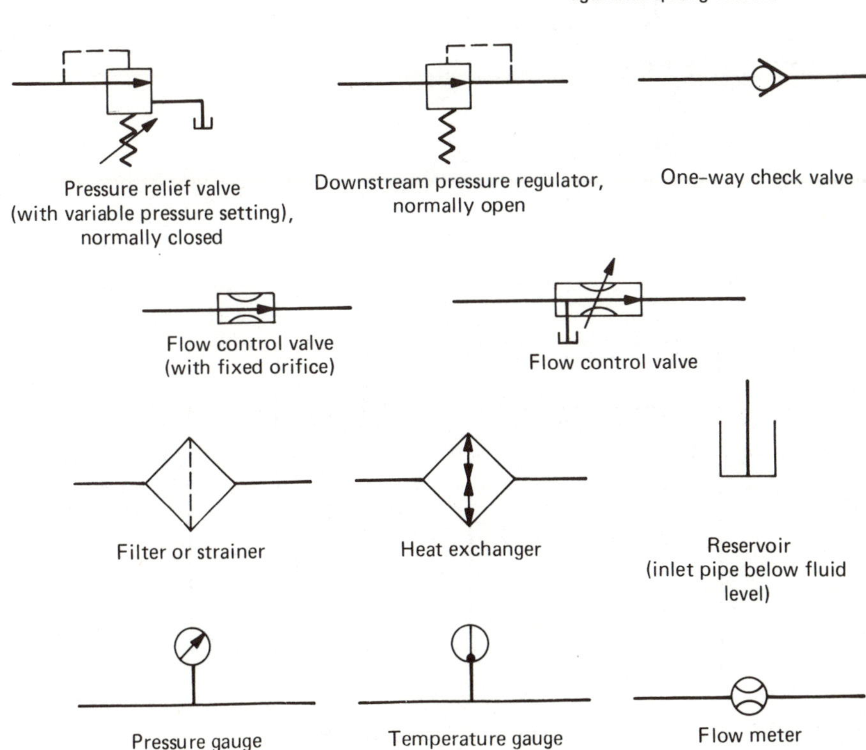

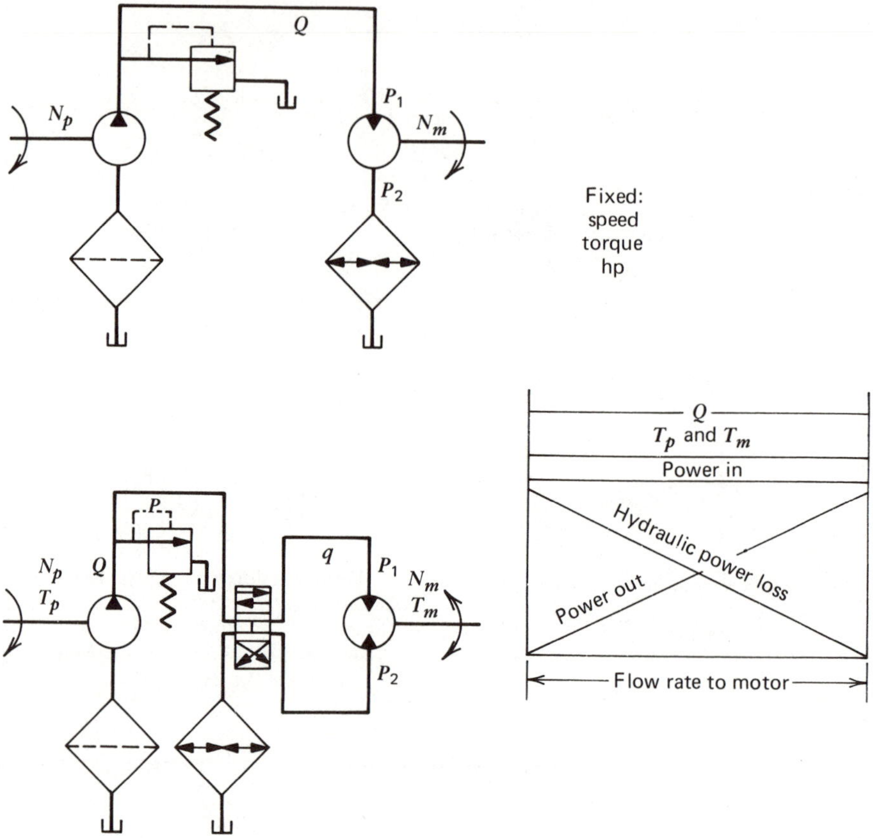

Q
P1
P2
Np
Nm
Fixed:
speed
torque
hp
Q
Tp and Tm
Power in
Hydraulic power loss
Power out
Flow rate to motor
Np
Tp
Q
P
q
P1
Nm
Tm
P2

Figure 10.1 Open loop circuits with fixed-displacement pump and motor.

Case cooling
Np
Tp
HPp
Q
P1
Nm
Tm
HPm
P2

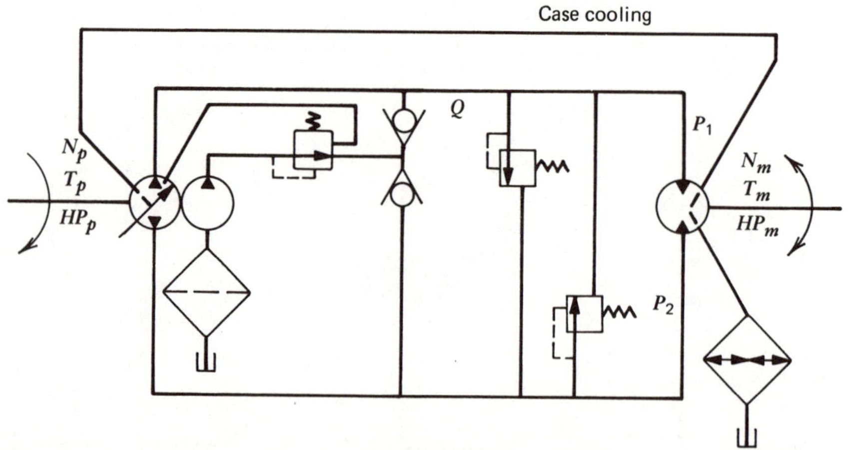

Figure 10.2 Closed loop hydrostatic drive hydraulic circuit.

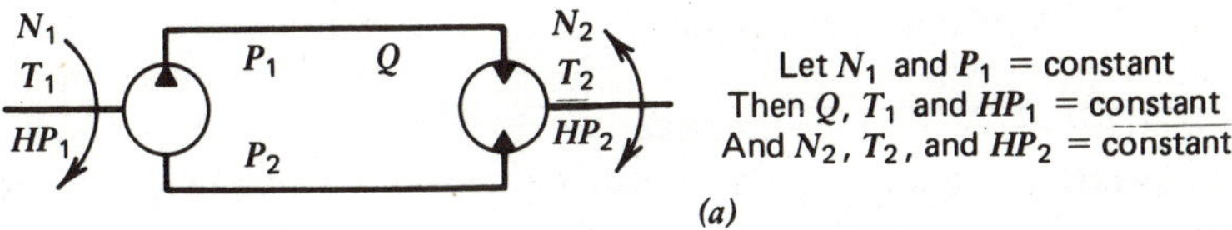

(a)

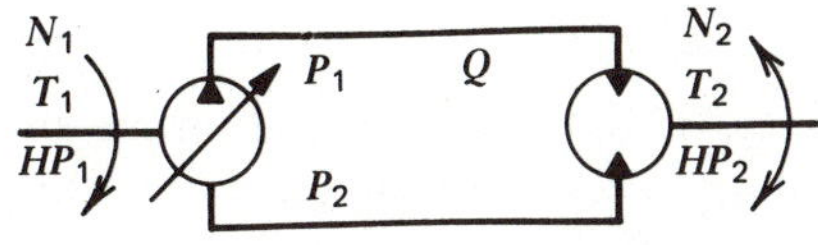

Let N_1 and P_1 = constant

See section of graph for maximum motor displacement

(b)

N_1 T_1 HP_1 P_1 Q P_2

Let N_1 and P_1 = constant

See section of graph for maximum pump displacement

(c)

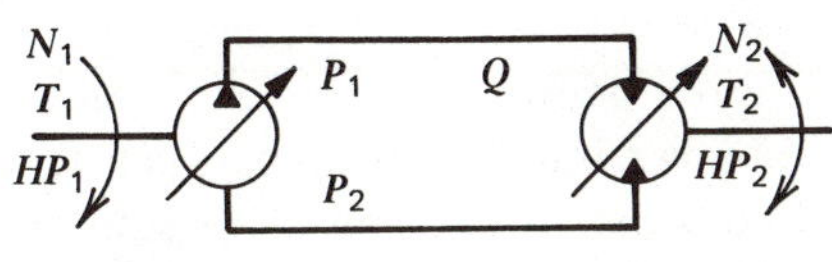

Let N_1 and P_1 = constant

See graph

(d)

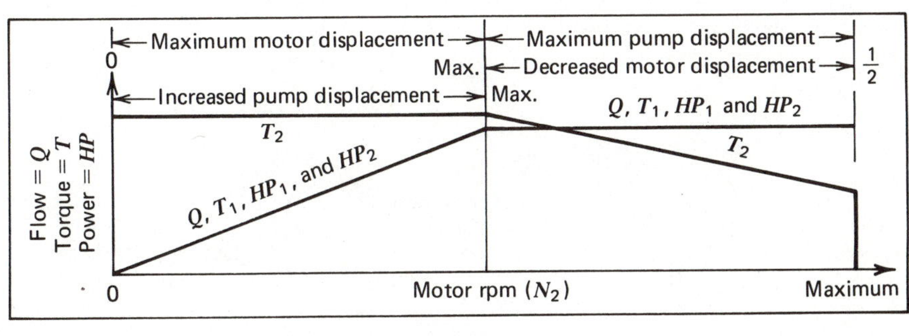

(e)

Figure 10.3 Types of closed loop hydrostatic drive circuits. (*a*) Fixed displacement pump, fixed displacement motor. (*b*) Variable displacement pump, fixed displacement motor. (*c*) Fixed displacement pump, variable displacement motor. (*d*) Variable displacement pump, variable displacement motor. (*e*) Graph of pump and motor performance versus displacements.

load. They also have the ability to provide braking for overrunning loads, such as a cable winch drive or a vehicle rolling downhill. As the load speed on the motor shaft increases above the speed driven by the flow of hydraulic fluid, the motor begins to produce a pumping action that attempts to drive up the speed of the pump unit. That action, in turn, attempts to increase the speed of the primary power unit to the pump shaft.

10.2 HYDRAULIC PUMPS

Three basic types of hydraulic pumps convert mechanical power of torque and rotation into hydraulic fluid power. The three types are designated according to the mechanical construction and operation of the device that displaces a quantity of fluid from the pump inlet to the outlet for each rotation of the pump shaft.

The type of pumps discussed here are the gear, vane, and piston pumps. They are classed as positive displacement pumps, because there is a theoretical volume of displacement associated with the physical dimensions of the hardware.

10.2.1 Gear Pumps

The gear pump has a simple construction. Its principal components are two gears that mesh with each other, an input power shaft to turn the drive gear, and a housing with inlet and outlet ports for fluid flow. Gear pumps can be either internal or external mesh. The pumping action of an external-mesh pump is shown in Fig. 10.4. Hydraulic fluid flows into the space between the gear teeth as the two gears unmesh. The fluid is carried around the periphery of both gears. At the outlet side of the pump, the fluid is forced out of the space between each tooth as the gears mesh. The theoretical displacement ($in.^3$/rev) of the pump is equal to the volume of the meshing teeth times the number of teeth per revolution.

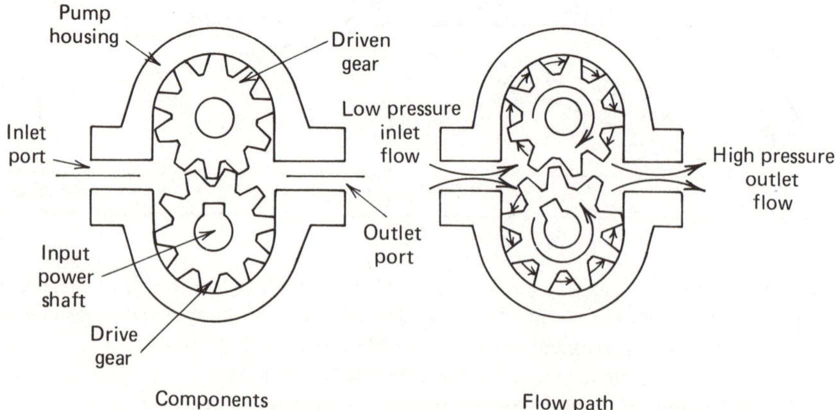

Figure 10.4 Gear pump components and hydraulic flow path.

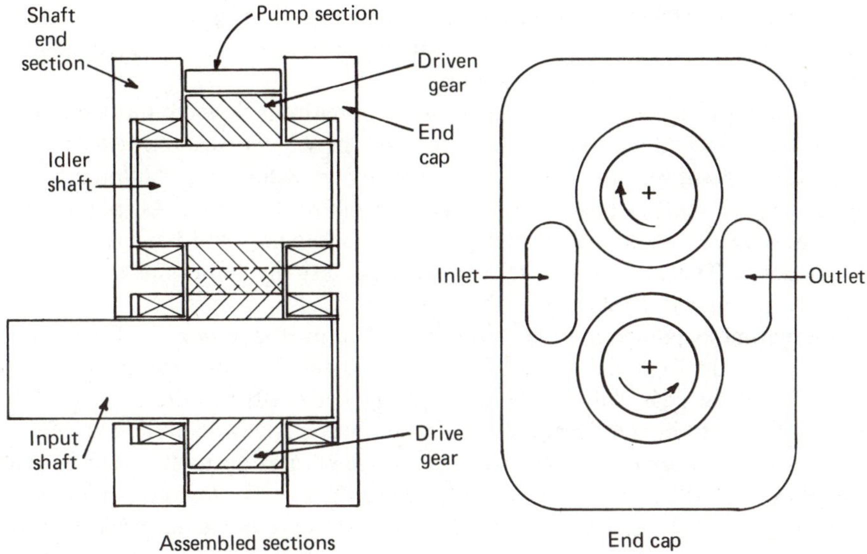

Figure 10.5 Gear pump construction and assembly.

Figure 10.5 shows the usual construction of a gear pump as a sandwich section of the pumping unit held between two end-cap sections. Bearings support the input power shaft and the drive gear connected to it. The driven gear is mounted on a jack shaft, which is supported and held in its relative position by bearings in the end sections. The inlet and outlet ports are through the end-section housings. Close mechanical construction tolerances prevent the flow of fluid from the outlet side back to the inlet side. The radial clearance between the tips of the gear teeth and the housing wall allow only a film of oil to pass between them. The same is true for the clearance between the sides of the gears and the end cap sections.

A common manufacturing process to obtain the closest side tolerances is to keep the two gears and the housing for the pumping unit as a set. All three are put on the lapping machine together, and the finished thickness dimension is nearly identical. Thus, a thin-film spacer can be inserted between the pumping-section housing and the end sections to provide the clearance required during thermal expansion for the gear to turn in the housing.

Manufacturers publish pump displacement data in terms of gear width. The pumping unit section can be replaced with a wider unit to provide a pump with a larger displacement per revolution. Gear pumps are available with gear widths ranging up to 3 in. wide; most are available in incremental sizes of 1/4 in. The manufacturer's rated delivery is published as gallons per minute, per inch of gear width, at a rated speed. Gear pumps are available with rated delivery flows, ranging from 1 to over 100 gal/min. Normal continuous duty speeds allowed are compatible with available power units. Rated speeds are usually within 2000 to 3000 rpm, rated continuous discharge pressures range from 2000 to 3000 psi. Gear pumps give good service at pressure applications of 1500 psi. Internal leakage increases rapidly as high pressures are produced, and the overall efficiency of power transfer drops off at high pressures. The overall efficiency is also low at low operating speeds because the internal leakage paths are independent of the rotating speed.

Gear pumps can be stacked together to form multiple flow circuits from one input shaft. Special sections are used in place of the end section between the two stacked pump sections. These have spline connections for continuation of the drive-gear shaft to turn the drive gear of the next section. All pump sections may have one common inlet port, but each pump will have individual outlet ports. The allowable torque for a shaft usually limits the number of pump units that can be stacked to three or four.

Internal-mesh pumps are also available. The most common is the gerotor pump shown in Fig. 10.6. The inner gear has one less tooth than the outer gear. The inner gear is driven by the input power shaft, and the outer gear is externally constrained to rotate with its rotation axis eccentric to the axis of the inner gear. Gerotor pumps are available for flow rates up to 10 gal/min and pressures up to 2000 psi; however, they are usually used for low pressure applications on low volume flows. Typical applications are power-steering circuits and recharging pumps for closed-loop fluid power transmission systems, which will be discussed later.

10.2.2 Vane Pump

Vane pumps contain a rotating drum mounted within a stationary ring with an axis eccentric to the axis of the rotating member. Figure 10.7 shows the pumping action of a vane pump. Vanes are mounted in radial slots of the rotating

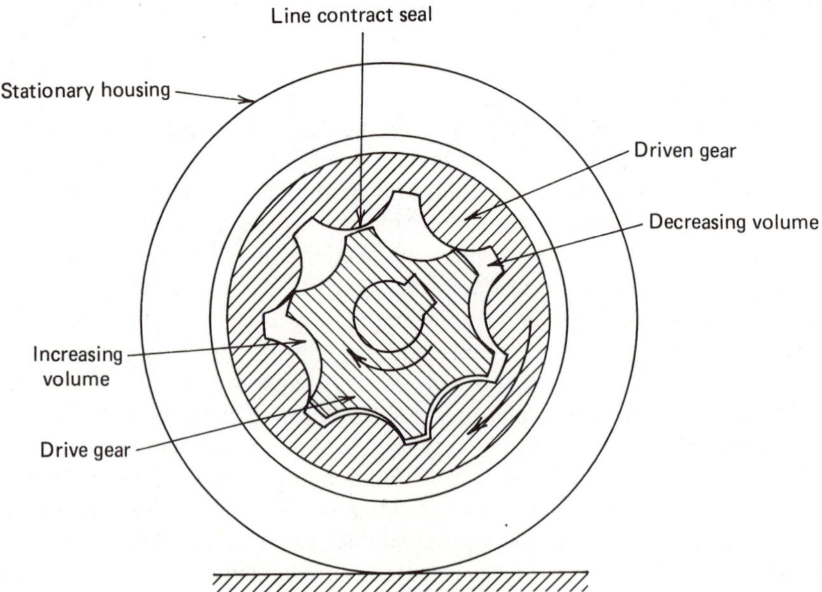

Figure 10.6 Gerotor pump components and parts.

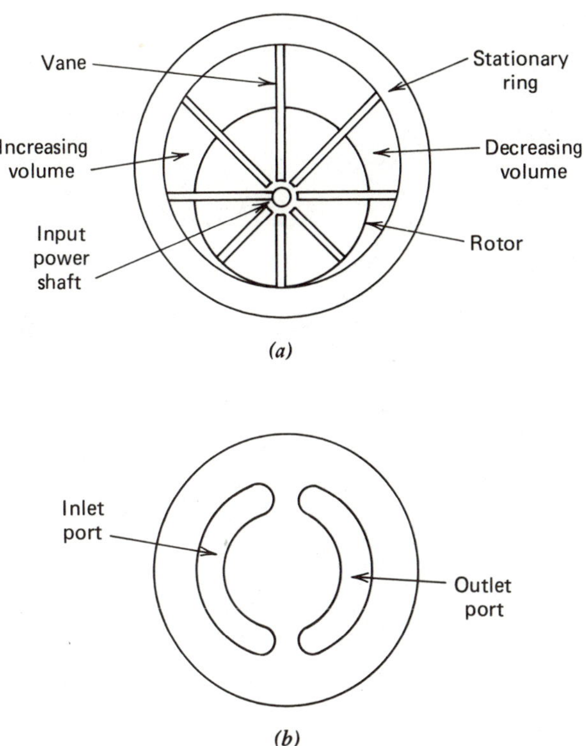

Figure 10.7 Vane pump components. (*a*) Component parts. (*b*) Part plate.

drum, which is called a rotor. Springs and hydraulic pressure force the vanes radially outward to remain in sliding contact with the wall of the stationary ring. As the input power shaft turns the rotor, the volume of the space between each vane increases for 180° of rotation. This volume decreases during the second half of the pump rotation. Port plates located in the side housings provide inlet ports to the area where the volume is increasing and a separate outlet port for the area where the high pressure flow is being forced from the pump.

The high outlet pressure on the discharge side of the pump creates a high-side load force on the bearings that support the shaft. A balanced vane pump, shown in Fig. 10.8, is made by putting two lobes on the inside of the stationary ring. The two pressure quadrants oppose each other, and the forces on the shaft are balanced. The vane pumps used in industrial systems that are constant volume displacement generally have a balanced design.

The operating limitations of vane pumps are similar to those of gear pumps. The operating efficiency of the vane pump tends to be somewhat better because of the sealing action of the sliding vanes on the housing wall, as opposed to the radial clearance of the gear teeth. Vane pumps can also be stack mounted, as shown for the gear pumps. Usually only two vane pumps are stacked together with one common inlet and separate discharge ports.

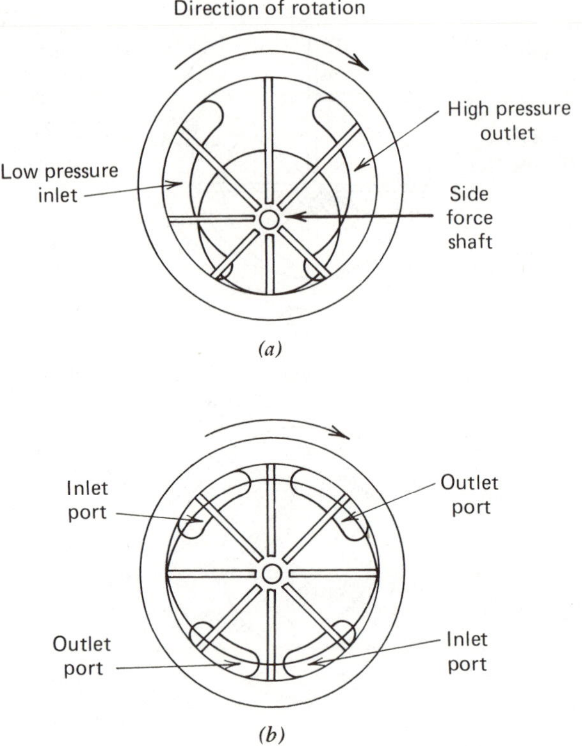

Figure 10.8 Vane pump types.
(*a*) Unbalanced vane pump. (*b*) Balanced vane pump.

10.2.3 Piston Pump

Piston pumps generate the flow of hydraulic fluid by the reciprocating motion of the piston within a cylinder bore. The components of an axial piston pump are shown in Fig. 10.9. This type of pump derives its name from the fact that the pistons reciprocate parallel to the shaft. The input power shaft, the rotating cylinder block, and the swash plate are constrained to their relative positions by the pump housing. The input shaft forces the cylinder block to rotate. The port plate connects the inlet and outlet flows from the rotating block to the respective passages in the end cap. The port plate is held against the end cap and does not rotate. The hydraulic pressure within the cylinder bores clamps the rotating block against the stationary port plate. A thin oil film separates the two components at this dynamic sealing surface. Therefore, both the port-plate and cylinder block end are lapped to precise tolerances of flatness and surface roughness in order to keep the flow from this leak path negligible.

Axial piston pumps are available with cylinder blocks that contain from five to nine pistons, sometimes more. An odd number is usually used to minimize hydraulic oil pressure pulses in the inlet and outlet ports. The sequential

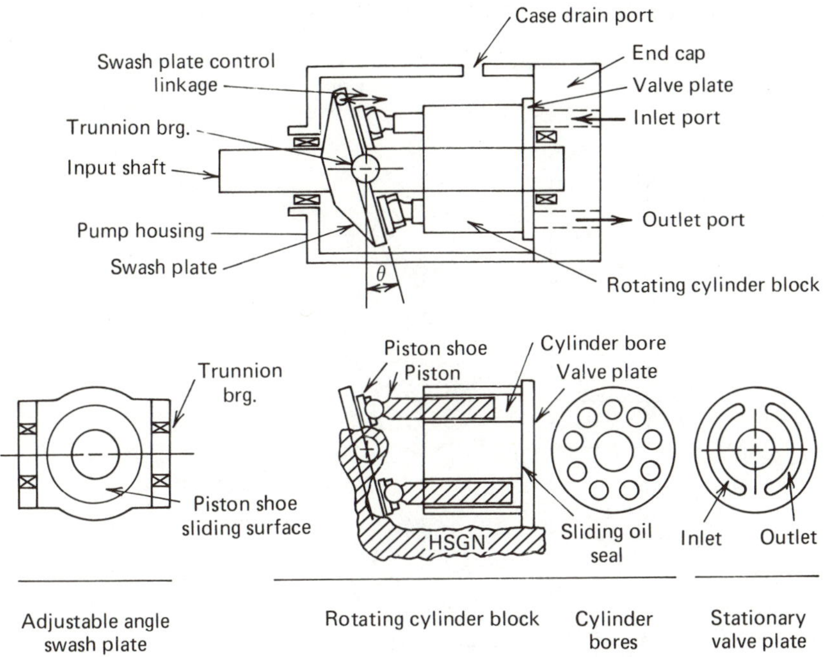

Figure 10.9 Axial piston pump layout of functioning parts.

discharge of the pistons produces a continuous flow from the pump. Light spring forces and hydraulic fluid pressure force the pistons out of the cylinder bores and hold the piston shoes against the stationary sliding surface on the swash plate. As the shaft is turned, the shoes slide in a circular path on the swash plate.

The volume of fluid displaced by each piston from the inlet side to the outlet side during one revolution of the cylinder block is depicted in Fig. 10.10. The piston shown at the top of the figure is at its maximum position inside the cylinder bore–only a clearance volume of fluid is contained in the cylinder bore. During 180° of shaft rotation, the piston shoe traveled in a circular arc on the swash plate. The piston moved axially from the bottom dead-center position to the top dead-center position. Hydraulic fluid was drawn into the pump to fill the space created by the moving piston. That same volume of fluid is forced out of the pump on the discharge side as the shaft is turned the second 180° of rotation. The displacement volume for one piston is the cross-sectional area of the piston times the axial travel distance. The total pump displacement (in.^{3}/rev) is that volume times the number of pistons in the cylinder block. The swash plate can be rotated around the trunnion axis, which is perpendicular to the input shaft axis. Rotation of the swash plate changes the cam angle of the piston shoe sliding surface. Thus the displacement per revolution can be changed by changing this angle. All pumping action stops when the cam angle is zero, that is, when the swash plate surface is perpendicular to the pistons.

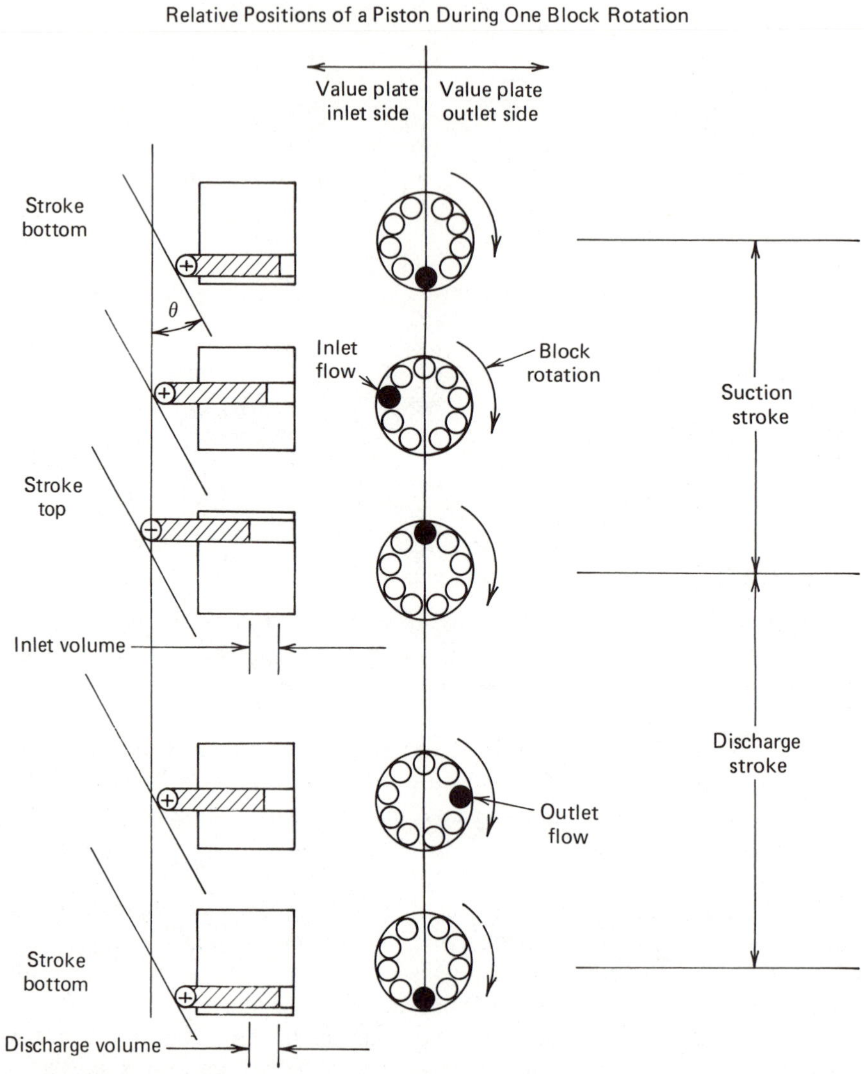

Figure 10.10 Axial piston pump fluid displacement strokes.

Continued rotation about the trunnion axis will move the cam to a negative angle. Reviewing the discussion of Fig. 10.10 illustrates that this rotation changes the direction of flow through the pump. The inlet now becomes the outlet.

A fixed displacement pump operates in the same manner, but it is made without a movable swash plate. A fixed angle cam plate is fastened within the pump housing. Because it does not require the trunnion, bearings, and components to change the angle, this pump can be placed in a much smaller housing and manufactured at a lower cost.

High pressure applications, where internal pump sealing becomes a problem, requires the use of piston pumps. Piston pumps are more complex and require closer design tolerances than do gear or vane pumps. Therefore, piston pumps cost more than gear or vane pumps with equal displacement.

Piston pumps are used in a broad range of applications. Hydrostatic drives on self-propelled farm equipment and high pressure fluid supply for backhoes and wheel loaders are examples. Piston pumps are typically rated for 3500 psi continuous duty and for short intervals at pressures of 5000 or 6000 psi. The rated speed is a function of the dimensions of the rotating block. Manufacturers limit the allowable speed to protect against the inertial forces of the reciprocating pistons. Smaller unit size pumps can be rated at 3600 rpm, whereas the larger volume pumps with larger diameter blocks may be limited to 1800 rpm.

Radial piston pumps are constructed with the pistons in a radial pattern perpendicular to the input power shaft. Piston shoes slide on a cam at the center of the cylinder block. These pumps can be made with a cam that is stationary and a rotating cylinder block, or the cylinder block can be held stationary and the cam rotated with the shaft.

10.2.4 Mountings and Drives

Hydraulic pumps and motors can be base mounted or flange mounted. A base mounted unit requires a flexible coupling on the drive shaft and mounting adjustments to align the pump shaft with the drive shaft. Flange mountings, shown in Fig. 10.11, are more common. Flange mounts have a concentric ring to align the shaft to the input coupling. Standards have been established to match mounting flanges and shaft sizes that are compatible with pump horsepower input and motor horsepower output. SAE J744C, "Hydraulic Power Pumps," has tables for the recommended flange mountings and shaft sizes required for shaft torques and horsepowers.

Flange mountings are defined as a two-bolt or four-bolt type flange, so named for the number of mounting bolts used. Standardized dimensions for the pilot ring and the radial distance to the mounting bolt holes are specified in the SAE standard. Letter designations from A to F define the size of the flange, and all dimensions for a given letter size are compatible with the allowable torque input for that size mounting.

The input shaft can usually be ordered with a choice of input drive couplings. The most common styles of input shafts are spline drive, key-way drive, and taper-lock drive shafts. The customer usually has the option of requesting the style of input shaft needed for the power-train connection to the pump.

10.2.5 Multiple Drives

Two or more variable displacement pumps can be driven from one input power supply by tandem mounting pumps in-line or by using a splitter box between the power source and the input drives of the pumps. In both styles of

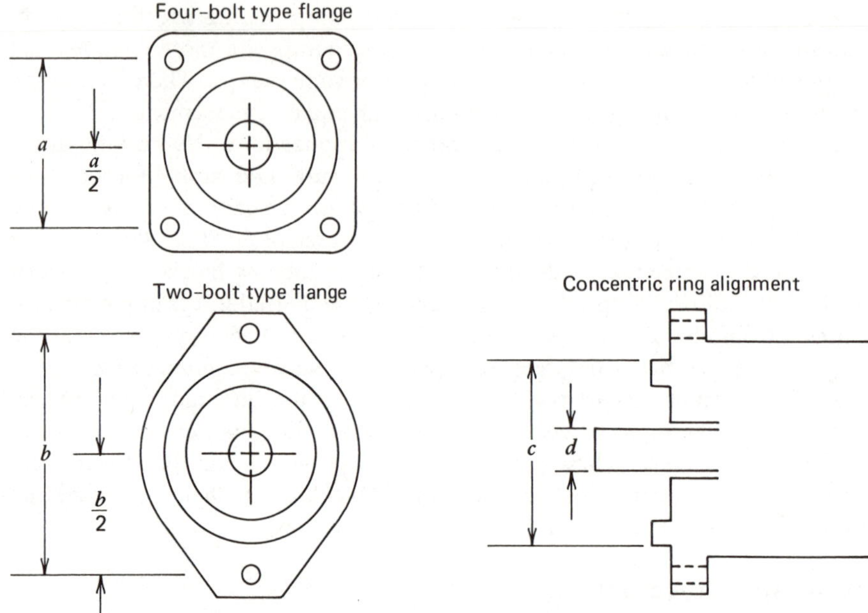

Typical SAE Mounting Flange Dimensions

Mounting Flange Size	*Shaft*					*Pilot Dimensions*		
	Torque IN·LB	*HP at 1000 RPM*	*Splined 30° involute*		*Keyed diameter d*	*a*	*b*	*c*
A-A	260	4.25	9T	20/40 DP	0.500	—	3.25	2.00
A	517	8.25	9T	16/32	0.625	—	4.188	3.25
B	1852	29.3	13T	16/32	0.875	3.536	5.750	4.00
B-B	2987	47.5	15T	16/32	1.000	3.536	5.750	4.00
C	5677	90.0	14T	12/24	1.250	4.508	7.125	5.00
C-C	10777	171.0	17T	12/24	1.500	4.508	7.125	5.00
D	15057	239.0	13T	8/16	1.750	6.364	9.00	6.00
E	15057	239.0	13T	8/16	1.750	8.839	12.50	6.50
F	24245	285.0	15T	8/16	—	9.745	13.781	7.00

Figure 10.11 Hydraulic pump and motor flange mountings and drive shafts.

multiple pump drives, the individual pressure and flow of each pump is controlled independently from the other pump.

The tandem arrangement is accomplished by using a pump with a machined mounting flange pad on the end away from the input drive shaft. A second pump can be bolted to this mounting pad and a shaft coupling connects the shaft of the first pump to the input shaft of the second pump.

Tandem drives are quite useful for applications on skid steer loaders, hay windrowers, and similar vehicles where each side of the vehicle has a separate but equal variable speed power drive. The skid steer loader can be made to spin in place by driving one side forward and the other side in reverse. Often a third small gear pump will be driven from the end of the tandem pump as a supply of oil for hydraulic cylinders or for small hydraulic motor power.

A splitter box is an extra gear box that has an input shaft driven from the main power source. Two or three output shafts can be gear driven from the input shaft. A flange mounting pattern at each of the output shaft locations provides means for multiple pump drives.

Splitter boxes are commonly found on applications for backhoes, cranes, trenchers, and front-end loaders. Power circuits for vehicle propelling, bucket lift, boom swing, and winch drives can be driven from the pumps mounted to the splitter box. Each pump circuit can operate independently from the others. Since full horsepower capability is usually not required from each of the pump drives at the same time, the summation of the maximum horsepower capability of each of the pumps can be three or four times the available horsepower from the engine of the vehicle.

10.2.6 Pump Performance and Ratings

The performance of a pump is stated in terms of the volume of flow and the hydraulic horsepower delivered by the pump. The equation for theoretical pump flow is:

$$Q_{\text{theo}} = \frac{D \times N}{231} \text{ gal/min} \qquad [10.1]$$

and

$$Q_{\text{theo}} = D \times N \text{ liters/min} \qquad [10.2]$$

where Q_{theo} = theoretical flow from the pump in gallons per minute or liters per minute

D = volume of displacement of the pump in one shaft revolution in cubic inches (in.^{3}/rev) or in liters (liters/rev)

N = shaft speed in revolutions per minute (rpm)

231 = constant equating a cubic-inch volume to gallons (in.^{3}/gal)

The actual pump flow is calculated as

$$Q_{\text{act}} = \frac{D \times N}{231} \times V_{\text{eff}} \text{ gal/min} \qquad [10.3]$$

and

$$Q_{\text{act}} = D \times N \times V_{\text{eff}} \text{ liters/min} \qquad [10.4]$$

where V_{eff} = the volumetric efficiency of the pump in percent, expressed as a decimal fraction

The volumetric efficiency is a function of both the internal leakage and the compressibility of the oil. The volumetric efficiency of a variable displacement pump is reduced as the magnitude of displacement is reduced. At a lower displacement, the internal leakage represents a larger percent of the total displacement.

Also, at reduced piston stroke for a variable-displacement pump, the carryover of compressed oil in the clearance volume from the high pressure side back to the low pressure side is larger as well as being a higher percentage of the displacement.

The torque required to drive the input shaft of a pump is

$$T_{\text{theo}} = \frac{P \times D}{2\,\pi} \qquad (\text{in} \cdot \text{lb}) \qquad [10.5]$$

and

$$T_{\text{theo}} = \frac{P \times D}{2\,\pi \times 100} \qquad (\text{N} \cdot \text{m}) \qquad [10.6]$$

where T_{theo} = theoretical input torque to create the given pressure rise through the pump

P = pump outlet pressure minus pump inlet pressure (psi) or (bar)

The actual torque required is

$$T_{\text{act}} = \frac{P \times D}{2\,\pi} \times \frac{1}{T_{\text{eff}}} \qquad (\text{in} \cdot \text{lb}) \qquad [10.7]$$

and

$$T_{\text{act}} = \frac{P \times D}{2\,\pi \times 100} \times \frac{1}{T_{\text{eff}}} \qquad (\text{N} \cdot \text{m}) \qquad [10.8]$$

where T_{eff} = the torque efficiency of the pump in percent, expressed as a decimal fraction

The torque efficiency is a function of sliding friction between parts, oil shear, churning losses within the pump housing, and flow forces required to move oil into and out of the pump.

The horsepower required to drive a pump is

$$\text{power HP} = \frac{\text{T (in} \cdot \text{lb)} \times \text{N (rpm)} \times \text{ft/12 in.} \times 2\,\pi}{33000 \text{ ft} \cdot \text{lb/min} \cdot \text{hp}} \qquad [10.9]$$

Therefore,

$$\text{HP}_{\text{in.}} = \frac{T \times N}{63{,}025} \text{ (hp)} \qquad \text{in English units} \qquad [10.10]$$

and

$$\text{HP}_{\text{in.}} = T \times N \times 1.05 \times 10^{-4} \text{ (kW)} \qquad \text{in SI units} \qquad [10.11]$$

where T = input shaft torque (in · lb) or (N · m)

N = shaft speed (rpm)

$\text{HP}_{\text{in.}}$ = input horsepower

The hydraulic horsepower delivered by the pump is

$$HP_{\text{hyd}} = \frac{P\ (lb/\text{in.}^2) \times Q\ (\text{gal/min}) \times 231 \text{ in.}^3/\text{gal}}{33{,}000 \text{ ft} \cdot \text{lb/min} \cdot \text{hp} \times 12 \text{ in./ft}} \qquad [10.12]$$

Therefore,

$$HP_{\text{hyd}} = \frac{P \times Q}{1714} \ (hp) \tag{10.13}$$

where P = pressure difference across the pump (psi)
Q = actual volume of pump discharge (gal/min)

The overall efficiency of the pump is

$$OA\text{eff} = \frac{HP_{\text{hyd}}}{HP_{\text{in}}} \times 100\,(\%) \tag{10.14}$$

which is also equal to the product of the volumetric and torque efficiencies.

$$OA\text{eff} = V\text{eff} \times T\text{eff} \tag{10.15}$$

To assist in calculations that involve these equations, sales literature specifies the displacement of the pump at rated operating conditions of speed and discharge pressure. The volumetric and overall efficiencies may also be stated at the rated operating condition.

An SAE hydraulic power pump test procedure is available to use as a recommended practice to establish the pump displacement and efficiencies of an existing pump. The SAE volumetric rating of the pump is the measured flow-of-output delivery when tested at 1000 rpm and 100 psi delta pressure between inlet and discharge pressure. The theoretical displacement per revolution can be calculated from that volumetric flow rate by assuming a 100% volumetric efficiency at that condition. It is thus not necessary to determine the physical dimensions of the pumping components in order to define a theoretical displacement. The pump characteristics of efficiencies can be established by comparisons between the pump performance at no load and loaded at a given condition of speed, pressure, and displacement. Typical pump performance curves for a variable displacement piston pump are shown in Fig. 10.12.

10.3 HYDRAULIC MOTORS

A hydraulic actuator is a component in the hydraulic power circuit that converts energy in the form of flow of a pressurized fluid into a mechanical force and motion. If the motion is in the form of a rotating shaft, the actuator is defined as a hydraulic motor. The motor is a positive displacement device. The volume of displacement may be a constant amount, as in a fixed displacement motor, or it may be adjustable, as in a variable displacement motor. For any given displacement setting, the motor speed remains relatively constant for a constant flow of input oil. This is so regardless of a change in motor shaft torque, which would require a corresponding change in delta pressure between the input and outlet of the motor. The speed of the motor can therefore be controlled by regulating the flow of oil to the motor, and the oil pressure required to drive the motor must increase or decrease as the motor shaft torque changes. Figure 10.13 shows an open loop circuit with multiple motor drives

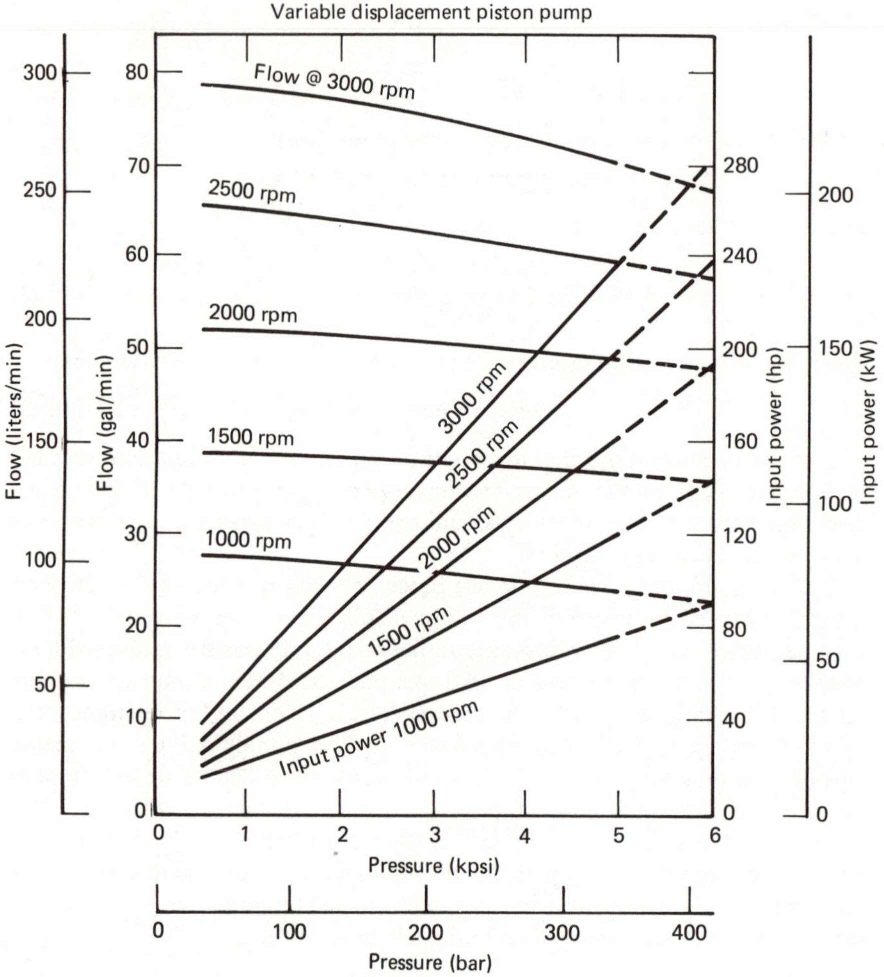

Figure 10.12 Typical pump performance curves.

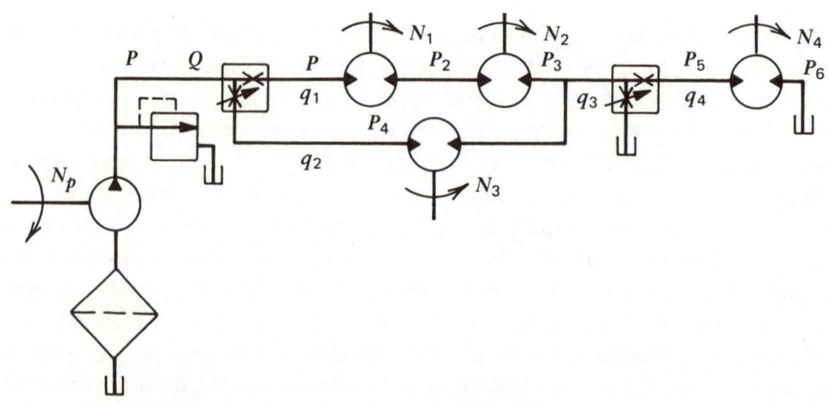

Figure 10.13 Open loop multiple motor drive circuit.

that receive hydraulic power from the same pump. The motors are connected in series and in parallel. Various combinations of motor shaft speed and torque can be obtained by selecting motors with different displacements.

10.3.1 Gear Motors

The gear motor has the same hardware as that of a gear pump, as shown in Fig. 10.4. The gear motor is driven by high-pressure oil flowing into the motor. The delta pressure between the inlet and outlet causes the gear to rotate. The shaft speed is a function of the volume of oil flowing to the motor and the quantity of oil that the gear mesh will displace in one revolution.

Gear motors have simple manufacturing requirements, are reasonably efficient at moderate pressures, and can tolerate some dirt particles in the oil. They can be used on applications with light starting loads and for high running loads. These characteristics make this a popular motor drive on agricultural, construction, and mining equipment.

10.3.2 Vane Motors

The vane pump, shown in Figs. 10.7 and 10.8, is used as a motor when hydraulic power is supplied to the inlet port and a torque load is driven by the shaft. This motor is generally rated for high speeds and can handle medium to hard starting loads. It is not as tolerant of dirt particles as a gear pump and requires filtration to a smaller particle size. This motor can operate at a significantly lower noise dBA level than can a gear motor or a piston motor because of the means of operation and smooth flow of fluid. It is widely used in industrial applications where the duration of service and lower noise level justify the higher initial cost per horsepower rating.

10.3.3 Gerotor Motors

The gerotor shown in Fig. 10.14 is widely used on agricultural equipment. This motor can also be constructed as a combination roller vane and gerotor drive. It provides a gear reduction by the gerotor action. The inner gear member has one less tooth than the outer gear member, thus the hydraulic fluid must force the outer member to turn N_o/N_i times for one revolution of the inner gear, which is connected to the motor shaft. The fluid displacement rating for this motor is the volume required to turn the output shaft one turn. The motor therefore can develop a much higher output torque level for a given physical size than a comparable gear or vane pump. The hydraulic pressure supplied to drive the motor can be significantly lower, and a higher volume of low pressure oil can be used to provide the energy for the work function.

Gear, vane, and gerotor motors are used in open loop circuit drives where one pump can provide a fluid supply for several motors. These motors have a fixed displacement, which requires a fixed volume of fluid for one revolution. The speed of these motors must be regulated by controlling the volume of fluid supplied to the motor.

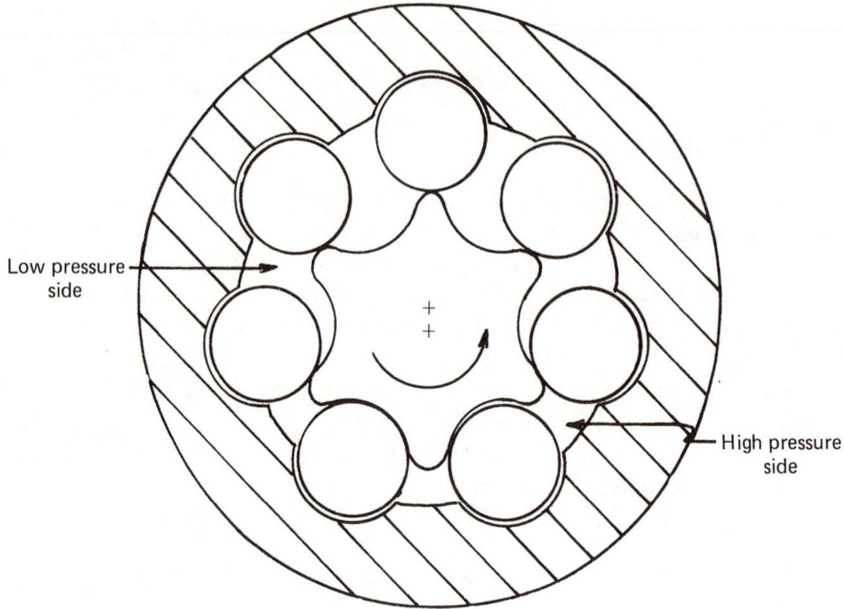

Figure 10.14 Roller vane gerotor motor.

10.3.4 Piston Motors

A piston pump functions as a motor when it is driven by a flow of high pressure fluid. Fig. 10.10 illustrates the rotary motion created when the unit is motoring. A flow of fluid under pressure into the clearance volume of a piston all the way into the bore of the rotating block will force the piston to move out of the block. This motion is only possible if the cylinder block rotates as the piston shoes slide in a circular pattern on the cam surface. The angle of the cam surface can be a permanent machined surface, making the unit a fixed displacement motor. Motors that have swash plates that can be moved to vary the cam angle are variable displacement motors. As the cam angle is changed to reduce displacement, the speed of the motor increases with a constant supply of flow to the inlet port.

10.3.5 Motor Performance and Ratings

The speed of a hydraulic motor is a function of motor displacement and fluid flow to the motor. The equation for motor speed is

$$N_{\text{theo}} = \frac{231\,Q}{D}\ \text{(rpm)} \qquad \text{or} \qquad = \frac{Q}{D}\ \text{(rpm)} \qquad [10.16]$$

where Q = fluid flow to the motor in (gal/min) or (liters/min)

D = volume of displacement of the motor for one rotation of the shaft (in.^{3}/rev) or (liters/rev)

The actual motor speed is

$$N_{act} = \frac{231\,Q}{D} \times V\text{eff (rpm)} \qquad \text{or} \qquad \frac{Q}{D} \times V\text{eff (rpm)} \qquad [10.17]$$

Veff = the volume efficiency of the motor in percent expressed as decimal fraction

The volumetric efficiency of a motor is affected by the same variables as that of a pump. The volumetric efficiency increases as the motor displacement is increased and decreases as pressure increases.

The torque that is developed at the motor shaft is a function of motor displacement and pressure drop across the motor. The equation for theoretical motor torque is

$$T_{theo} = \frac{P \times D}{2\pi}\ (\text{in} \cdot \text{lb}) \qquad \text{or} \qquad \frac{P \times D}{2\pi \times 100}\ (\text{N} \cdot \text{m}) \qquad [10.18]$$

T_{theo} = theoretical torque that can be produced
P = motor inlet pressure minus motor outlet pressure in psi or bar

The actual motor shaft torque is

$$T_{act} = \frac{P \times D}{2\pi} \times T\text{eff (in} \cdot \text{lb)} \qquad \text{or} \qquad \frac{P \times D}{2\pi \times 100} \times T\text{eff (N} \cdot \text{m)} \qquad [10.19]$$

where Teff = torque efficiency of the motor expressed as a decimal percent

The horsepower that is provided by a motor is

$$HP_{out} = \frac{T \times N}{63{,}025}\ (\text{hp}) \qquad \text{or} \qquad = T \times N \times 1.05 \times 10^{-4} \qquad (kW) \qquad [10.20]$$

where T = output shaft torque (in·lb) or (N·m)
N = shaft speed (rpm)
HP_{out} = output horsepower

The overall efficiency of a motor is

$$OA_{eff} = \frac{HP_{out}}{HP_{hyd}} \times 100\% \qquad [10.21]$$

Thus, the OA eff = Teff× Veff

An SAE hydraulic motor test procedure details recommended practice steps to determine a theoretical displacement per revolution of the motor. The test requires instrumentation to measure the flow of hydraulic fluid to the motor as it is driven at 1000 rpm. No coupling is connected to the motor shaft therefore the motor operates at zero output torque. The theoretical displacement is taken as the measured flow divided by 1000 rpm. This theoretical displacement can be used to predict motor speed and torque at 100% efficiency. The difference between the predicted performance and the actual speed of the motor under load will determine the efficiencies for that particular operating condition. Typical motor performance curves for a fixed displacement piston motor are shown in Fig. 10.15.

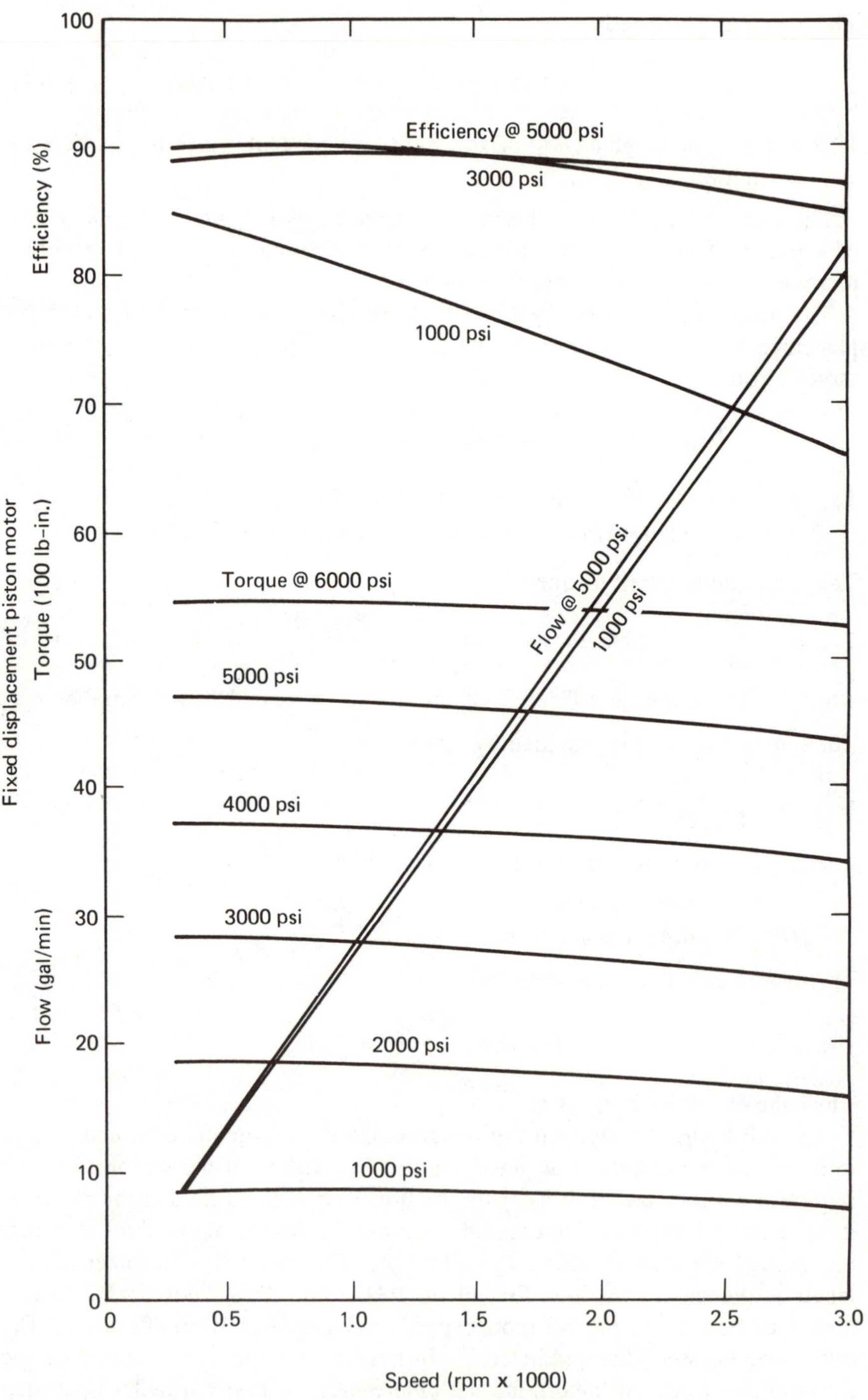

Figure 10.15 Typical motor performance curves.

10.4 SIZING HYDROSTATIC TRANSMISSIONS

Sizing the drive train for a vehicle requires selecting or designing each of the components in the unit to handle the speeds and torques required for the vehicle's proposed duty cycle. The transmission ratio must match up the speed and torque provided by the engine to the same parameters required at the vehicle's drive wheels. The transmission may be as simple as a one-speed gear box, or it may be complicated, to provide a changing transmission ratio to improve the vehicle performance as drive wheel torque and speeds change. Component efficiency and life expectancy play major roles in the acceptance of the individual components.

The transmission is normally located in the drive train, as shown in Fig. 10.16. Because the transmission is located in the middle of the drive train, the imposed parameters are dictated from either end. The performance requirements of the vehicle define the speed and tractive effort required at the vehicle drive wheel.

The tractive effort necessary at the drive wheel is defined as that required for the vehicle to exert a given drawbar pull, to traverse across a surface at a stated rolling resistance, or to climb a given grade. The speed that the vehicle must travel while providing that tractive effort must also be defined. These two items determine the horsepower required at the drive wheel. They also establish the horsepower required from the engine (see Fig. 10.17).

The maximum tractive effort required at the drive wheels, the fastest speed required for the vehicle, and the horsepower capability of the engine must be compatible, as shown in Fig. 10.18. If the parameters are defined for the vehicle, they will dictate minimum requirements for engine selection.

If the engine has been sized for other applications on the vehicle, then the engine parameters are available and the transmission must be sized to accept them and to transfer that power to the drive wheels. The engine speeds at rated

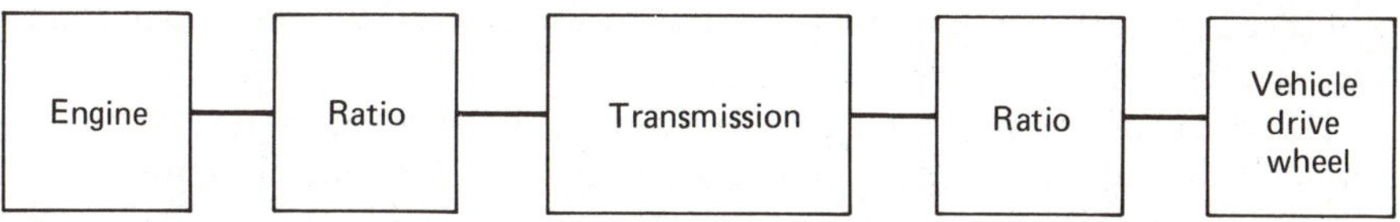

Figure 10.16 Vehicle drive train block schematic.

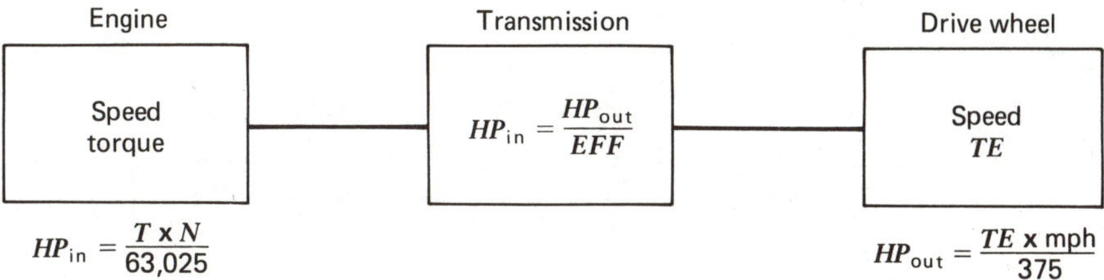

Figure 10.17 Parameters imposed on transmission.

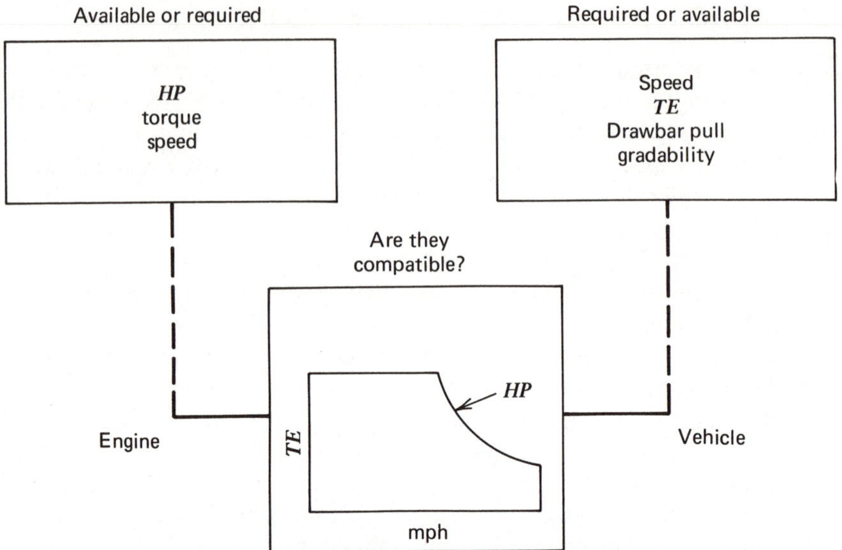

Figure 10.18 Compatibility of parameters imposed on transmissions.

horsepower for most mobile equipment applications fall into a narrow range of revolutions per minute. Once an engine has been selected to provide the necessary horsepower, the transmission must be sized to provide the proper speed and torque ratios and match the engine parameters to the requirements of the drive wheel.

10.4.1 Hydrostatic Transmission

The hydrostatic transmission conveniently provides a varied transmission ratio with a continuous range of working conditions–from slow speed, high torque to high speed, low torque. The goal of transmission design is to make this output speed range at constant input speed as wide as practical and to avoid shifting interruptions.

This discussion of hydrostatic transmissions is limited to a system with a single load for each hydraulic pump, commonly called a closed circuit system, (see Fig. 10.19). In mobile equipment, the prime mover is usually an internal combustion engine; for industrial applications, it is usually an electric motor. The work load can be the driving wheels of a vehicle, a cable winch, or a conveyor belt.

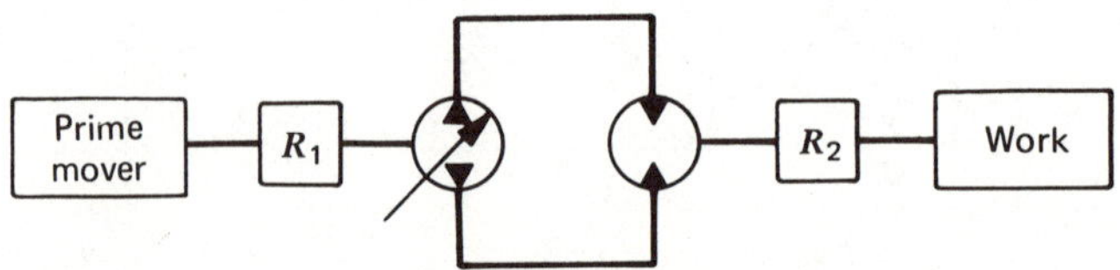

Figure 10.19 Closed loop hydrostatic transmission circuit.

A gear-mesh or belt-drive speed ratio is shown as R_1 to match the rated pump speed to the rated engine speed. The final drive ratio is shown as R_2 and is a fixed ratio between the motor output shaft and the work load.

Because the hydraulic pump or motor can be a unit of fixed or variable displacement, four combinations can be used as a hydrostatic transmission. The four combinations are:

1. PF-MF: a fixed displacement pump that drives a fixed displacement motor. Given the condition that the input speed and power remain constant, the output speed and torque will also remain constant. The drive is useful to power a constant-speed fan or conveyor that is remotely located from the power source.
2. PF-MV: a fixed displacement pump and a variable displacement motor. The flow from the pump and the pump discharge pressure will remain constant while driven by a constant horsepower input. As the motor displacement is varied, the output speed can be varied. The output torque changes inversely. Industrial drive applications include winding machinery.
3. PV-MF: a variable displacement pump and a fixed displacement motor. This combination has a wide range of applications on mobile equipment drives, from garden tractors to self-propelled farm equipment. It is also a common drive for industrial equipment such as fork lifts and excavators. The output speed is a function of the pump displacement, and the forward and reverse speeds are easily controlled by changing the pump displacement. If the transmission is operating at a constant system pressure, the drive is one of increasing horsepower as the pump displacement is increased. The output torque thus remains constant, but the output speed will increase during this change.
4. PV-MV: the displacement of both units can be varied. Full displacement of the motor with the pump at a low displacement position will give maximum output torque capabilities at low speeds to start a load. The advantage of the variable displacement motor is the extension of the total speed range after the pump has reached full stroke by decreased motor displacement. Of course, the output torque is reduced at the faster speed because the motor displacement is smaller. This system requires more complicated controls to ensure that the displacements of the two units are sequenced properly and to safeguard against the possibility of overspeeding one of the hydraulic units, either in the driving or braking modes.

This discussion applies specifically to the PV-MF system, although the concepts in sizing the transmission apply to all four systems. The closed loop hydrostatic drive that propels a mobile equipment vehicle is depicted in Fig. 10.20. The pump speed is proportional to the engine speed, which is assumed to be at the rated value for the sizing of the transmission. The input torque to the pump is inversely affected by R_1. Changing the pump displacement

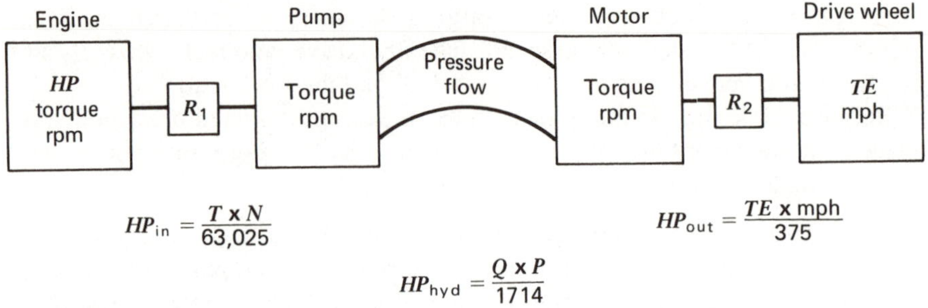

Figure 10.20 Hydrostatic transmission parameters throughout the drive train.

changes the pump flow; either the system delta pressure or the input horsepower to the pump also changes. The horsepower transmitted to the drive wheel is a function of the tractive effort and the vehicle speed.

10.4.2 Vehicle Parameters

Definition of the required vehicle performance is shown in Fig. 10.21. The tractive effort of the vehicle must be sufficient to cause the drive wheels to slip or the operator may feel that the vehicle does not have enough power. Thus the maximum tractive effort is set at that value. The tractive effort can be calculated when the gross vehicle weight and the coefficient of adhesion for the given traction surface are shown.

The corner horsepower concept is the amount of power required to produce both maximum tractive effort and maximum output speed at the same time. Because it is usually not necessary to spin the wheels at the fastest speed, using the corner horsepower requirement would result in an oversize engine. The horsepower demand can be greatly reduced if that tractive effort demand is relaxed at the top of the speed range.

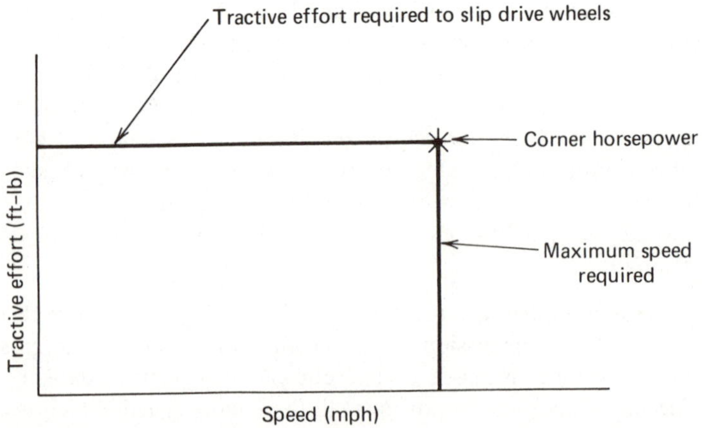

Figure 10.21 Vehicle speed and traction requirements.

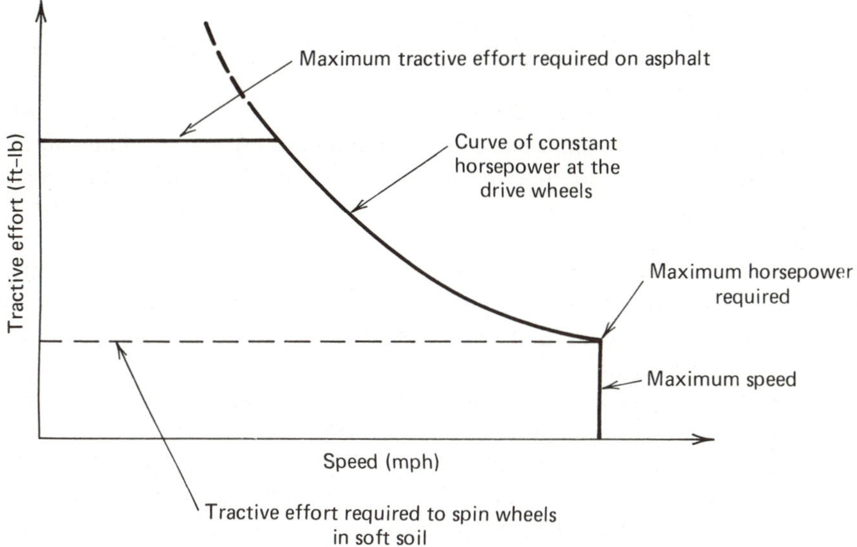

Figure 10.22 Modified vehicle speed and traction requirements.

For example, one requirement for the design of a tractor transmission is to spin the wheels on an asphalt surface at the slowest speed. Another requirement is to spin the wheels in soft soil at a field operation speed. This vehicle-performance curve resembles Fig. 10.22. The required engine size can be much smaller than that required in Fig. 10.21.

Corresponding hydraulic unit curves for a variable-pump fixed-motor hydrostatic transmission are shown in Fig. 10.23. The pump speed is fixed as a constant by a direct drive from the engine. Because the motor displacement is fixed and stays constant, the output torque is a function of the system's delta pressure. The hydraulic units have upper physical limits of speed and system pressure set by the manufacturer. They must operate below that limit in order

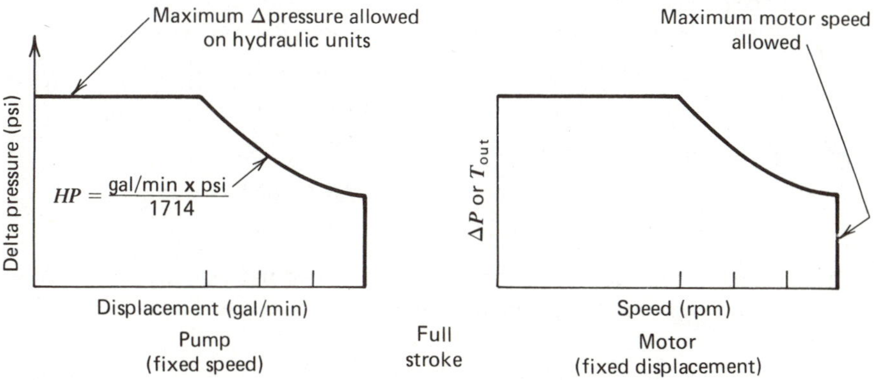

Figure 10.23 Hydraulic unit speed, pressure, and displacement limitations.

to provide the advertised life of the unit. The manufacturer's specified duty cycle may allow a 5000-psi system pressure for a short duration but will restrict that pressure to 3500-psi or less for constant use. The motor must be sized to provide the required tractive effort without exceeding that system pressure limit. Violation of pressure limits causes oil at a high pressure to dump across safety relief valves. The maximum delta pressure portion of the curve represents the maximum tractive effort that the vehicle will be allowed to develop.

A good way to ensure that the hydraulic safety relief valve will not dump high pressure oil back to the low pressure side and create a buildup of heat in the oil system is to size the unit so that the drive wheels break traction below this maximum allowed pressure.

Hydraulic horsepower is a function of the fluid displacement and delta system pressure across the pump. The pump input horsepower is larger than hydraulic horsepower by a factor of the overall efficiency of the pump. The pump input torque and the pump input horsepower increase as the pump stroke is increased. When the pump flow reaches the point of requiring maximum engine horsepower, the rest of the curve of the maximum performance available from the hydraulic units is at constant horsepower input and decreasing system pressure as the pump is stroked to full stroke. The motor output torque follows the system pressure and is calculated as

$$\text{torque} = \frac{\text{displacement} \times \text{delta pressure}}{2\pi} \qquad [10.22]$$

The motor displacement size must be selected so that maximum flow from the pump will not cause the motor speed to exceed the manufacturer's stated limit.

10.4.3 Sizing the Hydrostatic Units

The sizing process works from the required vehicle performance back to the hydrostatic transmission and, at the same time, from the engine toward the transmission. The designer can use the corner horsepower or the reduced corner horsepower, determined from the vehicle performance curve, to select the first trial hydraulic motor. The final drive ratio is determined by

$$R_2 = \frac{\text{max motor rpm}}{\text{max wheel rpm}} = \frac{\text{max motor rpm} \times \text{loaded radius}}{168.06 \times \text{mph}} \qquad [10.23]$$

The torque required at the motor output shaft will then be

$$\text{motor torque} = \frac{\text{wheel torque}}{R_2} = \frac{TE \times \text{loaded radius}}{R_2} \qquad [10.24]$$

If the output torque for the hydraulic motor does not match this required torque with a reasonable system pressure, the designer will have to select a more appropriate motor displacement.

The pump size selected is fixed by the flow required to operate the motor at the maximum speed used in the final drive calculation. The pump speed should be geared to operate near the rated speed to obtain the most cost effective size selection. The pump input horsepower can now be calculated because the flow required at the various system pressure levels has been determined.

Example 10.1 Tractor Hydrostatic Drive

Given:

GVW = 10,000 lb
loaded tire radius = 32 in.
Required vehicle parameters (low speed range).

At slow speeds, maximum tractive effort to spin tires on asphalt is required.
Maximum speed = 6 mph
At 6 mph, tractive effort (TE) is required to spin tires on firm soil.

From Table 10.2, select the coefficients of traction for asphalt and firm soil of 1.0 and 0.5, respectively. Then

$$TE \text{ max} = 1 \times 10{,}000 \text{ lb} = 10{,}000 \text{ lb}$$

and at 6 mph

$$TE = 0.5 \times 10{,}000 \text{ lb} = 5{,}000 \text{ lb}$$

The drive wheel horsepower would be

$$\frac{TE \times \text{mph}}{375} = \frac{5{,}000 \times 6}{375} = 80 \text{ hp} \qquad [10.25]$$

The fastest speed available for the maximum tractive effort with 80-hp wheel power is

$$S = \frac{\text{hp} \times 375}{\text{max TE}} = \frac{80 \times 375}{10{,}000} = 3 \text{ mph} \qquad [10.26]$$

TABLE 10.2 Coefficient of Rolling Resistance for Off-the-Road Vehicle Tires on the Given Material

Surface	*Coefficient* (C_{rr})
Concrete	0.02
Asphalt	0.02
Smooth, hard gravel	0.025
Dirt roadway (2-in. tire penetration)	0.05
Hard soil	0.06
Firm soil	0.08
Loose-tilled soil	0.10
Soft, sandy soil	0.120
Muddy, rutted road	0.10–0.20

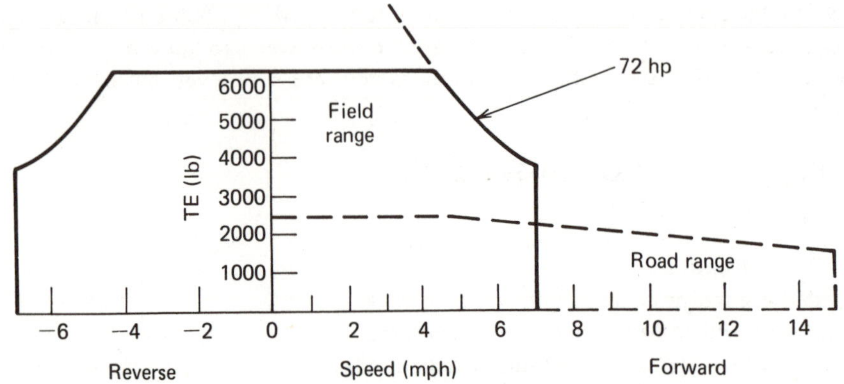

Figure 10.24 Two range combine vehicle speed and traction parameters.

The values of maximum speed, maximum tractive effort, and constant wheel horsepower can be laid out on a curve similar to that in Fig. 10.24.

To size the hydrostatic transmission, assume that a fixed motor speed of 2400 rpm will provide overspeed allowance for the hydraulic units. The final drive ratio is determined by Eq. 10.23 to be

$$R_2 = \frac{\text{motor speed} \times \text{wheel radius}}{168.06 \times \text{mph}} = \frac{2400 \times 32}{168.06 \times 6} = 76.2$$

and the motor torque required is calculated from Eq. 10.24 as

$$T = \frac{\max TE \times \text{wheel radius}}{\mathrm{R}_2} = \frac{10{,}000 \times 32}{76.2} = 4200 \text{ lb} \cdot \text{in}$$

To provide this motor shaft torque, the required motor displacement is calculated from Eq. 10.19.

$$D = \frac{2\,\pi \times T}{P \times \text{torque eff}} = \frac{2\,\pi \times 4200}{5000 \times 0.95} = 5.56 \text{ in.}^3/\text{rev}$$

for an assumed 95% efficiency and a pressure limit of 5000 psi delta pressure.

The engine horsepower can be estimated as

$$HP_{\text{in}} = \frac{HP_{\text{out}}}{\text{eff}} \qquad [10.27]$$

where the efficiency of the hydrostatic transmission and drive train can be approximated as 80% for the first trial solution. Hence the engine power is approximately 80 hp/0.80 = 100 hp.

The designer should select hydraulic units in the 5.56 in.^{3}/rev range and engine power of 100 hp. For improved accuracy of the hydrostatic transmission design, the designer should use the manufacturer's values for speed, torque, and efficiency ratings for the hydraulic units chosen.

The vehicle parameters available with that selection are then recalculated. Other hydraulic units are chosen and the vehicle parameters redefined as necessary.

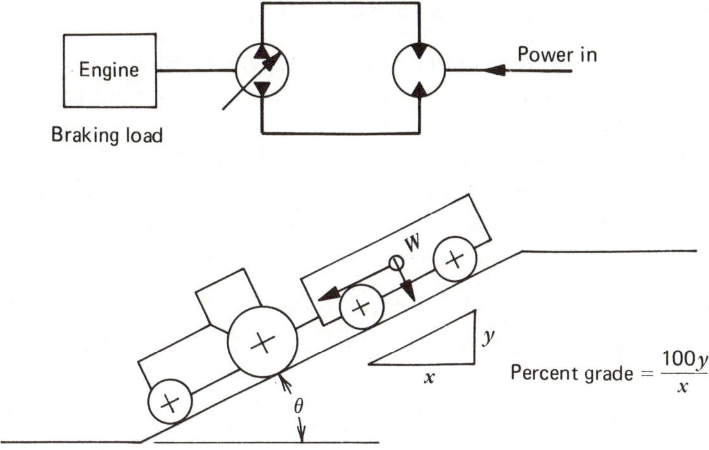

Figure 10.25 Closed loop hydrostatic transmission braking mode.

Figure 10.25 shows a vehicle with the transmission in a braking mode. The same effect would occur with an overrunning load caused by the inertia of a flywheel. To slow the drive wheels, the energy is transmitted to the engine, where it is absorbed. The drive wheels thus power the hydraulic motor unit, which becomes a pump. Flow from the fixed displacement unit drives the variable displacement hydraulic unit, which, in turn, forces the engine speed to increase. The engine speed may be driven up to as much as 10% overspeed during braking.

The volumetric efficiencies of both the pump and motor now work to a disadvantage in the size of the hydraulic unit. Due to the volumetric efficiencies of the hydraulic units, the motor will be allowed to turn faster while providing the hydraulic fluid flow required to drive the pump at the same speed. Engine overspeeding and the change in volumetric efficiencies of the hydraulic units in the braking mode allow the hydraulic motor speed to increase. The maximum motor speed allowed in sizing for the driving mode has to be limited to at least 25% below the manufacturer's rated motor speed to protect from overspeeding the hydraulic unit in the braking mode.

The designer usually selects the hydrostatic transmission and other components of the drive train from available hardware and will thus select an engine, hydraulic pump, hydraulic motor, and, possibly, gear reductions that closely approximate the design requirements. The designer can calculate backward or forward from the hydrostatic transmission after these units have been selected, using the engine horsepower available to obtain a new set of vehicle parameters of tractive effort and speed. The available components might thus allow a slightly wider speed range, or they might exceed the tractive effort required at certain points in the vehicle performance characteristics. If the tractive effort required were just a little lower, the designer might be able to select hydraulic units one size smaller. Consequently, the designer would examine the original vehicle performance requirement to see if it could be slightly altered to take advantage of a smaller unit size.

10.4.4 Sizing for a Vehicle without a Drawbar Load

The tractive effort requirements for a self-propelled vehicle, such as a grain combine or a forage windrower, provide a specific gradability for the implement. For example, the tractive effort requirement for a grain combine might be to overcome the rolling resistance of a machine with a fully loaded grain tank while climbing a slight grade. The total tractive effort required would be a sum of the rolling resistance and the force required to move the vehicle weight up the slope. The tractive effort is calculated as

$$TE = RR + GR = C_{rr} \times GVW\,(\cos\theta) + GVW\,(\sin\theta) \qquad [10.28]$$

where RR = rolling resistance
GR = grade resistance
C_{rr} = coefficient of rolling resistance
GVW = gross vehicle weight

and the percent grade and the slope angle θ are calculated by

$$G\% = 100\,(\tan\theta) = \frac{\text{ft rise}}{\text{ft horizontal distance}} \qquad [10.29]$$

Because small slope angles are involved, the tractive effort can be closely approximated as

$$TE = C_{rr} \times GVW + 0.01 \times \%G \times GVW \qquad [10.30]$$

A three- or four-speed gear box in the final drive ratio will provide for different tractive effort requirements and speed ranges. The low speed tractive effort could be selected to slip the wheels over a small speed range to provide maximum traction to get out of a bad traction spot in the field. The middle gear ratio could be required to operate from zero to the maximum field operation speed and provide a tractive effort equal to the rolling resistance of normal harvesting conditions and to traverse typical field grades with a loaded tank. The road speed gear ratio might give a speed range from zero to the maximum road speed.

The tractive effort at low speeds must be able to start and accelerate the vehicle. The tractive effort required at the top speed must be sufficient to overcome the rolling resistance of the combine with an empty grain tank on a gravel road and must also be sufficient to maintain the vehicle speed while going up normal road slopes.

The vehicle performance curve for a vehicle with a two-speed gear box in the final drive is shown in Fig. 10.24. One hydrostatic transmission cannot be selected to provide the exact performance characteristics as stated in both speed ranges. Some of the parameters will be exceeded to meet the requirements at other conditions. Therefore, the final designed hydrostatic transmission will exceed some of the maximum values required and will provide more tractive effort at the fastest speed or a wider speed range in the field operation mode or some other condition.

Example 10.2 Two-range hydrostatic transmission for a grain combine

Given:

$GVW = 20{,}000$ lb (empty grain tank)

$GVW = 27{,}500$ lb (full grain tank)

130-hp engine, 2500 rpm
(50 hp required to operate combine; therefore, 90-hp engine power required to propel combine)
Loaded wheel radius = 27 in.
Field speed range: maximum speed = 7 mph
Tractive effort is required to propel loaded combine up a 15% slope on firm, dry soil.
Road speed range: maximum speed = 15 mph
Tractive effort is required to start the combine on a gravel road with a 10% grade.
At maximum speed tractive effort for a 5% grade is required.

Solution:

Field speed range = 7 mph maximum

$$TE \text{ max} = \text{rolling resistance} + \text{grade resistance} = RR + GR$$
$$= C_{rr} + GWV + .01\%\text{G} \times GWV$$

from Eq.10.30.
From Table 10.2, select $C_{rr} = 0.08$ for firm soil. Therefore,

$$TE \text{ max} = (0.08 \times 27{,}500) + (0.01 \times 15 \times 27{,}500)$$
$$= 2200 + 4125 = 6325 \text{ lb}$$

The maximum values of 7 mph and 6325 lb tractive effort can be put on the low range curve of Fig. 10.24. The maximum performance curve for the low range can be completed as follows.

Assume an overall efficiency of 80% for the hydrostatic units and drive-train. The drive-wheel power available with 90-hp engine power will be approximately

$$\text{wheel hp} = 90 \text{ hp} \times 0.80 = 72 \text{ hp}$$

TABLE 10.3 **Coefficient of Traction Between Vehicle Tire and Given Surface**

Material	*Coefficient* (C_t)
Concrete and asphalt	0.8 – 1.0
Dry clay loam	0.55
Firm soil	0.50
Loose soil	0.45
Soft, sandy soil	0.40
Loose gravel road	0.35
Dry sand	0.20

Therefore, using Eq. 10.25, the tractive effort available at 7 mph would be

$$TE = \frac{72 \text{ hp} \times 375}{7 \text{ mph}} = 3857 \text{ lb}$$

and the fastest speed at which the maximum required tractive effort could be obtained is found from Eq. 10.26 to be

$$S = \frac{72 \text{ hp x } 375}{6325 \text{ lb}} = 4.27 \text{ mph}$$

which provides the required information to draw in the constant 72-hp portion of the curve. To size the hydrostatic units, assume that a motor will be selected that can operate at 3200 rpm. Gear the motor to operate at 2400 rpm to allow 25% overspeed during braking. The required final drive ratio will be

$$R_2 = \frac{2400 \text{ rpm} \times 27 \text{ in.}}{168.06 \times 7 \text{ mph}} = 55.1$$

The maximum motor torque for this drive ratio at maximum tractive effort is

$$T = \frac{6325 \text{ lb} \times 27 \text{ in.}}{55.1} = 3100 \text{ in} \cdot \text{lb}$$

and the motor displacement required would be

$$D = \frac{2\,\pi \times \text{T}}{\text{P} \times \text{torque eff}} = \frac{2\,\pi \times 3100 \text{ in} \cdot \text{lb}}{5000 \text{ psi} \times 0.95} = 4.1 \text{ in.}^3/\text{rev}$$

for an assumed 95% torque efficiency and a delta pressure limit of 5,000 psi.

Road speed range: $TE \text{ max} = RR + GR$

Select $C_{rr} = 0.025$ from Table 10.2 for gravel road

$$\begin{aligned} TE \text{ max} &= (0.025 \times 20{,}000) + (0.01 \times 10 \times 20{,}000) \\ &= 500 + 2000 = 2500 \text{ lb} \end{aligned}$$

The required final drive ratio for road speed is

$$R_2 = \frac{2400 \text{ rpm} \times 27 \text{ in.}}{168.06 \times 15 \text{ mph}} = 25.71$$

and the motor torque required is

$$T = \frac{2500 \text{ lb} \times 27 \text{ in.}}{25.71} = 2626 \text{ in} \cdot \text{lb}$$

The 4.1 in.^{3}/rev motor previously selected for field range will provide this torque at a delta pressure of

$$P = \frac{2\,\pi \times T}{D \times \text{torque eff}} = \frac{2\,\pi \times 2626 \text{ in} \cdot \text{lb}}{4.1 \text{ in}^3/\text{rev} \times 0.95} = 4236 \text{ psi}$$

at 15 mph the tractive effort required is

$$TE = (0.025 \times 20{,}000) + (0.01 \times 5 \times 20{,}000) = 500 + 1000 = 1500 \text{ lb}$$

The wheel power would be

$$\frac{1500 \text{ lb} \times 15 \text{ mph}}{375} = 60 \text{ hp}$$

and the motor delta pressure would be

$$P = \frac{2\pi \times 1500 \text{ lb} \times 27 \text{ in.}}{25.71 \times 4.1 \text{ in.}^3/\text{rev} \times 0.95} = 2541 \text{ psi}$$

This information can be used to complete Fig. 10.24. Select hydrostatic units that meet or exceed 4.1 in.^{3}/rev requirements. Check the vehicle performance using data for those units. Resize hydraulic units or alter vehicle performance as required.

A properly sized hydrostatic transmission must provide the vehicle requirements of maximum tractive effort at one condition and of maximum speed at another. The available horsepower to drive the hydraulic units does not have to provide both parameters at the same time. The allowable hydraulic motor speed and final drive ratio must provide the required vehicle speed. The motor displacement must be of sufficient size to provide the maximum tractive effort required at the vehicle drive wheels. The hydraulic unit displacement must be large enough to develop the required torque without exceeding the allowable system pressure. The properly sized hydrostatic transmission will provide the required vehicle parameters with the smallest size of hydraulic units that will satisfy those requirements.

10.5 HYDRAULIC CYLINDERS

A hydraulic cylinder is a hydraulic actuator that converts fluid power into linear mechanical force. A single acting cylinder is one that can have fluid force applied to only one side of the movable element. Hydraulic fluid can force the rod end out of the piston bore. Gravity or a mechanical spring force must be used to retract the piston into the bore and force the hydraulic fluid out of the cylinder.

A double-acting cylinder is shown in Fig. 10.26. A fluid force can be applied to either side of the piston to force the piston rod to move in either direction or to provide the counteracting force necessary to hold the piston stationary.

The force F that the hydraulic cylinder can exert against a load is equal to a pressure applied to the cross-sectional area of the piston. Summation of forces of the piston in Fig. 10.26 in the X direction gives

$$F = P_1A_1 - P_2A_2 - mX^2 \qquad [10.31]$$

$$A_1 = \frac{\pi}{4} D^2 \qquad [10.32]$$

$$A_2 = \frac{\pi}{4} (D^2 - d^2) \qquad [10.33]$$

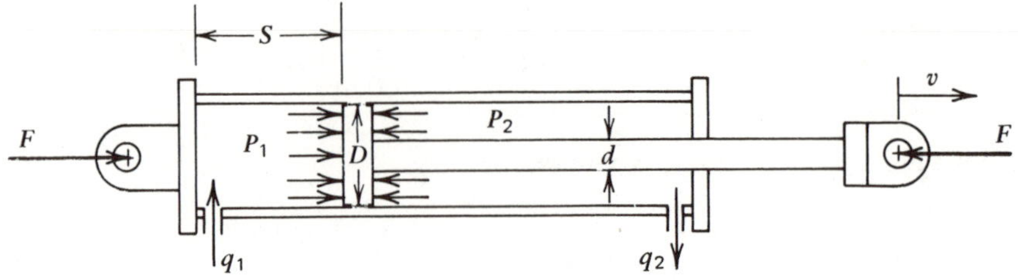

Figure 10.26 Double acting hydraulic cylinder.

where P_1, P_2 = hydraulic pressures (psi or bar)
D = piston diameter (in. or mm)
d = rod diameter (in. or mm)
S = stroke (in. or mm)
A_1 = piston area of the cap end (in.^{2} or m^2)
A_2 = piston area at the rod end (in.^{2} or mm^2)
m = mass of piston, rod, and fluid that is accelerated
X^2 = acceleration of the piston in the X direction (in./min or mm/min)

Assuming that the inertia force is negligible, the equation for hydrostatic force transmission is

$$F = \frac{\pi}{4} D^2 P_1 - \frac{\pi}{4} (D^2 - d^2) P_2 \qquad [10.34]$$

$$F = \frac{\pi}{4} [P_1 D^2 - P_2 (D^2 - d^2)] \qquad (\text{lb or } 10^{-1} \text{ N}) \qquad [10.35]$$

For cases where the return oil line from the rod end is ported back to the reservoir at atmospheric pressure, the force equation is simplified to

$$F = \frac{\pi}{4} P_1 D^2 \qquad (\text{lb or } 10^{-1} \text{ N}) \qquad [10.36]$$

However, the hydrostatic force transmission is only

$$F = \frac{\pi}{4} P_2 (D^2 - d^2) \qquad (\text{lb or } 10^{-1} \text{ N}) \qquad [10.37]$$

when the supply is to the rod end and the cap end is vented.

The speed of the rod movement is obtained by dividing the volume of fluid flow into the cylinder by the cross-sectional area of the cylinder bore. The flow required to move the piston at a given speed is

$$Q \text{ (gal/min)} = \frac{A \text{ (in.}^2\text{)}}{231} \times V \text{ (in./min)} \qquad \text{and}$$

$$Q \text{ (liters/min)} = 10^{-6} A \text{ (mm}^2\text{)} \times V \text{ (mm/min)}$$

where Q = inlet or outlet flow
A = cross-sectional area as described for force
V = velocity of the piston

Examination of the equation will show that the flow to the cylinder is unequal to the return flow from the cylinder of a piston with a rod from only one end.

10.5.1 Special Application Cylinders

A cylinder with a single piston and a piston rod extending from each end of the bore is defined as a double-rod cylinder. The inlet and outlet flow is equal for the double-rod cylinder.

A telescoping cylinder provides a long working stroke but occupies a short space in the retracted position. It is composed of a series of multiple tubular-rod segments that retract sequentially within each larger cylinder bore, as seen in Fig. 10.27.

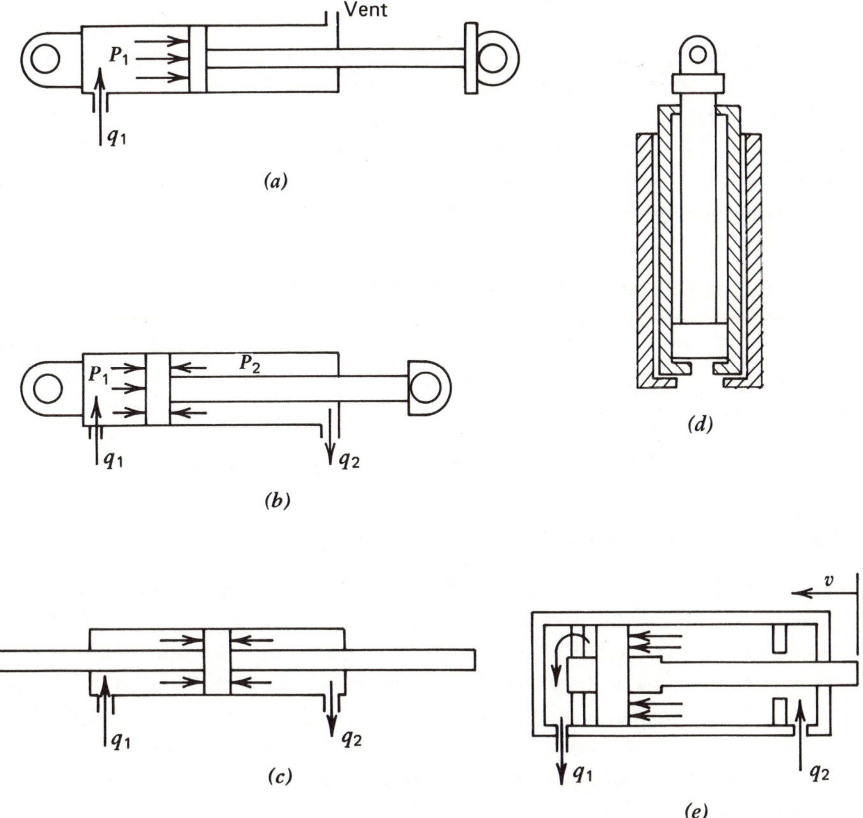

Figure 10.27 Hydraulic cylinders. (*a*) Single acting. (*b*) Double acting. (*c*) Double acting, double rod. (*d*) Telescoping. (*e*) Cushioned top.

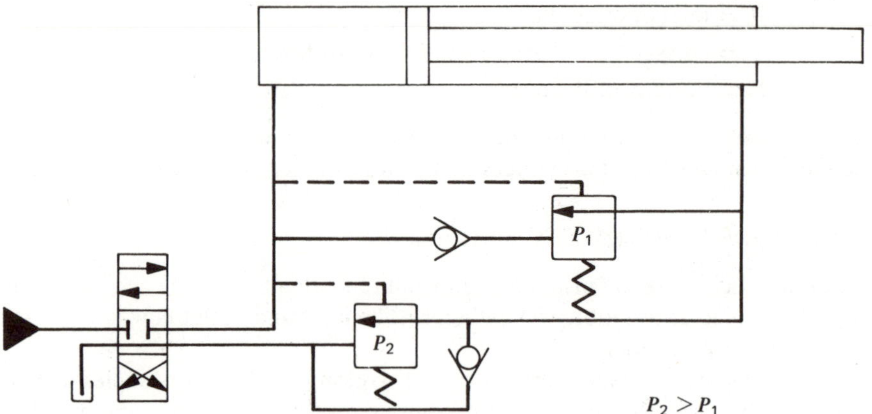

Figure 10.28 Regenerative hydraulic cylinder circuit.

Figure 10.28 shows a schematic of a double acting cylinder that is capable of regenerative flow. Flow from the rod end is ported to the cap end to be added to the pump flow. The piston is forced to move by the force developed as a result of the differential areas on the piston faces. When the rod end encounters an increase in force resistance, the pressure regulator at P_1 closes and the remaining stroke distance is traveled at a slower velocity, depending on the pump flow. The return stroke depends on pump flow to the rod end. This circuit allows the use of a relatively small displacement pump capable of developing high pressure.

10.6 HYDRAULIC CONTROL VALVES

Hydraulic power is controlled by pressure regulating and directional flow valves. The ability to control the power supplied to several different actuators at remote locations makes the selection of hydraulic power the preferred choice in many applications. Valves for specific functions provide methods to control one or several flow paths from a single pump with relative ease.

10.6.1 Pressure Regulating Valves

The function of pressure-regulating valves is to protect the components within the hydraulic circuit from damage from excessive pressure. These valves are usually called safety relief valves. The most basic type is a poppet held against a cone seat by a spring force. When the pressure force on the end of the poppet exceeds that spring force, the poppet is lifted off the seat and fluid can escape back to the reservoir. This valve does not provide a flat cutoff pressure limit with flow rate because the spring must be deflected more to open the cone passageway as the rate of flow is increased.

Curves of pressure versus flow for regulating valves with a nearly flat pressure cut-off limit can be obtained by a pilot operated pressure regulator valve.

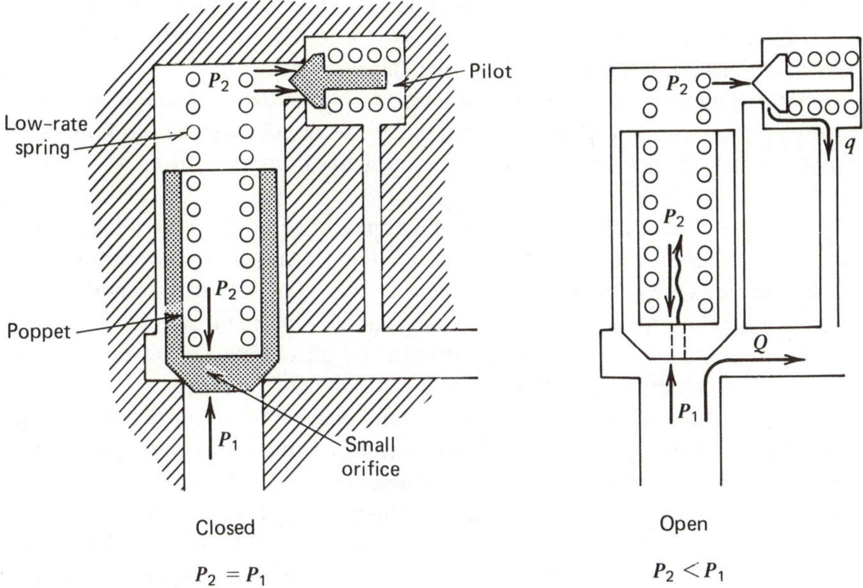

Figure 10.29 Pilot operated relief valve.

Figure 10.29 shows the action of this valve. The main poppet is clamped into the seat by hydraulic pressure acting on opposing areas. The small cone valve is held in place by a spring force adjusted for the limiting pressure. The small valve is lifted off its seat when the system pressure exceeds this set pressure. Fluid is allowed to escape past the cone. Fluid flows through the small orifice in the larger poppet into this chamber, but a large pressure drop develops across that orifice. The poppet is no longer hydraulically held on the seat, and it opens to the widest position against the relatively small spring force. As the system pressure drops back below the set pressure, the small cone valve closes, and once again a hydraulic clamping force is produced to close the main relief port.

Pilot relief valves provide fast action to respond to flow changes caused by rapidly closing valves and, hence, can prevent excess pressure spikes within the circuit. The system pressure can be closely regulated at the same value without increasing proportionately with the flow through the valve. Also, the poppet is securely held in the seat at pressures just below the set point, which prevents high leakage rates just prior to the opening of the valve.

10.6.2 Directional Control Valves

Directional control valves regulate the rate of flow that is permitted to operate individual hydraulic actuators. Various forms of directional control valves that function in one or more ways are check valves, remotely actuated poppet valves, and spool valves.

Check valves permit free flow in only one direction. They are constructed of a ball, flapper, or poppet held lightly against a seat by a small spring force. Their only purpose is to prohibit flow reversals within the circuit.

Remotely operated poppet valves are used to regulate flow through the valve by moving the poppet position proportionately to obtain the desired flow rate. Remote control to the valve is provided by a hydraulic piston or an electric solenoid to move the poppet.

Spool valves derive their name from the appearance of the related hardware. They are basically a shaft sliding in a bore. The shaft has grooves around the circumference, which make it look like a spool. The grooves provide fluid passageways to interconnect holes entering the bore perpendicular to the axial motion of the spool. Spool valves are classed according to the number of hydraulic lines connected to the valve, the number of operating positions for the spool, and how the hydraulic lines are interconnected when the spool is in the neutral position. A spool-type directional flow valve is shown in Fig. 10.30.

A spool valve with two, three, or four connected hydraulic lines is defined as a two-way, three-way, or four-way valve. The two-way spool valve would only operate as an on or off control valve inserted in a hydraulic line. A three-way valve switches the interconnection of one hydraulic line with either of the other two attached lines. As the spool is moved between its two positions, the connecting passageway from the first line to the second line is closed, and a passageway between the first and third is created. A four-way valve controls the flow interconnections between four hydraulic flow paths.

The number of operating positions of a spool valve refers to the distinct axial positions in which the spool can be placed within the bore of the valve body. Movement from one position to another will close off or change the interconnecting passageways. A two-way valve can only have two positions, corresponding to the open or closed position. Three-position valves have a

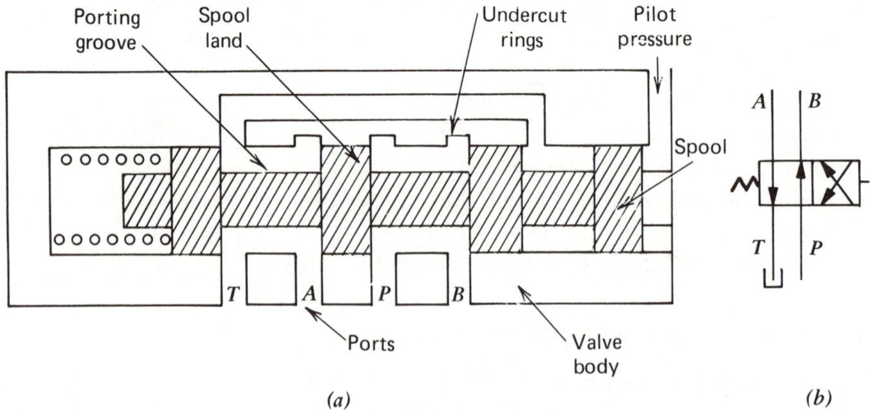

Figure 10.30 Four way two position directional flow valve. (*a*) Spool valve. (*b*) Symbol.

center neutral position and distinct interconnections for spool movement to the right or left from the central position. A two- or three-position valve that can be incrementally moved axially to slowly open the flow path between two hydraulic lines is called a throttling valve. These valves control the rate of flow as well as an on-off or gating function.

Four-way, three-position valves are further defined by the service that they provide in the neutral position. The three principle center conditions are shown in Fig. 10.31. The open center valve has all four parts interconnected to the tank return line (open position) at the center neutral position. The use of an open center valve in a pump discharge line ports all the pump flow to the tank with a low pressure drop through the valve while in the center neutral position. The pump can therefore operate at full flow and little pressure during standby conditions and requires low power input. When the spool is moved to the right or left, the *A* and *B* ports are connected to pressure or drain, and the pump flow to tank is closed off.

The closed center valve blocks all four lines while in the center neutral position. The advantage of this is that a load held by a hydraulic cylinder connected to *A* and *B* ports would be held in its relative position while the spool valve was in the neutral position. This condition is also necessary if more than one directional control valve is supplied from a single pump and if they are to function independently. It would not be possible to use an open center valve and operate the second valve in the neutral position. The disadvantage is that all pump flow must go over a pressure relief valve while in the standby position. Variable displacement pumps that sense the circuit pressure and destroke the pump to provide only the flow required to maintain the desired pressure and avoid flow over the relief valve are normally used.

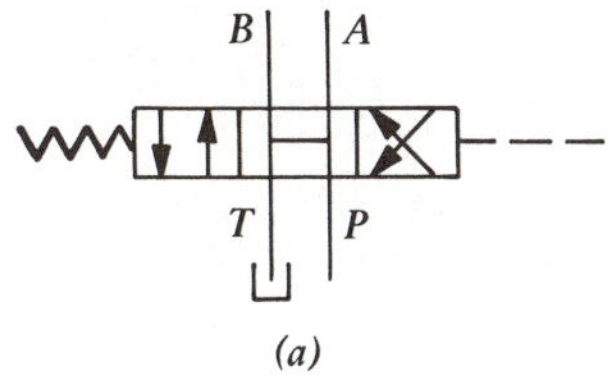

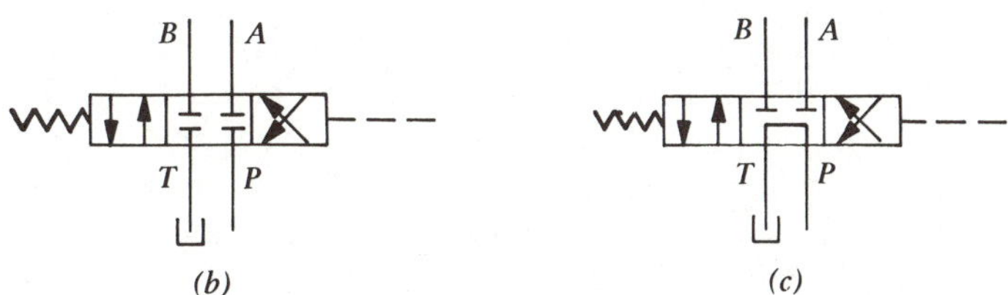

Figure 10.31 Four way three position directional control valve. (*a*) Open center spool. (*b*) Closed center spool. (*c*) Tandem center spool.

A tandem center valve is effective if only one directional flow valve is supplied by a pump or if it is a priority valve that must be displaced from neutral before any other directional flow valve can be used. The tandem center blocks the *A* and *B* ports while in the neutral position but unloads the pump by porting all pump flow to tank.

10.7 HYDRAULIC HOSES AND FITTINGS

The hydraulic circuit fluid flow connection between components can be hydraulic hose or steel tubing. Steel tubing has relatively thin walls and can be formed into smooth bends that provide streamlined hydraulic flow with low pressure drop over the length of the tube. Circuit connections are simplified by bending the tube around obstructions between two connection points within the circuit. By bending and forming, the tube can be made to lie unobtrusively alongside the frame of a machine and to transmit hydraulic power a long distance from the pump. The installation and removal for component repair is simplified because the turns made by tube bends make fewer fittings necessary and every fitting can serve as a union to open the circuit at locations of easy access by wrenches.

Tubing is sized by the outside diameter of the tube, regardless of the tube-wall thickness. Thus one fitting size can be used for the various wall thicknesses. To have sufficient hoop stress for the various internal pressures and duty cycles expected, a specified wall thickness is required for a given pressure and service factor. For a given tube size, the inside diameter is the outside diameter minus two times the wall thickness. Thus the internal cross-sectional area decreases as the wall thickness is increased. The internal size of a tube in a hydraulic circuit should be selected so that the maximum velocity of the hydraulic fluid within the conduit does not exceed 25 ft/sec. This practice prevents excessive turbulence in the fluid, which in turn, would increase the heat dissipated into the oil and increase the loss of usable power.

Hydraulic hoses are used where a flexible line is required in the circuit. They permit relative motion between two components, facilitate alignment and length changes between two points, and dampen vibrations caused by hydraulic pulses.

10.7.1 Hydraulic Hose Selection

Hose sizes are designated by the nominal inside diameter of the hose, expressed in fractions of inches. Thus all classes of hoses of the same size have equal internal areas, regardless of the construction. Hose sizes are often designated by a dash number which signifies the number of 1/16 in. increments in the internal diameter for example (-10 designates a 5/8-in. inside diameter hose).

Hose size selection is made on the basis of allowable velocity of the oil within the hose. The recommended suction side of a pump is to not exceed 4 ft/sec. Higher inlet velocities cause increased pressure losses and create pump cavitation unless the reservoir is pressurized to force the fluid to the pump inlet. The inlet hose size selection is a compromise between higher velocities and the additional cost of providing larger diameter hoses for low flow rates. A hose size selection nomograph is shown in Fig. 10.32. The recommended velocity for intake lines is shown to lie between 2 and 4 ft/sec.

A similar range is shown for hydraulic hoses used for the pump discharge and other pressurized lines. Oil velocity above the recommended range causes excessive turbulence in the oil and creates power losses and heating of the oil.

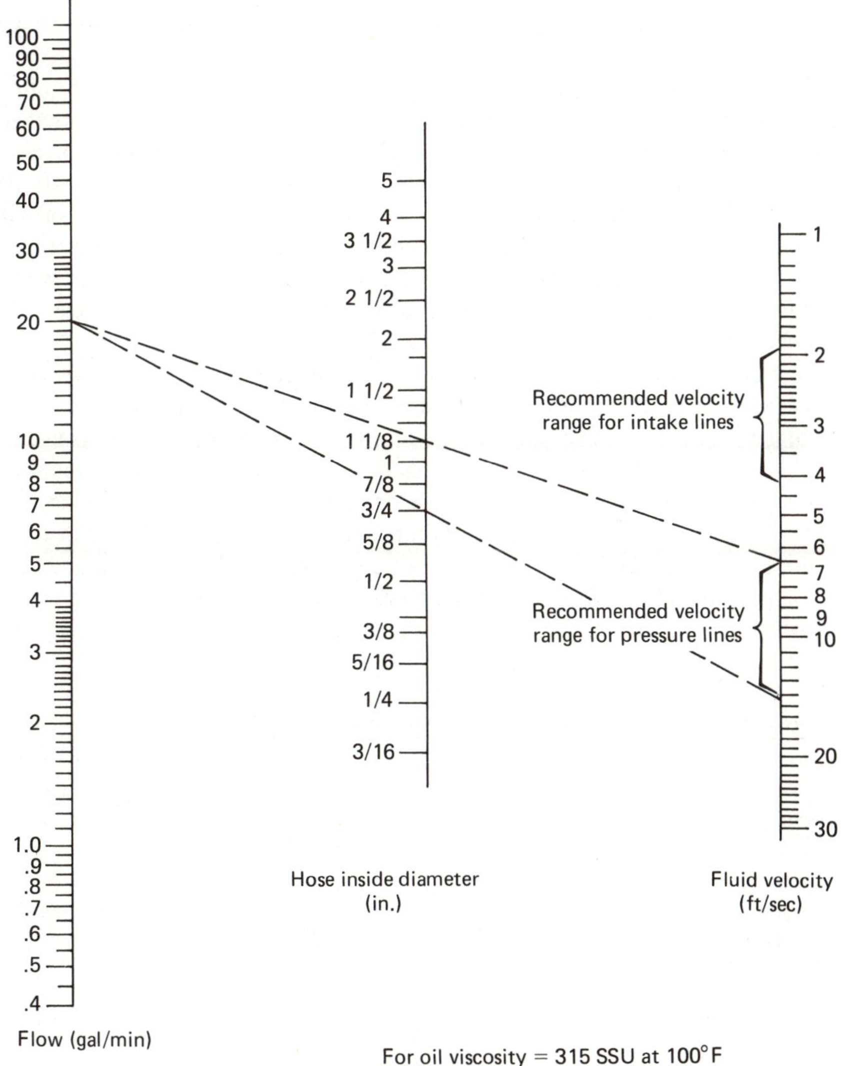

Figure 10.32 Hose size selection nomograph.

High oil temperature and high oil velocity cause hose deterioration and erosion of the internal rubber tube of the hose. The nomograph shows an example of selecting the hose size for a 20 gal/min pressurized hose. Dashed lines are drawn on the nomograph from that point to both extremes of the recommended velocity range. Selection of any standard size hose within this range from the center column will provide the most economical hose for this application. In this case, the smallest acceptable hose would have a 3/4 in. inside diameter.

Petroleum based hydraulic-fluid power hoses are constructed of a seamless inner tube of oil resistant synthetic rubber, a middle layer of steel wire braid or spiral plies reinforcement for strength, and an oil and weather resistant synthetic rubber cover.

Hydraulic hoses are classed according to the pressure rating, the steel or fiber reinforcement, the number of braids or spiral plies, and the material used for the cover. An SAE classification of SAE 100 R1 designates a medium pressure hose that is constructed with a single wire braid reinforcement. An SAE 100 R2A or SAE 100 R2B designates high pressure hoses with two braids of wire reinforcement or two spiral plies plus one braid of wire reinforcement, respectively. The SAE 100 R4 classification is for a hydraulic suction line with a spiral wire reinforcement. Heavy duty hoses with four and six spiral steel reinforcement layers are classed SAE 100 R9, 10, or 11.

Hose standards specify that a hose must be sufficiently constructed so that a new section of hose will not leak or fail below a minimum burst pressure listed for that class of hose. The maximum operating pressure for the hose is usually rated equal to one-fourth the burst pressure rating to ensure a reasonable hose life as aging and pressure fluctuations deteriorate the strength of the hose. The pressure ratings for hoses are a function of the wall strength and thus decrease as the hose diameter increases, as shown in Table 10.4. Operating pressures that exceed the maximum operating pressure rating will result in a service life below the manufacturer's design life of the hose.

TABLE 10.4 **Typical Hydraulic Hose Specifications**

Nominal hose i.d., in.	¼	½	¾	1	1½
Hose dash number	−4	−8	−12	−16	−24
Minimum bend radius, in.	4	7	9.5	12	20
Medium pressure hose					
Minimum burst pressure, psi	11,000	8,000	5,000	4,000	2,000
Maximum operating pressure, psi	2,750	2,000	1,250	1,000	500
High pressure hose					
Minimum burst pressure, psi	20,000	14,000	9,000	8,000	5,000
Maximum operating pressure, psi	5,000	3,500	2,250	2,000	1,250

Factors that reduce hose life expectancy are:

1. Flexing or bending the hose to a smaller radius than the minimum allowed.
2. Twisting or kinking the hose.
3. Crushing the steel reinforcement.
4. Abrasion or damage to the hose cover.
5. Pulling the hose–causing strain at the connections.
6. Operating above 200°or below −40°F hydraulic fluid temperature.
7. Surge pressures or pressure spikes above the maximum operating pressure.

All hydraulic circuits have pressure peaks above the indicated gauge pressure. High frequency pressure ripples and rapidly rising pressure peaks caused by closing valves will not be detected by most pressure gauges. Electronic measuring instrumentation is required to determine the severity of these pressure surges.

10.7.2 Hydraulic Hose and Tube Fittings

Tube and hose fittings are available in assorted types as well as sizes, corresponding to tube o.d. or hose i.d. Five of the most common types of fittings for hydraulic power usage are shown in Fig. 10.33. The first type shown is a dryseal pipe connection. A mechanical seal is formed by interference of the pipe threads as the connection is tightened. This connection is acceptable for brass or low carbon steels. However, brittle metals and high strength steel would, at best, seal only the first application and would require thread sealants to prevent leaking each time the connection is opened and retightened.

The second type shown uses straight threads with an elastomer gasket in the form of an O-ring. The clamping force of the threads squeeze the O-ring seal into a chamfer on the mating part, which is called an O-ring boss. This seal can be recreated each time the connection is disassembled by installing a new O-ring.

The third and fourth types show flared fittings with 37° and 45° cones. The seal is formed by a mechanical circular line contact of metal to metal as the cone on the nose of the fitting is forced against the flare of the steel tube or hose connection. These can be reused and will reseal when the connection is remade. A sleeve is usually inserted over the outside of a tube to permit the fastening nut to turn without twisting the hose or tube as shown for the third type. Thus these fittings function as unions in the line and provide a means to remove a tube or hose without removing one of the components at the end of a line.

The fifth type shown is for high-pressure lines and is most often used for the discharge lines from high pressure pumps. The pump discharge face would have a smooth machined surface with a drilled discharge port and a four-bolt

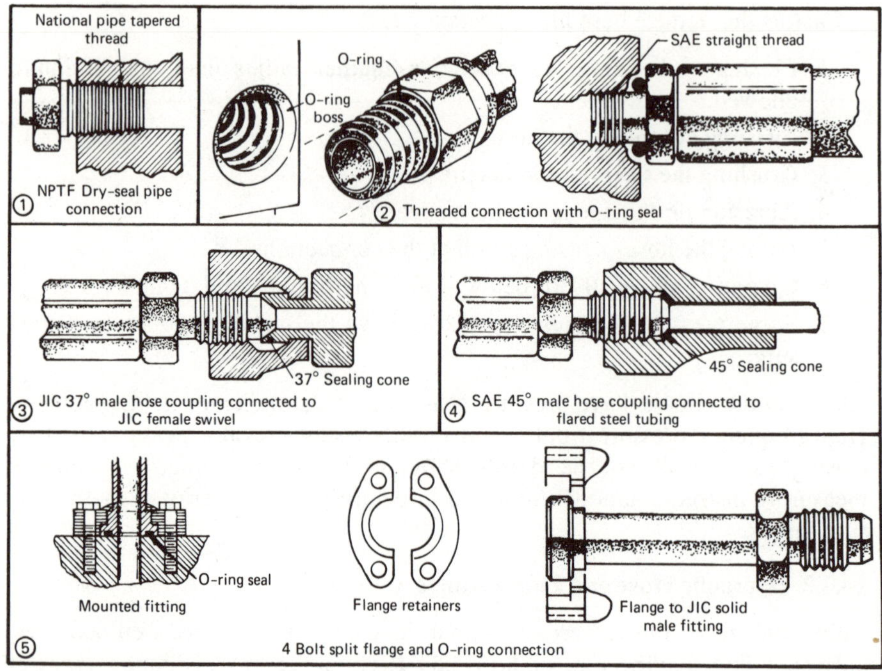

Figure 10.33 Five types of hydraulic fittings.

pattern for the fitting. The seal is formed as the flange retainers are clamped against the flange of the tube. The bolts are tightened until the tube flange makes metal-to-metal contact to the pump discharge port face. The O-ring is squeezed in the groove and creates a seal that prohibits leakage between the two metal faces.

Fitting elbows, bulkhead connectors, reducers, hose ends, and tubing connectors are available to make the transition between these types of connections. It is common to supply hoses with straight thread O-ring fittings on one end to mount on a valve or pump and with 37° flare fitting on the other to connect to steel tubing. Care must be taken to specify the type of fitting on both ends of each piece in the hydraulic circuit.

10.8 FILTERS, COOLERS, AND RESERVOIRS

The design of hydrostatic transmissions also includes auxiliary equipment for fluid filtration, hydraulic oil reservoir, and a heat exchanger to cool the oil. Closed loop hydrostatic systems use a charge or replenishing pump to make up loop leakage and provide extra flow for loop cooling. Cool oil is pumped from the reservoir into the low pressure side of the loop. Some of the hot oil flowing in the low pressure side is vented into the motor case. The motor case drain is ported into the pump case and the pump case is drained to the reservoir. This flow of oil with the flow from internal leakage absorbs the heat generated within the hydraulic loop and is passed through a heat exchanger to cool the oil before it is returned to the reservoir.

A filter is placed between the charge pump inlet and the reservoir. Thus the exchange flow is constantly cooling and filtering the loop fluid. The particle size filtration required for a piston pump is much smaller than that for gear pumps, which can tolerate more dirt without degrading pump performance. The recommended filter particle size for gear pumps is usually 40μ whereas piston pumps normally require filters that will pass only 10μ or smaller dirt particles. The charge pump inlet filter must be sized for the flow capacity of the charge pump. Excessive restriction at the filter will cause pump cavitation. A bypass in the suction line around the filter should be avoided and the operator should notice poor hydrostatic transmission performance and replace a plugged filter caused by contaminated oil. A bypass would allow the operation of the transmission with the filter plugged, and the contaminated oil would rapidly cause the pump and motors to fail.

The reservoir should be able to contain a quantity of oil sufficient so that the oil flow through the reservoir would have a 30 sec dwell time. Holding the oil for this period of time allows the gas and air bubbles that are entrained in the fluid to float to the surface. The liquid capacity of the reservoir should be designed to hold at least one-half the maximum charge pump flow. The return oil should enter the reservoir below the liquid surface to prevent entraining air with the fluid flow, which would cause oil foaming and pump cavitation. The suction line to the pump must always be below the oil surface so that air will not be drawn into the suction line. Placement of the suction line near the bottom of the reservoir, but above the sediment level, prevents contaminated particles from being drawn into the suction line.

HOMEWORK PROBLEMS

10.1 What input horsepower is required to drive a variable displacement hydraulic pump with 6 in.^{3} displacement? The maximum discharge pressure is 5000 psi, and the pump is to be driven at 2400 rpm. Assume a 97% torque efficiency and 88% volumetric efficiency.

10.2 Determine the torque required to drive a 0.1 liter pump at 2000 rpm if the delta pressure across the pump is 240 bar.

10.3 What would be the volumetric efficiency of the pump in Fig. 10.12: 1. at 1000 psi and 1000 rpm and, 2. at 4000 psi and 3000 rpm? Assume that the pump displacement is 6.25 in.^{3}/rev.

10.4 Use Fig. 10.12 and determine the volumetric efficiency of that pump for: 1. 3000 rpm and 100 bar and, 2. 1500 rpm and 300 bar. Assume that the pump displacement is 102 ml/rev.

10.5 Determine the power required to drive the pump in Fig. 10.12 at 3000 rpm and 4000 psi. Assume the pump displacement to be 6.25 in.^{3}/rev.

10.6 Calculate the input power required to drive the 102 ml/rev displacement pump in Fig. 10.12 at 2000 rpm and 200 bar pressure.

10.7 To drive a combine threshing cylinder, 1500 in · lb torque and a variable speed range from 800 to 1600 rpm are required. Specify a hydraulic motor to be used to drive this cylinder. What flow rate must be provided if the maximum pressure is 3500 psi?

10.8 Determine the overall efficiency of a transmission made up of the pump in Problem 10.3 and the motor used for Fig. 10.15. Assume pump speed of 3000 rpm and a 4000 psi delta pressure in the system loop.

10.9 A self-propelled windrower weighs 8000 lb. The drive wheel tire size is 16.5×16.1. The maximum forward speed required is 12 mph. The windrower is powered by an engine with a maximum torque of 190 ft · lb and a rated speed of 3200 rpm.
The ground speed drive consists of two hydrostatic drive systems (two pumps mounted in tandem to the engine, each providing power to one hydraulic wheel motor).
Select the size of hydraulic pump and motor required to propel one side of this windrower.

10.10 A hydraulic circuit consists of one variable displacement pump and three load functions. The three load functions are double-acting hydraulic cylinders. One cylinder operates independently of the other two. Of the remaining cylinders, one will only operate if the other is in neutral.
Draw a schematic to represent the components required for this circuit.

10.11 Make a schematic of the hydraulic system required for Problem 10.9.

10.12 A hydraulic cylinder is to be used on a front-end loader to lift the bucket or force it down into the pile. The maximum force required of the cylinder is 8000 lb. Size the cylinder for a safe operating pressure of 2500 psi.

10.13 What hydraulic horsepower is required in Problem 10.6 to extend the cylinder 18 in. in 5 sec?

10.14 A pump delivers 40 liters/min to a 100-mm diameter cylinder that is required to produce 2200 N force. What is required to extend the cylinder? What hydraulic power is required?

10.15 Specify the hydraulic hoses required for the inlet and discharge of the pump in Problem 10.1. Draw a schematic of the hydraulic circuit and make a parts list of components and fittings that would be required.

REFERENCES

1. "Analyzing Hydraulic Systems," Bulletin No. 0222-B1, Parker Hannifin Corporation, Cleveland, Ohio, 1976.
2. *Fluid Power Design Engineers Handbook*, Parker Hannifin Corporation, Cleveland, Ohio, 1981.
3. "Industrial Hydraulic Technology," Bulletin No. 0221-B1, Parker Hannifin Corporation, Cleveland, Ohio, 1978.
4. Merritt, H.E. *Hydraulic Control Systems*, John Wiley and Sons, Inc., New York, N.Y., 1967.
5. *Mobile Hydraulics Manual*, M2990A, Sperry Vickers, Troy Michigan, 1967.
6. "SAE Recommended Practices: J744 Hydraulic Power Pumps, J745 Hydraulic Power Pump Test Procedure, J746 Hydraulic Motor Test Procedure, J747 Hydraulic Control Valve Test Procedure," *SAE Handbook*, Society of Automotive Engineers, Inc., Warrendale, Pa., 1981.
7. Stringer, J. *Hydraulic System Analysis*, John Wiley and Sons, Inc., New York, N.Y., 1976.

APPENDIX

COMMON CROSS-SECTIONAL PROPERTIES
HOT-ROLLED LOW-CARBON STEEL PLATE
COMMON PIPE DIMENSIONS
ANGLE IRON
RECTANGULAR HOLLOW STRUCTURAL TUBING
SQUARE HOLLOW STRUCTURAL TUBING
HOT FINISHED STRUCTURAL HOLLOW SECTIONS
HOT-ROLLED SQUARE STEEL BARS
HOT-ROLLED ROUND STEEL BARS
COLD-ROLLED FLAT STOCK
HOT-ROLLED T-STEEL WITH SHARP CORNERS
HOT-ROLLED RECTANGULAR STEEL BARS
SEAMLESS AND WELDED STEEL TUBES
CLEVIS PINS
SLOTTED SPRING PIN DIMENSIONS
CLEVIS PINS WITH HEAD
AMERICAN NATIONAL STANDARD FINISHED HEXAGON CASTLE NUTS
HEXAGON BOLTS
RECOMMENDED TORQUE VALUES FOR METRIC FASTENERS
AMERICAN NATIONAL STANDARD COTTER PINS
SPLIT COTTER PINS
AMERICAN NATIONAL STANDARD SET SCREWS
AMERICAN NATIONAL STANDARD LOCK WASHERS
SPLIT LOCK WASHER
SERRATED WASHER
PLAIN WASHERS: METRIC AND ENGLISH
PROPERTY CLASS IDENTIFICATION SYMBOLS FOR BOLTS, SCREWS, AND STUDS

COMMONLY USED CROSS SECTIONAL PROPERTIES

Section of beam I = moment of inertia

$$Z = \text{section modulus} = \frac{I}{C}$$

$$I_{1-1} = \frac{bh^3}{12} \qquad I_{2-2} = \frac{b^3h}{12}$$

$$Z_{1-1} = \frac{bh^2}{6} \qquad Z_{2-2} = \frac{b^2h}{6}$$

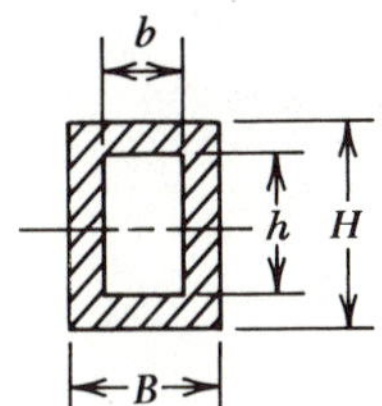

$$I = \frac{BH^3 - bh^3}{12}$$

$$Z = \frac{BH^3 - bh^3}{6H}$$

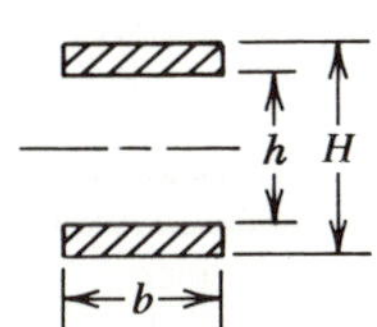

$$I = \frac{b}{12}(H^3 - h^3)$$

$$Z = \frac{b}{6H}(H^3 - h^3)$$

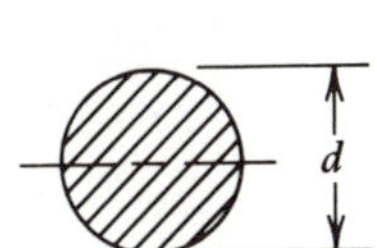

$$I = \frac{\pi d^3}{64} = 0.0491d^4$$

$$Z = \frac{\pi d^3}{32} = 0.0982d^4$$

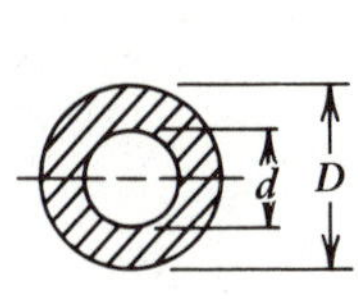

$$I = \frac{\pi}{64}(D^4 - d^4) = 0.0491(D^4 - d^4)$$

$$Z = \frac{\pi}{32D}(D^4 - d^4) = 0.0983\left(\frac{D^4 - d^4}{D}\right)$$

Hot-Rolled Low-Carbon Steel Plate, Sheet, and Strip (ISO 630, 1052, 3573)

		1	*2*	*3*	*4*
		Ductile Grades			
Quality		*Commercial*	*Drawing*	*Deep Drawing*	*Deep Drawing Special Killed*
Physical	Tensile (MPa)		−430	−370	−390
Properties	Yield (MPa)				
Properties	Elongation (%)		25−	28−	28−
	Rockwell HR30T				
Chemical	Carbon (%)	C = −0.15	C = −0.12	C = −0.10	C = −0.08
Composition	Manganese (%)	Mn = −0.60	Mn = −0.50	Mn = −0.45	Mn = −0.45
(ISO)	Silicon (%)				
	Phosphore (%)	P = −0.05	P = −0.04	P = −0.03	P = −0.03
	Sulfur (%)	S = −0.05	S = −0.04	S = −0.03	S = −0.03
		ISO 3573	ISO 3573	ISO 3573	ISO 3573
ISO		HR 1	HR 2	HR 3	HR 4
North America		ASTM- A569 (AISI-CQ)	ASTM- A621 (AISI-DQ)		ASTM- A622 (AISI-DQ SPEC.) (KILLED)

5	*6*	*7*	*8*	*9*	*10*	*11*
			Structural Grades			
Commercial						
330-440	360-470	410-520	510-610	190-620	590-750	690-870
195-	210-	230-	250-	290-	330-	360
22-	20-	17-	17-	15-	11-	7-
C = −0.15	C = −0.17	C = −0.20	C = −0.20	C = −0.30	C = −0.40	C = −0.50
P = −0.05	P = −0.05	P = −0.05	P = −0.05	P = −0.06	P = −0.06	P = −0.06
S = −0.05	S = −0.05	S = −0.05	S = −0.05	S = −0.05	S = −0.05	S = −0.05
ISO 630	ISO 630	ISO 630	ISO 630	ISO 1052	ISO 1052	ISO 1052
Fe 33	Fe 37-A Fe 37-B Fe 37 C Fe 37 D	Fe 42 A Fe 42 B Fe 42 C Fe 42 D	Fe 52-B Fe 52-C Fe 52-D	Fe 50-1 Fe 50-2	Fe 60-1 Fe 60-2	Fe 70-2
ASTM- A570 GRADE A	ASTM- A570 GRADE B	ASTM- A570 GRADE C	ASTM- A570 GRADE D	ASTM- A570 GRADE E	AISI 1040	AISI 1050

Common Pipe Dimensions

Nominal Sizes (in.)	*OD (in.)*	*Wall Thickness (in.)*	*Weight per Foot (lb.)*	*Standard or Extra Strong*	*Schedule Number*	*Area ($in.^2$)*	*Moment of Inertia ($in.^4$)*	*Section Modulus ($in.^3$)*	*Radius of Gyration (in.)*
½	0.840	0.109	0.85	Std.	40	0.250	0.017	0.041	0.26
		0.147	1.09	XS	80	0.320	0.020	0.048	0.25
¾	1.050	0.113	1.13	Std.	40	0.333	0.037	0.071	0.33
		0.154	1.47	XS	80	0.433	0.045	0.085	0.32
1	1.315	0.133	1.68	Std.	40	0.494	0.087	0.133	0.42
		0.179	2.17	XS	80	0.639	0.106	0.161	0.41
2	2.375	0.154	3.65	Std.	40	1.075	0.666	0.561	0.79
		0.218	5.02	XS	80	1.477	0.868	0.731	0.77
3	3.500	0.156	5.58			1.641	2.299	1.314	1.18
		0.188	6.63			1.951	2.685	1.534	1.17
		0.216	7.58	Std.	40	2.228	3.017	1.724	1.16
		0.300	10.25	XS	80	3.016	3.894	2.225	1.14
4	4.500	0.156	7.25			2.132	5.035	2.238	1.54
		0.188	8.64			2.540	5.917	2.630	1.53
		0.219	10.00			2.942	6.759	3.004	1.52
		0.237	10.79	Std.	40	3.174	7.233	3.214	1.51
		0.337	14.98	XS	80	4.407	9.610	4.271	1.48

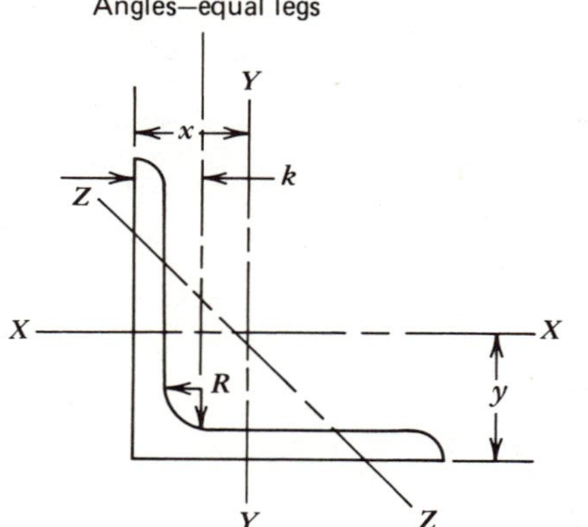

Section number and size in.	Thickness in.	Weight per foot lb	Area of section in.^{2}	k in.	Axis X-X and Axis Y-Y				Axis Z-Z
					$I_{x,y}$ in.^{4}	$Z_{x,y}$ in.^{3}	$r_{x,y}$ in.	x or y in.	r_z in.
L4 × 4 ×	3/4	18.5	5.44	1 1/8	7.67	2.81	1.19	1.27	0.778
R = 3/8	5/8	15.7	4.61	1	6.66	2.40	1.20	1.23	0.779
	1/2	12.8	3.75	7/8	5.56	1.97	1.22	1.18	0.782
	7/16	11.3	3.31	13/16	4.97	1.75	1.23	1.16	0.785
	3/8	9.8	2.86	3/4	4.36	1.52	1.23	1.14	0.788
	5/16	8.2	2.40	11/16	3.71	1.29	1.24	1.12	0.791
	1/4	6.6	1.94	5/8	3.04	1.05	1.25	1.09	0.795

$L3 \times 3 \times$	1/2	9.4	2.75	13/16	2.22	1.07	0.898	0.932	0.584
$R = 5/16$	7/16	8.3	2.43	3/4	1.99	0.954	0.905	0.910	0.585
	3/8	7.2	2.11	11/16	1.76	0.833	0.913	0.888	0.587
	5/16	6.1	1.78	5/8	1.51	0.707	0.922	0.865	0.589
	1/4	4.9	1.44	9/16	1.24	0.577	0.930	0.842	0.592
	3/16	3.71	1.09	1/2	0.962	0.441	0.939	0.820	0.596
$L2 \times 2 \times$	3/8	4.7	1.36	11/16	0.479	0.351	0.594	0.636	0.389
$R = 9/32$	9/16	3.92	1.15	5/8	0.416	0.300	0.601	0.614	0.390
	1/4	3.19	0.938	9/16	0.348	0.247	0.609	0.592	0.391
	3/16	2.44	0.715	1/2	0.272	0.190	0.617	0.569	0.394
	1/8	1.65	0.484	7/16	0.190	0.131	0.626	0.546	0.398
$L1 \times 1 \times$	1/4	1.49	0.438	3/8	0.037	0.056	0.290	0.339	0.196
$R = 1/8$	3/16	1.16	0.340	5/16	0.030	0.044	0.297	0.318	0.195
	1/8	0.80	0.234	1/4	0.022	0.031	0.304	0.296	0.196

Source: Bethlehem Structural Shapes Catalog

Rectangular Hollow Structural Tubing

Size (in.)			Weight per Foot (lb)	Area of Metal (in.^{3})	Axis X-X			Axis Y-Y		
Axis Y-Y	Axis X-X	Wall (in.)	W	A	Moment of Inertia I	Section Modulus Z	Radius of Gyration r	Moment of Inertia I	Section Modulus Z	Radius of Gyration r
		0.141	4.32	1.2720	1.4972	0.9981	1.0849	0.7951	0.7951	0.7906
3	2	0.1875	5.59	1.6438	1.8551	1.2367	1.0623	0.9758	0.9758	0.7704
		0.250	7.10	2.0890	2.2030	1.4687	1.0269	1.1466	1.1466	0.7409
		0.155	5.78	1.7015	3.3477	1.6738	1.4027	1.1230	1.1230	0.8124
4	2	0.1875	6.86	2.0188	3.8654	1.9327	1.3837	1.2849	1.2849	0.7978
		0.250	8.80	2.5890	4.6893	2.3447	1.3458	1.5321	1.5321	0.7692
		0.156	6.88	2.0240	4.5198	2.2599	1.4944	2.8949	1.9299	1.1959
4	3	0.1875	8.14	2.3938	5.2291	2.6146	1.4780	3.3404	2.2269	1.1813
		0.250	10.50	3.0890	6.4498	3.2249	1.4450	4.0988	2.7326	1.1519
		0.3125	12.69	3.7329	7.4838	3.7169	1.4112	4.7000	3.1333	1.1221
		0.1875	9.31	2.7383	8.8629	3.5452	1.7091	4.0118	2.6746	1.2104
		0.250	12.02	3.5354	10.949	4.3797	1.7598	4.9195	3.2797	1.1795
5	3	0.3125	14.52	4.2720	12.612	5.0448	1.7182	5.6255	3.7504	1.1475
		0.375	16.84	4.9543	13.907	5.5628	1.6754	6.1552	4.1084	1.1146
		0.500	20.88	6.1416	15.355	6.1418	1.5812	6.6889	4.4559	1.0432
		0.1875	11.86	3.4883	17.160	5.7198	2.2179	9.1952	4.5976	1.6236
		0.250	15.42	4.5354	21.574	7.1913	2.1810	11.509	5.7544	1.5930
6	4	0.3125	18.77	5.5220	25.346	8.4487	2.1424	13.463	6.7313	1.5614
		0.375	21.94	6.4543	28.553	9.5178	2.1033	15.097	7.5486	1.5294
		0.500	27.68	8.1416	33.213	11.071	2.0198	17.400	8.7002	1.4619

Square Hollow Structural Tubing (Actual Properties with Rounded Corners)

Size (in.)	*Wall (in.)*	*Weight per Foot (lb)* W	*Area of Metal ($in.^3$)* A	*Moment of Inertia* I	*Section Modulus* Z	*Radius of Gyration* r
2 × 2	0.110	2.69	0.7914	0.4574	0.4574	0.7603
	0.125	3.04	0.8934	0.5079	0.5079	0.7540
	0.154	3.65	1.0750	0.5911	0.5911	0.7514
	0.1875	4.31	1.2688	0.6667	0.6667	0.7249
2½ × 2½	0.141	4.32	1.2720	1.1498	0.9198	0.9507
	0.1875	5.59	1.6438	1.4211	1.1369	0.9298
	0.250	7.10	2.0890	1.6849	1.3479	0.8981
3 × 3	0.155	5.78	1.7015	2.2509	1.5006	1.1502
	0.1875	6.86	2.0188	2.5977	1.7318	1.1344
	0.250	8.80	2.5890	3.1509	2.1006	1.1032
4 × 4	0.1875	9.31	2.7383	6.4677	3.2338	1.5369
	0.250	12.02	3.5354	7.9880	3.9940	1.5031
	0.3125	14.52	4.2720	9.2031	4.6016	1.4677
	0.375	16.84	4.9543	10.152	5.0760	1.4315
	0.500	20.88	6.1416	11.234	5.6169	1.3524
6 × 6	0.1875	14.41	4.2383	23.496	7.8322	2.3545
	0.250	18.82	5.5354	29.845	9.9482	2.3220
	0.3125	23.02	6.7720	25.465	11.822	2.2884
	0.375	27.04	7.9543	40.436	13.479	2.2547
	0.500	34.48	10.142	48.379	16.126	2.1841

Hot Finished Structural Hollow Sections

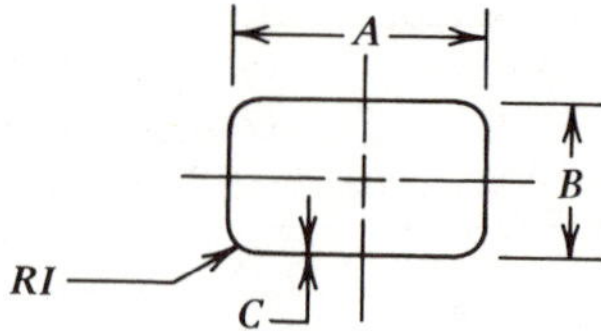

Millimeters					Inches				
A	× B	× C		RI	A	× B	× C		RI
20 ×	20 ×	2.0		4.0	0.787 ×	0.787 ×	0.079		0.16
		2.6		5.2			0.102		0.20
30 ×	30 ×	2.0		4.0	1.181 ×	1.181 ×	0.079		0.16
		2.6		5.2			0.102		0.20
		3.2		6.4			0.126		0.25
60 ×	60 ×	3.2		6.4	2.362 ×	2.362 ×	0.126		0.25
		4.0		8.0			0.157		0.31
		5.0		10.0			0.197		0.39
80 ×	80 ×	3.2		6.4	3.150 ×	3.150 ×	0.126		0.25
		3.6		7.2			0.142		0.28
		4.0		8.0			0.157		0.31
		5.0		10.0			0.197		0.39
		6.3		12.6			0.248		0.50
150 ×	150 ×	4.0		8.0	5.906 ×	5.906 ×	0.157		0.31
		5.0		10.0			0.197		0.39
		6.3		12.6			0.248		0.50
		8.0		16.0			0.315		0.63
		10.0		20.0			0.394		0.79
		12.5		25.0			0.492		0.98
		16.0		32.0			0.630		1.26
70 ×	40 ×	3.2		6.4	2.756 ×	1.575 ×	0.126		0.25
		4.0		8.0			0.157		0.31
		5.0		10.0			0.197		0.39

Millimeters				Inches			
A ×	B ×	C	RI	A ×	B ×	C	RI
100 ×	60 ×	3.2	6.4	3.937 ×	2.362 ×	0.126	0.25
		3.6	7.2			0.142	0.28
		4.0	8.0			0.157	0.31
		5.0	10.0			0.197	0.39
		6.3	12.6			0.248	0.50
120 ×	80 ×	3.2	6.4	4.724 ×	3.150 ×	0.126	0.25
		4.0	8.0			0.157	0.31
		5.0	10.0			0.197	0.39
		6.3	12.6			0.248	0.50
		8.0	16.0			0.315	0.63
		10.0	20.0			0.394	0.79

Notes:

1. Reference: ISO 657/XIV, 1978
2. Hot-finish structural hollow sections ISO 657/XIV, basis 1 in. = 25.4 mm.
3. RI is the maximum corner radius which is specified in ISO 657/14 standard to RI = 2C up to sizes $A \leqslant 100$ or $A \times B \leqslant 100 \times 150$ mm. The corner radii for larger sizes is according to BS 4848/2.
4. *Cold finish:* Tubing is specified in ISO 4019 standard to the same, with a few exceptions, outside dimensions as shown here and with wall thicknesses of 1.2, 1.6, 2, 2.6, 3.2, 4, 5, 6.3, 7.1, 8, 10, 12.5 mm.

Hot-Rolled Square Steel Bars (ISO 1035/II)

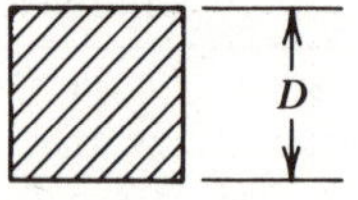

Nominal Size = D mm	in.	Nominal Size = D mm	in.
3	0.118	58	2.283
4	0.157	60	2.362
5	0.197	64	2.520
6	0.236	65	2.559
7	0.276	68	2.677
8	0.315	70	2.756
9	0.354	74	2.913
10	0.394	75	2.953
11	0.433	80	3.150
12	0.472	85	3.346
13	0.512	90	3.543
14	0.551	95	3.740
15	0.591	100	3.937
16	0.630	105	4.134
17	0.669	110	4.331
18	0.709	115	4.528
19	0.748	120	4.724
20	0.787	125	4.921
21	0.827	130	5.118
22	0.866	140	5.512
23	0.906	150	5.906
24	0.945	160	6.299
25	0.984	180	7.087
26	1.024	200	7.874
28	1.102	220	8.661
30	1.181	250	9.843
32	1.260	280	11.024
34	1.339	300	11.811
35	1.378	320	12.598
36	1.417		
38	1.496		
40	1.575		
42	1.654		
45	1.772		
48	1.890		
50	1.969		
52	2.047		
54	2.126		
55	2.165		
56	2.205		

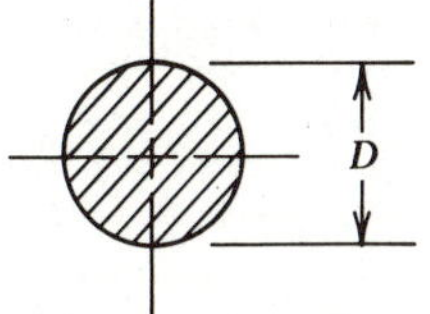

Hot-Rolled Round Steel Bars (ISO 1035/I)

Nominal Size = D		*Nominal Size = D*		*Nominal Size = D*	
mm	*in.*	*mm*	*in.*	*mm*	*in.*
5	0.197	26.5	1.043	66	2.598
5.5	0.217	27	1.063	67	2.638
6	0.236	27.5	1.083	68	2.677
6.5	0.256	28	1.102	70	2.756
7	0.276	28.5	1.122	73	2.874
7.5	0.295	29	1.142	75	2.953
8	0.315	29.5	1.161	78	3.071
8.5	0.335	30	1.181	80	3.150
9	0.354	31	1.220	83	3.268
9.5	0.374	32	1.260	85	3.346
10	0.394	33	1.299	88	3.465
10.5	0.413	34	1.339	90	3.543
11	0.433	35	1.378	95	3.740
11.5	0.453	36	1.417	100	3.937
12	0.472	37	1.457	105	4.134
12.5	0.492	38	1.496	110	4.331
13	0.512	39	1.535	115	4.528
13.5	0.531	40	1.575	120	4.724
14	0.551	41	1.614	125	4.921
14.5	0.571	42	1.654	130	5.118
15	0.591	43	1.693	135	5.315
15.5	0.610	44	1.732	140	5.512
16	0.630	45	1.772	145	5.709
16.5	0.650	46	1.811	150	5.906
17	0.669	47	1.850	155	6.102
17.5	0.689	48	1.890	160	6.299
18	0.709	49	1.929	165	6.496
18.5	0.728	50	1.969	170	6.693
19	0.748	51	2.008	175	6.890
19.5	0.768	52	2.047	180	7.087
20	0.787	53	2.087	185	7.283
20.5	0.807	54	2.126	190	7.480

(Continued)

Hot-Rolled Round Steel Bars (ISO 1035/I) (Continued)

Nominal Size = D *mm*	*in.*	*Nominal Size = D* *mm*	*in.*	*Nominal Size = D* *mm*	*in.*
21	0.827	55	2.165	195	7.677
21.5	0.846	56	2.205	200	7.874
22	0.866	57	2.244	210	8.268
22.5	0.886	58	2.283	220	8.661
23	0.906	59	2.323	230	9.055
23.5	0.925	60	2.362	240	9.449
24	0.945	61	2.402	250	9.843
24.5	0.965	62	2.441	260	10.236
25	0.984	63	2.480	280	11.024
25.5	1.004	64	2.520	300	11.811
26	1.024	65	2.559		

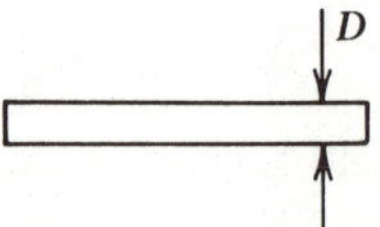

Cold-Rolled Flat Stock Steel Sheet and Strip Thicknesses (ANSI B32.1)

Nominal Size = D *mm*	*in.*	*Nominal Size = D* *mm*	*in.*	*Nominal Size = D* *mm*	*in.*
0.1	0.0039	2	0.0787	22	0.866
0.12	0.0047	2.2	0.0866	25	0.984
0.14	0.0055	2.5	0.0984	28	1.102
0.16	0.0063	2.8	0.1102	30	1.181
0.18	0.0071	3	0.1181	32	1.260
0.2	0.0079	3.2	0.1260	35	1.378
0.22	0.0087	3.5	0.1378	36	1.417
0.25	0.0098	3.8	0.1496	38	1.496
0.28	0.0110	4	0.1575	40	1.575
0.3	0.0118	4.5	0.177	45	1.772
0.35	0.0138	4.8	0.189	50	1.969
0.4	0.0157	5	0.197	55	2.165
0.45	0.0177	5.5	0.217	60	2.362
0.5	0.0197	6	0.236	70	2.756
0.55	0.0217	7	0.276	80	3.150
0.6	0.0236	8	0.315	90	3.543

Nominal Size = D mm	in.	Nominal Size = D mm	in.	Nominal Size = D mm	in.
0.65	0.0256	9	0.354	100	3.937
0.7	0.0276	10	0.394	110	4.331
0.8	0.0315	11	0.433	120	4.724
0.9	0.0354	12	0.472	130	5.118
1	0.0394	14	0.551	140	5.512
1.1	0.0433	15	0.591	150	5.906
1.2	0.0472	16	0.630	160	6.299
1.4	0.0551	18	0.709		
1.5	0.0591	20	0.787		
1.6	0.0630				
1.8	0.709				

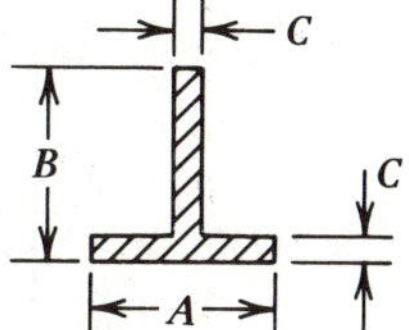

Hot-Rolled T-Steel With Sharp Corners (DIN 59051)

Nominal Size Millimeters A × B	C	Inches A × B	C
20 × 20	3.0	0.787 × 0.787	0.12
	4.0		0.16
25 × 25	3.5	0.984 × 0.984	0.14
	4.5		0.18
30 × 30	4.0	1.181 × 1.181	0.16
	5.0		0.20
35 × 35	4.5	1.378 × 1.378	0.18
	5.5		0.22
40 × 40	5.0	1.575 × 1.575	0.20
	6.0		0.24
45 × 45	6.5	1.772 × 1.772	0.26
50 × 50	7.0	1.969 × 1.969	0.28
60 × 60	8.0	2.362 × 2.362	0.31
70 × 70	9.0	2.756 × 2.756	0.35
80 × 80	10.0	3.150 × 3.150	0.39
100 × 100	11.0	3.937 × 3.937	0.43

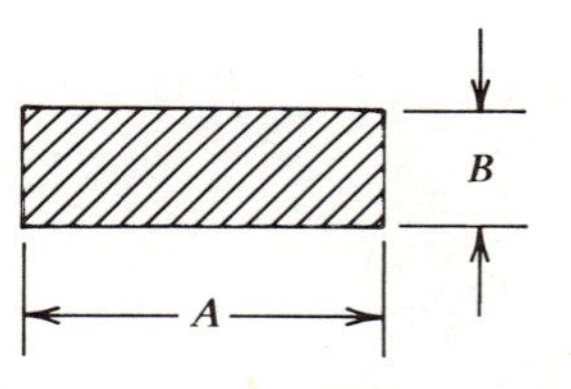

Hot-Rolled Rectangular Steel Bars (ISO 1035/III)

Nominal Size = A × B mm	in.
10 × 3	0.394 × 0.118
4	× 0.157
5	× 0.197
6	× 0.236
7	× 0.276
12 × 3	0.472 × 0.118
4	× 0.157
5	× 0.197
6	× 0.236
7	× 0.276
8	× 0.315
14 × 3	0.551 × 0.118
4	× 0.157
5	× 0.197
6	× 0.236
7	× 0.276
8	× 0.315
16 × 3	0.630 × 0.118
4	× 0.157
5	× 0.197
6	× 0.236
7	× 0.276
8	× 0.315
10	× 0.394

Nominal Size = A × B mm	in.
22 × 7	0.866 × 0.276
8	× 0.315
10	× 0.394
12	× 0.472
15	× 0.591
18	× 0.709
25 × 3	0.984 × 0.118
4	× 0.157
5	× 0.197
6	× 0.236
7	× 0.276
8	× 0.315
10	× 0.394
12	× 0.472
14	× 0.551
15	× 0.591
16	× 0.630
18	× 0.709
28 × 3	1.102 × 0.118
4	× 0.157
5	× 0.197
6	× 0.236
7	× 0.276
8	× 0.315
10	× 0.394
12	× 0.472

Nominal Size = A × B mm	in.
32 × 10	1.260 × 0.394
12	× 0.472
15	× 0.591
16	× 0.630
18	× 0.709
20	× 0.787
25	× 0.984
35 × 3	1.378 × 0.118
4	× 0.157
5	× 0.197
6	× 0.236
7	× 0.276
8	× 0.315
10	× 0.394
12	× 0.472
14	× 0.551
15	× 0.591
16	× 0.630
18	× 0.709
20	× 0.787
25	× 0.984
30	× 1.181
32	× 1.260
40 × 3	1.575 × 0.118
4	× 0.157
5	× 0.197

Nominal Size = A × B mm	in.
45 × 18	1.772 × 0.709
20	× 0.787
25	× 0.984
30	× 1.181
32	× 1.260
50 × 3	1.969 × 0.118
4	× 0.157
5	× 0.197
6	× 0.236
7	× 0.276
8	× 0.315
10	× 0.394
12	× 0.472
14	× 0.551
15	× 0.591
16	× 0.630
18	× 0.709
20	× 0.787
22	× 0.866
25	× 0.984
30	× 1.181
32	× 1.260
35	× 1.378
40	× 1.575
55 × 4	2.165 × 0.157
5	× 0.197

Nominal Size = A × B mm	in.
60 × 25	2.362 × 0.984
30	× 1.181
32	× 1.260
35	× 1.378
40	× 1.575
45	× 1.772
65 × 4	2.559 × 0.157
5	× 0.197
6	× 0.236
7	× 0.276
8	× 0.315
10	× 0.394
12	× 0.472
15	× 0.591
16	× 0.630
18	× 0.709
20	× 0.787
25	× 0.984
30	× 1.181
32	× 1.260
35	× 1.378
40	× 1.575
45	× 1.772
70 × 4	2.756 × 0.157
5	× 0.197
6	× 0.236

Nominal Size = A × B mm	in.
75 × 16	2.953 × 0.630
18	× 0.709
20	× 0.787
25	× 0.984
30	× 1.181
32	× 1.260
35	× 1.378
40	× 1.575
45	× 1.772
50	× 1.969
80 × 4	3.150 × 0.157
5	× 0.197
6	× 0.236
7	× 0.276
8	× 0.315
10	× 0.394
12	× 0.472
14	× 0.551
15	× 0.591
16	× 0.630
18	× 0.709
20	× 0.787
25	× 0.984
30	× 1.181
32	× 1.260
35	× 1.378
40	× 1.575

18 × 3	0.709 × 0.118	15	× 0.591	6	× 0.236	6	× 0.236	7	× 0.276	45	× 1.772
4	× 0.157	18	× 0.709	7	× 0.276	7	× 0.276	8	× 0.315	50	× 1.969
5	× 0.197			8	× 0.315	8	× 0.315	10	× 0.394	60	× 2.362
6	× 0.236	30 × 3	1.181 × 0.118	10	× 0.394	10	× 0.394	12	× 0.472		
7	× 0.276	4	× 0.157	12	× 0.472	12	× 0.472	15	× 0.591	90 × 5	3.543 × 0.197
8	× 0.315	5	× 0.197	14	× 0.551	15	× 0.591	16	× 0.630	6	× 0.236
10	× 0.394	6	× 0.236	15	× 0.591	18	× 0.709	18	× 0.709	7	× 0.276
		7	× 0.276	16	× 0.630	20	× 0.787	20	× 0.787	8	× 0.315
20 × 3	0.787 × 0.118	8	× 0.315	18	× 0.709	25	× 0.984	25	× 0.984	10	× 0.394
4	× 0.157	10	× 0.394	20	× 0.787	30	× 1.181	30	× 1.181	12	× 0.472
5	× 0.197	12	× 0.472	25	× 0.984	32	× 1.260	32	× 1.260	14	× 0.551
6	× 0.236	14	× 0.551	30	× 1.181			35	× 1.378	15	× 0.591
7	× 0.276	15	× 0.591	32	× 1.260	60 × 4	2.362 × 0.157	40	× 1.575	16	× 0.630
8	× 0.315	16	× 0.630			5	× 0.197	45	× 1.772	18	× 0.709
10	× 0.394	18	× 0.709	45 × 4	1.772 × 0.157	6	× 0.236	50	× 1.969	20	× 0.787
12	× 0.472	20	× 0.787	5	× 0.197	7	× 0.276			25	× 0.984
15	× 0.591	25	× 0.984	6	× 0.236	8	× 0.315	75 × 4	2.953 × 0.157	30	× 1.181
16	× 0.630			7	× 0.276	10	× 0.394	5	× 0.197	32	× 1.260
18	× 0.709	32 × 3	1.260 × 0.118	8	× 0.315	12	× 0.472	6	× 0.236	35	× 1.378
		4	× 0.157	10	× 0.394	14	× 0.551	7	× 0.276	40	× 1.575
22 × 3	0.866 × 0.118	5	× 0.197	12	× 0.472	15	× 0.591	8	× 0.315	45	× 1.772
4	× 0.157	6	× 0.236	14	× 0.551	16	× 0.630	10	× 0.394	50	× 1.969
5	× 0.197	7	× 0.276	15	× 0.591	18	× 0.709	12	× 0.472	60	× 2.362
6	× 0.236	8	× 0.315	16	× 0.630	20	× 0.787	15	× 0.591		

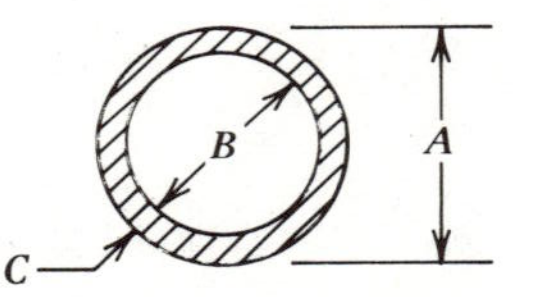

Seamless and Welded Steel Tubes (ISO 134, 221, 336, 2937)

Millimeters			Inches			Millimeters			Inches			Millimeters			Inches		
A	*C*	*B*	*A*	*C*	*B*	*A*	*C*	*B*	*A*	*C*	*B*	*A*	*C*	*B*	*A*	*C*	*B*
10.2	0.5	9.2	0.402	0.020	0.363	26.9	0.5	25.9	1.059	0.020	1.020	42.4	4.5	33.4	1.669	0.177	1.315
	1.6	7.0		0.063	0.276		1.8	23.3		0.071	0.917		5.0	32.4		0.197	1.275
	2.0	6.2		0.079	0.245		2.0	22.9		0.079	0.902		6.3	29.8		0.248	1.173
	2.3	5.6		0.091	0.221		2.3	22.3		0.091	0.878		10.0	22.4		0.394	0.882
	2.6	5.0		0.102	0.197		2.6	21.7		0.102	0.854		11.0	20.4		0.433	0.803
							2.9	21.1		0.114	0.831						
12.0	0.5	11.0	0.472	0.020	0.433		3.2	20.5		0.126	0.807	44.5	0.5	43.5	1.750	0.020	1.711
	1.6	8.8		0.063	0.346		3.6	19.7		0.142	0.776		2.0	40.5		0.079	1.593
							4.0	18.9		0.157	0.744		2.3	39.9		0.091	1.569
13.5	0.5	12.5	0.531	0.020	0.492		5.6	15.7		0.220	0.618		2.6	39.3		0.102	1.545
	1.6	10.3		0.063	0.405		8.0	10.9		0.315	0.429		12.5	19.5		0.492	0.766
	1.8	9.9		0.071	0.389												
	2.0	9.5		0.079	0.374	30.0	0.5	29.0	1.181	0.020	1.142	48.3	0.5	47.3	1.900	0.020	1.861
	2.3	8.9		0.091	0.350		1.6	26.8		0.063	1.055		2.3	43.7		0.091	1.719
	2.9	7.7		0.114	0.303		2.0	26.0		0.079	1.024		2.6	43.1		0.102	1.695
							2.3	25.4		0.091	1.000		2.9	42.5		0.114	1.672
16.0	0.5	15.0	0.630	0.020	0.591		2.6	24.8		0.102	0.976		3.2	41.9		0.126	1.648
	1.8	12.4		0.071	0.488		8.0	14.0		0.315	0.551		3.6	41.1		0.142	1.617
													4.0	40.3		0.157	1.585
17.2	0.5	16.2	0.677	0.020	0.638	31.8	0.5	30.8	1.250	0.020	1.211		4.5	39.3		0.177	1.546
	1.6	14.0		0.063	0.551		2.0	27.8		0.079	1.093		5.0	38.3		0.197	1.506
	1.8	13.6		0.071	0.535		2.6	26.6		0.102	1.045		5.6	37.1		0.220	1.459

	2.0	13.2		0.079	0.520
	2.3	12.6		0.091	0.496
	2.9	11.4		0.114	0.449
	3.2	10.8		0.126	0.425
19.0	0.5	18.0	0.750	0.020	0.711
20.0	0.5	19.0	0.787	0.020	0.748
	1.6	16.8		0.063	0.661
	2.0	16.0		0.079	0.630
21.3	0.5	20.3	0.840	0.020	0.801
	1.8	17.7		0.071	0.698
	2.0	17.3		0.079	0.683
	2.3	16.7		0.091	0.659
	2.6	16.1		0.102	0.635
	2.9	15.5		0.114	0.612
	3.2	14.9		0.126	0.588
	3.6	14.1		0.142	0.557
	5.0	11.3		0.197	0.446
	7.1	7.1		0.280	0.281
25.0	0.5	24.0	0.984	0.020	0.945
	1.6	21.8		0.063	0.858
	2.0	21.0		0.079	0.827
	6.3	12.4		0.248	0.488

	8.0	15.8		0.315	0.620
33.7	0.5	32.7	1.327	0.020	1.288
	2.0	29.7		0.079	1.170
	2.3	29.1		0.091	1.146
	2.6	28.5		0.102	1.122
	2.9	27.9		0.114	1.099
	3.2	27.3		0.126	1.075
	3.6	26.5		0.142	1.044
	4.0	25.7		0.157	1.012
	4.5	24.7		0.177	0.973
	5.0	23.7		0.197	0.933
	6.3	21.1		0.248	0.831
	8.8	16.1		0.346	0.634
38.0	0.5	37.0	1.500	0.020	1.461
	2.0	34.0		0.079	1.343
	2.3	33.4		0.091	1.319
	2.6	32.8		0.102	1.295
	10.0	18.0		0.394	0.713
42.4	0.5	41.4	1.669	0.020	1.630
	2.0	38.4		0.079	1.512
	2.3	37.8		0.091	1.488
	2.6	37.2		0.102	1.464
	2.9	36.6		0.114	1.441
	3.2	36.0		0.126	1.417
	3.6	35.2		0.142	1.386
	4.0	34.4		0.157	1.354

	7.1	34.1		0.280	1.341
	10.0	28.3		0.394	1.113
	12.5	23.3		0.492	0.916
51.0	0.5	50.0	2.000	0.020	1.961
	2.3	46.4		0.091	1.819
	2.6	45.8		0.102	1.795
	14.2	22.6		0.559	0.882
54.0	0.5	53.0	2.125	0.020	2.086
	1.6	50.8		0.063	1.999
	2.0	50.0		0.079	1.968
	2.3	49.4		0.091	1.944
	2.6	48.8		0.102	1.920
	2.9	48.2		0.114	1.897
	14.2	25.6		0.559	1.007
57.0	0.5	56.0	2.250	0.020	2.211
	1.6	53.8		0.063	2.124
	2.0	53.0		0.079	2.093
	2.3	52.4		0.091	2.069
	2.6	51.8		0.102	2.045
	2.9	51.2		0.114	2.022
	16.0	25.0		0.630	0.990
60.3	0.5	59.3	2.375	0.020	2.336
	2.3	55.7		0.091	2.194
	2.9	54.4		0.114	2.147
	3.2	53.9		0.126	2.123

(Continued)

Seamless and Welded Steel Tubes (ISO 134, 221, 336, 2937) *(Continued)*

Millimeters			*Inches*			*Millimeters*			*Inches*			*Millimeters*			*Inches*		
A	*C*	*B*	*A*	*C*	*B*	*A*	*C*	*B*	*A*	*C*	*B*	*A*	*C*	*B*	*A*	*C*	*B*
82.5	0.5	81.5	3.250	0.020	3.211	60.3	3.6	53.1	2.375	0.142	2.092	114.3	4.5	105.3	4.500	0.177	4.146
	2.6	77.3		0.102	3.045		4.0	52.3		0.157	2.060		5.0	104.3		0.197	4.106
	3.2	76.1		0.126	2.998		4.5	51.3		0.177	2.021		5.4	103.5		0.213	4.075
	22.2	38.1		0.874	1.502		5.0	50.3		0.197	1.981		5.6	103.1		0.220	4.059
							5.6	49.1		0.220	1.934		5.9	102.5		0.232	4.035
88.9	0.5	87.9	3.500	0.020	3.461		6.3	47.7		0.248	1.879		6.3	101.7		0.248	4.004
	2.0	84.9		0.079	3.343		7.1	46.1		0.280	1.816		7.1	100.1		0.280	3.941
	2.9	83.1		0.114	3.272		8.8	42.7		0.346	1.682		8.0	98.3		0.315	3.870
	3.2	82.5		0.126	3.248		11.0	38.3		0.433	1.509		8.8	96.7		0.346	3.807
	3.6	81.7		0.142	3.217		16.0	28.3		0.630	1.115		11.0	92.3		0.433	3.634
	4.0	80.9		0.157	3.185								14.2	85.9		0.559	3.382
	4.5	79.9		0.177	3.146	63.5	0.5	62.5	2.500	0.020	2.461		17.5	79.3		0.689	3.122
	5.0	78.9		0.197	3.106		2.0	59.5		0.079	2.343		32.0	50.3		1.260	1.980
	5.4	78.1		0.213	3.075		2.3	58.9		0.091	2.319						
	5.6	77.7		0.220	3.059		2.6	58.3		0.102	2.295	127.0	0.6	125.8	5.000	0.024	4.953
	6.3	76.3		0.248	3.004		2.9	57.7		0.114	2.272		3.2	120.6		0.126	4.748
	7.1	74.7		0.280	2.941		16.0	31.5		0.630	1.240		4.0	119.0		0.157	4.685
	8.0	72.9		0.315	2.870								36.0	55.0		1.417	2.165
	11.0	66.9		0.433	2.634	70.0	0.5	69.0	2.750	0.020	2.711						
	16.0	56.9		0.630	2.240		2.0	66.0		0.079	2.593	133.0	0.6	131.8	5.250	0.024	5.203
	25.0	38.9		0.984	1.531		2.6	64.8		0.102	2.545		3.6	125.8		0.142	4.967
							2.9	64.2		0.114	2.522		4.0	125.0		0.157	4.935
101.6	0.5	100.6	4.000	0.020	3.961		3.2	63.6		0.126	2.498		36.0	61.0		1.417	2.415
	2.0	97.6		0.079	3.843		3.6	62.8		0.142	2.467						
	2.9	95.8		0.114	3.771		17.5	35.0		0.689	1.372	139.7	0.6	138.5	5.500	0.024	5.453

	3.2	95.2		0.126	3.748								3.6	132.5		0.142	5.217
	3.6	94.4		0.142	3.717	73.0	0.5	72.0	2.875	0.020	2.836		4.0	131.7		0.157	5.185
	4.0	93.6		0.157	3.685		2.0	69.0		0.079	2.718		4.5	130.7		0.177	5.146
	4.5	92.6		0.177	3.646		2.9	67.2		0.114	2.647		5.0	129.7		0.197	5.106
	5.0	91.6		0.197	3.606		3.2	66.6		0.126	2.623		5.4	128.9		0.213	5.075
	5.6	90.4		0.220	3.559		3.6	65.8		0.142	2.592		5.6	128.5		0.220	5.059
	6.3	89.0		0.248	3.504		4.0	65.0		0.157	2.560		6.3	127.1		0.248	5.004
	7.1	87.4		0.280	3.441		4.5	64.0		0.177	2.521		8.0	123.7		0.315	4.870
	8.0	85.6		0.315	3.370		5.0	63.0		0.197	2.481		10.00	119.7		0.394	4.713
	28.0	45.6		1.102	1.795		5.6	61.8		0.220	2.434		40.0	59.7		1.575	2.350
							6.3	60.4		0.248	2.379						
108.0	0.5	107.0	4.250	0.020	4.211		7.1	58.8		0.280	2.316	141.3	0.6	140.1	5.563	0.024	5.516
	2.0	104.0		0.079	4.093		10.0	53.0		0.394	2.088		2.0	137.3		0.079	5.406
	2.6	102.8		0.102	4.045		14.2	44.6		0.559	1.757		3.2	134.9		0.126	5.311
	2.9	102.2		0.114	4.022		20.0	33.0		0.787	1.300		4.0	133.3		0.157	5.248
	3.2	101.6		0.126	3.998								4.5	132.3		0.177	5.209
	3.6	100.8		0.142	3.967	76.1	0.5	75.1	3.000	0.020	2.961		5.0	131.3		0.197	5.169
	30.0	48.0		1.181	1.888		2.6	70.9		0.102	2.795		5.6	130.1		0.220	5.122
							2.9	70.3		0.114	2.772		6.3	128.7		0.248	5.067
114.3	0.5	113.3	4.500	0.020	4.461		3.2	69.7		0.126	2.748		7.1	127.1		0.280	5.004
	2.0	110.3		0.079	4.343		3.6	68.9		0.142	2.717		8.8	123.7		0.346	4.870
	2.6	109.1		0.102	4.295		4.5	67.1		0.177	2.646		10.0	121.3		0.394	4.776
	2.9	108.5		0.114	4.272		5.4	65.3		0.213	2.575		12.5	116.3		0.492	4.579
	3.2	107.9		0.126	4.248		6.3	63.5		0.248	2.504		16.0	109.3		0.630	4.303
	3.6	107.1		0.142	4.217		7.1	61.9		0.280	2.441		20.0	101.3		0.787	3.988
	4.0	106.3		0.157	4.185		20.0	36.1		0.787	1.425		40.0	61.3		1.575	2.413

Clevis Pins (ISO 2340)

Type A—
Without Holes

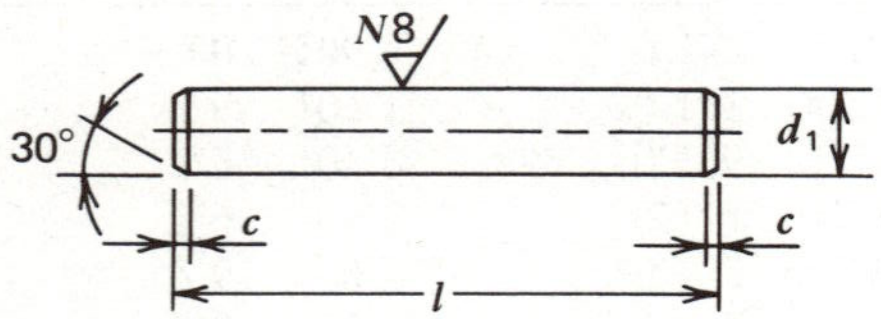

Type B—
With Holes

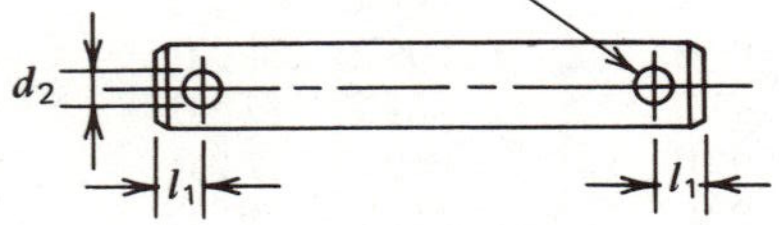

30° N8 d_1 c l

Break edges of holes d_2 l_1

d_1		3	4	5	6	8	10	12	14	16	18	20	22	24	27	30	33	36	40	45	50
d_2		0.8	1	1.2	1.6	2	3.2	3.2	4	4	5	5	5	6.3	6.3	8	8	8	8	10	10
c	max.	1	1	2	2	2	2	3	3	3	3	4	4	4	4	4	4	4	4	4	4
l_1	min.	1.6	2.2	2.9	3.2	3.5	4.5	5.5	6	6	7	8	8	9	9	10	10	10	10	12	12
l	from	6	8	10	12	16	20	25	30	35	40	40	45	50	55	60	65	70	80	90	100
	to	30	40	50	60	80	100	120	140	160	180	200	200	200	200	200	200	200	200	200	200

Notes: All dimensions are in mm.

Slotted Spring Pin Dimensions (IFI-512-S)

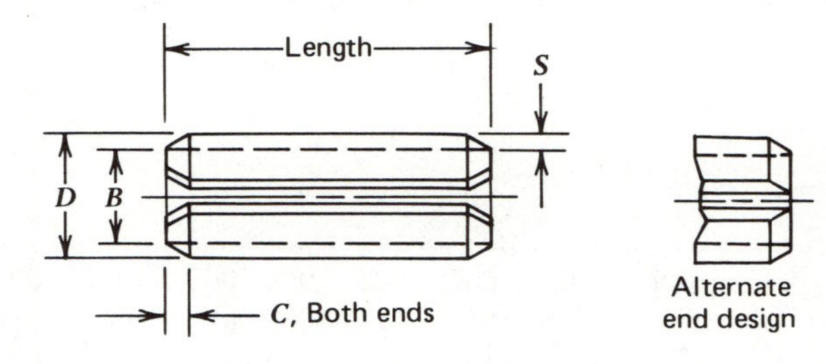

Nominal Pin Size	D Diameter Max.	D Diameter Min.	B Chamfer Diameter Max.	C Chamfer Length Max.	C Chamfer Length Min.	S Stock Thickness Nom.	Recommended Hole Size Max.	Recommended Hole Size Min.
1.5	1.68	1.60	1.4	0.7	0.15	0.3	1.57	1.50
2	2.20	2.12	1.9	0.8	0.2	0.4	2.07	2.00
2.5	2.72	2.63	2.4	0.9	0.2	0.5	2.58	2.50
3	3.25	3.15	2.9	1.0	0.2	0.6	3.10	3.00
4	4.28	4.15	3.9	1.2	0.3	0.8	4.10	4.00
5	5.33	5.17	4.8	1.4	0.3	1.0	5.12	5.00
6	6.36	6.20	5.8	1.6	0.4	1.2	6.13	6.00
8	8.40	8.22	7.8	2.0	0.4	1.6	8.15	8.00
10	10.43	10.25	9.7	2.4	0.5	2.0	10.15	10.00
12	12.48	12.28	11.7	2.8	0.6	2.5	12.18	12.00

Notes: All dimensions are in mm.

Clevis Pins With Head (ISO 2341)

Type A—
Without Holes

Type B—
With Holes

d_1		3	4	5	6	8	10	12	14	16	18	20	22	24	27	30	33	36	40	45	50
D		5	6	8	10	14	18	20	22	25	28	30	33	36	40	44	47	50	55	60	66
d_2		0.8	1	1.2	1.6	2	3.2	3.2	4	4	5	5	5	6.3	6.3	8	8	8	8	10	10
c	max.	1	1	2	2	2	2	3	3	3	3	4	4	4	4	4	4	4	4	4	4
e	approx.	0.5	0.5	1	1	1	1	1.6	1.6	1.6	1.6	2	2	2	2	2	2	2	2	2	2
k		1	1	1.6	2	3	4	4	4	4.5	5	5	5.5	6	6	8	8	8	8	9	9
l_1	min.	1.6	2.2	2.9	3.2	3.5	4.5	5.5	6	6	7	8	8	9	9	10	10	10	10	12	12
R		0.6	0.6	0.6	0.6	0.6	0.6	0.6	0.6	0.6	1	1	1	1	1	1	1	1	1	1	1
l		6	8	10	12	16	20	25	30	35	40	40	45	50	55	60	65	70	80	90	100
		30	40	50	60	80	100	120	140	160	180	200	200	200	200	200	200	200	200	200	200

Notes: All dimensions are in millimeters.

American National Standard Finished Hexagon Castle Nuts

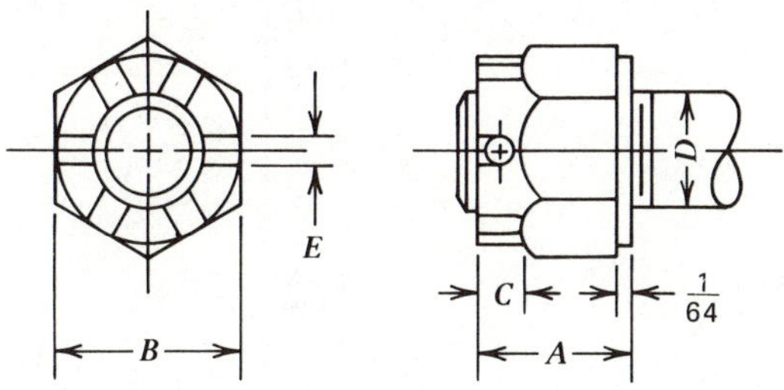

Nominal Size— Diameter D		*Threads per Inch* UNC	UNF	*Thickness* A	*Width Across Flats* B	*Slot* *Depth* C	*Width* E	*Diameter of Cylindrical Part (min.)*
0.250	1/4	20	28	9/32	7/16	0.094	0.078	0.371
0.3125	5/16	18	24	21/64	1/2	0.094	0.094	0.425
0.375	3/8	16	24	13/32	9/16	0.125	0.125	0.478
0.4375	7/16	14	20	29/64	11/16	0.156	0.125	0.582
0.500	1/2	13	20	9/16	3/4	0.156	0.156	0.637
0.5625	9/16	12	18	39/64	7/8	0.188	0.156	0.744
0.625	5/8	11	18	23/32	15/16	0.219	0.188	0.797
0.750	3/4	10	16	13/16	1 1/8	0.250	0.188	0.941
0.875	7/8	9	14	29/32	1 5/16	0.250	0.188	1.097
1.000	1	8	12	1	1 1/2	0.281	0.250	1.254
1.125	1 1/8	7	12	1 5/32	1 11/16	0.344	0.250	1.411

ANSI B18.2.2-1965 Dimensions are in inches

Thread may be coarse- or fine-thread series, class 2B tolerance; unless otherwise specified, fine-thread series will be furnished.

Hexagon Bolts (IFI-506)

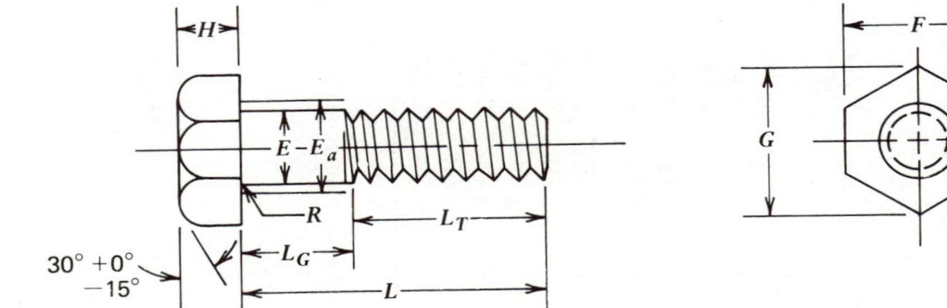

Nominal Bolt Size and Thread Pitch	E		F		G		H		E_a	R	L_T (Ref)		
	Body Diameter		*Width Across Flats*		*Width Across Corners*		*Fillet Head Height*		*Radius Transition Diameter*	*Radius of Fillet*	*Thread Length (Basic)*		
	Max.	*Min.*	*Max.*	*Min.*	*Max.*	*Min.*	*Max.*	*Min.*	*Max.*	*Min.*	*Bolt Lengths ≤125*	*Bolt Lengths >125 &, ≤200*	*Bolt Lengths >200*
M5 × 0.8	5.48	4.52	8.00	7.75	9.24	8.84	3.88	3.35	5.8	0.2	16	22	35
M6 × 1	6.48	5.52	10.00	9.69	11.55	11.05	4.70	4.13	7.0	0.3	18	24	37
M8 × 1.25	8.58	7.42	13.00	12.60	15.01	14.36	5.73	5.10	9.2	0.4	22	28	41
M10 × 1.5	10.58	9.42	15.00	14.50	17.32	16.53	6.86	6.17	11.2	0.4	26	32	45

M12 × 1.75	12.70	11.30	18.00	17.40	20.78	19.84	7.99	7.24	13.2	0.4	30	36	49
M14 × 2	14.70	13.30	21.00	20.30	24.25	23.14	9.32	8.51	15.2	0.6	—	40	53
M16 × 2	16.70	15.30	24.00	23.20	27.71	26.45	10.56	9.68	17.8	0.6	—	44	57
M20 × 2.5	20.84	19.16	30.00	29.00	34.64	33.06	13.12	12.12	22.4	0.8	—	52	65
M24 × 3	24.84	23.16	36.00	34.80	41.57	39.67	15.68	14.56	26.4	0.8	—	60	73
M30 × 3.5	30.84	29.16	46.00	44.50	53.12	50.73	19.48	17.92	33.6	1.2	66	72	85
M36 × 4	37.00	35.00	55.00	53.20	63.51	60.65	23.38	21.72	39.6	1.2	78	84	97
M42 × 4.5	43.00	41.00	65.00	62.90	75.06	71.71	26.97	25.03	45.6	1.2	90	96	109
M48 × 5	49.00	47.00	75.00	72.60	86.60	82.76	31.07	28.93	52.6	1.5	102	108	121
M56 × 5.5	57.20	54.80	85.00	82.20	98.15	93.71	36.20	33.80	62.0	2.0	—	124	137
M64 × 6	65.52	62.80	95.00	91.80	109.70	104.65	41.32	38.68	70.0	2.0	—	140	153
M72 × 6	73.84	70.80	105.00	101.40	121.24	115.60	46.45	43.55	78.0	2.0	—	156	169
M80 × 6	82.16	78.80	115.00	111.00	132.79	126.54	51.58	48.42	86.0	2.0	—	172	185
M90 × 6	92.48	88.60	130.00	125.50	150.11	143.07	57.74	54.26	96.0	2.0	—	192	205
M100 × 6	102.80	98.60	145.00	140.00	167.43	159.60	63.90	60.10	107.0	2.5	—	212	225

Notes: All dimensions are in mm.

Recommended Torque Values for Metric Fasteners

Nominal Size	Stress Area (mm^2)	Class 4.6 400 MPa Minimum Tensile Strength (N·m)	(lb·ft)	Class 8.8 830 MPa Minimum Tensile Strength (N·m)	(lb·ft)	Class 9.8 900 MPa Minimum Tensile Strength (N·m)	(lb·ft)	Class 10.9 1040 MPa Minimum Tensile Strength (N·m)	(lb·ft)	Class 12.9 1220 MPA Minimum Tensile Strength (N·m)	(lb·ft)
M 2.0	2.1	0.14	0.10	0.37	0.27	0.40	0.30	0.52	0.38	0.61	0.45
M 2.5	3.4	0.28	0.21	0.76	0.56	0.82	0.61	1.06	0.78	1.24	0.92
M 3.0	5.0	0.51	0.37	1.35	1.00	1.47	1.08	1.88	1.39	2.21	1.63
M 3.5	6.8	0.80	0.59	2.13	1.57	2.31	1.70	2.96	2.18	3.47	2.56
M 4.0	8.8	1.18	0.87	3.15	2.32	3.41	2.52	4.38	3.23	5.14	3.79
M 5.0	14.2	2.39	1.76	6.36	4.69	6.90	5.09	8.86	6.54	10.39	7.67
M 6.0	20.1	4.05	2.99	10.81	7.97	11.72	8.65	15.05	11.10	17.66	13.02
M 6.3	22.6	4.78	3.53	12.76	9.41	13.84	10.21	17.77	13.11	20.84	15.37
M 8.0	36.6	9.84	7.26	26.25	19.36	28.46	20.99	36.54	29.95	42.87	31.62
M10.0	58.0	19.49	14.37	51.99	38.35	56.38	41.58	72.38	53.39	84.91	62.63
M12.0	84.3	33.99	25.07	90.68	66.88	98.33	72.52	126.25	93.12	148.10	109.23
M14.0	115.0	54.10	39.90	144.32	106.45	156.49	115.42	200.93	148.20	235.70	173.85
M16.0	157.0	84.40	62.25	225.18	166.08	244.17	180.09	313.50	231.22	367.76	271.24
M20.0	245.0	164.64	121.43	439.24	323.96	476.28	351.29	611.52	451.03	717.36	529.10
M24.0	353.0	284.66	209.95	759.43	560.13	823.48	607.37	1057.31	779.83	1240.30	914.80
M30.0	561.0	565.49	417.08	1508.64	1112.72	1635.88	1206.56	2100.38	1549.16	2463.91	1817.29
M36.0	817.0	988.24	728.89	2636.49	1944.58	2858.85	2108.58	3670.62	2707.31	4305.92	3175.88
M42.0	1120.0	1580.54	1165.75	4216.66	3110.05	4572.29	3372.35	5870.59	4329.93	6886.66	5079.34

Notes:

1. 1 lb (force) × ft = 1.355818 N · m (Newtons × meters).
2. The minimum recommended torque values shown are valid for zinc-coated fasteners assembled in rigid joints to 75% of proof loads. For maximum or 100% proof torques, multiply table values by 1.33. Reduce torque values for fasteners with less friction or nonrigid (gasket) joints). See Chapt. 6.

American National Standard Cotter Pins*

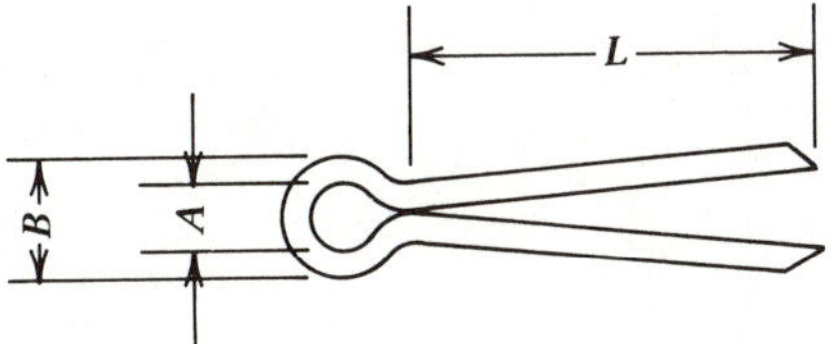

Pin Diameter				*Eye Diameter*		*Drill Size Recommended (hole/diameter)*	*Clevis Pin or Shaft Diameter*
Nominal		*Max*	*Min*	*Inside, A*	*Outside, B*		
0.031	1/32	0.032	0.028	1/32	1/16	−0.0469	1/8
0.047	3/64	0.048	0.044	3/64	3/32	−0.0625	3/16
0.062	1/16	0.060	0.056	1/16	1/8	−0.0781	1/4
0.078	5/64	0.076	0.072	5/64	5/32	−0.0938	5/16
0.094	3/32	0.090	0.086	3/32	3/16	−0.1094	3/8
0.125	1/8	0.120	0.116	1/8	1/4	−0.1406	1/2
0.156	5/32	0.150	0.146	5/32	5/16	−0.1719	5/8
0.188	3/16	0.176	0.172	3/16	3/8	−0.2031	
0.219	7/32	0.207	0.202	7/32	7/16	−0.2344	
0.250	1/4	0.225	0.220	1/4	1/2	−0.2656	

*ANSI B5.20 – 1958.

For shafts up to 5/8 in. diameter select a cotter pin that is approximately equal to one-fourth of the shaft diameter. For larger sizes, use a cotter pin that is from one-fourth to one-sixth of the shaft diameter. All dimensions are in inches.

Split Cotter Pins (ISO 1234)

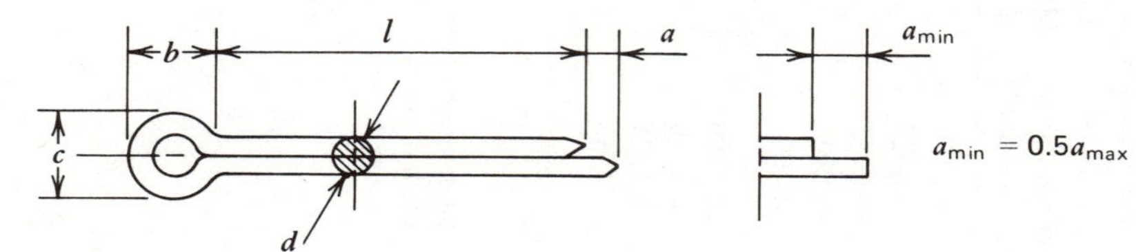

Nominal Size		0.6	0.8	1	1.2	1.6	2	2.5	3.2	4	5	6.3	8	10
d	max.	0.5	0.7	0.9	1	1.4	1.8	2.3	2.9	3.7	4.6	5.9	7.5	9.5
	min.	0.4	0.6	0.8	0.9	1.3	1.7	2.1	2.7	3.5	4.4	5.7	7.3	9.3
a	max.	1.6	1.6	1.6	2.5	2.5	2.5	1.5	3.2	4	4	4	4	6.3
b	≈	2	2.4	3	3	3.2	4	4	6.4	8	10	12.6	15	20
c	max.	1	1.4	1.8	2	2.8	3.6	4.6	5.8	7.4	9.2	11.8	15	19
	min.	0.9	1.2	1.6	1.7	2.4	3.2	4	5.1	6.5	8	10.3	13.1	16.6
Bolts	over	—	2.5	3.5	4.5	5.3	7	9	11	14	20	27	39	56
	to	2.5	3.5	4.5	5.5	7	9	11	14	20	27	39	56	80
Clevis pins	over	—	2	3	4	5	6	8	9	12	17	23	29	44
	to	2	3	4	5	6	8	9	12	17	23	29	44	69
Length	from	4	5	6	8	8	10	12	14	18	22	32	40	45
	to	12	16	20	25	32	40	50	63	80	100	125	160	200

Notes: All dimensions are in millimeters.

American National Standard Set Screws*

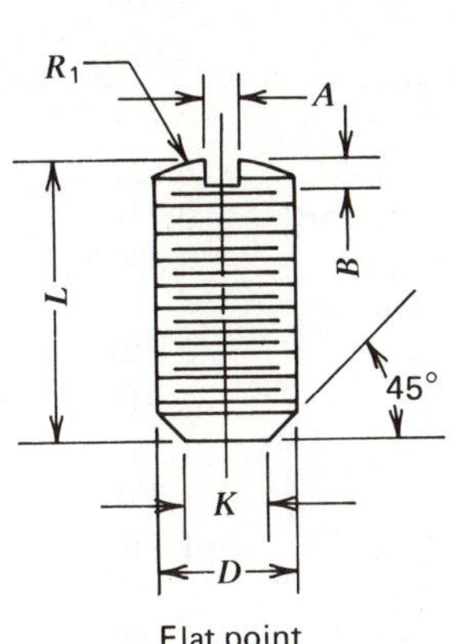

Flat point

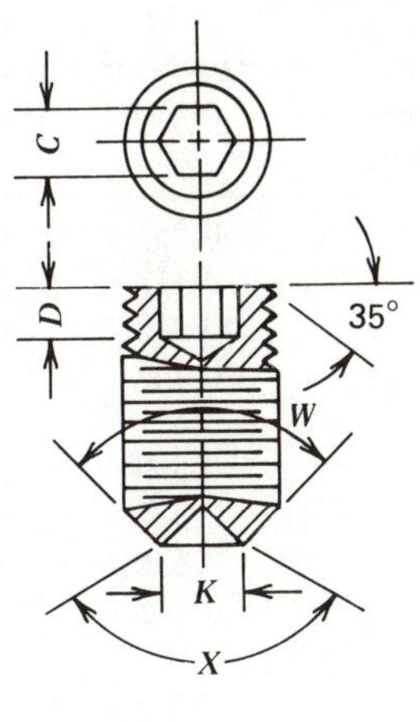

Cup point

Nominal Size—Diameter, D		*Slotted Headless* A	B	R_1	*Hexagonal Socket (min)* C	D	*Points* *Cup and Flat* (K_{max}) K	W	X
0.125	5	0.023	0.031	0.125	1/16	0.050	0.067		
0.138	6	0.025	0.035	0.138	1/16	0.050	0.074		
0.164	8	0.029	0.041	0.164	5/64	0.062	0.087		
0.190	10	0.032	0.048	0.190	3/32	0.075	0.102		
0.216	12	0.036	0.054	0.216	3/32	0.075	0.115		
0.250	1/4	0.045	0.063	0.250	1/8	0.100	0.132		
0.3125	5/16	0.051	0.078	0.313	5/32	0.125	0.172		
0.375	3/8	0.064	0.094	0.375	3/16	0.150	0.212		
0.4375	7/16	0.072	0.109	0.438	7/32	0.175	0.252		
0.500	1/2	0.081	0.125	0.500	1/4	0.200	0.291	W = 80° – 90°	X = 118° ± 5°
0.5625	9/16	0.091	0.141	0.563	1/4	0.200	0.332		
0.625	5/8	0.102	0.156	0.625	5/16	0.250	0.371		
0.750	3/4	0.129	0.188	0.750	3/8	0.300	0.450		
0.875	7/8				1/2	0.400			
1.000	1				9/16	0.450			
1.125	1 1/8				9/16	0.450			
1.250	1 1/4				5/8	0.500			
1.375	1 3/8				5/8	0.500			
1.500	1 1/2				3/4	0.600			

Note: Dimensions apply to screws 1 diameter in length or longer and are in inches.

American National Standard Lock Washers

Washer Size—Nominal		Regular (Light) Inside Diameter (min) C	Regular (Light) Outside Diameter (max) D	Regular (Light) Thickness (min) T	Extra-Duty (Heavy) Inside Diameter (min) C	Extra-Duty (Heavy) Outside Diameter (max) D	Extra-Duty (Heavy) Thickness (min) T
0.250	1/4	0.255	0.489	0.062	Same as for regular lock washers	0.535	0.084
0.312	5/16	0.318	0.586	0.078		0.622	0.108
0.375	3/8	0.382	0.683	0.094		0.741	0.123
0.438	7/16	0.446	0.779	0.109		0.839	0.143
0.500	1/2	0.509	0.873	0.125		0.939	0.162
0.562	9/16	0.572	0.971	0.141		1.041	0.182
0.625	5/8	0.636	1.079	0.156		1.157	0.202
0.750	3/4	0.763	1.271	0.188		1.361	0.241
0.875	7/8	0.890	1.464	0.219		1.576	0.285
1.000	1	1.017	1.661	0.250		1.799	0.330
1.125	1 1/8	1.144	1.853	0.281		2.019	0.375
1.250	1 1/4	1.271	2.045	0.312		2.231	0.417
1.375	1 3/8	1.398	2.239	0.344		2.439	0.458
1.500	1 1/2	1.525	2.430	0.375		2.638	0.496

Note: Lock washers are specified by giving nominal size and series (3/8 regular lock washer).
All dimensions are in inches.
ANSI B27.1, 1965

Split Lock Washer

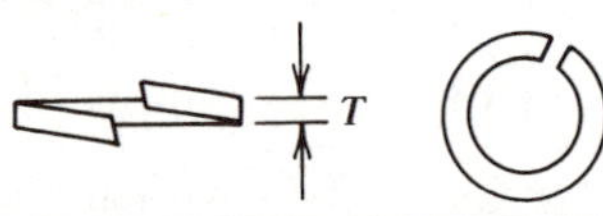

Nominal Size (mm) Screw Diameter	Data (mm) i.d.	o.d.	T
3	3.1	6.2	0.8
4	4.1	7.6	0.9
5	5.1	9.2	1.2
6	6.1	11.8	1.6
7	7.1	12.8	1.6
8	8.1	14.8	2.0
10	10.2	18.1	2.2
12	12.2	21.1	2.5
14	14.2	24.1	3.0
16	16.2	27.4	3.5
18	18.2	29.4	3.5
20	20.2	33.6	4.0
22	22.5	35.9	4.0
24	24.5	40.0	5.0
27	27.5	43.0	5.0
30	30.5	48.2	6.0

Serrated Washer

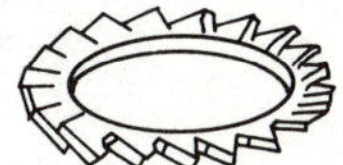

External-A

Nominal Size (mm) Screw Diameter	Data (mm)		
	i.d.	*o.d.*	T
2	2.2	4.5	0.3
2.5	2.7	5.5	0.4
3	3.2	6.0	0.4
4	4.3	8.0	0.5
5	5.3	10.0	0.6
6	6.4	11.0	0.7
8	8.4	15.0	0.8
10	10.5	18.0	0.9
12	12.5	20.5	1.0

Plain Washers: Metric and English

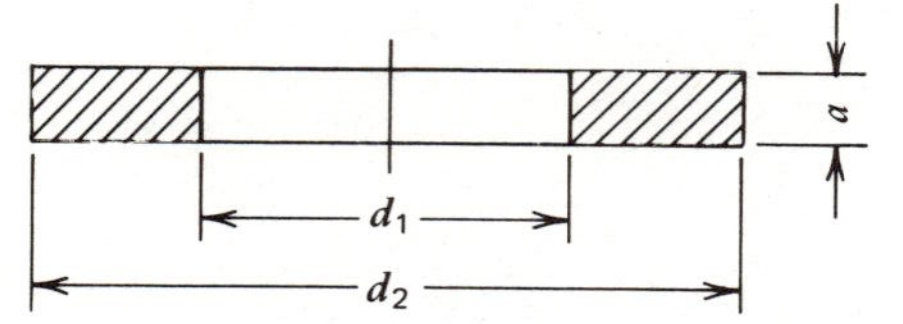

Clearance Hole d_1		Diameters of the Washer d_2 for Widths across Flats			Thickness of the Washer	Washer Size—		English Standard Plate		
Fine Series (mm)	Medium Series (mm)	Normal (mm)	Small (mm)	Large (mm)	a (mm)	Nominal (in.)		d_1 (in.)	d_2 (in.)	a (in.)
1.7		4			0.3	0.250	1/4	0.312	0.734	0.065
2.2		5			0.3	0.312	5/16	0.375	0.875	0.083
2.7		6.5			0.5	0.375	3/8	0.438	1.000	0.083
3.2		7			0.5	0.438	7/16	0.500	1.250	0.083
4.3		9			0.8	0.500	1/2	0.562	1.375	0.109
5.3	5.5	10			1	0.562	9/16	0.625	1.469	0.109
6.4	6.6	12.5			1.6	0.625	5/8	0.688	1.750	0.134
7.4	7.6	14			1.6	0.750	3/4	0.812	2.000	0.148
8.4	9	17	15.5	21	1.6	0.875	7/8	0.938	2.250	0.165

10.5	11	21	18	24	2	1.000	1	1.062	2.500	0.165
13	14	24	21	28	2.5	1.125	1⅛	1.250	2.750	0.165
15	16	28	24	30	2.5	1.250	1¼	1.375	3.000	0.165
17	18	30	28	34	3	1.375	1⅜	1.500	3.250	0.180
19	20	34	30	37	3	1.500	1½	1.625	3.500	0.180
21	22	37	34	39	3					
23	24	39	37	44	3					
25	26	44	39	50	4					
28	30	50	44	56	4					
31	33	56	50	60	4					
34	36	60	56	66	5					
37	39	66	60	72	5					
40	42	72	66	78	6					

Notes:
1. There are also Light-SAE washers.
2. These clearance holes conform to ISO 273.

Property Class Identification Symbols for Bolts, Screws and Studs

Property Class	*Bolt Head Markings*	*Studs Smaller Than M12*	*Identification Symbols of SAE and ISO Bolt Grades*	
4.6	4.6	—		*SAE Bolt Head Markers*
4.8	4.8	—	Grades 0, 1, and 2, no markings	
5.8	5.8	—	Grade 3: 2 radial dashes 180° apart	
8.8	8.8	○	Grade 5: 3 radial dashes 120° apart	
9.8	9.8	+	Grades 6: 4 radial dashes 90° apart	
10.9	10.9	□	Grade 7: 5 radial dashes 72° apart	
12.9	12.9	△	Grade 8: 6 radial dashes 60° apart	

Median Ranks*

Rank order number	Sample Size N: 1	2	3	4	5	6	7	8	9	10	11	12	13	14	15	16	17	18	19	20	21	22	23	24	25
1	.5000	.2929	.2063	.1591	.1294	.1091	.0943	.0830	.0741	.0670	.0611	.0561	.0519	.0483	.0452	.0424	.0400	.0378	.0358	.0341	.0330	.0315	.0301	.0289	.0277
2		.7071	.5000	.3864	.3147	.2655	.2295	.2021	.1806	.1632	.489	.1368	.1266	.1178	.1101	.1034	.0975	.0922	.0874	.0831	.0797	.0761	.0729	.0698	.0671
3			.7937	.6136	.5000	.4218	.3648	.3213	.2871	.2594	.2366	.2175	.2013	.1873	.1751	.1644	.1550	.1465	.1390	.1322	.1264	.1207	.1156	.1108	.1064
4				.8409	.6853	.5782	.5000	.4404	.3935	.3557	.3244	.2982	.2760	.2568	.2401	.2254	.2125	.2009	.1905	.1812	.1731	.1654	.1583	.1518	.1458
5					.8706	.7345	.6352	.5596	.5000	.4519	.4122	.3789	.3506	.3263	.3051	.2865	.2700	.2553	.2421	.2302	.2198	.2100	.2010	.1927	.1852
6						.8909	.7705	.6787	.6065	.5481	.5000	.4596	.4253	.3958	.3700	.3475	.3275	.3097	.2937	.2793	.2665	.2546	.2437	.2337	.2245
7							.9057	.7979	.7129	.6443	.5878	.5404	.5000	.4653	.4350	.4085	.3850	.3641	.3453	.3283	.3132	.2992	.2864	.2747	.2639
8								.9170	.8194	.7406	.6756	.6211	.5747	.5347	.5000	.4695	.4425	.4184	.3968	.3774	.3599	.3438	.3291	.3156	.3032
9									.9259	.8368	.7634	.7018	.6494	.6042	.5650	.5305	.5000	.4728	.4484	.4264	.4066	.3885	.3719	.3566	.3426
10										.9330	.8511	.7825	.7240	.6737	.6300	.5915	.5575	.5272	.5000	.4755	.4533	.4331	.4146	.3976	.3819
11											.9389	.8632	.7987	.7432	.6949	.6525	.6150	.5886	.5516	.5245	.5000	.4777	.4573	4385	.4213
12												.9439	.8734	.8127	.7599	.7135	.6725	.6359	.6032	.5736	.5467	.5223	.5000	.4795	.4606
13													.9481	.8822	.8249	.7746	.7300	.6903	.6547	.6226	.5934	.5669	.5427	.5205	.5000
14														.9517	.8899	.8356	.7875	.7447	.7063	.6717	.6401	.6115	.5854	.5615	.5394
15															.9548	.8966	.8450	.7991	.7579	.7207	.6868	.6562	.6281	.6024	.5787
16																.9576	.9025	.8535	.8095	.7698	.7335	.7008	.6709	.6434	.6181
17																	.9600	.9078	.8610	.8188	.7802	.7454	.7136	.6844	.6574
18																		.9622	.9126	.8678	.8269	.7900	.7563	.7253	.6968
19																			.9642	.9169	.8736	.8346	.7990	.7663	.7361
20																				.9659	.9203	.8793	.8417	.8073	.7755
21																					.9670	.9239	.8844	.8482	.8248
22																						.9685	.9271	.8892	.8542
23																							.9699	.9302	.8936
24																								.9711	.9329
25																									.9723

(Continued)

Median Ranks* (Continued)

Rank order number	Sample Size N 26	27	28	29	30	31	32	33	34	35	36	37	38	39	40	41	42	43	44	45	46	47	48	49	50
1	0.267	.0257	.0248	.0239	.0231	.0224	.0217	.0210	.0204	.0198	.0193	.0187	.0182	.0178	.0173	.0169	.0165	.0161	.0158	.0154	.0151	.0147	.0144	.0141	.0139
2	.0645	.0622	.0600	.0579	.0560	.0542	.0525	.0509	.0495	.0481	.0467	.0455	.0443	.0432	.0421	.0411	.0401	.0392	.0383	.0374	.0366	.0358	.0351	.0344	.0337
3	.1024	.0986	.0952	.0919	.0888	.0860	.0834	.0809	.0785	.0763	.0742	.0722	.0703	.0685	.0668	.0652	.0637	.0622	.0608	.0595	.0582	.0569	.0558	.0546	.0535
4	.1403	.1351	.1304	.1259	.1218	.1179	.1142	.1108	.1076	.1045	.1017	.0989	.0964	.0935	.0916	.0894	.0873	.0852	.0833	.0815	.0797	.0780	.0764	.0749	.0734
5	.1718	.1716	.1656	.1599	.1547	.1497	.1451	.1408	.1367	.1328	.1291	.1257	.1224	.1993	.1163	.1135	.1108	.1083	.1058	.1035	.1013	.0991	.0971	.0951	.0932
6	.2160	.2081	.2008	.1939	.1876	.1816	.1760	.1707	.1657	.1610	.1566	.1524	.1484	.1447	.1411	.1377	.1344	.1313	.1284	.1255	.1228	.1202	.1178	.1154	.1131
7	.2539	.2446	.2360	.2279	.2204	.2134	.2068	.2006	.1948	.1892	.1841	.1792	.1745	.1701	.1658	.1618	.1580	.1544	.1509	.1476	.1444	.1413	.1384	.1356	.1329
8	.2917	.2811	.2712	.2620	.2533	.2453	.2377	.2306	.2209	.2175	.2116	.2059	.2003	.1954	.1906	.1860	.1816	.1774	.1734	.1696	.1659	.1624	.1591	.1559	.1528
9	.3296	.3176	.3064	.2960	.2862	.2771	.2685	.2605	.2529	.2458	.2390	.2326	.2266	.2208	.2153	.2101	.2052	.2005	.1959	.1916	.1875	.1835	.1797	.1761	.1726
10	.3675	.3541	.3416	.3300	.3191	.3089	.2994	.2904	.2820	.2740	.2665	.2594	.2526	.2462	.2401	.2343	.2288	.2235	.2185	.2136	.2090	.2046	.2004	.1963	1924
11	.4053	.3905	.3768	.3640	.3520	.3408	.3303	.3204	.3111	.3023	.2940	.2861	.2787	.2716	.2649	.2585	.2524	.2465	.2410	.2357	.2306	.2257	.2211	.2168	.2123
12	.4432	.4270	.4120	.3980	.3849	.3726	.3611	.3503	.3401	.3305	.3214	.3128	.3047	.2970	.2896	.2826	.2759	.2696	.2635	.2577	.2521	.2468	.2417	.2368	.2321
13	.4811	.4635	.4472	.4320	.4178	.4045	.3920	.3803	.3692	.3588	.3489	.3396	.3307	.3223	.3144	.3068	.2995	.2926	.2860	.2797	.2737	.2679	.2624	.2571	.2520
14	.5189	.5000	.4824	.4660	.4507	.4363	.4228	.4102	.3983	.3870	.3764	.3663	.3568	.3477	.3391	.3309	.3231	.3157	.3086	.3018	.2953	.2890	.2830	.2773	.2718
15	.5568	.5365	.5176	.5000	.4836	.4682	.4537	.4401	.4273	.4183	.4039	.3931	.3828	.3731	.3639	.3551	.3467	.3387	.3311	.3238	.3168	.3101	.3037	.2976	.2917
16	.5947	.5730	.5528	.5340	.5164	.5000	.4846	.4701	.4564	.4435	.4313	.4198	.4088	.3985	.3986	.3792	.3703	.3617	.3536	.3458	.3384	.3312	.3244	.3178	.3115
17	.6325	.6095	.5880	.5680	.5493	.5318	.5154	.5000	.4855	.4718	.4588	.4465	.4349	.4239	.4134	.4034	.3939	.3848	.3761	.3678	.3599	.3523	.3450	.3380	.3313
18	.6704	.6459	.6332	.6020	.5822	.5617	.5463	.5299	.5145	.5000	.4863	.4733	.4609	.4492	.4381	.4275	.4175	.4078	.3986	.3899	.3815	.3734	.3657	.3583	.3512
19	.7083	.6824	.6584	.6360	.6151	.5955	.5772	.5599	.5436	.5282	.5137	.5000	.4870	.4746	.4629	.4517	.4410	.4309	.4212	.4119	.4030	.3945	.3864	.3785	.3710
20	.7461	.7189	.6936	.6700	.6480	.6274	.6080	.5898	.5727	.5565	.5412	.5267	.5130	.5000	.4876	.4758	.4646	.4539	.4437	.4339	.4246	.4156	.4070	.3988	.3909
21	.7840	.7554	.7288	.7040	.6809	.6592	.6389	.6197	.6017	.5847	.4687	.5535	.5391	.5254	.5124	.5000	.4882	.4770	.4662	.4559	.4461	.4367	.4277	.4190	.4107
22	.8219	.7919	.7640	.7380	.7138	.6911	.6697	.6497	.6308	.6130	.5961	.5802	.5651	.5508	.5371	.5242	.5118	.5000	.4887	.4780	.4677	.4578	.4483	.4393	.4306
23	.8597	.8284	.7992	.7721	.7467	.7229	.7006	.6796	.6599	.6412	.6236	.6069	.5911	.5761	.5619	.5483	.5354	.5230	.5113	.5000	.4892	.4789	.4690	.4595	.4504
24	.8976	.8649	.8344	.8061	.7796	.7547	.7315	.7096	.6889	.6695	.6511	.6337	.6172	.6015	.5866	.5725	.5590	.5491	.5338	.5220	.5108	.5000	.4897	.4798	.4702
25	.9355	.9014	.8696	.8401	.8124	.7866	.7623	.7395	.7180	.6977	.6786	.6604	.6432	.6269	.6114	.5966	.5825	.5691	.5563	.5441	.5323	.5211	.5103	.5000	.4901

26	.9733	.9378	.9048	.8741	.8453	.8184	.7932	.7694	.7471	.7260	.7060	.6872	.6693	.6523	.6361	.6208	.6061	.5922	.5788	.5661	.5539	.5422	.5310	.5202	.5099
27		.9743	.9400	.9081	.8782	.8503	.8240	.7944	.7761	.7542	.7335	.7139	.6953	.6777	.6609	.6449	.6297	.6152	.6014	.5881	.5754	.5633	.5517	.5405	.5298
28			.9752	.9421	.9111	.8821	.8549	.8293	.8052	.7825	.7610	.7406	.7213	.7030	.6856	.6691	.6533	.6383	.6239	.6101	.5970	.5844	.5723	.5607	.5496
29				.9461	.9440	.9140	.8858	.8592	.8343	.8107	.7884	.7674	.7474	.7284	.7104	.6932	.6769	.6613	.6464	.6322	.6185	.6055	.5930	.5810	.5694
30					.9769	.9458	.9166	.8892	.8633	.8390	.8159	.7941	.7734	.7538	.7351	.7174	.7005	.6843	.6689	.6542	.6401	.6266	.6136	.6012	5893
31						.9776	.9474	.9191	.8924	.8672	.8434	.8208	.7995	.7792	.7599	.7415	.7241	.7074	.6914	.6762	.6616	.6477	.6343	.6215	.6091
32							.9783	.9491	.9215	.8955	.8909	.8476	.8255	.8046	.7847	.7657	.7476	.7304	.7140	.6982	.6832	.6688	.6550	.6417	.6290
33								.9790	.9505	.9237	.8983	.8743	.8516	.8299	.8094	.7899	.7712	.7535	.7365	.7203	.7047	.6899	.6756	.6620	.6488
34									.9796	.9519	.9258	.9011	.8776	.8553	.8342	.8140	.7948	.7765	.7590	.7423	.7263	.7110	.6963	.6822	.6687
35										.9802	.9533	.9278	.9036	.8807	.8589	.8382	.8184	.7995	.7815	.7643	.7479	.7321	.7170	.7024	.6885
36											.9807	.9545	.9297	.9061	.8837	.8623	.8420	.8226	.8041	.7864	.7694	.7532	.7376	.7227	.7083
37												.9813	.9557	.9315	.9084	.8865	.8656	.8456	.8266	.8084	.7910	.7743	.7583	.7429	.7282
38													.9818	.9568	.9332	.9106	.8892	.8687	.8491	.8304	.8125	.7954	.7789	.7632	.7480
39														.9822	.9579	.9348	.9127	.8917	.8716	.8524	.8341	.8165	.7996	.7834	.7679
40															.9827	.9589	.9363	.9148	.8942	.8745	.8556	.8376	.8203	.8037	.7877
41																.9831	.9599	.9378	.9167	.8965	.8772	.8587	.8409	.8239	.8076
42																	.9835	.9608	.9392	.9185	.8987	.8798	.8616	.8541	.8274
43																		.9839	.9617	.9405	.9203	.9009	.8822	.8644	.8472
44																			.9842	.9626	.9418	.9220	.9029	.8846	.8671
45																				.9846	.9634	.9431	.9236	.9049	.8869
46																					.9849	.9642	.9442	.9251	.9068
47																						.9853	.9649	.9454	.9266
48																							.9856	.9656	.9465
49																								.9859	.9663
50																									.9861

Source: Statistical Design and Analysis of Engineering Experiments, by Lipson and Sheth, McGraw-Hill Book Company, New York, copyright © 1973. Reprinted by permission.

Two-Sided Confidence Band Factors on WEIBULL Plots

Sample Size	*1*						*2*					
Confidence	*90%*		*95%*		*99%*		*90%*		*95%*		*99%*	
% Rank	*5.0%*	*95.0%*	*2.5%*	*97.5%*	*0.5%*	*99.5%*	*5.0%*	*95.0%*	*2.5%*	*97.5%*	*0.5%*	*99.5%*
1	5.0	95.0	2.5	97.5	0.50	99.5	2.5	77.7	1.3	84.2	0.25	92.9
2							22.4	97.5	15.8	98.7	7.1	99.8

Sample Size	*3*						*4*					
Confidence	*90%*		*95%*		*99%*		*90%*		*95%*		*99%*	
% Rank	*5.0%*	*95.0%*	*2.5%*	*97.5%*	*0.5%*	*99.5%*	*5.0%*	*95.0%*	*2.5%*	*97.5%*	*0.5%*	*99.5%*
1	1.7	63.2	0.84	70.8	0.17	82.9	1.3	52.7	0.63	60.2	0.12	73.4
2	13.5	86.5	9.4	90.6	4.1	95.9	9.8	75.1	6.8	80.6	2.9	88.9
3	36.8	98.3	29.2	99.2	17.1	99.8	24.9	90.2	19.4	93.2	11.1	97.1
4							47.3	98.7	39.8	9.4	26.6	99.9

Sample Size	*5*						*6*					
Confidence	*90%*		*95%*		*99%*		*90%*		*95%*		*99%*	
% Rank	*5.0%*	*95.0%*	*2.5%*	*97.5%*	*0.5%*	*99.5%*	*5.0%*	*95.0%*	*2.5%*	*97.5%*	*0.5%*	*99.5%*
1	1.0	45.1	0.50	52.2	0.10	65.3	0.85	39.3	0.42	45.9	0.08	58.6
2	7.6	65.7	5.3	71.6	2.3	81.5	6.3	58.2	4.3	64.1	1.9	74.6
3	18.9	81.1	14.7	85.3	8.3	91.7	15.3	72.9	11.8	77.7	6.6	85.6
4	34.3	92.4	28.4	94.7	18.5	97.7	27.1	84.7	22.3	88.2	14.4	93.4
5	54.9	99.0	47.8	99.5	34.7	99.9	41.8	93.7	35.9	95.7	25.4	98.1
6							60.7	99.1	54.1	99.6	41.4	99.9

Sample Size				7						8		
Confidence	*90%*		*95%*		*99%*		*90%*		*95%*		*99%*	
% Rank	*5.0%*	*95.0%*	*2.5%*	*97.5%*	*0.5%*	*99.5%*	*5.0%*	*95.0%*	*2.5%*	*97.5%*	*0.5%*	*99.5%*
1	0.73	34.8	0.36	41.0	0.07	53.1	0.64	31.2	0.13	36.9	0.06	48.4
2	5.3	42.1	3.7	57.9	1.6	68.5	4.6	47.1	2.0	52.7	1.4	63.1
3	12.9	65.9	9.9	71.0	5.5	79.7	11.1	60.0	6.1	65.1	4.7	74.2
4	22.5	77.5	18.4	81.6	11.8	88.2	19.3	71.1	12.1	75.5	10.0	83.0
5	34.1	87.1	29.0	90.1	20.3	94.5	28.9	80.7	19.8	84.3	17.0	90.0
6	47.9	94.7	42.1	96.3	31.5	98.4	40.0	88.9	29.3	91.5	25.8	95.2
7	65.2	99.3	59.0	99.6	46.9	99.9	52.9	95.4	41.0	96.8	36.8	98.6
8							68.8	99.4	56.2	79.7	51.6	99.9

Sample Size				9						10		
Confidence	*90%*		*95%*		*99%*		*90%*		*95%*		*99%*	
% Rank	*5.0%*	*95.0%*	*2.5%*	*97.5%*	*0.5%*	*99.5%*	*5.0%*	*95.0%*	*2.5%*	*97.5%*	*0.5%*	*99.5%*
1	0.57	28.3	0.28	33.6	0.06	44.5	0.51	25.9	0.25	30.8	0.05	41.1
2	4.1	42.9	2.8	48.2	1.2	58.5	3.7	39.4	2.5	44.5	1.1	54.4
3	9.8	55.0	7.5	60.0	4.2	69.3	8.7	50.7	6.7	55.6	3.7	64.8
4	16.9	65.5	13.7	70.1	8.7	78.1	15.0	60.7	12.2	65.2	7.7	73.5
5	25.1	74.9	21.2	78.8	14.6	85.4	22.2	69.6	18.7	73.8	12.8	80.9
6	34.5	83.1	29.9	86.3	21.9	91.3	30.4	77.8	26.2	81.3	19.1	87.2
7	45.0	90.2	40.0	92.5	30.7	95.8	39.3	85.0	34.8	87.8	26.5	92.3
8	57.1	95.9	51.8	97.2	41.5	98.8	49.3	91.3	44.4	93.3	35.2	96.3
9	71.7	99.4	66.4	99.7	55.5	99.9	60.6	96.3	55.5	97.5	45.6	98.9
10							74.1	99.5	69.2	99.7	58.9	99.9

(Continued)

Two-Sided Confidence Band Factors on WEIBULL Plots (Continued)

Sample Size			*11*						*12*			
Confidence	*90%*		*95%*		*99%*		*90%*		*95%*		*99%*	
% Rank	*5.0%*	*95.0%*	*2.5%*	*97.5%*	*0.5%*	*99.5%*	*5.0%*	*95.0%*	*2.5%*	*97.5%*	*0.5%*	*99.5%*
1	0.46	23.8	0.23	28.5	0.05	38.2	0.43	22.1	0.21	26.5	0.04	35.7
2	3.3	36.4	2.3	41.3	1.0	50.9	3.0	33.9	2.1	38.5	0.9	47.7
3	7.9	47.0	6.0	51.8	3.3	60.8	7.2	43.8	5.5	48.4	3.0	57.3
4	13.5	56.4	10.9	61.0	6.9	69.3	12.3	52.7	9.9	57.2	6.2	65.5
5	20.0	65.0	16.7	69.2	11.4	76.7	18.1	60.9	15.2	65.1	10.3	72.8
6	27.1	72.9	23.4	76.6	16.9	83.1	24.5	68.5	21.1	72.3	15.2	79.1
7	35.0	80.0	30.8	83.3	23.3	88.6	31.5	75.5	27.7	78.9	20.9	84.9
8	43.6	86.5	39.0	89.1	30.7	93.1	39.1	81.9	34.9	84.8	27.2	89.7
9	53.0	92.1	48.2	94.0	39.2	96.7	47.3	87.7	42.8	90.1	34.5	93.8
10	63.6	96.7	58.7	97.7	49.1	99.0	56.2	92.8	51.6	94.5	42.7	97.0
11	76.2	99.5	71.5	99.8	61.8	99.9	66.1	97.0	61.5	97.9	52.3	99.1
12							77.9	99.6	73.5	99.8	64.3	99.9

Sample Size			*13*						*14*			
Confidence	*90%*		*95%*		*99%*		*90%*		*95%*		*99%*	
% Rank	*5.0%*	*95.0%*	*2.5%*	*97.5%*	*0.5%*	*99.5%*	*5.0%*	*95.0%*	*2.5%*	*97.5%*	*0.5%*	*99.5%*
1	0.39	20.6	0.20	24.7	0.04	33.5	0.37	19.3	0.18	23.2	0.04	31.5
2	2.8	31.6	1.9	36.0	0.8	44.9	2.6	29.7	1.8	33.9	0.76	42.4
3	6.6	41.0	5.0	45.4	2.8	54.1	6.1	38.5	4.7	42.8	2.6	51.2
4	11.3	49.5	9.1	53.8	5.7	62.1	10.4	46.6	8.4	50.8	5.3	58.9
5	16.6	57.3	13.9	61.4	9.4	69.1	15.3	54.0	12.8	58.1	8.7	65.8
6	22.4	64.5	19.2	68.4	13.8	75.5	20.6	61.0	17.7	64.9	12.7	72.0
7	28.7	71.3	25.1	74.9	18.9	81.1	26.4	67.5	23.0	71.1	17.2	77.7

8	35.5	77.6	31.6	80.8	24.5	86.2	32.5	73.6	28.9	77.0	22.3	82.8
9	42.7	83.4	38.6	86.1	30.9	90.6	39.0	79.4	35.1	82.3	28.0	87.3
10	50.5	88.7	46.2	90.9	37.9	94.3	46.0	84.7	41.9	87.2	34.2	91.3
11	59.0	93.4	54.6	95.0	45.9	97.2	53.4	89.6	49.2	91.6	41.1	94.7
12	68.4	97.2	64.0	98.1	55.1	99.2	61.5	93.9	57.2	95.3	48.8	97.4
13	79.4	99.6	75.3	99.8	66.5	99.9	70.3	97.4	66.1	98.2	57.6	99.2
14							80.7	99.6	76.8	99.8	68.5	99.9

Sample Size			*15*						*16*			
Confidence	*90%*		*95%*		*99%*		*90%*		*95%*		*99%*	
% Rank	*5.0%*	*95.0%*	*2.5%*	*97.5%*	*0.5%*	*99.5%*	*5.0%*	*95.0%*	*2.5%*	*97.5%*	*0.5%*	*99.5%*
1	0.34	18.1	0.17	21.8	0.03	29.8	0.32	17.1	0.16	20.5	0.03	28.2
2	2.4	27.9	1.7	31.9	0.7	40.2	2.3	26.4	1.6	30.2	0.67	38.1
3	5.7	36.3	4.3	40.5	2.4	48.6	5.3	36.4	4.0	38.3	2.2	46.3
4	9.7	44.0	7.8	48.1	4.9	56.1	9.0	41.7	7.3	45.6	4.5	53.4
5	14.2	51.1	11.8	55.1	8.0	62.7	13.2	48.4	11.0	52.4	7.5	59.9
6	19.1	57.7	16.3	61.6	11.7	68.8	17.8	54.8	15.2	58.7	10.9	65.8
7	24.4	64.0	21.3	67.7	15.9	74.4	22.7	60.9	19.8	64.6	14.7	71.3
8	30.0	70.0	26.6	73.4	20.5	79.5	27.9	66.7	24.7	70.1	19.0	76.4
9	36.0	75.6	32.3	78.7	25.6	84.1	33.3	72.1	29.9	75.3	23.6	81.0
10	42.3	80.9	38.4	83.7	31.2	88.3	39.1	77.3	35.4	80.2	28.7	85.3
11	48.9	85.8	44.9	88.2	37.3	92.0	45.2	82.2	41.3	84.8	34.2	89.1
12	56.0	90.3	51.9	92.2	43.9	95.1	51.6	86.8	47.6	89.0	40.1	92.5
13	63.7	94.3	59.5	95.7	51.4	97.6	58.3	91.0	54.4	92.7	46.6	95.5
14	72.1	97.6	68.1	98.3	59.8	99.3	65.6	94.7	61.7	96.0	53.7	97.8
15	81.9	99.7	78.2	99.8	70.2	99.9	73.6	97.7	69.8	98.4	61.9	99.3
16							82.9	99.7	79.4	99.8	71.8	99.9

(Continued)

Two-Sided Confidence Band Factors on WEIBULL Plots (Continued)

Sample Size	*17*						*18*					
Confidence	*90%*		*95%*		*99%*		*90%*		*95%*		*99%*	
% Rank	*5.0%*	*95.0%*	*2.5%*	*97.5%*	*0.5%*	*99.5%*	*5.0%*	*95.0%*	*2.5%*	*97.5%*	*0.5%*	*99.5%*
1	0.30	16.2	0.16	19.5	0.03	26.8	0.28	15.3	0.14	18.5	0.03	25.5
2	2.1	25.0	1.5	28.7	0.63	36.3	2.0	23.8	1.4	27.3	0.59	34.6
3	5.0	32.6	3.8	36.4	2.1	44.1	4.7	31.0	3.6	34.7	2.0	42.2
4	8.5	39.6	6.8	43.4	4.3	51.0	8.0	37.7	6.4	41.4	4.0	48.8
5	12.4	46.1	10.3	49.9	7.0	57.3	11.6	43.9	9.7	47.6	6.5	54.9
6	16.6	52.2	14.2	56.0	10.1	63.1	15.6	49.8	13.3	53.5	9.5	60.5
7	21.2	58.0	18.4	61.7	13.7	68.5	19.9	55.4	17.3	59.0	12.8	65.8
8	26.0	63.6	23.0	67.1	17.6	73.4	24.4	60.8	21.5	64.3	16.5	70.7
9	31.1	68.9	27.8	72.2	21.9	78.1	29.1	65.9	26.0	69.2	20.5	75.3
10	36.4	74.0	32.9	77.0	26.6	82.4	34.1	70.9	30.8	74.0	24.7	79.5
11	42.0	78.8	38.3	81.6	31.5	86.3	39.2	75.6	35.7	78.4	29.3	83.5
12	47.8	83.4	44.0	85.8	36.9	89.9	44.6	80.1	41.0	82.7	34.2	87.2
13	53.9	87.6	50.1	89.7	42.7	93.0	50.2	84.4	46.5	86.7	39.5	90.5
14	60.4	91.5	56.6	93.2	49.0	95.7	56.1	88.4	52.4	90.3	45.1	93.5
15	67.4	95.0	63.6	96.2	55.9	97.9	62.3	92.0	58.6	93.6	51.2	96.0
16	75.0	97.9	71.3	98.5	63.7	99.4	69.0	95.3	65.3	96.4	57.8	98.0
17	83.8	99.7	80.5	99.9	73.2	99.9	76.2	98.0	72.7	98.6	65.4	99.4
18							84.7	99.7	81.5	99.9	74.5	99.9

Sample Size	19						20					
Confidence	90%		95%		99%		90%		95%		99%	
% Rank	*5.0%*	*95.0%*	*2.5%*	*97.5%*	*0.5%*	*99.5%*	*5.0%*	*95.0%*	*2.5%*	*97.5%*	*0.5%*	*99.5%*
1	0.27	14.6	0.13	17.6	0.03	24.3	0.26	13.9	0.13	16.8	0.02	23.3
2	1.9	22.6	1.3	26.0	0.56	33.1	1.8	21.6	1.2	24.9	0.53	31.7
3	4.4	29.6	3.4	33.1	1.9	40.4	4.2	28.3	3.2	31.7	1.8	38.7
4	7.5	35.9	6.1	39.6	3.8	46.8	7.1	34.4	5.7	37.9	3.6	44.9
5	11.0	41.9	9.1	45.6	6.2	52.7	10.4	40.1	8.7	43.7	5.8	50.7
6	14.7	47.6	12.6	51.2	9.0	58.2	14.0	45.6	11.9	49.1	8.5	56.0
7	18.8	53.0	16.3	56.6	12.1	63.3	17.7	50.8	15.4	54.3	11.4	61.0
8	23.0	58.2	20.3	61.6	15.5	68.7	21.7	55.8	19.1	59.2	14.6	65.7
9	27.4	63.2	24.4	66.5	19.2	72.6	25.9	60.6	23.1	63.9	18.1	70.1
10	32.0	68.0	28.9	71.1	23.2	76.8	30.2	65.3	27.2	68.5	21.8	74.3
11	36.8	72.6	33.5	75.6	27.4	80.8	34.7	69.8	31.5	72.8	25.7	78.2
12	41.8	77.0	38.4	79.7	31.9	84.5	39.4	74.1	36.1	76.9	29.9	81.9
13	47.0	81.2	43.4	83.7	36.7	87.9	44.2	78.3	40.8	80.9	34.3	85.4
14	52.4	85.3	48.8	87.4	41.8	91.0	49.2	82.3	45.7	84.6	39.0	88.6
15	58.1	89.0	54.4	90.9	47.3	93.8	54.4	86.0	50.9	88.1	44.0	91.5
16	64.1	92.5	60.4	93.9	53.2	96.2	59.9	89.6	56.3	91.3	49.3	94.2
17	70.4	95.6	66.9	96.6	59.6	98.1	65.6	92.9	62.1	94.3	55.0	96.4
18	77.4	98.1	74.0	98.7	66.9	99.4	71.7	95.8	68.3	96.8	61.3	98.2
19	85.4	99.7	82.4	99.9	75.7	99.9	78.4	98.2	75.1	98.8	68.3	99.5
20							86.1	99.7	83.2	99.9	76.7	99.9

(Continued)

Two-Sided Confidence Band Factors on WEIBULL Plots (Continued)

Sample Size	*21*						*22*					
Confidence	*90%*		*95%*		*99%*		*90%*		*95%*		*99%*	
% Rank	*5.0%*	*95.0%*	*2.5%*	*97.5%*	*0.5%*	*99.5%*	*5.0%*	*95.0%*	*2.5%*	*97.5%*	*0.5%*	*99.5%*
1	0.24	13.2	0.12	16.1	0.02	22.2	0.23	12.7	0.11	15.4	0.02	21.4
2	1.7	20.6	1.1	23.8	0.50	30.4	1.6	19.8	1.1	22.8	0.48	29.2
3	4.0	27.0	3.0	30.3	1.6	37.1	3.8	25.9	2.9	29.1	1.5	35.7
4	6.7	32.9	5.4	36.3	3.3	43.2	6.4	31.5	5.1	34.9	3.2	41.6
5	9.8	38.4	8.2	41.9	5.5	48.7	9.4	36.9	7.8	40.2	5.2	46.9
6	13.2	43.6	11.2	47.1	8.0	53.9	12.6	41.9	10.7	45.3	7.6	52.0
7	16.8	48.7	14.5	52.1	10.7	58.7	15.9	46.8	13.8	50.2	10.2	56.7
8	20.5	53.5	18.1	56.9	13.8	63.3	19.5	51.5	17.1	54.8	13.0	61.2
9	24.4	58.2	21.8	61.5	17.0	67.7	23.2	56.0	20.7	59.3	16.1	65.4
10	28.5	62.8	25.7	65.9	20.5	71.8	27.1	60.4	24.3	63.6	19.4	69.5
11	32.8	67.1	29.7	70.2	24.2	75.7	31.1	64.7	28.2	67.7	22.9	73.4
12	37.1	71.4	34.0	74.2	28.1	79.4	35.2	68.8	32.2	71.7	26.5	77.0
13	41.7	75.5	38.4	78.1	32.2	82.9	39.5	72.8	36.3	75.6	30.4	80.5
14	46.4	79.4	43.0	81.8	36.6	86.1	43.9	76.7	40.6	79.2	34.5	83.8
15	51.2	83.1	47.8	85.4	41.2	89.2	48.4	80.4	45.1	82.8	38.7	86.9
16	56.3	86.7	52.8	88.7	46.0	91.9	53.1	84.0	49.7	86.1	43.2	89.7
17	61.5	90.1	58.0	91.7	51.2	94.4	58.0	87.3	54.6	89.2	47.9	92.3
18	67.0	93.2	63.6	94.5	56.7	96.6	63.0	90.5	59.7	92.1	53.0	94.7
19	72.9	95.9	69.6	96.9	62.8	98.3	68.4	93.5	65.0	94.8	58.3	96.7
20	79.3	98.2	76.1	98.8	69.5	99.4	74.0	96.1	70.8	97.0	64.2	98.4
21	86.7	99.7	83.8	99.8	77.7	99.9	80.1	98.3	77.1	98.8	70.7	99.5
22							87.2	99.7	84.5	99.8	78.5	99.9

Sample Size	23						24					
Confidence	*90%*		*95%*		*99%*		*90%*		*95%*		*99%*	
% Rank	*5.0%*	*95.0%*	*2.5%*	*97.5%*	*0.5%*	*99.5%*	*5.0%*	*95.0%*	*2.5%*	*97.5%*	*0.5%*	*99.5%*
1	0.22	12.2	0.11	14.8	0.02	20.5	0.21	11.7	0.10	14.2	0.02	19.8
2	1.5	19.0	1.0	21.9	0.45	28.1	1.5	18.2	1.0	21.1	0.44	27.1
3	3.6	24.9	2.7	28.0	1.5	34.4	3.4	23.9	2.6	26.9	1.4	33.2
4	6.1	30.3	4.9	33.5	3.0	40.1	5.9	29.2	4.7	32.3	2.9	38.7
5	8.9	35.4	7.4	38.7	5.0	45.3	8.5	34.1	7.1	37.3	4.7	43.7
6	12.0	40.3	10.2	43.7	7.2	50.2	11.4	38.9	9.7	42.1	6.9	48.5
7	15.2	45.0	13.2	48.4	9.7	54.8	14.5	43.4	12.6	46.7	9.2	53.0
8	18.6	49.6	16.3	52.9	12.4	59.2	17.7	47.8	15.6	51.0	11.8	57.3
9	22.1	54.0	19.7	57.2	15.3	63.3	21.1	52.1	18.7	55.3	14.6	61.3
10	25.8	58.3	23.1	61.4	18.4	67.3	24.6	56.2	22.1	59.4	17.5	65.3
11	29.6	62.4	26.8	65.5	21.7	71.1	28.2	60.3	25.5	63.3	20.6	69.0
12	33.5	66.4	30.5	69.4	25.2	74.7	31.9	64.2	29.1	67.1	23.9	72.6
13	37.5	70.3	34.4	73.1	28.8	78.2	35.7	68.0	32.8	70.8	27.3	76.0
14	41.6	74.1	38.5	76.8	32.6	81.5	39.6	71.7	36.6	74.4	30.9	79.3
15	45.9	77.8	42.7	80.2	36.6	84.6	43.7	75.3	40.5	77.8	34.6	82.4
16	50.3	81.3	47.0	83.6	40.7	87.5	47.8	78.8	44.6	81.2	38.6	85.3
17	54.9	84.7	51.5	86.7	45.1	90.2	52.1	82.2	48.9	84.3	42.6	88.1
18	59.6	87.9	56.2	89.7	49.7	92.7	56.5	85.4	53.2	87.3	46.9	90.7
19	64.5	91.0	61.2	92.5	54.6	94.9	61.0	88.5	57.8	90.2	51.4	93.0
20	69.6	93.8	66.4	95.0	59.8	96.9	65.8	91.4	62.6	92.8	56.2	95.2
21	75.0	96.3	71.9	97.2	65.5	98.4	70.7	94.0	67.6	95.2	61.2	97.0
22	80.9	98.4	78.0	98.9	71.8	99.5	76.0	96.5	73.0	97.3	66.7	98.5
23	87.7	99.7	85.1	99.8	79.4	99.9	81.7	98.4	78.8	98.9	72.8	99.5
24							88.2	99.7	85.7	99.8	80.1	99.9

(Continued)

Two-Sided Confidence Band Factors on WEIBULL Plots (Continued)

Sample Size	25						26					
Confidence	*90%*		*95%*		*99%*		*90%*		*95%*		*99%*	
% Rank	*5.0%*	*95.0%*	*2.5%*	*97.5%*	*0.5%*	*99.5%*	*5.0%*	*95.0%*	*2.5%*	*97.5%*	*0.5%*	*99.5%*
1	0.20	11.2	0.10	13.7	0.02	19.0	0.19	10.8	0.09	13.2	0.01	18.4
2	1.4	17.6	0.98	20.3	0.42	26.1	1.3	16.9	0.94	19.6	0.40	25.2
3	3.3	23.1	2.5	26.0	1.3	32.1	3.2	22.2	2.4	25.1	1.3	31.0
4	5.6	28.1	4.5	31.2	2.8	37.4	5.4	27.1	4.3	30.1	2.7	36.2
5	8.2	32.9	6.8	36.0	4.5	42.8	7.8	31.8	6.5	34.8	4.4	40.9
6	11.0	37.5	9.3	40.7	6.6	46.9	10.5	36.2	8.9	39.3	6.3	45.5
7	13.9	41.9	12.0	45.1	8.8	51.3	13.3	40.5	11.5	43.6	8.5	49.7
8	17.0	46.2	14.9	49.3	11.3	55.5	16.3	44.6	14.3	47.7	10.8	53.8
9	20.2	50.3	17.9	53.5	13.9	59.5	19.3	48.7	17.2	51.7	13.3	57.7
10	23.5	54.3	21.1	57.4	16.7	63.3	22.5	52.6	20.2	55.6	16.0	61.5
11	26.9	58.3	24.4	61.3	19.7	67.0	25.8	56.4	23.3	59.4	18.8	65.1
12	30.5	62.1	27.7	65.0	22.8	70.5	29.2	60.1	26.5	63.0	21.8	68.5
13	34.1	65.8	31.3	68.6	26.0	73.9	32.6	63.7	29.9	66.6	24.8	71.9
14	37.8	69.4	34.9	72.2	29.4	77.1	36.2	67.3	33.3	70.0	28.0	75.1
15	41.6	73.0	38.6	75.5	32.9	80.2	39.8	70.7	36.9	73.4	31.4	78.1
16	45.6	76.4	42.5	78.8	36.6	83.2	43.5	74.1	40.5	76.6	34.8	81.1
17	49.6	79.7	46.5	82.0	40.4	86.0	47.3	77.4	44.3	79.7	38.5	83.9
18	53.7	82.9	50.6	85.0	44.4	88.6	51.3	80.6	48.2	82.7	42.2	86.6
19	58.0	86.0	54.8	87.9	48.6	91.1	55.3	83.6	52.2	85.6	46.1	89.1
20	62.4	88.9	59.2	90.6	53.0	93.3	59.4	86.6	56.3	88.4	50.2	91.4

21	67.0	91.7	63.9	93.1	57.6	95.4	63.7	89.4	60.6	91.0	54.5	93.6
22	71.8	94.3	68.7	95.4	62.5	97.1	68.1	92.1	65.1	93.4	59.0	95.5
23	76.8	96.6	73.9	97.4	67.8	98.6	72.8	94.5	69.8	95.6	63.7	97.2
24	82.3	98.5	79.6	99.0	73.8	99.5	77.7	96.7	74.8	97.5	68.9	98.6
25	88.7	99.7	86.2	99.8	80.9	99.9	83.0	98.6	80.3	99.0	74.7	99.5
26							89.1	99.8	86.7	99.9	81.5	99.9

Sample Size	27						28					
Confidence	*90%*		*95%*		*99%*		*90%*		*95%*		*99%*	
% Rank	*5.0%*	*95.0%*	*2.5%*	*97.5%*	*0.5%*	*99.5%*	*5.0%*	*95.0%*	*2.5%*	*97.5%*	*0.5%*	*99.5%*
1	0.19	10.5	0.09	12.7	0.01	17.8	0.18	10.1	0.09	12.3	0.01	17.2
2	1.3	16.3	0.91	18.9	0.39	24.4	1.2	15.8	0.87	18.3	0.37	23.6
3	3.0	21.5	2.3	24.2	1.2	30.0	2.9	20.8	2.2	23.5	1.2	29.1
4	5.2	26.2	4.1	29.1	2.6	35.0	5.0	25.4	4.0	28.2	2.5	33.9
5	7.5	30.7	6.3	33.7	4.2	39.7	7.3	29.7	6.0	32.6	4.0	38.5
6	10.1	35.0	8.6	38.0	6.0	44.1	9.7	33.9	8.2	36.8	5.8	42.8
7	12.8	39.2	11.1	42.2	8.1	48.2	12.3	37.9	10.6	40.9	7.8	46.8
8	15.6	43.2	13.7	46.2	10.4	52.2	15.0	41.8	13.2	44.8	10.0	50.7
9	18.6	47.1	16.5	50.1	12.8	56.0	17.9	45.6	15.8	48.6	12.3	54.4
10	21.6	50.9	19.4	53.9	15.3	59.7	20.8	49.3	18.6	52.3	14.7	58.0
11	24.7	54.6	22.3	57.6	18.0	63.2	23.8	52.9	21.5	55.9	17.3	61.5
12	28.0	58.2	25.4	61.2	20.8	66.6	26.9	56.5	24.4	59.4	20.0	64.9
13	31.3	61.8	28.6	64.6	23.8	69.9	30.0	59.9	27.5	62.8	22.8	68.1
14	34.6	65.3	31.9	68.0	26.8	73.1	33.3	63.3	30.6	66.1	25.7	71.2
15	38.1	68.6	35.3	71.3	30.0	76.1	36.6	66.6	33.8	69.3	28.7	74.2

(Continued)

Two-Sided Confidence Band Factors on WEIBULL Plots (Continued)

Sample Size	27						28					
Confidence	*90%*		*95%*		*99%*		*90%*		*95%*		*99%*	
% Rank	*5.0%*	*95.0%*	*2.5%*	*97.5%*	*0.5%*	*99.5%*	*5.0%*	*95.0%*	*2.5%*	*97.5%*	*0.5%*	*99.5%*
16	41.7	71.9	38.7	74.5	33.3	79.1	40.0	69.9	37.1	72.4	31.8	77.1
17	45.3	75.2	42.3	77.6	36.7	81.9	43.4	73.0	40.5	75.5	35.0	79.9
18	49.0	78.3	46.0	80.5	40.2	84.6	47.0	76.1	44.0	78.4	38.4	82.6
19	52.8	81.3	49.8	83.4	43.9	87.1	50.6	79.1	47.6	81.3	41.9	85.2
20	56.7	84.3	53.7	86.2	47.7	89.5	54.3	82.0	51.3	84.1	45.5	87.6
21	60.7	87.1	57.7	88.8	51.7	91.8	58.1	84.9	55.1	86.7	49.2	89.9
22	64.9	89.8	61.9	91.3	55.8	93.9	62.0	87.6	59.0	89.3	53.1	92.1
23	69.2	92.4	66.2	93.7	60.2	95.7	66.0	90.2	63.1	91.7	57.1	94.1
24	73.7	94.7	70.8	95.8	64.9	97.3	70.2	92.6	67.3	93.9	61.4	95.9
25	78.4	96.9	75.7	97.6	69.9	98.7	74.5	94.9	71.7	95.9	66.0	97.4
26	83.6	98.6	81.0	99.0	75.5	99.6	79.1	97.0	76.4	97.7	70.8	98.7
27	89.4	99.8	87.2	99.9	82.1	99.9	84.1	98.7	81.6	99.1	76.3	99.6
28							89.8	99.8	87.6	99.9	82.7	99.9

Sample Size	29						30					
Confidence	*90%*		*95%*		*99%*		*90%*		*95%*		*99%*	
% Rank	*5.0%*	*95.0%*	*2.5%*	*97.5%*	*0.5%*	*99.5%*	*5.0%*	*95.0%*	*2.5%*	*97.5%*	*0.5%*	*99.5%*
1	0.17	9.8	0.08	11.9	0.01	16.6	0.17	9.5	0.08	11.5	0.01	16.1
2	1.2	15.3	0.84	17.7	0.36	22.9	1.1	14.8	0.81	17.2	0.35	22.2
3	2.8	20.1	2.1	22.7	1.2	28.2	2.7	19.5	2.1	22.0	1.1	27.4
4	4.8	24.6	3.8	27.3	2.4	32.9	4.6	23.8	3.7	26.5	2.3	32.0
5	7.0	28.8	5.8	31.6	3.9	37.4	6.8	27.9	5.6	30.7	3.7	36.3

6	9.4	32.8	7.9	35.7	5.6	41.5	9.0	31.8	7.7	34.7	5.4	40.4
7	11.9	36.8	10.2	39.7	7.5	45.5	11.4	35.7	9.9	38.5	7.2	44.2
8	14.5	40.5	12.7	43.5	9.6	49.3	14.0	39.3	12.2	42.2	9.2	47.9
9	17.2	44.2	15.2	47.2	11.8	52.9	16.6	42.9	14.7	45.8	11.4	51.5
10	20.0	47.9	17.9	50.8	14.1	56.5	19.3	46.5	17.2	49.3	13.6	55.0
11	22.9	51.4	20.6	54.3	16.6	59.9	22.1	49.9	19.9	52.8	16.0	58.3
12	25.8	54.8	23.5	57.7	19.2	63.1	24.9	53.3	22.6	56.1	18.5	61.5
13	28.9	58.2	26.4	61.0	21.9	66.3	27.8	56.6	25.4	59.3	21.0	64.6
14	32.0	61.5	29.4	64.3	24.6	69.4	30.8	59.8	28.3	62.5	23.7	67.7
15	35.2	64.8	32.5	67.4	27.5	72.4	33.8	63.0	31.2	65.6	26.4	70.6
16	38.4	67.9	35.6	70.5	30.5	75.3	36.9	66.1	34.3	68.7	29.3	73.5
17	41.7	71.0	38.9	73.5	33.6	78.0	40.1	69.1	37.4	71.6	32.2	76.2
18	45.1	74.1	42.2	76.4	36.8	80.7	43.3	72.1	40.6	74.5	35.3	78.9
19	48.5	77.0	45.6	79.3	40.0	83.3	46.6	75.0	43.8	77.3	38.4	81.4
20	52.0	79.9	49.1	82.0	43.4	85.8	50.0	77.8	47.1	80.0	41.6	83.9
21	55.7	82.7	52.7	84.7	47.0	88.1	53.4	80.6	50.6	82.7	44.9	86.3
22	59.4	85.4	56.4	87.2	50.6	90.3	57.0	83.3	54.1	85.2	48.4	88.5
23	63.2	88.0	60.2	89.7	54.4	92.4	60.6	85.9	57.7	87.7	52.0	90.7
24	67.1	90.5	64.2	92.0	58.4	94.3	64.2	88.5	61.4	90.0	55.7	92.7
25	71.1	92.9	68.3	94.1	62.5	96.0	68.1	90.9	65.2	92.2	59.5	94.5
26	75.3	95.1	72.6	96.1	67.0	97.5	72.0	93.1	69.2	94.3	63.6	96.2
27	79.8	97.1	77.2	97.8	71.7	98.8	76.1	95.3	73.4	96.2	67.9	97.6
28	84.6	98.7	82.2	99.1	77.0	99.6	80.4	97.2	77.9	97.8	72.6	98.8
29	90.1	99.8	88.0	99.9	83.3	99.9	85.1	98.8	82.7	99.1	77.7	99.6
30							90.4	99.8	88.4	99.9	83.8	99.9

(Continued)

Two-Sided Confidence Band Factors on WEIBULL Plots (Continued)

Sample Size	31						32					
Confidence	90%		95%		99%		90%		95%		99%	
% Rank	5.0%	95.0%	2.5%	97.5%	0.5%	99.5%	5.0%	95.0%	2.5%	97.5%	0.5%	99.5%
1	0.16	9.2	0.08	11.2	0.01	15.7	0.16	8.9	0.07	10.8	0.01	15.2
2	1.1	14.4	0.79	16.7	0.33	21.6	1.1	13.9	0.76	16.2	0.32	21.0
3	2.6	18.9	2.0	21.4	1.1	26.6	2.6	18.3	1.9	20.8	1.0	25.8
4	4.5	23.1	3.6	25.7	2.2	31.1	4.3	22.4	3.5	25.0	2.1	30.2
5	6.5	27.1	5.4	29.8	3.6	35.3	6.3	26.3	5.2	28.9	3.5	34.3
6	8.7	30.9	7.4	33.7	5.2	39.2	8.4	30.0	7.2	32.7	5.0	38.2
7	11.1	34.6	9.5	37.4	7.0	43.0	10.7	33.6	9.2	36.4	6.8	41.9
8	13.5	38.2	11.8	41.0	8.9	46.7	13.0	37.1	11.4	39.9	8.6	45.4
9	16.0	41.7	14.2	44.6	11.0	50.2	15.5	40.6	13.7	43.4	10.6	48.9
10	18.6	45.1	16.6	48.0	13.1	53.5	18.0	43.9	16.1	46.7	12.7	52.2
11	21.3	48.5	19.2	51.3	15.4	56.8	20.6	47.2	18.5	50.0	14.9	55.4
12	24.0	51.8	21.8	54.6	17.8	60.0	23.2	50.4	21.1	53.1	17.2	58.5
13	26.8	55.0	24.5	57.8	20.2	63.0	25.9	53.5	23.6	56.3	19.5	61.5
14	29.7	58.2	27.3	60.9	22.8	66.0	28.7	56.6	26.3	59.3	22.0	64.4
15	32.6	61.3	30.1	63.9	25.4	68.9	31.5	59.6	29.0	62.3	24.5	67.3
16	35.6	64.3	33.0	66.9	28.2	71.7	34.4	62.6	31.8	65.2	27.1	70.1
17	38.6	67.3	36.0	69.8	31.0	74.5	37.3	65.5	34.7	68.1	29.8	72.8
18	41.7	70.2	39.0	72.6	33.9	77.1	40.3	68.4	37.6	70.9	32.6	75.4
19	44.9	73.1	42.1	75.4	36.9	79.7	43.3	71.2	40.6	73.6	35.5	77.9
20	48.1	75.9	45.3	78.1	39.9	82.1	46.4	74.0	43.6	76.3	38.4	80.4

21	51.4	78.6	48.6	80.7	43.1	84.5	49.5	76.7	46.8	78.9	41.4	82.7
22	54.8	81.3	51.9	83.3	46.4	86.8	52.7	79.3	49.9	81.4	44.5	85.0
23	58.2	83.9	55.3	85.7	49.7	88.9	56.0	81.9	53.2	83.8	47.7	87.2
24	61.7	86.4	58.9	88.1	53.2	91.0	59.3	84.4	56.5	86.2	51.0	89.3
25	65.3	88.8	62.5	90.4	56.9	92.9	62.8	86.9	60.0	88.5	54.5	91.3
26	69.0	91.2	66.2	92.5	60.7	94.7	66.3	89.2	63.5	90.7	58.0	93.1
27	72.8	93.4	70.1	94.5	64.6	96.3	69.9	91.5	67.2	92.7	61.7	94.9
28	76.8	95.4	74.2	96.3	68.8	97.7	73.6	93.6	71.0	94.7	65.6	96.4
29	81.0	97.3	78.5	97.9	73.3	98.8	77.5	95.6	74.9	96.4	69.7	97.8
30	85.5	98.8	83.2	99.2	78.3	99.6	81.6	97.3	79.1	98.0	74.1	98.9
31	90.7	99.8	88.7	99.9	84.2	99.9	86.0	98.8	83.7	99.2	78.9	99.6
32							91.0	99.8	89.1	99.9	84.7	99.9

Sample Size	*33*						*34*					
Confidence	*90%*		*95%*		*99%*		*90%*		*95%*		*99%*	
% Rank	*5.0%*	*95.0%*	*2.5%*	*97.5%*	*0.5%*	*99.5%*	*5.0%*	*95.0%*	*2.5%*	*97.5%*	*0.5%*	*99.5%*
1	0.15	8.6	0.07	10.5	0.01	14.8	0.15	8.4	0.07	10.2	0.01	14.4
2	1.0	13.5	0.74	15.7	0.31	20.4	1.0	13.2	0.72	15.3	0.30	19.8
3	2.5	17.8	1.9	20.2	1.0	25.1	2.4	17.3	1.8	19.6	1.0	24.5
4	4.2	21.8	3.4	24.3	2.1	29.4	4.1	21.2	3.3	23.6	2.0	28.7
5	6.1	25.6	5.1	28.2	3.4	33.4	5.9	24.9	4.9	27.4	3.3	32.6
6	8.2	29.2	6.9	31.8	4.9	37.2	7.9	28.4	6.7	31.0	4.7	36.3
7	10.4	32.7	8.9	35.4	6.5	40.8	10.0	31.8	8.7	34.5	6.3	39.8
8	12.6	36.1	11.0	38.9	8.3	44.3	12.2	35.2	10.7	37.8	8.1	43.2
9	15.0	39.5	13.3	42.2	10.2	47.6	14.5	38.4	12.8	41.1	9.9	46.5
10	17.4	42.7	15.5	45.5	12.3	50.9	16.9	41.6	15.0	44.3	11.9	49.7

(Continued)

Two-Sided Confidence Band Factors on WEIBULL Plots (Continued)

Sample Size	33						34					
Confidence	*90%*		*95%*		*99%*		*90%*		*95%*		*99%*	
% Rank	*5.0%*	*95.0%*	*2.5%*	*97.5%*	*0.5%*	*99.5%*	*5.0%*	*95.0%*	*2.5%*	*97.5%*	*0.5%*	*99.5%*
11	19.9	45.9	17.9	48.7	14.4	54.0	19.3	44.7	17.3	47.4	13.9	52.7
12	22.5	49.0	20.4	51.8	16.6	57.1	21.7	47.8	19.7	50.5	16.0	55.7
13	25.1	52.1	22.9	54.8	18.9	60.0	24.3	50.8	22.1	53.5	18.2	58.6
14	27.7	55.1	25.4	57.8	21.2	62.9	26.8	53.7	24.6	56.4	20.5	61.5
15	30.4	58.1	28.1	60.7	23.7	65.7	29.5	56.6	27.1	59.3	22.9	64.3
16	33.2	61.0	30.7	63.6	26.2	68.5	32.1	59.5	29.7	62.1	25.3	66.9
17	36.0	63.9	33.5	66.4	28.8	71.1	34.8	62.3	32.4	64.8	27.8	69.6
18	38.9	66.7	36.3	69.2	31.4	73.7	37.6	65.1	35.1	67.5	30.3	72.1
19	41.8	69.5	39.2	71.8	34.2	76.2	40.4	67.8	37.8	70.2	33.0	74.6
20	44.8	72.2	42.1	74.5	37.0	78.7	43.3	70.4	40.6	72.8	35.7	77.0
21	47.8	74.8	45.1	77.0	39.9	81.0	46.2	73.1	43.5	75.3	38.4	79.4
22	50.9	77.4	48.1	79.6	42.8	83.3	49.1	75.6	46.4	77.8	41.3	81.7
23	54.0	80.0	51.2	82.0	45.9	85.5	52.1	78.2	49.4	80.2	44.2	83.9
24	57.2	82.5	54.4	84.4	49.0	87.6	55.2	80.6	52.5	82.6	47.2	86.0
25	60.4	84.9	57.7	86.7	52.3	89.7	58.3	83.0	55.6	84.9	50.3	88.0
26	63.8	87.3	61.0	88.9	55.6	91.6	61.5	85.4	58.8	87.1	53.4	90.0
27	67.2	89.5	64.5	91.0	59.1	93.4	64.7	87.7	62.1	89.2	56.7	91.8
28	70.7	91.7	68.1	93.0	62.7	95.0	68.1	89.9	65.4	91.2	60.1	93.6
29	74.3	93.8	71.7	94.8	66.5	96.5	71.5	92.0	68.9	93.2	63.6	95.2
30	78.1	95.7	75.6	96.5	70.5	97.8	75.0	94.0	72.5	95.0	67.3	96.6
31	82.1	97.4	79.7	98.0	74.8	98.9	78.7	95.8	76.3	96.7	71.2	97.9
32	86.4	98.9	84.2	99.2	79.5	99.6	82.6	97.5	80.3	98.1	75.4	98.9
33	91.3	99.8	89.4	99.9	85.1	99.9	86.7	98.9	84.6	99.2	80.1	99.6
34							91.5	99.8	89.7	99.9	85.5	99.9

Sample Size	*35*						*36*					
Confidence	*90%*		*95%*		*99%*		*90%*		*95%*		*99%*	
% Rank	*5.0%*	*95.0%*	*2.5%*	*97.5%*	*0.5%*	*99.5%*	*5.0%*	*95.0%*	*2.5%*	*97.5%*	*0.5%*	*99.5%*
1	0.14	8.2	0.07	10.0	0.01	14.0	0.14	7.9	0.07	9.7	0.01	13.6
2	1.0	12.8	0.70	14.9	0.30	19.3	0.99	12.5	0.68	14.5	0.29	18.8
3	2.3	16.9	1.8	19.1	0.98	23.8	2.3	16.4	1.7	18.6	0.96	23.2
4	3.9	20.6	3.2	23.0	1.9	27.9	3.8	20.1	3.1	22.4	1.9	27.2
5	5.8	24.2	4.8	26.7	3.2	31.8	5.6	23.6	4.6	26.0	3.1	31.0
6	7.7	27.7	6.5	30.2	4.6	35.4	7.5	27.0	6.3	29.4	4.4	34.5
7	9.7	31.0	8.4	33.6	6.1	38.8	9.4	30.2	8.1	32.8	6.0	37.9
8	11.9	34.3	10.4	36.9	7.8	42.1	11.5	33.4	10.1	36.0	7.6	41.2
9	14.1	37.4	12.4	40.1	9.6	45.4	13.7	36.5	12.1	39.1	9.3	44.3
10	16.3	40.5	14.6	43.2	11.5	48.5	15.9	39.5	14.2	42.2	11.1	47.4
11	18.7	43.6	16.8	46.3	13.5	51.5	18.1	42.5	16.3	45.1	13.0	50.3
12	21.1	46.6	19.1	49.2	15.5	54.4	20.4	45.4	18.5	48.1	15.0	53.2
13	23.5	49.5	21.4	52.2	17.6	57.3	22.8	48.3	20.8	50.9	17.1	56.0
14	26.0	52.4	23.8	55.0	19.8	60.1	25.2	51.1	23.1	53.7	19.2	58.8
15	28.5	55.2	26.3	57.8	22.1	62.8	27.7	53.9	25.5	56.5	21.4	61.4
16	31.1	58.0	28.8	60.6	24.4	65.5	30.2	56.6	27.9	59.2	23.7	64.1
17	33.7	60.8	31.3	63.3	26.8	68.1	32.7	59.3	30.4	61.9	26.0	66.6
18	36.4	63.5	33.9	66.0	29.3	70.6	35.3	62.0	32.9	64.5	28.4	69.1
19	39.1	66.2	36.6	68.6	31.8	73.1	37.9	64.6	35.4	67.0	30.8	71.5
20	41.9	68.8	39.3	71.1	34.4	75.5	40.6	67.2	38.0	69.5	33.3	73.9

(Continued)

Two-Sided Confidence Band Factors on WEIBULL Plots (Continued)

Sample Size	35						36					
Confidence	*90%*		*95%*		*99%*		*90%*		*95%*		*99%*	
% Rank	*5.0%*	*95.0%*	*2.5%*	*97.5%*	*0.5%*	*99.5%*	*5.0%*	*95.0%*	*2.5%*	*97.5%*	*0.5%*	*99.5%*
21	44.7	71.4	42.1	73.6	37.1	77.8	43.3	69.7	40.7	72.0	35.8	76.2
22	47.5	73.9	44.9	76.1	39.8	80.1	46.0	72.2	43.4	74.4	38.5	78.5
23	50.4	76.4	47.7	78.5	42.6	82.3	48.8	74.7	46.2	76.8	41.1	80.7
24	53.3	78.8	50.7	80.8	45.5	84.4	51.6	77.1	49.0	79.1	43.9	82.8
25	56.3	81.2	53.6	83.1	48.4	86.4	54.5	79.5	51.8	81.4	46.7	84.9
26	59.4	83.6	56.7	85.3	51.4	88.4	57.4	81.8	54.8	83.6	49.6	86.9
27	62.5	85.8	59.8	87.5	54.5	90.3	60.4	84.0	57.7	85.8	52.6	88.8
28	65.6	88.0	63.0	89.5	57.8	92.1	63.4	86.2	60.8	87.8	55.6	90.6
29	68.9	90.2	66.3	91.5	61.1	93.8	66.5	88.4	63.9	89.8	58.8	92.3
30	72.2	92.2	69.7	93.4	64.5	95.3	69.7	90.5	67.1	91.8	62.0	94.0
31	75.7	94.1	73.2	95.1	68.2	96.7	72.9	92.4	70.5	93.6	65.4	95.5
32	79.3	96.0	76.9	96.7	72.0	98.0	76.3	94.3	73.9	95.3	68.9	96.8
33	83.0	97.6	80.8	98.1	76.1	99.0	79.8	96.1	77.5	96.8	72.7	98.0
34	87.1	98.9	85.0	99.3	80.6	99.7	83.5	97.6	81.3	98.2	76.7	99.0
35	91.7	99.8	89.9	99.9	85.9	99.9	87.4	99.0	85.4	99.3	81.1	99.7
36							92.0	99.8	90.2	99.9	86.3	99.9

Sample Size	*37*						*38*					
Confidence	*90%*		*95%*		*99%*		*90%*		*95%*		*99%*	
% Rank	*5.0%*	*95.0%*	*2.5%*	*97.5%*	*0.5%*	*99.5%*	*5.0%*	*95.0%*	*2.5%*	*97.5%*	*0.5%*	*99.5%*
1	0.13	7.7	0.06	9.4	0.01	13.3	0.13	7.5	0.06	9.2	0.01	13.0
2	0.96	12.1	0.66	14.1	0.28	18.4	0.94	11.8	0.64	13.8	0.27	17.9
3	2.2	16.0	1.7	18.1	0.93	22.7	2.1	15.6	1.6	17.7	0.90	22.1
4	3.7	19.6	3.0	21.9	1.8	26.6	3.6	19.1	2.9	21.3	1.8	26.0
5	5.4	23.0	4.5	25.4	3.0	30.2	5.3	22.4	4.4	24.8	2.9	29.5
6	7.3	26.3	6.1	28.7	4.3	33.7	7.1	25.6	6.0	28.0	4.2	32.9
7	9.2	29.5	7.9	32.0	5.8	37.0	8.9	28.8	7.7	31.2	5.6	36.2
8	11.2	32.6	9.8	35.1	7.4	40.2	10.9	31.8	9.5	34.3	7.2	39.3
9	13.3	35.6	11.7	38.2	9.0	43.3	12.9	34.7	11.4	37.3	8.8	42.3
10	15.4	38.6	13.7	41.1	10.8	46.3	15.0	37.6	13.4	40.2	10.5	45.2
11	17.6	41.5	15.8	44.1	12.7	49.2	17.1	40.5	15.4	43.1	12.3	48.1
12	19.8	44.3	18.0	46.9	14.6	52.0	19.3	43.3	17.5	45.9	14.2	50.9
13	22.1	47.1	20.2	49.7	16.6	54.8	21.5	46.0	19.6	48.6	16.1	53.6
14	24.5	49.9	22.4	52.5	18.6	57.5	23.8	48.7	21.8	51.3	18.1	56.3
15	26.9	52.6	24.7	55.2	20.8	60.1	26.1	51.4	24.0	54.0	20.1	58.9
16	29.3	55.3	27.0	57.9	22.9	62.7	28.4	54.0	26.3	56.6	22.3	61.4
17	31.7	58.0	29.4	60.5	25.2	65.2	30.8	56.6	28.6	59.1	24.4	63.9
18	34.2	60.6	31.9	63.0	27.5	67.7	33.2	59.2	30.9	61.7	26.6	66.3
19	36.8	63.1	34.4	65.6	29.8	70.1	35.7	61.7	33.3	64.1	28.9	68.7
20	39.3	65.7	36.9	68.0	32.2	72.4	38.2	64.2	35.8	66.6	31.2	71.0

(Continued)

Two-Sided Confidence Band Factors on WEIBULL Plots (Continued)

Sample Size	*37*						*38*					
Confidence	*90%*		*95%*		*99%*		*90%*		*95%*		*99%*	
% Rank	*5.0%*	*95.0%*	*2.5%*	*97.5%*	*0.5%*	*99.5%*	*5.0%*	*95.0%*	*2.5%*	*97.5%*	*0.5%*	*99.5%*
21	41.9	68.2	39.4	70.5	34.7	74.7	40.7	66.7	38.2	69.0	33.6	73.3
22	44.6	70.6	42.1	72.9	37.2	77.0	43.3	69.1	40.8	71.3	36.0	75.5
23	47.3	73.0	44.7	75.2	39.8	79.1	45.9	71.5	43.3	73.6	38.5	77.6
24	50.0	75.4	47.4	77.5	42.4	81.3	48.5	73.8	45.9	75.9	41.0	79.8
25	52.8	77.8	50.2	79.7	45.1	83.3	51.2	76.1	48.6	78.1	43.6	81.8
26	55.6	80.1	53.0	81.9	47.9	85.3	53.9	78.4	51.3	80.3	46.3	83.8
27	58.4	82.3	55.8	84.1	50.7	87.2	56.6	80.6	54.0	82.4	49.0	85.7
28	61.3	84.5	58.8	86.2	53.6	89.1	59.4	82.8	56.8	84.5	51.8	87.6
29	64.3	86.6	61.7	88.2	56.6	90.9	62.3	84.9	59.7	86.5	54.7	89.4
30	67.3	88.7	64.8	90.1	59.7	92.5	65.2	87.0	62.6	88.5	57.6	91.1
31	70.4	90.7	67.9	92.0	62.9	94.1	68.1	89.0	65.6	90.4	60.6	92.7
32	73.6	92.6	71.2	93.8	66.2	95.6	71.1	91.0	68.7	92.2	63.7	94.3
33	76.9	94.5	74.5	95.4	69.7	96.9	74.3	92.8	71.9	93.9	67.0	95.7
34	80.3	96.2	78.0	96.9	73.3	98.1	77.5	94.6	75.1	95.5	70.4	97.0
35	83.9	97.7	81.8	98.2	77.2	99.0	80.8	96.3	78.6	97.0	73.9	98.1
36	87.8	99.0	85.8	99.3	81.5	99.7	84.3	97.8	82.2	98.3	77.8	99.0
37	92.2	99.8	90.5	99.9	86.6	99.9	88.1	99.0	86.1	99.3	82.0	99.7
38							92.4	99.8	90.7	99.9	86.9	99.9

Sample Size				39						40		
Confidence	*90%*		*95%*		*99%*		*90%*		*95%*		*99%*	
% Rank	*5.0%*	*95.0%*	*2.5%*	*97.5%*	*0.5%*	*99.5%*	*5.0%*	*95.0%*	*2.5%*	*97.5%*	*0.5%*	*99.5%*
1	0.13	7.3	0.06	9.0	0.01	12.7	0.12	7.2	0.06	8.8	0.01	12.4
2	0.91	11.5	0.62	13.4	0.26	17.5	0.89	11.3	0.61	13.1	0.26	17.1
3	2.1	15.2	1.6	17.3	0.88	21.6	2.0	14.9	1.5	16.9	0.86	21.1
4	3.5	18.6	2.8	20.8	1.7	25.4	3.4	18.2	2.7	20.3	1.7	24.8
5	5.1	21.9	4.2	24.2	2.8	28.9	5.0	21.4	4.1	23.6	2.7	28.2
6	6.9	25.0	5.8	27.4	4.1	32.2	6.7	24.5	5.7	26.8	4.0	31.5
7	8.7	28.1	7.5	30.5	5.5	35.4	8.5	27.4	7.3	29.8	5.3	34.6
8	10.6	31.0	9.2	33.5	7.0	38.4	10.3	30.3	9.0	32.7	6.8	37.6
9	12.6	33.9	11.1	36.4	8.5	41.4	12.2	33.2	10.8	35.6	8.3	40.5
10	14.6	36.8	13.0	39.3	10.2	44.3	14.2	35.9	12.6	38.4	9.9	43.3
11	16.6	39.6	15.0	42.1	12.0	47.1	16.2	38.7	14.6	41.1	11.6	46.1
12	18.8	42.3	17.0	44.8	13.8	49.8	18.3	41.3	16.5	43.8	13.4	48.8
13	20.9	45.0	19.0	47.5	15.6	52.5	20.4	44.0	18.5	46.5	15.2	51.4
14	23.1	47.6	21.2	50.2	17.6	55.1	22.5	46.6	20.6	49.1	17.1	53.9
15	25.4	50.3	23.3	52.8	19.6	57.6	24.7	49.1	22.7	51.6	19.0	56.5
16	27.6	52.8	25.5	55.3	21.6	60.1	26.9	51.7	24.8	54.1	21.0	58.9
17	30.0	55.4	27.8	57.9	23.7	62.6	29.1	54.2	27.0	56.6	23.0	61.3
18	32.3	57.9	30.0	60.3	25.8	65.0	31.4	56.6	29.2	59.1	25.1	63.7
19	34.7	60.4	32.4	62.8	28.0	67.3	33.7	59.1	31.5	61.5	27.2	66.0
20	37.1	62.8	34.7	65.2	30.3	69.6	36.1	61.5	33.8	63.8	29.4	68.3

(Continued)

Two-Sided Confidence Band Factors on WEIBULL Plots (Continued)

Sample Size	*39*						*40*					
Confidence	*90%*		*95%*		*99%*		*90%*		*95%*		*99%*	
% Rank	*5.0%*	*95.0%*	*2.5%*	*97.5%*	*0.5%*	*99.5%*	*5.0%*	*95.0%*	*2.5%*	*97.5%*	*0.5%*	*99.5%*
21	39.5	65.2	37.1	67.5	32.6	71.9	38.4	63.8	36.1	66.1	31.6	70.5
22	42.0	67.6	39.6	69.9	34.9	74.1	40.8	66.2	38.4	68.4	33.9	72.7
23	44.5	69.9	42.1	72.1	37.3	76.2	43.3	68.5	40.8	70.7	36.2	74.8
24	47.1	72.3	44.6	74.4	39.8	78.3	45.7	70.8	43.3	72.9	38.6	76.9
25	49.6	74.5	47.1	76.6	42.3	80.3	48.2	73.0	45.8	75.1	41.0	78.9
26	52.3	76.8	49.7	78.7	44.8	82.3	50.8	75.2	48.3	77.2	43.4	80.9
27	54.9	79.0	52.4	80.9	47.4	84.3	53.3	77.4	50.8	79.3	46.0	82.8
28	57.6	81.1	55.1	82.9	50.1	86.1	55.9	79.5	53.4	81.4	48.5	84.7
29	60.4	83.3	57.8	84.9	52.8	87.9	58.6	81.6	56.1	83.4	51.1	86.5
30	63.1	85.3	60.6	86.9	55.6	89.7	61.2	83.7	58.8	85.3	53.8	88.3
31	66.0	87.3	63.5	88.8	58.5	91.4	64.0	85.7	61.5	87.3	56.6	90.0
32	68.9	89.3	66.4	90.7	61.5	92.9	66.7	87.7	64.3	89.1	59.4	91.6
33	71.8	91.2	69.4	92.4	64.5	94.4	69.6	89.6	67.2	90.9	62.3	93.1
34	74.9	93.0	72.5	94.1	67.7	95.8	72.5	91.4	70.1	92.6	65.3	94.6
35	78.0	94.8	75.7	95.7	71.0	97.1	75.4	93.2	73.1	94.2	68.4	95.9
36	81.3	96.4	79.1	97.1	74.5	98.2	78.5	94.9	76.3	95.8	71.7	97.2
37	84.7	97.8	82.6	98.3	78.3	99.1	81.7	96.5	79.6	97.2	75.1	98.2
38	88.4	99.0	86.5	99.3	82.4	99.7	85.0	97.9	83.0	98.4	78.8	99.1
39	92.6	99.8	90.9	99.9	87.2	99.9	88.6	99.1	86.8	99.3	82.8	99.7
40							92.7	99.8	99.1	99.9	87.5	99.9

Weld Metal Required for Various Joint Designs

Table 5-7- Weld Metal Required for Various Joint Designs

Joint Type	Base Material Thickness, in.	Width of Bead or Groove, in.	Max. Root Spacing, in.	Approx Amount of Metal Deposited per Linear Ft: cu in.	Approx Amount of Metal Deposited per Linear Ft: lb
Square Butt (Reinforcement .03"-.06"; Removeable Copper Backing)	0.037	1/8	0	0.07	0.02
	0.050	5/32	0	0.13	0.04
	0.062	3/16	0	0.13	0.04
	0.093	3/16-1/4	1/32	0.18	0.06
	0.125	1/4	1/16	0.22	0.07
Square Butt (Reinforcement .03"-.07")	1/8	1/4	1/32	0.35	0.11
	3/16	3/8	1/16	0.74	0.24
	1/4	7/16	3/32	0.97	0.31
V Groove (Reinf. .04"-.08"; Removeable Copper Backing)	3/16	0.35	1/8	0.72	0.227
	1/4	0.51	3/16	1.39	0.443
	5/16	0.61	3/16	1.84	0.582
	3/8	0.71	3/16	2.36	0.745
	1/2	0.91	3/16	3.68	1.16
	5/8	1.16	3/16	5.10	1.61
V Groove (Reinf. .04"-.08"; No Backing Used. Under Side of Weld Chipped and Welded)	1/4	0.41	3/32	1.33	0.42
	5/16	0.51	3/32	1.71	0.54
	3/8	0.65	1/8	2.30	0.73
	1/2	0.85	1/8	3.85	1.21
	5/8	1.06	1/8	4.63	1.46
Double V Groove	1/2	0.40	1/8	2.65	0.89
	5/8	0.49	1/8	3.45	1.08
	3/4	0.62	1/8	4.60	1.46
	1	0.81	1/8	7.70	2.42
	1-1/4	1.03	1/8	9.26	2.92
U Groove (Reinf. .04"-.08")	1/2	0.679	1/8	3.27	1.03
	5/8	0.745	1/8	4.37	1.38
	3/4	0.813	1/8	5.33	1.68
	1	0.957	1/8	8.35	2.63
	1-1/4	1.073	1/8	11.48	3.62
	1-1/2	1.215	1/8	15.16	4.79
	1-3/4	1.349	1/8	18.90	5.98
	2	1.485	1/8	23.45	7.40
Double U Groove (Reinf. .08")	1	0.679	1/8	6.54	2.06
	1-1/4	0.745	1/8	8.74	2.76
	1-1/2	0.813	1/8	10.66	3.36
	2	0.957	1/8	16.66	5.26
	2-1/2	1.073	1/8	22.96	7.24

Joint Type	Base Material Thickness, in.	Width of Groove (W), in.	Approx Amount of Metal Deposited per Linear Ft: cu in.	Approx Amount of Metal Deposited per Linear Ft: lb
Bevel Groove (Reinf. .04"-.08")	1/4	0.125	0.22	0.07
	5/16	0.188	0.40	0.13
	3/8	0.250	0.61	0.19
	1/2	0.375	1.21	0.38
	5/8	0.500	1.98	0.63
	3/4	0.625	2.95	0.93
	1	0.875	5.57	1.77
Double J Groove (Reinf. .04"-.08")	1	0.500	4.67	1.48
	1-1/4	0.563	6.90	1.90
	1-1/2	0.594	8.10	2.56
	1-3/4	0.625	9.83	3.11
	2	0.656	12.06	3.81
	2-1/4	0.688	14.29	4.51
	2-1/2	0.750	16.68	5.27
Double Bevel Groove (Reinf. .04"-.08")	1/2	0.188	0.78	0.25
	5/8	0.250	1.24	0.39
	3/4	0.313	1.78	0.56
	1	0.438	3.13	0.99
	1-1/4	0.563	4.87	1.54
	1-1/2	0.688	7.00	2.21
	1-3/4	0.813	9.47	3.00
	2	0.938	12.33	3.90
J Groove (Reinf. .04"-.08")	1	0.625	5.64	1.78
	1-1/4	0.719	7.91	2.50
	1-1/2	0.781	10.20	3.23
	1-3/4	0.875	12.95	4.09
	2	0.969	15.60	4.93
	2-1/4	1.031	18.35	5.80
	2-1/2	1.094	21.95	6.94
Corner (LAP)	1/16	—	0.05	0.02
	3/32	—	0.09	0.03
	1/8	—	0.15	0.05
	3/16	—	0.33	0.10
	1/4	—	0.59	0.19
	5/16	—	0.92	0.29
	3/8	—	1.32	0.42
	1/2	—	2.35	0.74
Fillet	—	1/8	0.09	0.03
	—	3/16	0.22	0.07
	—	1/4	0.38	0.12
	—	5/16	0.59	0.19
	—	3/8	0.84	0.27
	—	1/2	1.50	0.47
	—	5/8	2.34	0.74
	—	3/4	3.38	1.07
	—	1	6.00	1.90

IMP-75, The Integrated Mechanisms Program

```
REMARK/  FOLDING TOOLBAR
SYSTEM=  FOLDING TOOLBAR
GROUND=FRAME
REVLUT(FRAME,WING)=MHNG
    DATA/LINK(FRAM,MHNG)=0.0,0.0,0.0/0.0,0.0,1.0/1.0,0.0,0.0
    DATA/LINK(WING,MHNG)=0.0,0.0,0.0/0.0,0.0,1.0/1.0,0.0,0.0
REVLUT(FRAME,BCYL)=HNG1
    DATA/LINK(FRAM,HNG1)=-41.1,-1.24,0.0/-41.1,-1.24,1.0/-40.1,-1.24,0.0
    DATA/LINK(BCYL,HNG1)=-41.1,-1.24,0.0/-41.1,-1.24,1.0/-40.1,-1.24,0.0
REVLUT(FRAME,LNKA)=HNG2
    DATA/LINK(FRAM,HNG2)=-2.91,-5.75,0.0/-2.91,-5.75,1.0/-1.91,-5.75,0.0
    DATA/LINK(LNKA,HNG2)=-2.91,-5.75,0.0/-2.91,-5.75,1.0/-1.91,-5.75,0.0
REVLUT(WING,LNKB)=HNG3
    DATA/LINK(WING,HNG3)=4.25,-2.85,0.0/4.25,-2.85,1.0/5.25,-2.85,0.0
    DATA/LINK(LNKB,HNG3)=4.25,-2.85,0.0/4.25,-2.85,1.0/2.16,5.39,0.0
PRISM(BCYL,CRAM)=HCYL
    DATA/LINK(BCYL,HCYL)=-13.917,2.926,0.0/2.16,5.39,0.0/$
                         -13.917,2.926,1.0
    DATA/LINK(CRAM,HCYL)=-13.917,2.926,0.0/2.16,5.39,0.0/$
                         -13.917,2.926,1.0
REVLUT(CRAM,LNKB)=HNG4
    DATA/LINK(CRAM,HNG4)=2.16,5.39,0.0/2.16,5.39,1.0/-41.1,-1.24,0.0
    DATA/LINK(LNKB,HNG4)=2.16,5.39,0.0/2.16,5.39,1.0/4.25,-2.85,0.0
REVLUT(LNKB,LNKA)=HNG5
    DATA/LINK(LNKB,HNG5)=2.16,5.39,0.0/2.16,5.39,1.0/4.25,-2.85,0.0
    DATA/LINK(LNKA,HNG5)=2.16,5.39,0.0/2.16,5.39,1.0/-2.91,-5.75,0.0
REMARK/  CENTER OF GRAVITY OF FOLDING OUTRIGGER
POINT(WING)=CGOR
    DATA/POINT(CGOR,MHNG)=82.0,-21.5,0.0
REMARK/  WEIGHT OF FOLDING OUTRIGGER
    DATA/WEIGHT(WING,MHNG)=1440.0,82.0,-21.5,0.0
REMARK/  COUPLING POINT OF CYLINDER AND FOLDING LINKAGE
POINT(LNKB)=FLWR
    DATA/POINT(FLWR,HNG4)=0.0,0.0,0.0
REMARK/  GRAVITATIONAL CONSTANTS
    DATA/GRAV=0.0,-386.088,0.0
REMARK/  IMP TOLERANCES
ZERO(POSITN)=0.001
REMARK/  HYDRAULIC CYLINDER--OPERATION OF FOLDING TOOLBAR
    DATA/POSITN(HCYL)=0.0,-1.687,9
    DATA/VELCTY(HCYL)=-3.0
REMARK/  IMP REQUEST STATEMENTS
PRINT/POSITN(ALL)
PRINT/VELCTY(CGOR,FLWR)
PRINT/ACCL(CGOR,FLWR)
PRINT/FORCE(ALL)
EXECUTE/HOLD
```

Analysis of Folding Toolbar Kinematic Mode

```
***************************************************************************
*                              ANALYSIS OF                                *
*                          FOLDING TOOLBARLB                              *
*                            KINEMATIC MODE                               *
*                  ON       i                   AT    -                  *
***************************************************************************

POSITION    1

THE GENERALIZED COORDINATES ARE
    JOINT             SET BY          POSITION        VELOCITY      ACCELERATION
                                       DEG,IN        RAD,IN/SEC      RAD,IN/SEC2
    HCYL               USER             0.0            -3.000           0.0

DEGREE OF FREEDOM =   1.   QUALITY INDEX =   0.130E 05

POSITION RESULTS (DEG,IN)                X               Y               Z

  JNT. MHNG             0.001

  JNT. HNG1             0.000

  JNT. HNG2             0.000

  JNT. HNG3           104.232

  JNT. HCYL             0.0

  JNT. HNG4            95.520

  JNT. HNG5           -38.704

  PNT. CGOR            84.772         82.000         -21.499            0.0

  PNT. FLWR             5.807          2.160           5.390            0.0

VELOCITY RESULTS (RAD,IN/SEC)            X               Y               Z

  PNT. CGOR            55.577         14.095          53.760            0.0

  PNT. FLWR             3.585         -3.263           1.485            0.0

ACCELERATION RESULTS (RAD,IN/SEC2)       X               Y               Z

  PNT. CGOR            38.307        -32.246          20.680            0.0

  PNT. FLWR             1.200          0.094          -1.196            0.0

FORCE RESULTS (IN-LB,LB)                 X               Y               Z

  JNT. MHNG FROM FRAM ONTO WING
       FORC         34132.156       8639.867      -33020.551           ----
       TORQ             0.028          ----            ----            0.028

  JNT. HNG1 FROM FRAM ONTO BCYL
       FORC         26621.602     -26314.355       -4032.930           ----
       TORQ             0.023          ----            ----            0.023

  JNT. HNG2 FROM FRAM ONTO LNKA


       FORC         42377.184      17554.039       38570.477           ----
       TORQ             0.076          ----            ----           -0.076

  JNT. HNG3 FROM WING ONTO LNKB
       FORC         35631.309     -35631.313          -0.024           ----
       TORQ             0.144          ----            ----           -0.144

  JNT. HCYL FROM BCYL ONTO CRAM
       FORC         26621.605          ----           -0.078      -26621.609
       TORQ             1.399          1.399           ----            ----

  JNT. HNG4 FROM CRAM ONTO LNKB
       FORC         26621.609      -2560.603      -26498.180           ----
       TORQ             0.106          ----            ----            0.106

  JNT. HNG5 FROM LNKB ONTO LNKA
       FORC         42377.156      42377.156          -0.006           ----
       TORQ             0.056          ----            ----            0.056

*******************
```

(Continued)

Analysis of Folding Toolbar Kinematic Mode (Continued)

```
POSITION   2

THE GENERALIZED COORDINATES ARE
    JOINT            SET BY          POSITION        VELOCITY       ACCELERATION
                                     DEG, IN        RAD, IN/SEC     RAD, IN/SEC2
    HCYL             USER            -1.687          -3.000            0.0

DEGREE OF FREEDOM =  1.  QUALITY INDEX =  0.143E 05

POSITION RESULTS (DEG, IN)               X               Y                Z

  JNT. MHNG            21.923

  JNT. HNG1             1.263

  JNT. HNG2             9.063

  JNT. HNG3           101.358

  JNT. HCYL            -1.687

  JNT. HNG4           113.305

  JNT. HNG5           -48.689

  PNT. CGOR            84.772          84.097          10.671           0.0

  PNT. FLWR             6.059           0.342           6.050           0.0

VELOCITY RESULTS (RAD, IN/SEC)           X               Y                Z

  PNT. CGOR            58.644          -7.382          58.177           0.0

  PNT. FLWR             3.321          -3.201           0.882           0.0

ACCELERATION RESULTS (RAD, IN/SEC2)      X               Y                Z

  PNT. CGOR            40.569         -40.215          -5.351           0.0

  PNT. FLWR             0.975           0.121          -0.968           0.0

FORCE RESULTS (IN-LB, LB)                X               Y                Z

  JNT. MHNG  FROM FRAM ONTO WING
         FORC       32303.090        7001.387      -31535.223          ----
         TORQ           0.030            ----            ----           0.030

  JNT. HNG1  FROM FRAM ONTO BCYL
         FORC       27906.949      -27584.871       -4227.645          ----
         TORQ           0.028            ----            ----           0.028

  JNT. HNG2  FROM FRAM ONTO LNKA
         FORC       34121.578       14134.336       31056.449          ----
         TORQ           0.133            ----            ----          -0.133

  JNT. HNG3  FROM WING ONTO LNKB
         FORC       33565.906      -33565.910          -0.000          ----
         TORQ           0.096            ----            ----          -0.096

  JNT. HCYL  FROM BCYL ONTO CRAM
         FORC       27906.930            ----          -0.081      -27906.934
         TORQ           1.371           1.371            ----           ----

  JNT. HNG4  FROM CRAM ONTO LNKB
         FORC       27906.938      -11040.770      -25630.035          ----
         TORQ           0.095            ----            ----           0.095

  JNT. HNG5  FROM LNKB ONTO LNKA
         FORC       34121.590       34121.590          -0.028          ----
         TORQ           0.002            ----            ----           0.002

********************

POSITION   3

THE GENERALIZED COORDINATES ARE
    JOINT            SET BY          POSITION        VELOCITY       ACCELERATION
                                     DEG, IN        RAD, IN/SEC     RAD, IN/SEC2
    HCYL             USER            -3.374          -3.000            0.0

DEGREE OF FREEDOM =  1.  QUALITY INDEX =  0.158E 05

POSITION RESULTS (DEG, IN)               X               Y                Z

  JNT. MHNG            43.910

  JNT. HNG1             2.191
```

```
  JNT. HNG2          17.564

  JNT. HNG3          95.719

  JNT. HCYL          -3.374

  JNT. HNG4         128.725

  JNT. HNG5         -56.535

  PNT. CGOR          84.772         73.986         41.380          0.0

  PNT. FLWR           6.560         -1.438          6.401          0.0

VELOCITY RESULTS (RAD,IN/SEC)          X              Y              Z

  PNT. CGOR          56.488        -27.574         49.301          0.0

  PNT. FLWR           3.151         -3.128          0.379          0.0

ACCELERATION RESULTS (RAD,IN/SEC2)    X              Y              Z

  PNT. CGOR          38.167        -29.768        -23.887          0.0

  PNT. FLWR           0.845          0.137         -0.834          0.0

FORCE RESULTS (IN-LB,LB)              X              Y              Z

  JNT. MHNG FROM FRAM ONTO WING
       FORC       25559.066       3497.478     -25318.645          ----
       TORQ           0.026           ----           ----         0.026

  JNT. HNG1 FROM FRAM ONTO BCYL
       FORC       23220.543     -22952.555      -3517.704          ----
       TORQ           0.068           ----           ----         0.068

  JNT. HNG2 FROM FRAM ONTO LNKA
       FORC       21715.660       8995.359      19764.957          ----
       TORQ           0.026           ----           ----        -0.026

  JNT. HNG3 FROM WING ONTO LNKB
       FORC       26500.754     -26500.758          0.013          ----
       TORQ           0.028           ----           ----        -0.028

  JNT. HCYL FROM BCYL ONTO CRAM
       FORC       23220.559           ----         -0.067    -23220.563
       TORQ           1.426          1.426           ----          ----

  JNT. HNG4 FROM CRAM ONTO LNKB
       FORC       23220.574     -14526.309     -18115.785          ----
       TORQ           0.204           ----           ----         0.204

  JNT. HNG5 FROM LNKB ONTO LNKA
       FORC       21715.648      21715.652          0.006          ----
       TORQ           0.110           ----           ----         0.110

********************

POSITION   4

THE GENERALIZED COORDINATES ARE
     JOINT              SET BY         POSITION       VELOCITY     ACCELERATION
                                       DEG,IN      RAD,IN/SEC    RAD,IN/SEC2
     HCYL               USER           -5.061         -3.000          0.0

DEGREE OF FREEDOM =  1.   QUALITY INDEX =   0.172E 05

POSITION RESULTS (DEG,IN)             X              Y              Z

  JNT. MHNG          64.641

  JNT. HNG1           2.802

  JNT. HNG2          25.712

  JNT. HNG3          88.303

  JNT. HCYL          -5.061

  JNT. HNG4         141.428

  JNT. HNG5         -61.703

  PNT. CGOR          84.772         54.548         64.890          0.0

  PNT. FLWR           7.222         -3.175          6.487          0.0
```

(Continued)

Analysis of Folding Toolbar Kinematic Mode (Continued)

```
VELOCITY RESULTS (RAD,IN/SEC)          X              Y              Z
  PNT. CGOR          52.527        -40.208         33.799          0.0
  PNT. FLWR           3.049         -3.048         -0.066          0.0
ACCELERATION RESULTS (RAD,IN/SEC2)     X              Y              Z
  PNT. CGOR          33.324        -15.468        -29.517          0.0
  PNT. FLWR           0.771          0.146         -0.756          0.0
FORCE RESULTS (IN-LB,LB)               X              Y              Z
  JNT. MHNG FROM FRAM ONTO WING
       FORC       16991.180        655.591     -16978.535          ----
       TORQ           0.019           ----           ----         0.019
  JNT. HNG1 FROM FRAM ONTO BCYL
       FORC       15756.578     -15574.727      -2386.975          ----
       TORQ           0.046           ----           ----         0.046
  JNT. HNG2 FROM FRAM ONTO LNKA
       FORC       11157.328       4621.734      10155.074          ----
       TORQ           0.068           ----           ----         0.068
  JNT. HNG3 FROM WING ONTO LNKB
       FORC       17607.969     -17607.973         -0.000          ----
       TORQ           0.005           ----           ----        -0.005
  JNT. HCYL FROM BCYL ONTO CRAM
       FORC       15756.563           ----         -0.048     -15756.563
       TORQ           1.014          1.014           ----          ----
  JNT. HNG4 FROM CRAM ONTO LNKB
       FORC       15756.570     -12318.977      -9824.070          ----
       TORQ           0.122           ----           ----         0.122
  JNT. HNG5 FROM LNKB ONTO LNKA
       FORC       11157.324      11157.324          0.010          ----
       TORQ           0.134           ----           ----         0.134

********************

POSITION   5

THE GENERALIZED COORDINATES ARE
    JOINT            SET BY       POSITION       VELOCITY      ACCELERATION
                                   DEG,IN       RAD,IN/SEC     RAD,IN/SEC2
    HCYL             USER          -6.748         -3.000          0.0

DEGREE OF FREEDOM =  1.   QUALITY INDEX =   0.181E 05

POSITION RESULTS (DEG,IN)              X              Y              Z
  JNT. MHNG          83.885
  JNT. HNG1           3.090
  JNT. HNG2          33.665
  JNT. HNG3          79.646
  JNT. HCYL          -6.748
  JNT. HNG4         151.727
  JNT. HNG5         -64.337
  PNT. CGOR          84.772         30.112         79.243          0.0
  PNT. FLWR           7.986         -4.866          6.332          0.0
VELOCITY RESULTS (RAD,IN/SEC)          X              Y              Z
  PNT. CGOR          48.895        -45.706         17.368          0.0
  PNT. FLWR           3.003         -2.965         -0.480          0.0
ACCELERATION RESULTS (RAD,IN/SEC2)     X              Y              Z
  PNT. CGOR          28.737         -4.856        -28.324          0.0
  PNT. FLWR           0.738          0.150         -0.722          0.0
FORCE RESULTS (IN-LB,LB)               X              Y              Z
  JNT. MHNG FROM FRAM ONTO WING
       FORC        8567.164       -268.978      -8562.941          ----
       TORQ           0.008           ----           ----         0.008
```

```
  JNT. HNG1 FROM FRAM ONTO BCYL
       FORC       8001.047      -7908.707      -1212.084          ----
       TORQ          0.015         ----           ----           0.015

  JNT. HNG2 FROM FRAM ONTO LNKA
       FORC       4204.613       1741.706       3826.913          ----
       TORQ          0.049         ----           ----          -0.049

  JNT. HNG3 FROM WING ONTO LNKB
       FORC       8867.496      -8867.496          0.018          ----
       TORQ          0.041         ----           ----           0.041

  JNT. HCYL FROM BCYL ONTO CRAM
       FORC       8001.047         ----          -0.025      -8001.047
       TORQ          0.497          0.497         ----            ----

  JNT. HNG4 FROM CRAM ONTO LNKB
       FORC       8001.047      -7046.551      -3789.840          ----
       TORQ          0.024         ----           ----           0.024

  JNT. HNG5 FROM LNKB ONTO LNKA
       FORC       4204.613       4204.613         -0.010          ----
       TORQ          0.027         ----           ----          -0.027

********************

POSITION   6

THE GENERALIZED COORDINATES ARE
    JOINT             SET BY         POSITION        VELOCITY     ACCELERATION
                                     DEG, IN       RAD, IN/SEC     RAD, IN/SEC2
    HCYL              USER           -8.435          -3.000           0.0

DEGREE OF FREEDOM =  1.   QUALITY INDEX =   0.182E 05

POSITION RESULTS (DEG, IN)              X               Y               Z

  JNT. MHNG        101.966

  JNT. HNG1          3.026

  JNT. HNG2         41.571

  JNT. HNG3         69.920

  JNT. HCYL         -8.435

  JNT. HNG4        160.146

  JNT. HNG5        -64.786

  PNT. CGOR         84.772          4.031          84.676            0.0

  PNT. FLWR          8.818         -6.509           5.948            0.0

VELOCITY RESULTS (RAD, IN/SEC)          X               Y               Z

  PNT. CGOR         46.497        -46.444           2.211            0.0

  PNT. FLWR          3.013         -2.880          -0.886            0.0

ACCELERATION RESULTS (RAD, IN/SEC2)     X               Y               Z

  PNT. CGOR         25.666          1.672         -25.611            0.0

  PNT. FLWR          0.745          0.149          -0.730            0.0

FORCE RESULTS (IN-LB,LB)                X               Y               Z

  JNT. MHNG FROM FRAM ONTO WING
       FORC       1554.040        976.029      -1209.302          ----
       TORQ          0.000         ----           ----           0.000

  JNT. HNG1 FROM FRAM ONTO BCYL
       FORC        894.409       -884.086       -135.494          ----
       TORQ          0.001         ----           ----           0.001

  JNT. HNG2 FROM FRAM ONTO LNKA
       FORC        335.746        139.079        305.586          ----
       TORQ          0.002         ----           ----          -0.002

  JNT. HNG3 FROM WING ONTO LNKB
       FORC        984.277       -984.277          0.002          ----
       TORQ          0.002         ----           ----           0.002

  JNT. HCYL FROM BCYL ONTO CRAM
       FORC        894.408         ----          -0.003       -894.408
       TORQ          0.052          0.052         ----            ----
```

(Continued)

Analysis of Folding Toolbar Kinematic Mode (Continued)

```
  JNT. HNG4 FROM CRAM ONTO LNKB
       FORC         894.409        -841.249        -303.755           ----
       TORQ           0.001            ----            ----          0.001

  JNT. HNG5 FROM LNKB ONTO LNKA
       FORC         335.747         335.747          -0.002           ----
       TORQ           0.025            ----            ----         -0.025

********************

POSITION    7

THE GENERALIZED COORDINATES ARE
     JOINT              SET BY          POSITION        VELOCITY    ACCELERATION
                                        DEG, IN      RAD, IN/SEC    RAD, IN/SEC2
     HCYL               USER            -10.122          -3.000          0.0

DEGREE OF FREEDOM =  1.  QUALITY INDEX =  0.174E 05

POSITION RESULTS (DEG, IN)                 X               Y               Z

  JNT. MHNG         119.444

  JNT. HNG1           2.552

  JNT. HNG2          49.587

  JNT. HNG3          59.004

  JNT. HCYL         -10.122

  JNT. HNG4         167.183

  JNT. HNG5         -63.332

  PNT. CGOR          84.772         -21.586          81.977          0.0

  PNT. FLWR           9.702          -8.105           5.332          0.0

VELOCITY RESULTS (RAD, IN/SEC)             X               Y               Z

  PNT. CGOR          45.799         -44.289         -11.662          0.0

  PNT. FLWR           3.090          -2.798          -1.312          0.0

ACCELERATION RESULTS (RAD, IN/SEC2)        X               Y               Z

  PNT. CGOR          24.750           5.727         -24.079          0.0

  PNT. FLWR           0.808           0.142          -0.795          0.0

FORCE RESULTS (IN-LB, LB)                  X               Y               Z

  JNT. MHNG FROM FRAM ONTO WING
       FORC        6209.043        4278.422        4499.703           ----
       TORQ           0.001            ----            ----         -0.001

  JNT. HNG1 FROM FRAM ONTO BCYL
       FORC        5563.965        5499.750         842.886           ----
       TORQ           0.004            ----            ----         -0.004

  JNT. HNG2 FROM FRAM ONTO LNKA
       FORC        1381.203        -572.145       -1257.129           ----
       TORQ           0.058            ----            ----          0.058

  JNT. HNG3 FROM WING ONTO LNKB
       FORC        6045.238        6045.238          -0.001           ----
       TORQ           0.003            ----            ----         -0.003

  JNT. HCYL FROM BCYL ONTO CRAM
       FORC        5563.961            ----           0.020       5563.961
       TORQ           0.367          -0.367            ----           ----

  JNT. HNG4 FROM CRAM ONTO LNKB
       FORC        5563.961        5425.332        1234.276           ----
       TORQ           0.007            ----            ----         -0.007

  JNT. HNG5 FROM LNKB ONTO LNKA
       FORC        1381.203       -1381.203           0.003           ----
       TORQ           0.006            ----            ----         -0.006

********************

POSITION    8

THE GENERALIZED COORDINATES ARE
     JOINT              SET BY          POSITION        VELOCITY    ACCELERATION
                                        DEG, IN      RAD, IN/SEC    RAD, IN/SEC2
     HCYL               USER            -11.809          -3.000          0.0
```

```
DEGREE OF FREEDOM =  1.   QUALITY INDEX =  0.154E 05

POSITION RESULTS (DEG,IN)                X               Y               Z

  JNT. MHNG          137.082

  JNT. HNG1            1.565

  JNT. HNG2           57.922

  JNT. HNG3           46.419

  JNT. HCYL          -11.809

  JNT. HNG4          173.223

  JNT. HNG5          -60.050

  PNT. CGOR           84.772         -45.411          71.583          0.0

  PNT. FLWR           10.638          -9.657           4.462          0.0

VELOCITY RESULTS (RAD,IN/SEC)            X               Y               Z

  PNT. CGOR           47.525         -40.131         -25.459          0.0

  PNT. FLWR            3.263          -2.723          -1.799          0.0

ACCELERATION RESULTS (RAD,IN/SEC2)       X               Y               Z

  PNT. CGOR           27.333           9.122         -25.766          0.0

  PNT. FLWR            0.970           0.122          -0.962          0.0

FORCE RESULTS (IN-LB,LB)                 X               Y               Z

  JNT. MHNG FROM FRAM ONTO WING
       FORC        12538.930        9564.844        8107.930            ----
       TORQ            0.003            ----            ----          -0.003

  JNT. HNG1 FROM FRAM ONTO BCYL
       FORC        11859.668       11722.793        1796.619            ----
       TORQ            0.081            ----            ----          -0.081

  JNT. HNG2 FROM FRAM ONTO LNKA
       FORC         1615.295        -669.118       -1470.191            ----
       TORQ            0.038            ----            ----           0.038

  JNT. HNG3 FROM WING ONTO LNKB
       FORC        12583.227       12583.227          -0.006            ----
       TORQ            0.022            ----            ----           0.022

  JNT. HCYL FROM BCYL ONTO CRAM
       FORC        11859.668            ----           0.044       11859.668
       TORQ            0.844          -0.844            ----            ----

  JNT. HNG4 FROM CRAM ONTO LNKB
       FORC        11859.668       11776.793        1399.584            ----
       TORQ            0.035            ----            ----          -0.035

  JNT. HNG5 FROM LNKB ONTO LNKA
       FORC         1615.302       -1615.302           0.009            ----
       TORQ            0.136            ----            ----           0.136

********************

POSITION   9

THE GENERALIZED COORDINATES ARE
    JOINT              SET BY          POSITION        VELOCITY     ACCELERATION
                                        DEG,IN       RAD,IN/SEC      RAD,IN/SEC2
    HCYL               USER            -13.496          -3.000           0.0

DEGREE OF FREEDOM =  1.   QUALITY INDEX =  0.121E 05

POSITION RESULTS (DEG,IN)                X               Y               Z

  JNT. MHNG          156.140

  JNT. HNG1           -0.122

  JNT. HNG2           66.918

  JNT. HNG3           30.914

  JNT. HCYL          -13.496

  JNT. HNG4          178.464

  JNT. HNG5          -54.607
```

(Continued)

Analysis of Folding Toolbar Kinematic Mode (Continued)

```
   PNT. CGOR           84.772        -66.295        52.832        0.0

   PNT. FLWR           11.643        -11.171         3.282        0.0

VELOCITY RESULTS (RAD,IN/SEC)          X              Y              Z

   PNT. CGOR           53.937        -33.614       -42.181        0.0

   PNT. FLWR            3.613         -2.666        -2.438        0.0

ACCELERATION RESULTS (RAD,IN/SEC2)    X              Y              Z

   PNT. CGOR           39.236         14.983       -36.263        0.0

   PNT. FLWR            1.380          0.073        -1.378        0.0

FORCE RESULTS (IN-LB,LB)               X              Y              Z

   JNT. MHNG FROM FRAM ONTO WING
        FORC    19153.418      17056.938       8712.898          ----
        TORQ        0.000           ----           ----        -0.000

   JNT. HNG1 FROM FRAM ONTO BCYL
        FORC    18971.285      18752.348       2873.949          ----
        TORQ        0.099           ----           ----        -0.099

   JNT. HNG2 FROM FRAM ONTO LNKA
        FORC      623.859       -258.436       -567.813          ----
        TORQ        0.065           ----           ----         0.065

   JNT. HNG3 FROM WING ONTO LNKB
        FORC    19325.770      19325.773          0.000          ----
        TORQ        0.083           ----           ----         0.083

   JNT. HCYL FROM BCYL ONTO CRAM
        FORC    18971.270           ----          0.080     18971.277
        TORQ        1.409         -1.409           ----          ----

   JNT. HNG4 FROM CRAM ONTO LNKB
        FORC    18971.293      18964.477        508.576          ----
        TORQ        0.002           ----           ----        -0.002

   JNT. HNG5 FROM LNKB ONTO LNKA
        FORC      623.874       -623.874          0.019          ----
        TORQ        0.175           ----           ----         0.175

********************

POSITION   10

THE GENERALIZED COORDINATES ARE
      JOINT             SET BY        POSITION      VELOCITY     ACCELERATION
                                       DEG,IN      RAD,IN/SEC    RAD,IN/SEC2
      HCYL              USER          -15.183        -3.000         0.0

DEGREE OF FREEDOM =  1.  QUALITY INDEX =  0.694E 04

POSITION RESULTS (DEG,IN)              X              Y              Z

   JNT. MHNG          179.982

   JNT. HNG1           -2.923

   JNT. HNG2           77.307

   JNT. HNG3            8.185

   JNT. HCYL          -15.183

   JNT. HNG4          182.376

   JNT. HNG5          -45.330

   PNT. CGOR           84.772        -81.993        21.526        0.0

   PNT. FLWR           12.770        -12.664         1.644        0.0

VELOCITY RESULTS (RAD,IN/SEC)          X              Y              Z

   PNT. CGOR           77.472        -19.673       -74.932        0.0

   PNT. FLWR            4.403         -2.660        -3.508        0.0

ACCELERATION RESULTS (RAD,IN/SEC2)    X              Y              Z

   PNT. CGOR          115.084         45.440      -105.734        0.0

   PNT. FLWR            2.738         -0.088        -2.737        0.0
```

```
FORCE RESULTS ( IN-LB,LB)              X              Y              Z

  JNT. MHNG FROM FRAM ONTO WING
       FORC       25792.938      25654.512       2668.712           ----
       TORQ           0.000           ----           ----         -0.000

  JNT. HNG1 FROM FRAM ONTO BCYL
       FORC       27228.727      26914.484       4124.859           ----
       TORQ           0.099           ----           ----         -0.099

  JNT. HNG2 FROM FRAM ONTO LNKA
       FORC        1587.327        657.533       1444.735           ----
       TORQ           0.016           ----           ----          0.016

  JNT. HNG3 FROM WING ONTO LNKB
       FORC       26089.418      26089.418          0.005           ----
       TORQ           0.016           ----           ----          0.016

  JNT. HCYL FROM BCYL ONTO CRAM
       FORC       27228.715           ----          0.119      27228.715
       TORQ           2.122         -2.122           ----           ----

  JNT. HNG4 FROM CRAM ONTO LNKB
       FORC       27228.711      27205.305      -1128.864           ----
       TORQ           0.023           ----           ----         -0.023

  JNT. HNG5 FROM LNKB ONTO LNKA
       FORC        1587.316       1587.316         -0.012           ----
       TORQ           0.019           ----           ----         -0.019

*******************
```

Spring Data Chart

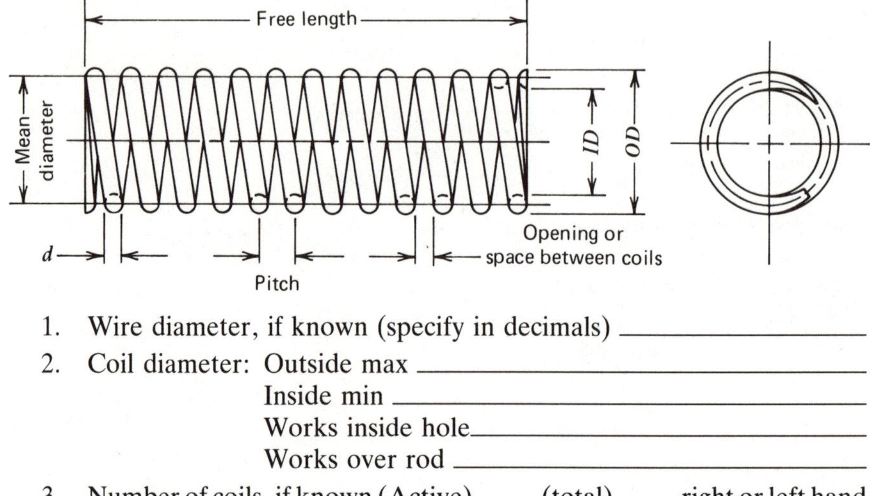

1. Wire diameter, if known (specify in decimals) ______________
2. Coil diameter: Outside max ______________
   Inside min ______________
   Works inside hole ______________
   Works over rod ______________
3. Number of coils, if known (Active) ____ (total) ____ right or left hand helix ____
4. Type of ends, (see below) ________ (squareness) ________
5. Free length (specify as approximate) ______________
6. Initial load _____ (min) _____ (max) ______ at ______ mm ht.
7. Final Load _____ (min) _____ (max) ______ at ______ mm ht.
8. Rate ______________
9. Max. solid height (if important) ______________
10. Type of finish ______________
11. Type of material (may be left to manufacturer) ______________

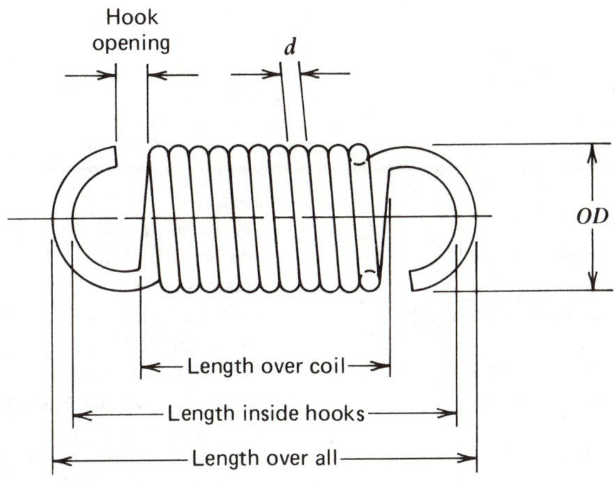

Material ____________________

Wire diameter (specify in decimals) ____________________

Outside diameter of coils ____________________

Number of coils ____________________

Length over coils (total coils plus one) multiplied by wire diameter ______

Type of ends (describe fully) ____________________

Rate ____________________

Initial installed length ____________________

Load at initial length ____________________

Final extended length ____________________

Load at final length ____________________

Additional data if significant: ____________________

Relative position of ends

Ambient temperature.

Frequency of operation.

If coil length is very long, will spring work in vertical or horizontal position?

INDEX